Wolfgang Detel/Claus Zittel (Hg.)

Wissensideale und Wissenskulturen in der frühen Neuzeit

Ideals and Cultures of Knowledge in Early Modern Europe

WISSENSKULTUR UND GESELLSCHAFTLICHER WANDEL

Herausgegeben vom Sonderforschungsbereich / Forschungskolleg 435
der Deutschen Forschungsgemeinschaft
»Wissenskultur und gesellschaftlicher Wandel«

Band 2

Wissensideale und Wissenskulturen in der frühen Neuzeit

Ideals and Cultures of Knowledge in Early Modern Europe

Herausgegeben von Wolfgang Detel und Claus Zittel

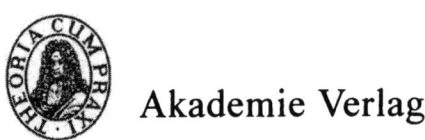 Akademie Verlag

Gedruckt mit Unterstützung der Deutschen Forschungsgemeinschaft.

Einbandgestaltung unter Verwendung von: Basilicae Philosophicae, in: Johann Daniel Mylius, Opus midico-chymicum, Frankfurt 1618.

ISBN 3-05-003713-X

Satz: Andreas Falke, Frankfurt/M.
Druck und Bindung: Druckhaus „Thomas Müntzer", Bad Langensalza
Einbandgestaltung: Dorén + Köster, Berlin

Printed in the Federal Republic of Germany

Contents

6

Wolfgang Detel / Claus Zittel

Introduction[1]:
Ideals and Cultures of Knowledge in Early Modern Europe

From a broad systematical and historical point of view this book examines some aspects of notions of knowledge entertained in 17th century Europe. Knowledge can be understood in different ways though. In particular, we might want to look at *ideals of knowledge* – at the notion of perfect knowledge that can never be false or unjustified; or we might be interested in *cultures of knowledge* – in practices and methods that are used to critically examine claims to knowledge. Traditionally, ideals of knowledge have been the subject of epistemology, while cultures of knowledge, in the sense just sketched, have been a topic of the philosophy of science. Ideals and cultures of knowledge were taken to constitute the core of scientific rationality, and it was widely assumed that this sort of scientific rationality has not substantially changed in the history of science since classical Greek antiquity, or at least since the age of Early Modern Europe.

This assumption has been increasingly attacked during the 20th century by historians like Canguilhem, Kuhn or Foucault, and more recently also by epistemologists like Michael Williams.[2] They reject the idea, so essential for traditional epistemology as well as philosophy and history of science, that the historical development of science is an evolutionary process that can be rationally reconstructed by applying universally valid criteria of rationality. And they insist that the history of science is full of contingencies and breaks that cannot be subsumed under a single story of evolutionary progress and increasing success. However, these far-reaching claims do not seem to be sufficiently supported by detailed historical studies. One of the aims of this collection of essays is to offer a number of such studies for the crucial period of the 17th century that extend the historical basis for deciding at some time in the future about the big claims concerning the history of science just sketched.

To do this, we suggest proceeding from a *broader notion of a culture of knowledge* than the one mentioned above. The idea is to conceive of a culture of knowledge as a set consisting firstly of practices and methods designed to produce scientific ideas and theories, secondly of background assumptions including an idea of what perfect knowledge and a perfected scien-

---

1 The essays collected in this volume originated in a conference on "Ideals and Cultures of Knowledge in Early Modern Europe: Concepts, Historical Background and Social Impact" (organized by W. Detel and C. Zittel), which took place in December 2000 at the Johann Wolfgang Goethe University in Frankfurt. This conference was linked to the research project "Cultures of Knowledge and Social Change," sponsored by the Deutsche Forschungsgemeinschaft and working at the Frankfurt University since 1999. Part of this research project comprises the investigation into the culture of knowledge in Early Modern Europe.
We are grateful to all participants of this conference for fruitful discussions and to the DFG for sponsoring our conference and this publication. Thanks too to Erna Mamane and Andreas Falke for giving support by the editorial work and to David Toalster for proofreading the Introduction.
2 Cfr. Williams (1991).

tific theory comes down to and thirdly of ways of organizing, learning and teaching established knowledge and theories. From this point of view, ideals of knowledge are but a part of a culture of knowledge. We might then ask whether there is a substantial change of cultures of knowledge in the development of science, but to answer such a big question, we have to look at the most important specific cultures of knowledge showing up in this development; in particular we have to adequately understand the different aspects and parts that are usually constitutive for influential cultures of knowledge. This is exactly what the essays presented in this volume attempt to do for cultures of knowledge in Early Modern Europe.

It does indeed seem obvious that the broader notion of a culture of knowledge outlined above can be applied to the situation of the sciences in 17th century Europe. Thus, there is no explicit distinction between epistemology and philosophy of science in this period of time. More importantly, it was not only the natural sciences that contributed to 17th century cultures of knowledge, but also political, religious, and aesthetic ideas and theories that were taken to also imply claims to knowledge or even to constitute specific cultures of knowledge themselves. Finally, there were a number of different ideas around about how one had to understand scientific practices. For instance, whether these practices were supposed to be part of a heuristic for scientific discoveries, or rather for laying out the logical and rational structure of established theories, or just for classifying and systematizing knowledge in a broader sense. In any case, the picture of 17th century cultures of knowledge in Europe looks extremely complex, and an adequate understanding of these cultures must be able to do justice to this complexity.

In fact, in recent times a number of essay volumes and monographies have attempted to look at a great number of thinkers[3], details and specific developmental stories about 17th century science[4] that have not attracted much attention from scholars so far. These studies stay away from, and sometimes criticize explicitly, big and unifying narratives.[5] But most of them still look at 17th century cultures of science from a specific methodological point of view and within the specific framework of the history of philosophy or the history of science.

The volume of essays presented here is different in at least two respects. Firstly, it offers an *interdisciplinary picture of 17th century cultures of knowledge*. Historians of science (Biagioli, Machamer, Steinle), and of philosophy (Detel, Garber, Wilson, Zittel), of general cultures and of scientific institutions (Brockliss, Neuber), of fine arts (Stumpfhaus) and of music (Becker, Vogel) contribute to this picture, as well as scholars working on the history of English or German literature, or in Comparative Literature (Reeves, Reichert). Secondly, these essays represent different *paradigmatic ways and methods in which cultures of knowledge can be historically explored*. Some of the essays for instance look at leading background assumptions or at practices and methods of such a culture, others at the ways claims to knowledge get accepted in scientific institutions. Some reveal the social forces driving developments in cultures of knowledge, some concentrate primarily on the internal history of scientific ideas. Other essays examine rather broad structures and longer processes or focus on very specific historical episodes.

---

3 Z. B. Osler (2000); Lindberg/Westman (1990); Field/James (1993).
4 Porter/Teich (1992); *Der Neue Ueberweg* (1988ff.).
5 Cfr. Cunningham/Williams (1993); Biagioli (1996); Shapin (1999); Golinski (1998).

One of the most characteristic elements of a culture of knowledge is the set of means and methods that are used to justify and to demonstrate claims to knowledge. Under this heading, *Machamer* and *Biagioli* examine an aspect of 17th century cultures of scientific knowledge that at first glance looks very specific, but turns out to be crucial and significant, namely the different ways parts of 17th century science relied on a display of pictures and illustrations.

Peter Machamer presents a grand picture of 17th century science that suggests a narrative unity among the many disparate strands of this period. The central claim is that in most branches of 17th century sciences as well as in many arts of the time, experience and a display of pictures, illustrations, and diagrams played a crucial role in demonstrating scientific positions and theories.

Machamer points out that there are a number of interesting aspects to these new kinds of demonstration. While there was certainly some sort of display and construction in ancient geometry, what is new in 17th century demonstrations is that the method of these demonstrations reacted to the major change towards an individualistic nature of epistemology that took place in the Renaissance (a sort of epistemology that Machamer calls "epistemic I" elsewhere). Demonstrations connected with a display of pictures, illustrations and diagrams were a way of making subjective experiments public – the basis of all knowledge. From this point of view, it basically consists of laying out a phenomenon before oneself and others. What is laid out provides an experience for those people seeing it. Moreover, this was a way of presenting and interpreting phenomena that anybody in principle could learn. One of the most important social aspects of this new mode of demonstrations was the new phenomenon of printing. Illustrations, tables and diagrams, being forms that were confined to two-dimensional page layouts, could be printed well and clearly and became a widespread form of demonstration in printed form. Tables allowed a reader to follow a procedure easily, diagrams and illustrations enabled the readers to understand, and to reproduce experiments; pictures, reproduced as engravings, made readers visually aware of new discoveries in different sciences. So these sorts of representations were open for public inspection, and they constituted a sort of externalization of memory.

One of the most interesting points of this picture is, as Machamer stresses that it does not only provide a new *concept* of demonstration that ties demonstration closely to experience, but also emphasizes the constructive and practical aspect of demonstrations: demonstrations are a way of doing something in a technical sense. Moreover, the display of pictures, illustrations, diagrams, and in general visible things that are constitutive of the new method of demonstration are at the same time the crucial basis of teaching and learning them. That is to say Machamer talks about the new method of demonstration in 17th century science exactly in terms of a culture of knowledge in the broad sense defined above. 17th century demonstrations as Machamer understands them, constituted a new culture of knowledge in Early Modern Europe.

Mario Biagioli presents an extensive essay on the status and function of illustrations as pictorial evidence in 17th century astronomy. One of the aims of his essay is to undermine the widely accepted distinction between schematic, diagrammatic pictures and realistic, mimetic pictures and the further distinction between nominalist and realist attitudes in 17th century astronomy. Biagioli shows that these distinctions often prove to be unstable and to produce, when applied to specific historical contexts, category mistakes. In addition, Biagioli denies the traditional claim that a common visual language was part of the ideal of knowledge proposed by 17th century astronomy. Illustrations served, in astronomical works of this time, very different purposes that were quite independent of representational accuracy. It is the epistemic context that determines the function of scientific illustrations as well as the criteria for accuracy.

Biagioli supports these claims by looking at the visual representations of, and the debates about, a new astronomical object, the sunspots, in 17th century astronomy. The main focus is on the discussion bewteen Galileo and Christoph Scheiner, a Jesuit mathematician. As Biagioli shows, the Jesuits procceeded from a sharp distinction between real and fiction. They were determined realists about the new discoveries because they wished to show that new phenomena could be categorized as belonging to well known kinds of objects und thus did not threaten the status quo in astronomy. This is why Scheiner painstakingly tried to show that the sunspots were satellites orbiting the sun. For this claim was consistent with the Aristotelian belief in the incorruptability of the heavens.

Galileo, however, tried to bypass the the Jesuit's real/fiction dichotomy. His concern was not to say what the real nature of sunspots is, but rather to prove that they were not observational artifacts. For Galileo, the real was just non-fiction. This allowed him to introduce new astronomical objects and to declare them as real without having to determine their true nature. Galileo rejected therefore Scheiner's claim about the nature of the sunspots as unwarranted by the observational evidence, thereby leaving open the dangerous possibility that they might be changes on the sun itself.

Interestingly, the illustrations and pictures utilized by Galileo and Scheiner to support their claims prove to be quite different modes of visual representation of sunspots. They relied on different observational systems; more importantly, the schematic illustrations produced by Scheiner were helpful for underscoring his realist claims, while Galileo's realist pictures supported his anti-realist and agnostic position concerning the real nature of the sunspots. So the usually accepted dichotomies turn out to be unhelpful in this particular example.

There is yet another historical background of this debate that has to be taken into account. Scheiner and Galileo were addressing different audiences that belonged to different epistemic cultures. Scheiner was speaking to his Jesuit superiors who needed to be shown that the work of a Jesuit mathematician was consistent with the accepted natural philosophy and theology, while Galileo was addressing primarily the courtiers at the Italian courts. These backgrounds shaped, as Biagioli shows, the different ways Scheiner and Galileo commented on the sunspots and used illustrations and pictures to support their claims.

There are hints, though, as Biagioli points out in conclusion, that Galileo too, not unlike Scheiner and the Jesuits, had substantial cosmological and ontological assumptions, but kept them secret. In particular, it was an organistic and realistic view of change which cast the sun as the center of life and let Galileo believe that the sunspots represent real changes on the surface of the sun.

If historians of science look not only at the development of scientific theories, but also at historical cultures of knowledge in the broad sense outlined above, then it seems natural to suppose that cultures of knowledge should be closely connected to all sorts of cultural and social context. *Reeves, Brockliss, Garber*, and *Wilson* reveal different strategies of tying the presentation, justification and development of scientific theories in Early Modern Europe intrinsically to cultural and social facts.

Eileen Reeves looks at the treatment of magnetism in Galileo's *Dialogue concerning the two chief world systems*. Her main interest is to examine in which way epistemological, cultural and scientific questions converge in this text. She detects an ambivalence in Galileo's discussion of the magnet. On the one hand, his spokesman in the *Dialogue,* Salviati is totally indifferent to the topos of navigation, exploration, and commerce, i. e. to the practical aspects of the use of magnets. On the other hand, one of the main conclusions in Galileo's report points to these practical aspects by claiming that the presence of hard gemstone in magnets which was of practical interest of artisans diminishes the attractive force and thus the worth of magnets. Reeves concludes that Galileo was uncertain about the status of the new scientific discipline that examines the attractive force of magnets. This had a considerable impact on Galileo's own professional status. For the recognition that magnets were closely connected to the commercial interests of any sea-going nation was already a commonplace notion in the 16th century and was reinforced by William Gilbert. In his essay, Galileo also presented himself as a mechanic eager to profit from magnetic devices. Another important aspect of Galileo's discussion of the magnet that Reeves focuses on is the way Galileo talks about the Aristotelian explanation of the magnetic attractive force. Simplicius takes the force to be an example of the overall sympathy that from an Aristotelian point of view governs the accord of all natural things. On the one hand, Salviati associates this doctrine with magic theories and practices well known at the time. A necromantic handbook of invocations and diagrams, the so-called *Clavicula Salomonis*, describes the practice of baptizing magnets for inducing great love and hatred into people subjected to these magic practices. On the other hand, Salviati mentions the ancient story of Diana and Acteon to exemplify his claim that the ancient theory of sympathetic forces in nature provides just empty names for describing a phenomenon that is by no means understood in ancient physics. One of the readings of this story sees Acteon as the tragic hunter who was consumed by his dogs after having spent all his substance on them. Reeves sees a hint here at Galileo's own personal situation after his brother had put a great financial and social burden on him.

Reeves concludes that Galileo's treatment of the magnet in the *Dialogue* and in particular the way he lets Salviati react to the Aristotelian theory reveals different levels of argument. First of all, there is the purely scientific examination of the attractive force of magnets that relies especially on Gilbert's work. Secondly, there are critical allusions to the connection between the ancient theory of magnets with magic doctrines and practises concerning the magnet. Thirdly, there is an allegorical hint at the social and financial situation Galileo was in; and this is all the more important because Galileo tried to profit from the magnetic devices he performed experiments on. As Reeves sees it, the presence of all these levels and aspects in Galileo's treatment of the magnet in the *Dialogue* does not only reveal his uncertainty about the scientific status of a new discipline of physics, but also his uncertainty about the way in which this new discipline might contribute to his status as a court scientist.

In a number of articles *Laurence Brockliss* has argued that in the 17th century the medical faculty at Paris was more open to many ideas from the new natural sciences of the time than contemporary satirists and later historians usually admitted. In general, 17th century Aristotelianism was a flexible natural philosophy that easily incorporated new material from new theories of physics – provided certain fundamentals remained untouched. In his essay for the present volume however, he observes that some aspects of the new science were accepted more quickly than others in institutions of higher education. Thus, Torricelli's work on air-pressure was received by medical faculty at Paris only three years after he had published his findings. On the other hand, it took nearly 40 years for the account of the circulation of the blood presented in Harvey's *De Motu Cordis* to be accepted in one of the leading medical faculties in Europe. In the remainder of his essay, Brockliss uses the reception of Torricelli and Harvey at Paris as a case study for examining some of the factors that shaped the appropriation of new ideas within institutions of higher education in 17th century France. He shows that there were primarily three factors that governed the professor's acceptance of any particular facts of the new sciences: Philosophical commitment, religious prejudice and peer-group pressure. If a new scientific fact could be incorporated into the Aristotelian or Aristotelo-Galenic world view without causing serious problems, and if it was not a threat to counter-reformation catholicism and if the university accepted it as harmless, then it could be taken into account. It is on the basis of these premises that Brockliss explains the different ways Torricelli's and Harvey's discoveries and were discussed by university professors.

Torricelli's work seemed at first glance to undermine one of the fundamentals of Aristotle's theory of the elements, since air-pressure was an Aristotelian nonsense. Nevertheless, his experimental findings could be incorporated into Aristotelian physics because the Aristotelian professors at Paris and elsewhere distinguished carefully between facts and conclusions drawn from them. They accepted the barometrical experiments, but they did not accept that these experiments demonstrated that the atmosphere had weight or spring. Instead, they were able to take Torricelli's experiments as further evidence of the traditional horror-vacui-theory. In this way, Aristotle's theory of the elements could be saved. Harvey's experiments however were different. They claimed to establish a new effect, namely that venal blood must flow from the extremities to the venal cava, not the other way around as the traditional theory has it. This new fact was inconsistent with traditional physiology. Therefore, Harvey's experimental work could not be detached from a new physiology that was inconsistent with the traditional physiological theory. It was thus very difficult for traditional Aristotelians to swallow the results of his experiments. In addition, Harvey's experiments and theory had serious therapeutic consequences. The Paris' medical faculties therapeutic identity was closely tied to the frequent use of the therapy of bleeding. Harvey's theory made the therapy of bleeding a nonsense and would therefore have demanded a profound change of therapeutic style. So there were deep professional reasons to stick to the traditional physiology. Brockliss concludes that a number of ideological and material factors constituted a major obstacle against a rapid assimilation of Harvey's work on the circulation of blood. Another aspect Brockliss mentions is the general interest of the Parisian university world in foreign scientific work. Thus, in the 1630ies and 40ies the faculty in Paris had little interest in scientific work performed in England, whereas Torricelli's work was widely discussed in this time.

One of the most important conclusions Brockliss draws from the material presented in his essay is that the widely accepted picture of 17th century Aristotelianism as a dynamic and plastic philosophy capable of embracing the new science and of constructing alternatives to chemical or mechanical philosophies may be an exaggeration. If we look at the internal and external factors shaping the way Harvey's and Torricelli's work was absorbed into Aristotelian physics in France, one can see that the professors of philosophy were on the back foot from the beginning of the century on. They were increasingly forced to give up all but the most fundamental Aristotelian principles.

Daniel Garber explores some of the social factors that influenced the rise of an anti-Aristotelian natural philosophy at the beginning of the seventeenth century. To do this, he looks at a specific case in the debate about Aristotelianism which took place in Paris in 1624. In this year, three young scholars, Jean Biteau, Antoine Vilon, and Estienne de Clave announced a public refutation of Aristotelian natural philosophy. The meeting was forbidden by the authorities, the theses were condemned and the three scholars were exiled from Paris. Moreover, the authorities ordered all persons to stay away from teaching any doctrines that go against the ancient and approved authors. The questions Garber addresses are, why did anyone care? Who objected to the theses of the three young scholars and why?

By examining written reactions of two young intellectuals, Marin Mersenne and Jean-Baptiste Morin, who were later to become pretty visible persons in the scientific life of early 17th-century Paris, Garber shows that both Mersenne and Morin, while they are certainly interested in the content of the theses proposed by the three anti-Aristotelians and in proving that their position is false and heretical, are, more or less implicitly, also concerned with issues having to do with truth and controversial debate in general. Both writers perceive the three young anti-Aristotelians not as disinterested seekers of truth, but rather as troublemakers and arrogant opponents of the authorities, daring to challenge the consensus of the learned word.

As Garber demonstrates, Mersenne and Morin are extremely nervous about heterodoxy and new controversial doctrines in general, because they worry that heterodoxy and new doctrines might lead straightforwardly to political and religious instability. From the late 1620[ies] on humanist free thinkers, the so-called libertins érudites began to flourish in France. In 1623 Mersenne already reports that there are fifty thousand atheists in Paris alone. And it is Mersenne, as Garber shows, who explicitly links the problem of heterodoxy with the problem of the stability of state. He perceives free thinkers and heretics as desiring to overthrow religion, the church and even the state itself. These reflections extend even to philosophical heterodoxy. Mersenne calls all people who reject Aristotle heretics.

But, as Garber points out, to understand why heresy and heterodoxy was such a pressing issue we have to turn to a larger historical context. The memory of the religious wars of 16th century France was still vivid in Paris from the middle of the 16th century on. Bloody wars and civic violence continued for more than thirty years. Stability had still not entirely returned in 1624, when the three young anti-Aristotelian scholars announced their refutation of Aristotle's natural philosophy. Garber concludes that heterodox belief was more than a purely intellectual issue in France in early 17th century, it was also an important social and political issue. The historical experience of the religious wars let the members of the Parlament, the doctors of the faculty of theology and thinkers like Mersenne and Morin conclude that difference in belief leads to political and social violence. In the context of early 17th century France, defending Aristotle against the new scientific ideas was seen as necessary for the stability of society itself.[6]

With a few exceptions, the preferred system of the modernists lay in the metaphysics and natural philosophy of the mid-17th century christianized Epicurianism and atomism. *Catherine Wilson* looks at a short episode in the 17th century reworking of Epicurean doctrine as a guiding background assumption of this culture of knowledge. She joins the critics of the traditional "two conversions"-theory according to which corpuscularianism and mechanism were first embraced and then rejected as experimentally inadequate. As Wilson points out, it is crucial to distinguish two different varieties of atomism: Firstly, qualitative corpuscularianism (QC), which teaches that many substances are not continuous, and many substances that appear to be one substance are not homogenous. Rather, they are composed of tiny, sub-visible particles or consists of a mixture of different kinds of particles stuck together. These tiny particles for instance compose fluids, fallids and also airs, vapors, mists and smokes. Second, there is reductive atomism (RA) that goes beyond QC by placing restrictions on the particles that are held to be the basic units, assigning them primary properties only and treating secundary properties as effects of the interaction between sentient beings and figured, moving particles.

According to Wilson it is important to see that 17th century RA indeed marks a major rupture with the eclectic systems of the Renaissance, whereas 17th century QC is continuous both with Lucretian atomism and with 18th century chemistry; moreover the experimental philosophy of the 17th century depended on QC, and not on RA. QC is, while RA is not a doctrine that is permanent in 16th, 17th and 18th century science in Europe. QC explained phenomena of dissolution and recovery easily, whereas Aristotelian theory dealt pretty poorly with this phenomenon. Contrary to RA, QC could be well supported by observations and experiments and because it was not a very demanding doctrine it was accepted wherever Aristotelian orthodoxy did not dominate exclusively. By distinguishing between RA and QC Wilson supports the traditional emphasis on the innovative character of 17th century science. Defenders of QC did usually not distinguish very sharply between invisible and light particles and active spirits and by entertaining a weaker version of mechanism undermined the Cartesian distinction between the intentional immaterial substance and the material inert substance.

Summing up her discussion about theories of corpuscular effluvia, Wilson concludes that the association of atomism with hedonism and skepticism continued to provide a threat to theology that was never really diminished. The detachment of atomism from morality was never really completed. However, from a scientific point of view, QC did not need to be transformed into RA because it had always been present since Lucretius wrote on atomism. Some of the leading scientists of 17th century science, like Robert Boyle, may have sometimes officially approved RA, but it remains still to be shown if this considerably influenced their scientific work.

One of the leading themes that pervade 17th century cultures of knowledge is the relation of form and diversity in nature. On a methodological level, this topic is reflected in the discussion about empiricism and systematization, and about skepticism and rational insights, among

---

6 For studying this interesting analysis it is very helpful that Garber in an appendix gives us a translation with commenting notes of the public thesis put forward by Biteau, Vilon and du Clave (the translation relies on a text that is taken from an original copy of broad side conserved in the Paris' Bibliothèque Nationale).

17th century scientists and philosophers. *Neuber, Steinle, Zittel, Reichert* and *Detel* explore different aspects and traditions that shaped this discussion. All contributions show how, depending on the context in different countries, traditional conceptions about the order of nature and about the corresponding systematization of kinds of knowledge were more or less taken over by skeptical or voluntaristic epistemologies.

In the 16th and 17th century there was a methodological competition between casuistic and systematical forms to provide an organization of scientific and empirical data. *Wolfgang Neuber's* diagnosis is that the method of systematization developed and worked out in the Ramistic tradition defeated the casuistic organization of knowledge. He sees this historical selection and victory as undermining and weakening the empiricism of science, i. e. the weight of an empirical foundation of scientific theories. His aim is to provide a better understanding of this process. On Neuber's view, one of the most important aspects of this development is that the Ramistic organization of established knowledge was a sort of standardization of a collective memory. This standardization was theologically justified, and supported, by the assumption that the structures of nature, intellect, and memory can be closely mapped to each other. Neuber stresses, among other things, the impact of this process for the rhetorical culture of knowledge which was still extremely influential in Early Modern Europe. The casuistic order of knowlegde lost its influence: Only *elocutio* and *pronunciatio*, the less important parts of a speech, remained a realm of the casuistic method, while crucial parts, i. e. inventio, dispositio, and memoria, were now reserved for moves of systematization in the Ramistic dialectical style. At the same time, Neuber emphasizes that the heuristic methods for making empirical discoveries vanished. This loss was a result of the Ramistic way of systematizing and memorizing established pieces of knowledge.

Friedrich Steinle explores the development of the ways the notion of natural laws was used in 17th century physics, in particular in theories presented and discussed in the Royal Society. The traditional conception of a natural law, as it was for instance still entertained by Descartes, relied on two basic features of a natural law. Firstly that it be authorized by God and be an expression of His Will and secondly that it be an abstract and fundamental principle in the framework of an axiomatized physics. Steinle points to the different ways this traditional conception was received and changed in the early history of the Royal Society. Some of the members, like Henry More, stick pretty much to the traditional use, although they already tend to broaden the meaning of the notion of a natural law a little by sometimes applying this notion to regularities that were labeled rules or propositions by traditional writers. This tendency was supported and extended by other members of the Royal Society, like Hooke and Boyle, and later Newton, who suggested to call all empirical regularities natural laws, quite independently of their position in a deductively organized scientific theory. Natural laws no longer had to be axiomatic principles. As Steinle points out, one of the consequences of broadening the notion of a natural law was an increasing metaphysical debate about the force of causality operating in natural laws and about the relation of this force to God's will. Another consequence was that the notion of natural laws became a lot more flexible, vague and heterogeneous and could therefore be taken to cover the results of very different scientific programs entertained in the Royal Society, in particular the discoveries of the new experimental philosophers. This impact proved to be helpful for defending the research program of the Royal Society against attacks from theologians, since the members of the Royal Society, by maintaining the voluntaristic character of all natural laws, were able to present their work as revealing more and more facts of the universe that were actually results of God's decisions. In this way, the new research program of the Royal Society could be seen to strongly support the admiration of God's actions.

Claus Zittel takes Francis Bacon's dictum "truth is the daughter of time" as the point of departure for discussing three different interpretations of Bacon's understanding of history: The skeptical-relativist, the optimistic-evolutionary and the eschatological one. The messianic interpretation is refuted by showing that for Bacon divinely revealed knowledge implies an ideal of knowledge which is useless for exploring nature insofar as it is accessible to human beings. By reading Bacon's metaphor that time is like a river as highlighting the fortuity and contingency of the transmission of knowledge, Zittel rejects the evolutionary interpretation of history as well. It is the relativist reading of Bacon's concept of history and the historical development of knowledge that turns out to be correct. Proceeding from the assumption that philosophical and scientific truths are changing in the course of history, Bacon reduces ancient scientific texts and natural histories to a mere source of material for further research. The status of this kind of inherited knowledge remained unclear, though. However, an encyclopedic collection of a vast number of empirical facts and data within the framework of a relativist view on the history of science and knowledge generates the difficult problem of ordering the collected data without using the traditional models of ordering. One of the methodological consequences of this difficulty is that Bacon begins to write aphoristically and attempts, when trying to order the diversity and chaos of the data, to make them somehow manageable without prematurely foisting illusory models of uniformity upon them. Zittel shows that Bacon, in selecting and gathering empirical material, does not exclude change, but instead uses it productively. Therefore, the scientific method developed by Bacon according to Zittel by no means functions like a "machine". When Bacon talks about form this should not be understood as a sort of structure or substance, but as an indication of similar effects. The discovery of form, understood in this way, does not guarantee a stable unified order of nature. Relying on Bacon's late work, *Sylva Sylvarum*, Zittel demonstrates that even an order as clear and rigid as that of the Centurions did not for Bacon instantiate a Platonic encyclopedic ideal. Actually, the Centuries in *Sylva Sylvarum* no longer fulfill the function of ordering things. Paradoxically, it was precisely the most highly ordered concept of form ever to serve as the foundation for one of his works that made possible the most extensive presentation of diversity and chaos. Here too, Bacon did not want to eliminate the "woods of experience," but to move within them. Bacon describes this procedure as a higher type of natural magic that does not aim at ordering nature, but rather at breaking it into "great and strange works." In this way Zittel shows that Bacon's understanding of history, his methodology and his encyclopedia, which are usually taken to be three different areas of Bacon's philosophy, are intrinsically related to each other.

The leading question *Klaus Reichert* raises is whether there is a connection between apocalyptic thinking and science in 17th century culture. Reichert proceeds from the observation that it is primarily in periods of cultural crisis that theological apocalyptic thinking is crucial for attempts to understand cultural developments. He firstly looks at what one might call the culmination of the close relation between theology and science in the 17th century, namely at the way Isaac Newton connects his physics to his interpretation of the book of the books. It is well-known that Newton wanted to make sure that the sayings of the prophets turn out to be clear, true and consistent with modern physics. Reichert stresses though that

Newton suggested a sharp methodological distinction between doing physics and interpreting the old scriptures. Reichert shows that Newton's conviction – although science without theology is impossible, science and theology are based on different methods – was entertained by many leading writers and scientists in the 17th century before Newton. In particular, this was not an assumption that was primarily proposed by Puritans. Many Catholics in the Royal Society for example worked on problems of physics in the way just outlined. One of the most important consequences of this scientific habitus in the 17th century was that a fallibilist notion of scientific knowledge could be taken to be consistent with infallible claims to theological knowledge. As Reichert sees it, Bacon's work was crucial for the development of this situation. In the overall research program proposed by Bacon every single science and discipline was supposed to rely on its own methods: In this way, the traditional ideal of knowledge was split into many different ideals. In particular, Bacon reserved the possibility of perfect knowledge for theological knowledge. However as Reichert shows extensively, Bacon related his vision of natural sciences and their evolutionary development. Bacon looked at the constant improvement of scientific knowledge as instantiation of an eschatological program, as fulfillment of theological apocalyptic prophecies. From this apocalyptic point of view the most important mission of the sciences was, for Bacon, the discovery of the deepest secrets of nature and of God's actions. This final discovery was supposed to be simultanious with the apocalyptic end of the world. It is in this way that Bacon tried, as Reichert sees it, to theologically legitimize a social and institutional support of the improvement of all sciences.

Wolfgang Detel looks at the relation between skepticism and scientific method in Gassendi's physics by way of examining, not only what Gassendi says about epistemology and scientific method, but also how he proceeds in doing physics himself. From a closer look at the way Gassendi examines the horror-vacui-theory and the famous ship-experiment on inertia conducted several times by Gassendi himself, Detel concludes several facts from Gassendi's picture of scientific method. They include the claims that theories should be approved mainly insofar as they explain and are confirmed by empirical observations and experiments, that statements attributing perceptible properties to things in the external world must form the empirical basis for the tests of theories, that theoretical assumptions about causes of perceptual external states and about their intrinsic nature remain at most probable and in principle fallible. And above all that these three claims are a specification and result of moderate skepticism for the specific case of the natural sciences.

At the same time, Gassendi seems – as Detel points out in accordance with most scholars – to defend the claim that the infallible empirical basis of our claims to knowledge consists of judgements that state that it appears to a person that fact p is the case. This claim however is inconsistent with the picture of scientific method just outlined. As Detel shows, Gassendi seems to make two more claims though: Firstly, that only if it appears to a person that fact p is the case *and* if there are no obstacles for judging about perceivable external things, the appearance of p is essential evidence for p being the case; secondly, that claiming that appearances of p are essential evidence for p under the mentioned conditions is itself a matter of reason and is therefore defeasible. This more sophisticated way of talking about the empirical basis of our knowledge proves to be fully consistent with Gassendi's view of science and applies skeptical probabilism to the empirical basis of scientific knowledge.

However, in determining whether normal conditions for perception and observation have been obtained, reason must itself rely on perception and observation. Detel tries to show that this problem of epistemic circularity is at least indirectly acknowledged by Gassendi. In some of Gassendi's works Detel finds a strategy that can be read as addressing this problem by introducing a version of an externalist epistemology. From all this it follows that Gassendi did not simply try to rehabilitate Epicurean philosophy; he rather looks at Epicurean philosophy as a philosophical position that could most easily be transformed into a promising physics guided methodologically by an empiricist fallibilism.

In paradigmatically different ways *Becker, Vogel* and *Stumpfhaus* explore relations between art and knowledge in Early Modern Europe. While Becker looks at forms of a knowledge of music and at their relations and historical development in the 16th and 17th century, Vogel shows how notions of truth and knowledge can be connected with causal effects of pieces of music for the emotional states of listeners. Finally, Stumpfhaus examines aesthetic ways of representing truths that turn out to be competitors with – and even superior to – representing things linguistically.

Alexander Becker proceeds from the observation that there are at least three different kinds of knowledge in the area of music: Firstly, a technical knowledge of the production of music; secondly, an aesthetic knowledge of standards or criteria that we use to evaluate music and its effects on the listeners; and thirdly, a scientific knowledge of musical phenomena. According to Becker, between the 14th and the 18th century the relationship between these kinds of musical knowledge changed fundamentally since the medieval predominance of the scientific approach was superseded by the end of the 17th century by the predominance of aesthetic knowledge of music. In his paper, Becker examines a short period of this general development in the 16th century. For Becker, one reason for focussing on this period is that in the 16th century debates on music all three ways of knowing music were present and influential, that no one clearly dominated the others and that they were widely interrelated. The details of this situation seem to be fairly confusing: On the one hand, the traditional scientific and mathematical theory of music continues to be accepted as a general framework; there are no theoretical alternatives available. On the other hand, theorists debate intensely on a lot of major and minor problems; it seems as if any element of the traditional theory could be questioned. One of the most important innovations of this debate is, as Becker shows, a new technical understanding of musical theory. According to it, an adequate theory of music no longer has to explain, in mathematical terms, the given musical material present in nature; rather, the task of the theory is to offer solutions to problems of musical practice by developing new constructions of the musical space and to open up possibilities for composers to explore. Another crucial aspect of this development was a change in the standards of justification. In the traditional medieval view a theory of music was justified if it was able to integrate musical phenomena in an account of nature as a whole. The new technical understanding of musical theory in the 16th century allowed a theory to be subjected to the test of compositional practice and to the test of the effects of these practices on the listener's sensory experience. The effects of music were also in the focus of the aesthetic approach to theory. Followers of this approach took a theory of music to be justified if it permitted the composers to produce pieces of music that were able to express emotions and meanings of the texts that were set to music. The most important explanatory term of the aesthetic understanding of theory was the concept of imitation that united an account of compositorical activity with a causal account of the effects of music on listeners. Obviously these three ways of understanding music theory – the scientific, the technical, and the aesthetic – could easily get in conflict with each other. They did so when methods of falsifying a theory that are implied in the technical and aesthetic approach were employed to criticize the traditional scientific corpus of musical knowledge. On the other hand, as Becker argues, the debates in the late 16th century were characterized less by clear–cut distinctions of the different approaches to theory than by attempts to integrate traditional and new conceptions. Nevertheless, the new developments present a fallibilistic picture of musical theory that fits well with the rhetorical culture of knowledge that dominated the 16th century.

Matthias Vogel works out what the aesthetic knowledge of music comes down to in the specific case of Monteverdi, who developed a conception of music in which the scientific and practical knowledge of the composer has to serve a normative aesthetical idea: namely – as Monteverdi puts it – the idea of truth. In particular, he wonders how this knowledge could still be connected to the notion of truth. He takes Claudio Monteverdi's dictum that the composer is working on the foundations of truth as a starting point for a discussion of how the notion of truth can possibly be applied to pieces of music and accordingly of how this dictum can be rationally reconstructed. Vogel firstly looks at influential modern ways of relating art and truth proposed by Habermas, Goodman, Hegel and Adorno to conclude that neither the concept of truthfulness (Habermas), nor a broad notion of adequacy (Goodman) are satisfactory notions for discussing the connection between art and truth; the same goes for the proposal that art is a deficient (Hegel) or superior (Adorno) form of non-conceptual truth. Relying on a distinction between two functions, namely representation and expression that works of art and especially works of music, can realize, Vogel maintains that while we certainly can and should conceive of works of art often as representing something, this representational function is not specific for the arts, since it can also be satisfied by using natural languages. Vogel then shows that John Dewey's concept of aesthetical communication is systematically helpful for reconstructing the notion of truth that Monteverdi applied to pieces of music. Dewey's crucial idea is that an artist, when producing a work of art, anticipates the way a potential recipient understands this work and that the recipient in turn interprets this work as being intentionally produced by the artist. Within this framework, Vogel broadly speaking suggests to talk of a piece of music as being true if it is enabling listeners to identify those emotional affects that the music is intended to express by the effects of listening to the music. This cognitive relationship to music is possible, because music makes use of means that are marked by exaggeration and therefore do not directly evoke the experience of affects but have effects that are specific for the presented affects. Pieces of music that are true in this sense enable the listeners, as Vogel sees it, to identify the emotional affects that these pieces are supposed to express. In this way, the representational and the expressive function of arts and in particular of music, are seen to work together to create a realm of truth and falsity in the area of art and especially of music. And since the effects that are mentioned in this notion of musical truth are specific for works of music, this notion cannot be substituted by a linguistic notion of truth. By interpreting Monteverdi's late opera "Coronation of Poppea" Vogel demonstrates, in the last part of his paper, how Monteverdi put these ideas to work in his music.[7]

---

7 Vogel's reason for looking at this opera was that the participants in the Frankfurt conference, the basis of the collection of these essays, viewed a performance of the "Coronation of Poppea" presented in the Frankfurt opera house.

Bernhard Stumpfhaus suggests a new understanding of the paintings of Nicolas Poussin, the "philosopher painter" (*peintre philosophe*), as he was labeled in the 17th century. Traditionally, the work of Poussin is seen as a sort of cognitive painting that seeks to express, and to translate into the "language" of painting, the stories that are told in previously chosen texts (mainly religious ones). According to the traditional interpretation, in doing this Poussin tries to follow the rules for producing dramatic sequences of actions proposed in Aristotle's *Poetics*. Linearity, logic and clarity are taken to be the main virtues of Poussin's works. By interpreting in great detail two of the most prominent paintings produced by Poussin, *Mannalese* and *Thunder Landscape with Pyramus and Thisbe*, Stumpfhaus shows that the traditional picture is, to say the least, oversimplified. It is wrong simply to identify the logic of the texts Poussin is relying on and the logic of his paintings that are related to these texts. Accordingly, it is problematic to read Poussin's paintings in the same way as linguistic texts and to assume that Poussin himself wanted his works to be read this way. As Stumpfhaus points out, Poussin usually does not rely on just one text, but compiles passages from different texts in such a way that it fits with his aesthetic intentions. These compilations often imply a profanisation of religious sujets. Moreover, on a closer look Poussin often does not comply with Aristotelian poetic rules; for instance, he often does not want to describe causal sequences of actions as outlined in the texts he has in mind, but rather translates causal sequences into simultaneous contrasts that primarily serve aesthetic interests, for example in that they help to express the intensity of emotional states. And the insights Poussin seeks to offer in his paintings are, even if the underlying texts tell Christian religious stories, connected to stoic doctrines shaped by 16th and 17th century skepticism. More importantly, according to the interpretation suggested by Stumpfhaus, Poussin wants to demonstrate that in his paintings he can express certain truths and empirical facts better by using artistic and aesthetic means, than poets or philosophers by using linguistic means. In particular, Poussin proves to be skeptical about the unity and conformity of soul and body. Picking up contemporary philosophical (Cartesian) ideas about a sharp distinction between a mental and a corporeal substance, Poussin seems to deny that facial and bodily expressions of human beings always truly express their mental states. Rather, he seems to think that we have to take into account peoples' surroundings – the landscape, the weather, the plants, the animals, the buildings – to be able to better determine what states of minds they are in. As Stumpfhaus sees it, this is exactly what Poussin wants to bring out in his paintings. In short, in Poussin's work the compilation of different texts, the substitution of simultaneous contrasts for causal sequences of actions, and the holistic interpretation of emotions and other mental states of human beings are regulated, not by Aristotelian ideals, but by Poussin's own aesthetic standards, in order to provide the recipients with insights stemming from a mixture of Stoicism, skepticism, and contemporary philosophy. If his paintings have a cognitive status, then not in the sense that they are supposed to tell exactly the same stories that are told in the texts the painter is relying on, but rather in the sense that the painter uses and combines texts freely to express his own views and the way he understands the chosen texts, by aesthetic means that live up, not to Aristotelian doctrines, but to the painter's own aesthetic ideals.

References

Biagioli, Mario (1996): "From Relativism to Contingentism," in Peter Galison and David J. Stump (eds.), *The Disunity of science. Boundaries, Contexts, and Power.* Stanford UP, 189-206.

Cunningham, Andrew and Williams, Percy (1993): "De-centering the 'Big Picture': The Origins of Modern Science and the Modern Origins of Science," *British Journal for the History of Science* 26 (1993), 407-432.

Der Neue Ueberweg. Grundriss der Geschichte der Philosophie. Die Philosophie des 17. Jahrhunderts, Vol. IV, ed. by Jean-Pierre Schobinger, Basel 1988ff.

Field, Judith V. and James, Frank A. J. L. (eds.) (1993): *Renaissance and Revolution: Humanists, Scholars, Craftsmen and Natural Philosophers in Early Modern Europe.* Cambridge.

Golinski, Jan (1998): *Making Natural Knowledge. Constructivism and History of Science.* Cambridge UP.

Lindberg, David and Westman, Robert (eds.) (1990): *Reappraisals of the Scientific Revolution.* Cambridge UP.

Osler, Margaret J. (ed.) (2000): *Rethinking the Scientific Revolution.* Cambridge UP.

Porter, Roy and Teich, Mikulas (eds.) (1992): *The scientific revolution in national context.* Cambridge.

Shapin, Steven (1999): *Die wissenschaftliche Revolution.* Frankfurt/M.

Williams, Michael (1991): *Unnatural Doubts: Epistemological Realism and the Basis of Scepticism.* Oxford.

Peter Machamer

17th Century Demonstrations[1]

The historiography of 17th century studies and the scientific revolution is replete with images, schemata and slogans meant to provide perspective and unification into the goings on four centuries past. The very concept of the scientific revolution itself was meant to focus scholarly attention on certain facets of historic happenings. The mechanization of the world picture (Dijksterhuis 1959), the role of artist-engineers (Panofsky 1954, Rossi 1962), the view from technology and society (Merton 1938/1959), the continued influence of Aristotelian ways of thought (Schmitt 1983, Geymonat 1965) were but a few of the historiographic selection principles used to weave a narrative unity among disparate wiggling strands of the chronological period. Even chronological periodicity itself is but one such way of narrative as Jakob Burkhardt (1921) showed us all those years ago. In recent times the so called new historiography has sought to stress intellectual and social contexts of the courts, of patronage, of local societies and alternative grand visions of the universe, and to privilege the actors' categories (Osler 2000, p. 7).

Not being loath to enter into this grand fray of proper perspectives, let me provide yet another guiding aphorism through which to tie the old times together. Recall that Aristotle said, in *Metaphysics* (1025 b25): "all thought is either practical [πρακτική], productive [ποιητική] or theoretical [υεωρητκή]." This was the dictum that guided thought down through the late 14th century. Then, in ways too complex to tell here, the synthesis, such as it was, fell apart; the center could not hold. Somewhat later under the influence of the Renaissance humanism, the introduction of printing, the enfranchisement of the individual as epistemic and moral authority, and the economic shift of power from the Mediterranean to the North Atlantic, the practical, productive and theoretical models of thought collapsed into each other[2].

Rephrasing this historiographic slogan more aptly for the topic at hand, it becomes: *Apodeixis* [ἀπόδειξις] (and *episteme*, ἐπιστήμη), on the one hand, and *techne* [τέχνη], on the other, collapse into *empeiria* [ἐμπειρία].

That is, experience (*empeiria*), which is constitutive of the expert practitioner of any art, becomes central to the method of demonstration (*apodeixis*). And production or, as I shall call it later, construction (*techne*) – which itself constitutes an experience – similarly becomes a crucial (dare I say, essential, in this context) aspect of the concept and practice of demonstration (*apodeixis*).

---

1 Many thanks to Jim Lennox, J. E. McGuire and Ioli Patellis. Also thanks to the discussants in Frankfurt, December 2000, where a version of this paper was presented.
2 Aristotle himself foreshadowed this collapse in his Nicomachean Ethics.

These terms and their relations are all familiar from their Greek context, first in Aristotle but then importantly in medicine (in the fight between the rationalists and the empirics). And they have a history throughout the Middle Ages, coming into challenge in the 14ᵗʰ Century with the advent of nominalistic and voluntaristic ways of thinking.

Since my topic is 17ᵗʰ century demonstrations, what I shall to do is mine this nugget-like slogan, and see where it takes us towards understanding some of the changes in the concept of proof and demonstration that exhibit themselves in the 16ᵗʰ and 17ᵗʰ centuries.

Let us first review the concept of demonstration:.

> *Demonstration*: apodeixis, ἀπόδειξις I. A showing forth, exhibiting, Eur.; a setting
> forth, exposition, publication Hdt., Thuc.; a showing forth, proving, proofs, Hdt., Att.
> Pl. proofs, arguments in proof of *tinos*, Dem. II. A display, performance (Hdt.) [From
> Liddel and Scott's *Greek-English Lexicon*]

Now both meanings I. and II. have relevance for our considerations. Showing, pointing, and displaying are the ways we humans have of making things clear. They are also the major ways in which get the attention of others. We set things forth orally or, even more patently, in a two dimensional display so that in the exhibition or display itself connections may be seen. We achieve clarity not just about that to which we refer, but from the very structure of the display itself. It is worth noting that this quasi-spatial representation is one of the fundamental cognitive forms of human knowledge. There will be more on this below.

In Latin, "demonstration" carries this same meaning:

> *Demonstro, -are*: to indicate, to point out, show clearly (literally with the hand, by
> gesture, *figuram digito*)
> Tranf. To explain, describe show; demonstrator, one who points out or indicates. [Cas-
> sell's *New Latin Dictionary*]

And still again in English, we see "demonstrate" coming in, in the late 16ᵗʰ and early 17ᵗʰ Century:

> *Demonstrate*
> 1. To point out, indicate; to set forth
> (2. To manifest, show, display (3. To describe and explain by help of specimens, or by
> experiment (4. To show or make evident by reasoning; to establish the truth of by
> deduction; to prove indisputably [*Shorter Oxford English Dictionary*]

> *Demonstration*
> 1. The action of demonstrating; exhibition, manifestation; (b. An illustration; a sign
> (2. A display, show or manifestation (3. The action or process of making evident by
> reasoning; proving indisputably by deduction or practical proof [*Shorter Oxford En-
> glish Dictionary*]

Notice demonstrating is doing something. It is an activity, which issues in a product, the demonstration. Now, one relevant experience consists in this activity. The experience is, if you will, the attending to the construction of the proof. The construction is the process of producing the product that is the demonstration. It is this experience that convinces the demonstrator; this experience that brings him into a state of knowing.

This constructive aspect of 17th Century science and philosophy was noted well by Amos Funkenstein (1986), who called it *ergetic knowledge* in contrast to contemplative knowledge. He traces it back, as do I, to the constructive character of proofs in Greek geometry, and relates its cause to *metabasis*, or the breakdown of distinct bases or principles for knowledge in each discipline (the Aristotelian doctrine that each science has its own principles). He attributes *metabasis* interestingly to the incoming thoughts about intension and remission of forms in the 14th century. The shift, says Funkenstein, is from "philosophical questions of justification to logico-mathematical questions of representation." (309)

Funkenstein's focus of the change in representation is important, but he is too linguistically mired to grasp the important epistemological differences such a change carries. Briefly stated, the Late Renaissance into the 17th Century saw shift to the individual as the basic epistemological source. This I have called elsewhere the introduction of the Epistemic I (Machamer 2000). This shift made the subjective experience of the individual knower the font of all knowledge. So much has been noted before. But centering all on the individual's experience opens a large problem as to how inter-subjectivity among individuals or teaching and learning may occur. Put in another way, how can private, subjective experience, the basis of all knowledge, be made public and accessible to other people? It would be interesting to draw precisely contrast between this dilemma of modern philosophers and the lack of a similar problem earlier; e. g., in Thomas Aquinas who likewise held all knowledge to be based in sense experience (there is nothing in the intellect unless it was in the senses). Though this would provide an excellent way of seeing the changes than modernity has wrought, there is not time to develop this contrast now. Now the argument is that this taking of demonstration as part of experience and tying it with other forms of experience is seen as the way to make public the discovery and display of knowledge so that it maybe inter-subjective, and by teaching other individuals the proper method of demonstration, they too may come to have the same knowledge.

This form of constructive demonstration, as Funkenstein notes, was with us in early geometry and came to heights with the divine Archimedes, yet it was there too in Aristotle (which Funkenstein fails to remark.). Briefly, let me remind you of how the syllogism works.

Discovery, in syllogistic terms, is the search for the proper middle term of the syllogism, which when found, gets displayed in an appropriate syllogism.

Schematically, the simplest BARBARA form is:

All A is B
All B is C
All A is C

B of course is the middle term connecting A to C, and the reason it connects is because A is included in B, and B is included in C, so that A is included in C. The *included in* relation is spatial (Cf. George Lakoff 1987, container schema), by metaphorical extension, of course, it comes to apply much more. But here the spatial inclusion is displayed, or demonstrated, by the form of syllogism. Its constructive character comes from having found the right middle that displays the proper inclusive connection.

Constructive proofs in geometry work much the same way. In fact, most of Euclid proceeds by phrasing the propositions to be proved in an activity-based language; e. g. Proposition I. 9: To bisect a given rectilineal angle; or III. 25: Given a segment of a circle to describe the compete circle of which it is a segment; or again IX. 18: Given two numbers, whether it is possible to a find a third proportional to them. In many proofs the constructing of the diagram itself and reading off results from what has been constructed is cognitively necessary.

So display and construction in some form is not new to the 17th century. But there is this major change to the individualistic nature of epistemology, to the epistemic I. Historically this shift is the result of a multi-dimensional confluence of causes. We cannot develop these lines at any length here, but tantalizingly to say that the rise of capitalism and individualism, the technological changes brought along with the explorations and exploitations of the New World, the Reformation and Renaissance engineering led to new forms of life. One of which was science and mathematics. The role of printing, the book, tables, diagrams and illustrations – all new more permanent forms of demonstration and display – is a major aspect[3].

It is an important historical phenomenon that the same show and display features of publicly characterizing experience can be seen elsewhere than in literary endeavors. The specific purposes of these forms of life were different from those espoused by natural philosophy, but the way of teaching and goal of learning was the same, and they shared more general, pragmatic goals. The baroque and mannerist painting, both in Southern and Northern Europe, shows the same tendency to reveal and display the secular. One sees it in the new emphases on portraits, landscapes and material commodities. One function here is the crass display of wealth and commercial success. (This aspect was noted by John Berger following Walter Benjamin and placed in the 18th Century and about oil paintings, but I believe it is a credible thesis that it had begun a good century before.)

The same display tendency can be seen in early 17th Century music, most notably with the spectacles developed by Monteverdi and his new musical form, the opera. And in the home too, young women were trained to display their attractions by learning virginals and performing music at home so that suitors could be persuaded to relish their charms.

This latter emphasis has a counterpart in the burgeoning literary form of pornography, which often gets tied explicitly to both natural philosophy and the experimental method (see Franz 1989 and Findlen 1993). But these are just some of the forms that the new way of experience took. If there were time and space much more might be elaborated and detailed.

How such multi-complex domains fit into the theme of demonstration and display is hard to articulate. Such synchronic causal relations are hard to describe, for they are simultaneous and non-linear, complex parallel multi-feedback loops, occurring within non-independent systems. Perhaps this is what Michel Foucault was trying to describe in his archeology, but perhaps not. I also say "causal relations" because I am unsure that metaphysically there is

---

3 I have written somewhat more on some of these aspects in my Introduction to *The Cambridge Companion to Galileo* 1999.

anything else to be coherently committed to, but I do not wish to argue this question now, so if it pleases you more, choose whatever term you will to describe historical forces and their effects on institutions, peoples, cultural practices, ideas, economic systems, and the like. I like causes and I think it is as proper to describe *sui generis* causal social mechanisms (e. g. the institution of a legal system) as it is cognitive or neural causal mechanisms. Further, it should be noted that a commitment to causality and even a commitment to explanation by mechanisms does not commit one to any particular ontology and certainly not to any form of reductionism.

One hallmark of the Renaissance that persists well into the 17th Century was the explicit and often vociferous rejection of Aristotlianism in various forms and guises. As Charles Schmitt and others have pointed out, this was often much more in name than in doctrine. But the hue and cry against the syllogism was widespread and rhetorically effective. Schmitt (1988, p. 800), for example, attributes to Ramus a new mode of exposition founded on visual components, and often consisting in the frequent use of dichotomous tables. This mode was designed to simplify and make things more communicable. This "New Logic" whether from Petrus Ramus (1574) or his predecessors (Cf. Walter Ong 1958, Chap. IV), shows how, at that time, it was thought that some method of demonstration was required to take the place of the syllogism. Knowledge, in order to be taught, had to be codified in ways that would allow people to be trained. Thomas Blundeville's *The Art of Logike* (1599) is typical of the logic and method books that exist at the time. He uses presentations of his results in tables, and the tables themselves reflect the way in which one is to think through a problem. For example, Blundeville exhibits the layout of types of substances in a hierarchical layout, so that the hierarchy of genus types is clear to all.

If we look closely at the 17th Century discussions of method, we find something quite remarkable in their concept of demonstration. Consider William Harvey in his *De motu cordis* (1628) where in chapter 14 he claims: "Since all things, both argument and occular demonstration, show that the blood passes through the lungs and by the force of the ventricles, and is sent to all parts of the body...."(114). Note the phrase "occular demonstration". The anatomical tradition from Vesalius and Fabricius had made much of occular demonstrations, and the illustrated drawings in their printed books displayed and so demonstrated their findings. Harvey here shows himself clearly in this new tradition. But the anatomists' illustrations are the representation of their demonstrations, their activity dissecting bodies and revealing the inner structures and workings of the human fabric. In such activity lies the stuff of *empeiria*, experience. You need to know how to dissect and how to observe, and when done properly the result is a demonstration, performed in front of students and represented in books, of how the heart works. So it was that surgeons first became admitted into medical faculties, rather than being considered as "mechanical practitioners".

As an aside, if there were more time, I would argue at length that the insights in Shapin and Schaffer (1985) are quite consistent with the project outlined in this essay. Specifically, when they argue about Boyle's book "What [Boyle's] *New Experiments* did was to exemplify a working philosophy of scientific knowledge." (p. 49) Or again, about the public dimension they write later, "Boyle's social technology constituted an objectifying resource by making the production of knowledge viable as a collective enterprise: 'It is not I who says this: it is all of us'... In the official formulation of the Royal Society, the production of experimental knowledge commenced with the individuals' acts of seeing and believing, and was completed when all individuals voluntarily agreed with one another about what had been seen or ought to be believed." (p. 78) But while the collectivity is necessary for making knowledge public, and this is an important part of knowledge, it is insufficient to explain how it is that all these 17th Century "heros" believed they alone had the right and true method and each alone had accessed the truth. It does not explain the individualistic epistemology as combined with entrepreneurial vision of the new science. (Cf. Machamer 2000)

The common model for rational representation is described in terms of what is easily visible or what can be clearly and distinctly seen. As Thomas Hobbes, in *De Corpore*, put it, "Demonstration was understood by them [the geometers] for that sort of ratiocination that placed the thing they were to prove, as it were before men's eyes." (1655, p. 86) The representations of knowledge are always spatial displays, very often the preferred form proportional geometry. But tables laying out agreements and exclusions, definitions in terms of subjects being included in predicates, and even pictorial diagrams and illustrations also fall into the spatial modes of representations. Causal, and thus explanatory relations, are conceived in terms of spatial, often mechanical models and metaphors.

Necessity attends to these representations because they can be seen to be true by anyone who properly attends to them. They provide the means for the novice to have the requisite experiences to become knowers. One implication from this new pattern of display is so clear that it hardly needs be pointed out; spatial relations as primary mode of understanding lends itself well to an ontology of three dimensional bodies and their motions, for these are easily picturable, and so may be represented in pictures. Body and motion, despite subtle differences, was *the* ontology of the new methodologists (even I would say including Kepler). The 17th Century thought in spatial terms, in terms of proportion and ratio; this mode of understanding and representation they took to be the model of intelligibility[4]. This was Descartes' point in the *Regulae* when in Rule 14 he pronounced:

> "The problem should be re-expressed in terms of the real extension of bodies and should be picture din our imagination entirely by mean of bare figures. Thus it will be perceived much more distinctly by our intellect." (p. 438)

Experience, and derivatively the experiences that come from manipulative experiments, is crucial. Experience is essential for it is only in experience that an individual, cognizing subject can interact with the world. This is one aspect of the common theme about learning from the book of nature. More specifically, experience and the experiences attendant upon one's own actions play a part in making geometry clear through the process of geometrical construction. Action also is part of the goal of method and it lies in practices and the making of products that result in more commodious living. Recall that all the 17th Century "heroes" gave as a goal for their science, the practical production of commodities (better living through the new philosophy). Truly, the theoretical, practical and productive sciences have collapsed.

---

4 Timothy J. Reiss (1997) argues at length against the spatiality thesis. The most clear argument is that if spatiality had been that important then algebra would not have been developed and adopted so quickly. Now while it is true that in the mid 18th century algebra does bring yet another cognitive change to science, it still relies on primarily spatial manipulations for doing proofs and probably on spatial cognitive models for intelligibility. This, of course, would have to be argued at length.

As noted above, the scheme of demonstration that is needed cannot be logic, for, according to Francis Bacon (1620), history shows that the pursuit of logic left to its own course is not to be trusted, for logic is too weak for curing the old disease, i. e. inadequate to sort out the confusions in natural philosophy. "... logic ... though it may be very properly applied to civil business and to those arts which rest on discourse and opinion is not nearly subtle enough to deal with nature."

We need, says Bacon, demonstrations, but not of the usual syllogistic type. We do not need "that method of discovery and proof according to which the most general principles are first established, and then the intermediate axioms are tried and proved by them, (for this method) is the parent of error and the curse of all science..." (p. 67)

For Bacon "the best demonstration by far is experience..." but experiences must be obtained in an orderly and systematic manner. Note that experience for Bacon is an action, something to be done by an individual. The proper experiences do not just happen to a person. This is the haphazard method that he criticized.

Galileo's mechanical and common sense experiments are also described in such perceptual terms. A man of genius, a philosopher like Salviati or his Academician friend, takes up a phenomenon to be explained or as in *Dialogo* (1632) a controversy to be resolved. If the interlocutors do not follow the method of the philosopher, they will reach false conclusions like that illustrated in the story that Galileo tells in Day 2 about the famous doctor who carries out an anatomical dissection aimed quieting a controversy between the Galenists and the Peripatetics. The traditional philosopher observing this finally remarks: "You have made me see this matter so plainly and palpably that if Aristotle's text were not contrary to it, stating clearly that the nerves originate in the heart, I should be forced to admit it to be true." Recall that Galen was the advocate of *empeiria*.

Here is a theme that we have seen before in different ways. Demonstration (especially in anatomy) epistemologically (as well as etymologically) consists in laying a phenomenon before oneself and others. This "laying out" exhibits the structure of the phenomenon, exhibits its true nature. What is laid out provides an experience for those seeing it. It carries informational certainty that causes assent.

I have written elsewhere about the balance as the model for what is intelligible during this period in history (Machamer 1999). I mention it again in this context to show how it was a model for demonstration. Of course, the balance and all the Archimedian simple machines in their mathematical form are used to give equilibrium proofs. But physically, it is obviously observable when the balance was in equilibrium, when the weights and arms were equalized. Any individual could judge when a problem had been solved, when "things were right," in right proportion [*ratio rectus*]. This was a concept of correctness, proof or right reason that could be easily taught. It was a way of interpreting phenomena that anybody could learn, and the standard for success was patent. There was no question of whether you had a proof or not; it was easily seen. Those who would not accept this model of intelligibility would not open their eyes. Personal ambition (e. g., claims to priority over Galileo), dogmatism or authority blinded them.

These are the experiences that make for demonstrations. Baliani says it well to Galileo when he writes: "I intend ... to show that science does nothing but to seek the causes [which] belong to a different habit called wisdom ... and just as the principles of the sciences are customarily definitions, axioms, and postulates, ... in physical things these are for the most part experiences, on which are founded astronomy, music, mechanics, optics, and all the rest." (July 20, 1639) The point is that experiences are literal experiences, literally *seeing* mechanical, optical and astronomical objects *as* the idealized objects of geometry. They also are the experiences of constructing proofs about mechanical, optical, musical and astronomical objects. All experiences are seeing things as they are accordingly to your model of intelligibility.

Galileo's geometry is quite representative of the time. It is the geometry of inscribed and proscribed circles. It is the geometry of mean proportionals (*media proportionales*). This is just the geometry of co-alternate angles and similar triangles. Similar triangles are the way to represent the balance. (Cf. Hooper 1992, p. 348)

It was in this way that the *more geometrico* provided the model of intelligibility and proof for science. The geometry involved was not a pure geometry but the physical geometry of the mixed sciences; an equilibrium geometry tied to experience. The visual paradigm of equilibrium proofs for the simple machines brought together the Galilean tenets of experiment, long observation and rigorous demonstration.

Experiments for Galileo were ways of providing for himself and others first person experiences in order to discover or verify relationships holding in nature. They were ways of demonstrating that the phenomenon that had been geometrically described actually could produce the results claimed. These experiences made the terms of the natural mechanical explanations meaningful and real to the person constructing or discovering. Mechanical models, comparable production of effects, and commonplace experiences all played the same set of roles in Galileo's thought. They were not merely rhetorical devices invoked to convince others about the certainty of conclusions, they were also epistemic requirements necessary to give the proper experience to the individual scientist putting forward the explanation. So it is that in *Discorsi* (1638) Galileo has Salviati reply to Simplicio's query about falling bodies as to whether "this is the acceleration employed by nature in the motion of her falling heavy bodies"?:

> "Like a true scientist, you make a very reasonable demand, for this is usual and necessary in those sciences which apply mathematical demonstrations to physical conclusions, as may be seen among writers on optics, astronomers, mechanics, musicians and others who confirm their principles with sensory experiences that are the foundations of all the resulting structures ... Therefore as to experiments the Author [Galileo] has not failed to make them, and in order to be assured that the acceleration of heavy bodies falling naturally does follow the ratio expounded above, I have often made the following test [prova] in the following manner ..." (212; 169)

"prova" here does mean test, but it is used similarly as it is in the phrase "the proof (or test) of the pudding is in the eating". Even in his letter to Castelli (1613) Galileo speaks of "those natural conclusions of which the manifest meaning or the necessary demonstration have made certain and sure". (NE V 279-88)

For Descartes, this is the point of analysis, the reduction to simples. Analysis lays out the parts, the substructures. For Bacon, the laying out was done by constructing the tables that organized and analyzed (interpreted) experience. The tables spatially represented presence, absence and difference. While for Galileo the exhibition or laying out that commanded assent was the reduction and representation of the phenomenon in a mechanical model, and the geometrical description of how the machine worked. Descartes makes this point in more abstract terms in Rule 14 of the *Regulae* (1628):

"Accordingly, in all reasoning it is only by means of comparison that we attain an exact knowledge of the truth. This is the way syllogisms work. But as we have frequently insisted, the syllogistic forms are of no help in grasping the truth of things ... Comparisons are said to be simple and straight forward only when the things sought and the initial data participate equally in <a common nature>. The chief part of human endeavor is simply to reduce these proportions to the point where an equality between what we are seeking and what we already know is clearly visible."

Equations lay out parts of things, so this just repeats the general point about demonstration. But equations that make the known and the unknown 'visible' are not wholly metaphorical. Manipulating equations literally provides visual, spatial experiences that carry informational certainly. Descartes (1628) becomes clearly Galilean as he continues,

"Nothing can be reduced to such an equality except what admits of differences of degree, and everything covered by the term magnitude. So we understand that all we have to deal with here are magnitudes in general." (pp. 57-58)

Now, when dealing with motion Descartes, of course, makes a well known shift, moving from Archimedes simple machines to collisions (under the influence of Isaac Beeckmann). But the general form of equilibrium proof is the same, and it is interesting to note that it borrows uncritically Galileo's principle of the relativity of perceived motion. For example, consider Descartes' (1644) third law of nature:

"40. The third law: that a body, upon coming in contact with a stronger one, loses none of its motion; but that, upon coming in contact with a weaker one, it loses as much as it transfers to that weaker body.
     This is the third law of nature: when a moving body meets another, if it has less force to continue to move in a straight line than the other has to resist it, it is turned aside in another direction, retaining its quantity of motion and changing only the direction of that motion. If however, it has more force; it moves the other body with it, and loses as much of its motion as it gives to that other. Thus we know from experience that when any hard bodies which have been set in motion strike an unyielding body, they do not on that account cease moving, but are driven back in the opposite direction; on the other hand, however, when they strike a yielding body to which they can easily transfer all their motion they immediately come to rest ..."

Descartes is balancing bodies here, using speed and weight.
     Again we may see a Galilean form of equilibrium demonstration in Descartes' "follower" Christiaan Huygens. He has followed Descartes in changing the form of equilibrium proof from simple machines to collisions, but still keeps clearly the to a form of equilibrium bal-

ance model, and he uses the relativity of motion to great probative effect. Here are two of Huygens (1703) basic hypotheses concerning motion:

> "Hypotheses II: Whatever may be the cause of hard bodies rebounding from mutual contact when they collide with one another, let us suppose that when two bodies, equal to each other and having equal speed, directly collide with one another, each rebounds with the same speed which it had before the collision.
>
> Two bodies are said to directly collide when both their motion and their contact occur in the same straight line which passes through the center of gravity of each body.
>
> Hypothesis V: If two hard bodies collide with each other, the relative speed of the departing bodies is the same as the relative speed of the approaching bodies." (p. 579)

His way of proving this is by moving an observer on a ship, in such direction and speed so that the two colliding balls appear to rebound from each other with equal speeds. This is balancing out things to achieve an equilibrium state.

So these are some examples of demonstrations and there could be many, many others. But let's return to an aspect of the social that was mentioned above.

One theme with which to begin to unwind this age is the relatively new phenomenon of the printing. By the end of the 16th Century, the new printing culture provided readily accessible books, pamphlets, and broadsides through which both formal and informal education and communication occurred. These were read and used in old institutions (schools and universities), new ones (ateliers, academies and private tutorials), and, not the least, by individual readers alone in their homes.

It is not just, as Elizabeth Eisenstein (1983) has noted, the greater accessibility, and consequent greater and more widespread literacy, among various people and classes that is worth noting. The spread of printed material across Europe, of course, meant that more people and more different types and classes of people were reading than ever had before, but it also had the far reaching consequence that education became more standardized, with many people reading the same books (and so getting the same information). Needless to say, there could have been no image of the book of nature open to all if there had been no printed books that were basically the same everywhere. More relevant here, this also meant that the information was presented in new forms confined to two-dimensional page layouts, in part, because this is what could be printed well and clearly (illustrations, tables, diagrams, etc.).

This new reading situation needs to be contrasted with what existed previously with aural learners hearing individual lectures or sermons, or workers and artisans being individually tutored in the idiosyncratic style of the master by whom they were taught. For the first time, someone learning anatomy in Padua would learn from the same text and diagrams that were used by another student in Pisa or even one in Paris. Even the Bible became more standardized so that many people could read the same text in their separate homes, and, though there were different editions, it was now possible for each individual to learn on his or her own and discuss what the Bible actually said, rather than rely on aural memory and the ultimate textual authority of the local priest or pastor[5].

---

5 Cf. Eisenstein (1983).

Printed texts were not only strings of sentences, they also contained representations in the form of spatial layouts, displays, tables and pictures. These became a preferred form of demonstration. Tables allowed a reader to follow a procedure, whether it was instruction for writing a letter, holding a civil conversation, determining the volume of a wine barrel, or performing an inductive discovery. Pictures, reproduced as woodcuts and then as engravings and copperplates, provided visual awareness of fascinating new discoveries and forms of life from far away lands, as well as being a source of knowledge and pleasure about events and practices closer to home. Diagrams and illustrations served as mechanical drawings and detailed models (as "blue prints") for those who would learn to construct and perform. Alan Gross and his colleagues (2000) in a very brief sample of 100 papers, starting in the 1660s, from *The Philosophical Transactions* and *Journal des Scavans* found that 40% of the papers were illustrated in some way or another. (p. 384)

This is one strong reason why geometry took on new life and excitement in the 16ᵗʰ Century. It was being revived by the re-introduction of classical geometrical texts by the humanists and academicians, and concurrently by the mechanics who would use it in their work. This should not be surprising since geometry depended essentially on the construction of diagrams, which now could be clearly and easily reproduced on the printed page, and was, through printing, available to many in ways accessible to them. Reviel Netz (1999) rightly goes even further:

"Mathematics would explode exponentially only when storing and retrieval became much more written, and when the construction of the tool box [the mathematical techniques and proofs] was done methodically rather than though sheer exposure.
My claim in this section is a mirror image of Eisenstein's [1979] thesis on the role of printing in early modern science. Already in early modern science, monographs lead on to monographs, in the recursive pattern which we now associate with the development of science." (p. 238)

It cannot be over-emphasized how much the way of learning by reading was a new form of knowledge acquisition. Again, this is not just that printing provided the opportunity for more people to read, and so more people learned to read, and so literacy increased. Nor is it just that more books were printed in the vernacular so that they would appeal and sell to these newly literate audiences. These were important features of the new phenomenon, but even more importantly, something fundamentally different cognitively happens when a person sits down with a book in hand, and, concentrating on a sentence at a time, reads as an isolated act. The contrast is with the hearing of spoken words from another person; minimally a very different social settings with very different demands on attention. The very cognitive form of learning and memory was essentially changed. We, having grown up with books unlimited, can only imagine what new interior worlds were opened up for these new readers. I think it is as a result of this new interior space, that personal and private imagination became a topic of much speculation and discussion.

The act of reading requires connections among the visual systems and the linguistic system, and together they form new ways of problem solving. Also, the spatial memory system, which is connected to the visual, takes on different more general functions and becomes more intimately connected to the declarative, semantic memory system in frontal problem solving.

But how these systems work and change is still mostly unstudied. We even do not know the cognitive differences between literate and illiterate persons.

The thesis about printing that I have outlined above is not without its problems (Cf. Johns 1998). But I shall not dwell on them here. However, I need to emphasize again that the change in the form of the experience that comes from reading and the form of the information conveyed by the printed page cannot be understood as an isolated phenomenon. The post-Reformation and anti-Aristotelian context in which these changes were occurring also emphasized an anti-establishment, pro-individual ideology, and the need to devise a new, systematic way of codifying knowledge must be seen as part of an attempt to establish intellectual and social stability. For a moment, consider that the very things that comprised knowledge, that made up the inventory of stuff in the world; the very things that comprised the subjects to be known had grown and changed. The voyages of exploration, East and West, introduced the Europeans to multitudinous new kinds of things, plants, animals and peoples. These novelties caused a sense of wonder, awe, and awakened a desire to collect them. Every king and every court, and soon, every rich man would have to build his collection of rare and wondrous objects. But these novelties needed to be understood. They needed to be fit into the very categories of memory so that they could form the basis for experiences.

Such new kinds of things did not fit well into the old systems of knowledge (and some old things did not fit well with the new), and raised many painful and difficult questions. There were questions about the nature of God [Did He give divine grace to the Indians, even though they were not Christians?], about the nature of human beings [Were the Indians really men at all? and what of the man-like apes from the East?], and about the natures of the flora and fauna in the world. In the late 16th Century botanical and zoological gardens began to be established because they provided places where the newly awakened curiosity could be appeased, and as well, places to study the new natures. Such gardens were established at Padua and Parma in 1544, and in Bologna in 1568. Printed books and posters proliferated the images of the unusual for those who could not travel to see the live specimens, but they could still have an experience of them and so come to know.

Let me briefly end this essay by turning back to the printed book. With the availability of printing came many books and broadsheets (and diagrams and pictures) directed towards the newly literate, and growing, middle class. These were works written in the local vernacular, and had a practical bent. Looking at the catalogues for the late 16th Century bookfairs, one sees that most books and pamphlets offered for sale are 'do-it-yourself' style recipe books. This is part of the advent of a constructive epistemology. But one must remark here that this literature most often eschewed any "philosophical" explanations for their topics. All was to be written, as the Royal Society was later to rule, in plain speaking, without excessive metaphor. This anti-philosophical bent fit with two other currents in Renaissance humanism. First, the sense of doing something new and disavowing of the recent past (in favor sometimes the real Greece). Secondly, the development of plain and simple, though often eloquent, vocabulary. These tenets make their way into the 17th century in the form of anti-scholasticism and anti-occultism. Each one of the major figures of the "new science", the mechanical philosophy, holds that he is doing something radically new, that he avoids magic and occult entities and qualities, and that he rejects the older, misguided forms of Aristotelianism. So the new philosophy or science had to be something that was accessible to the newly literate, something that could be printed in an intelligible form, and written in terms, that were plain and could be simply understood. Mechanisms, and possibly only geometrical mechanical mechanisms, fulfilled these conditions. Finding the mechanism by which some-thing worked was how one explained. Displaying mechanism descriptions or pictures was *the* form of demonstration by which one taught, even oneself, in the sciences and most every-where.

This thesis about the centrality of the notion of experience and display has two more important epistemological implications. First, the representation of experience into books, notebooks, diagrams and printed displays constitutes an interesting externalization of mem-ory. This Descartes saw in the *Regulae* (1628) when he alluded to how written proofs allow us to recall, the steps that had been achieved before.

> "Moreover, as we said, we should not contemplate in one and the same visual or mental gaze, more than two of the innumerable different dimensions which it is possi-ble to depict in the imagination. It is therefore important to retain all the others in such a way that they readily come to mind whenever we need to recall them. It seems that memory has been ordained by nature for this very purpose. But because memory is often unreliable, and in order not to have to squander one jot of our attention on refreshing it while engaged in other thoughts, human ingenuity has given us that happy invention – the practice of writing. Relying on this as an aid, we shall leave absolutely nothing to memory but put down in paper whatever we have to retain ..." (Rule 16, p. 67)

The second and related point is that these representations are open for public inspection. They provide the basis for the necessary inter-subjectivity that arises once the epistemologi-cal ego is made paramount and central in the process of knowing. It is not just the standard-ization brought by printing that Eisenstein (1983) refers to. Bacon (1620) refers to this too: "... that type once set can provide an infinite number of copies, while letters written by hand yield only one." (Book I, 110.) But more it is the representation of an experience that may now be had by anyone who is properly trained. And by such experiences all men may come to know.

References

Aristotle (1924): *Metaphysics,* A Revised text with Introduction and Commentary, by W. D. Ross. Oxford: Oxford University Press.

Bacon, Francis (1620): *Novum Organon,* Translated and Edited by Peter Urbach and John Gibson. Chicago: Open Court 1994.

Berger, John (1972): *Ways of Seeing.* London: British Broadcasting Corporation and Penguin Books.

Blundeville, Thomas (1599): *The Art of Logike.* London, reprinted 1956.

Burckhardt, Jacob (1921): *The civilisation of the renaissance in Italy,* authorized translation by S. G. C. Middlemore, 8th ed. New York: Macmillan.

Descartes, René (1628): "Rules for the Direction of the Mind" [Regulae], in John Cottingham, Robert Stoothoff, Dugald Murdoch (eds. and translators), *The Philosophical Writings of Descartes*, Vol. 1. Cambridge University Press.

Descartes, René (1644): *Principles of Philosophy*, Translated by V. R. Miller and R. P. Miller. French translation with additions 1647.

Dijksterhuis, E. J. (1959): *The Mechanization of the World Picture*, Translated by C. Dikshoon. Princeton University Press 1986.

Eisenstein, Elizabeth L. (1983): *The Printing Revolution in Early Modern Europe*. Cambridge University Press.

Findlen, Paula (1993): "Humanism, Politics and Pornography in Renaissance Italy," in Lynn Hunt (ed.), *The Invention of Pornography*. New York: Zone Books.

Franz, David O. (1989): *Festum Voluptatis: A Study of Renaissance Erotica*. Ohio State University Press.

Foucault, Michel (1969): *L'archeologie du savoir*. Paris: Editions Gallimard.

Funkenstein, Amos (1986): *Theology and the Scientific Imagination from the Middle Ages to the Seventeenth Century*. Princeton University Press.

Galilei, Galileo (1613): "Letter to Castelli", in Maurice A. Finnochiaro (ed. and trans.), *The Galileo Affair*. University of California press 1989.

Galilei, Galileo (1638): *The Two New Sciences [Discorsi]*, Translated by Stillman Drake. University of Wisconsin Press 1974.

Gross, A., Harmon, J. E. and Reidy, M. (2000): "Argument and 17th-Century Science: A Rhetorical Analysis with Sociological implications," *Social Studies of Science* June, 384.

Huygens, Christiaan (1703): "The Motion of Colliding Bodies (De motu corporum ex percusione)," Composed mid-1650s (published 1703 Opuscula postuma), Translated by Richard J. Blackwell, *Isis* 68 (1977), 574 ff.

Hobbes, Thomas (1655): *De Corpore*. Molesworth edition.

Geymonat, Lodovico (1965): *Galileo Galilei: a biography and inquiry into his philosophy of science*. Foreword by Giorgio de Santillana. Text translated from the Italian with additional notes and appendix by Stillman Drake. New York: McGraw Hill.

Harvey, William (1628): *The Circulation of the Blood [De motu cordis]*, Translated Kenneth J. Franklin, as. London/Dent 1963.

Hooper, Wallace (1992): *Galileo and the Problems of Motion*. Ph.D. Dissertation. Indiana University.

Johns, Adrian (1998): *The Nature of the Book: Print and Knowledge in the Making*. University of Chicago Press.

Lakoff, George (1987): *Women Fire and Dangerous Things: What Categories Reveal about the Mind*. University of Chicago.

Machamer, Peter (1999): "Introduction," in Peter Machamer (ed.), *The Cambridge Companion to Galileo*. Cambridge University Press.

Machamer, Peter (2000): "Individualism and the idea(l) of Method," in P. Machamer, M. Pera and A. Baltas (eds.), *Scientific Controversies*. Oxford University Press.

Merton, Robert K. (1938): *Science, Technology and Society in Seventeenth Century England*. New Jersey: Humanities Press 1978.

Netz, Reviel (1999): *The Shaping of Deduction in Greek Mathematics*. Cambridge University Press.

Ong, Walter (1958): *Method and the Decay of Dialogue*. Harvard University Press.

Osler, Margaret J. (ed.) (2000): *Rethinking the Scientific Revolution*. Cambridge University Press.

Panofsky, Erwin (1954): *Galileo as a critic of the arts*. The Hague, M. Nijhoff.

Ramus, Petrus (1574): *The Logike of Petrus Ramus*, Translated by Roland MacIlmaine. Northridge, CA: San Fernando State College 1969.

Reiss, Timothy J. (1997): *Knowledge, discovery and imagination in early modern Europe; The rise of aesthetic rationalism*. Cambridge University Press.

Rossi, Paolo (1962): *Philosophy, Technology and the Arts in the Early Modern Era*, Translated by Salvator Attanasio. New York: Harper and Row 1970.

Schmitt, Charles B. (1988): "The Rise of the Philosophical textbook," in Charles B. Schmitt and Quentin Skinner (eds.), *The Cambridge History of Renaissance Philosophy*. Cambridge University Press.

Schmitt, Charles B. (1983): *Aristotle and the Renaissance*. Cambridge, Mass.: Published for Oberlin College by Harvard University Press.

Shapin, Steven and Schaffer, Simon (1985): *Leviathan and the Air Pump: Hobbes, Boyle and the Experimental Life*. Princeton University Press.

Mario Biagioli

Picturing Objects in the Making:
Scheiner, Galileo and the Discovery of Sunspots[1]

Troubled dichotomies

Historians of early modern science and medicine have distinguished two kinds of illustrations, and situated them at opposite ends of the spectrum of pictorial evidence: schematic, diagrammatic, and normative pictures on one side, and realistic, mimetic, descriptive illustrations on the other. The former images are said to abstract from the physical details of the object, while the latter are supposed to identify an object as a specific physical entity.[2]

This distinction between diagrammatic and identifying images finds support both in standard histories of early modern printed illustrations as well as in the recent historiography about notions of evidence in early modern natural philosophy – especially the distinction between the everyday "generic" evidence of the Aristotelians, and the singular, specific, and often instrument-produced evidence of the new natural and experimental philosophy.[3] The distinction between diagrammatic and identifying images is also found in recent narratives about the development of scientific disciplines and their representational styles. Van Helden and Winkler have argued that astronomy changed its disciplinary character and developed its own "visual language" in the first half of the 17th century.[4] This change was epitomized by a movement from geometrical diagrams and schematic illustrations (traditionally used by astronomers to discuss planetary motions, eclipses, or to measure the size and distance of celes-

1 This paper was outlined almost ten years ago as part of a volume of translations and commentaries on the dispute between Galileo and Christopher Scheiner concerning the discovery of sunspots. Initially, Albert van Helden and I had planned to be co-authors but, after many vicissitudes, the volume is approaching completion through the efforts of Albert van Helden and Eileen Reeves (*Galileo vs. Scheiner: The Sunspots Controversy of 1611-1613* (forthcoming)). I wish to thank Al for the many discussions on issues of pictorial evidence in general, and on the representation of sunspots in particular. This essay is conceptually and materially indebted to those discussions, his suggestions, and his published work (even when it challenges it) in ways that may be difficult to specify in acknowledgments or footnotes. I also wish to thank Sande Cohen and especially Claus Zittel and Kriss Ravetto-Biagioli for all the comments, criticism, and plain corrections they have provided. Unless otherwise noted, all translations utilized in this essay are from Albert van Helden's and Eileen Reeves' forthcoming book.
2 Literature that proposes some version of the typifying/identifying dichotomy include: Ackerman (1985), Ashworth (1991); Ivins (1952); Edgerton (1984 and 1991); Harcourt (1987).
3 For a forceful statement about the impact of the printing press on the production and dissemination of images, and the distinction between identifying images (mechanically produced) and typifying images (handmade) see Ivins (1953). Various dimensions of singular evidence are discussed in Dear (1987); Daston (1991); Shapin/ Schaffer (1984); Licoppe (1996); Jones (2001). A long-duree analysis of the interest in the singular, unusual, and wondrous is provided by Daston/Park (1998).
4 Winkler/van Helden (1992 and 1993). Other views of the relationship between discipline-formation and disciplinary pictorial styles are Rudwick (1976) and Lynch (1985).

tial bodies) to accurate pictorial representations of celestial bodies not as geometrical constructs but as physical objects (as, for instance, the increasingly "realistic" pictures of the lunar surface).

The astronomers' early use of diagrams and schematic pictures did not simply reflect the unavailability of telescopes, but also the socioprofessional status of astronomy. Given the traditional hierarchy between mathematics and philosophy institutionalized in late medieval and renaissance universities, astronomers were not supposed to inquire into the physical features of celestial bodies or the physical causes of their motions. These questions had traditionally fallen in the domain of philosophy. Therefore, the increasing use of pictorial representations in 17th-century astronomy helped to reconstitute astronomy as a "philosophical" discipline – one that dealt with the physical nature of celestial bodies and of the causes of their motions, not just with their geometrical description. In this regard, the dichotomy between schematic and realistic pictures touches on yet another central theme in the history of early modern science: the distinction between nominalist and realist positions concerning the epistemological status of mixed mathematics.[5]

I do not believe these historiographical narratives have much to gain either empirically or conceptually by sustaining these oppositions. Dichotomies like diagrammatic/realistic, generic/specific, mathematical/physical, normative/descriptive, realist/nominalist are dangerously comprehensive and deceptively clear. When applied to historical analyses, some of these dichotomies have triggered analogical slips that not only blur the details of specific contexts but also end up producing categorical mistakes. For example, I see no reason to define "specificity" or "physicality" according to the conventions of pictorial realism, or to assume that pictorial realism and philosophical realism need to share the same modes of representation. Often the "realistic look" of an image has very little to do with its role in conveying the physical features or the individual specificity of an object. This coding also implies a progressivist trajectory from schematic pictures (produced by poor telescopes or printing techniques) to pictorially and informationally rich ones (produced by increasingly better instruments, better printing technologies, and more evolved disciplinary styles of visual evidence). But even if we set aside the objections about the stability of dichotomies in general, the opposition between schematic and realistic illustrations runs quickly into empirical trouble in the case of the visual representations of new kinds of objects.

This essay analyzes one such a case: the discovery of sunspots. I start with a discussion of the use of illustrations in telescopic astronomy prior to the discovery of sunspots. I then analyze the different modes of visual representation of sunspots utilized by Galileo and the Jesuit mathematician Christopher Scheiner and try to link them to the debate about the epistemological status of early modern astronomy. I hope to show that the dichotomy between "typifying" and "identifying" images is as unstable and ultimately contradictory as the distinction between nominalism and realism that has been traditionally invoked within discussions about the epistemological status of early modern astronomy.

---

5 Duhem (1969); Westman (1980); Dear (1987); Jardine (1979 and 1984); Baldini (1992a); Cosentino (1970); Biagioli (1990).

The *Sidereus nuncius'* cinematic representations

News of the existence of sunspots began to circulate toward the end of 1611. Their discovery was an important coda to the first wave of astronomical findings that followed the introduction of the telescope in 1609. Not only were sunspots independently detected by a number of people involved in or mobilized by previous telescopic discoveries, but their interpretations and visual representations were closely related to those of the other new astronomical objects observed between 1609 and 1611.

Published in the Spring of 1610, Galileo's *Sidereus nuncius* was the most famous printed report of early telescopic observations. It was also a trend-setting text concerning the use of visual representations of celestial bodies. There Galileo described the discovery of four satellites of Jupiter and of many new fixed stars. He also argued that the Moon was not the pristine celestial body the Aristotelians claimed it was, and that its surface was as irregular as the Earth's. Although philosophical in nature, most of Galileo's claims were constructed and presented through images. Some of them (like the pictures of the lunar surface) seem quite realistic in traditional pictorial terms and testify to Galileo's training in the arts of *disegno*.[6] He used contemporary chiaroscuro techniques for his drawings of the Moon, which were then transferred to engravings. The illustrations of the movements of the satellites of Jupiter, instead, look diagramatic and were printed from woodcuts. Jupiter is represented by a capital "O" and its satellites are designated by simple asteriscs of four different sizes depending on the satellites' luminosity.

It is surprising that some of the *Nuncius'* most realistic-looking pictures are remarkably inaccurate by modern standards. Their effectiveness, however, was not affected by their flaws.[7] Galileo's anti-Aristotelian claim about the irregularities of the lunar surface was conveyed by a sequence of engravings recording how rugged the Moon appeared as it went through its phases.[8] He produced those pictures by looking through the telescope and making washdrawings which he then passed on to the engraver (Fig. 1, p. 42). But as the washdrawings were transformed into engravings, a very large crater emerged almost out of nowhere in the Moon's southern hemisphere (Fig. 2, p. 43). This addition may have been the result of the engraver's poor skills or of the rush Galileo imposed on him, but it may have also reflected Galileo's desire to stress as much as possible the earth-like appearance of the Moon.

The gap between the washdrawings and the engravings is so noticeable that, were Galileo a modern astronomer, he would have probably been charged with scientific misconduct.[9] However, none of his contemporaries seemed to have noticed or commented on Galileo's

---

6 Horst Bredekamp's (2000) "Gazing Hands and Blind Spots: Galileo as Draftsman" provides the latest and most comprehensive discussion on Galileo and his training in *disegno*.

7 Galileo himself complained about them and hoped to produce a new edition of the *Sidereus nuncius* with more extensive and better illustrations of the Moon (Galilei (1890-1909), Vol. X, pp. 299-300, henceforth cited as *GO*). Such an edition, however, never appeared.

8 Galileo's tactics, however, were not exclusively pictorial. He based other arguments for the existence of topographical features on the lunar surface on the size and length of shadows cast by those irregularities to calculate their height. Galilei (1989), pp.51-2.

9 Righini Bonelli (1975); Gingerich (1975); Edgerton (1984); Whitaker (1978); Winkler/van Helden (1992); Feyerabend (1975), pp.99-143; Biagioli (2000).

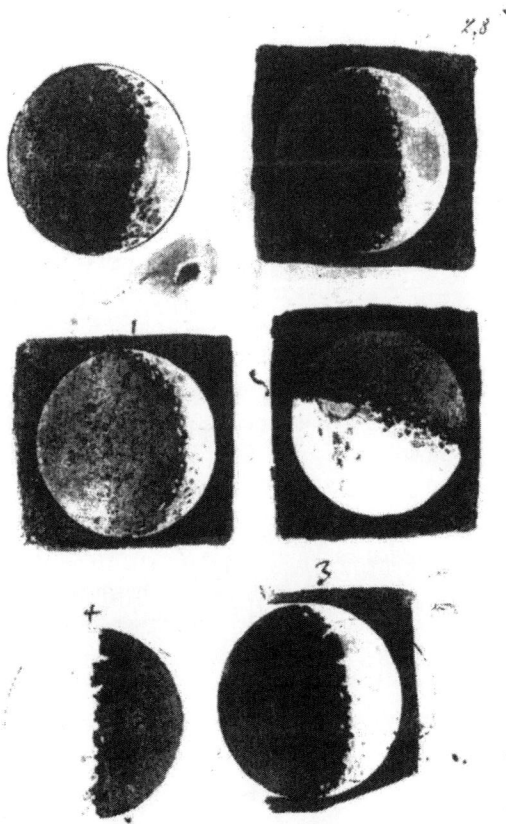

Figure 1: Galileo's 1609 washdrawings of the Moon, Ms. Gal 48, fol. 28r, courtesy of the Biblioteco Nazionale Centrale, Florence.

exaggerations (even when they were further amplified in the unauthorized reprint of the *Sidereus* in the Fall of 1610).[10] Galileo's pictures managed to convey a philosophical point about the physical nature of the Moon not by representing it "the way it was" (that is, in its specificity), but by exaggerating the irregularities of the lunar surface and making them more "generic." These pictures conveyed something about the physical nature of the moon without being mimetic.

Other pictures in *Sidereus nuncius,* however, seem to belong to a different genre: pictures that have the schematic look of diagrams but instead construe their objects as material and physical, not abstract and mathematical. The long sequence of illustrations mapping the movements of the satellites of Jupiter over forty-four days is an example of this visual genre (Fig. 3, p. 44). These pictures were constructed as a sequence showing that the bright spots around Jupiter displayed regular periodic motions which, in turn, indicated that they were neither optical artifacts nor fixed stars. Although they looked like diagrams, these illustrations conveyed a very physical, cosmological claim about very specific objects. Therefore, if we compare Galileo's "realistic" pictures of the lunar surface and the "diagrammatic" ones of the Medicean Stars we see that both these pictures make the *same kind of claim* about the existence of specific new objects but, at the same time, they seem to belong to *different pictorial genres*.

This suggests that the distinction between diagramatic and identifying images is not conceptually useful, and that empirical accuracy is a highly contextual notion that cannot be confused with the pictorial detail or fine texture one sees in a picture. To understand how physical claims could be conveyed either through diagramatic-looking pictures or through

---

10 Galilei, *Sidereus nuncius,* (Frankfurt: Paltheniano, 1610). Not only was the quality of the pictures of the Moon much worse in this edition, but some images were out of order and one was printed upside down. One of Galileo's interlocutors in the debate over the irregularities of the lunar surface, Johann Brengger, used the Frankfurt reprint but did not comment on the bad quality of the pictures (*GO*, Vol. X, p. 461).

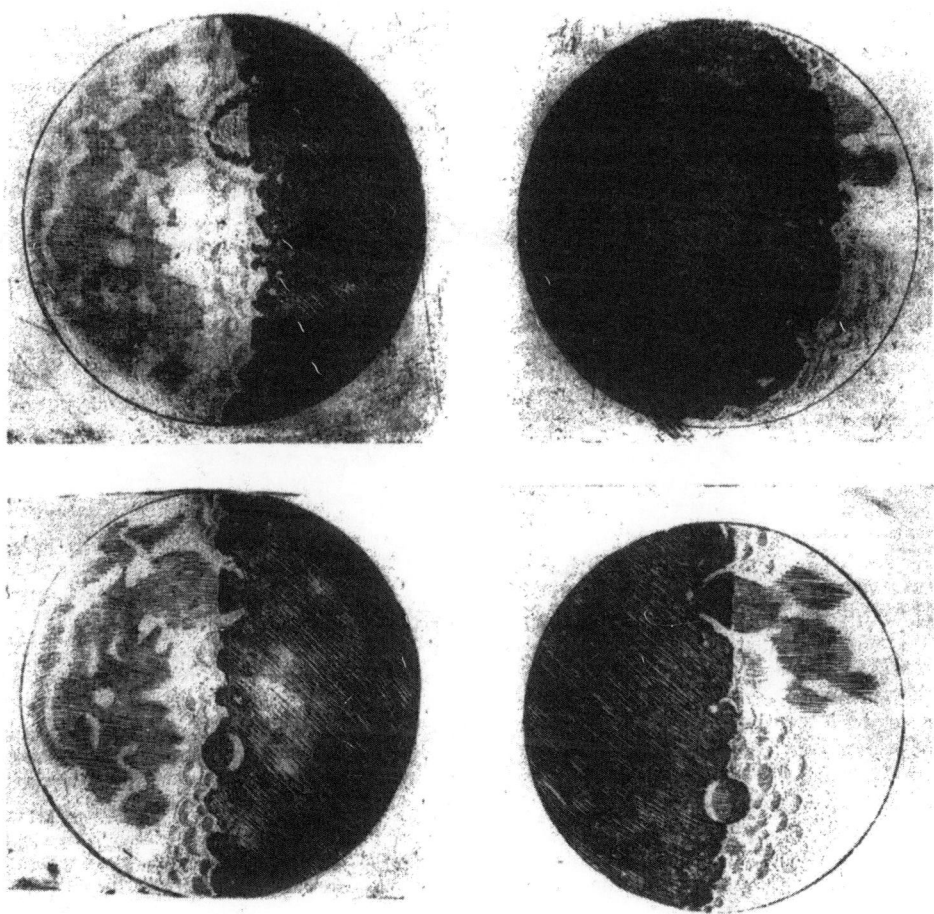

Figure 2: Galileo's engravings of the Moon in *Sidereus muncius* (1610), courtesy of Owen Gingerich. The crater is visible near the terminator in the two lower engravings.

realistic-looking (but substantially incorrect) illustrations we need to reconsider the notion of the referent. The referent of Galileo's pictures, I believe, is not just the object that is purported to be represented. Because an emerging object cannot, by virtue of being emerging, be reducible to an accepted kind, its representations cannot just refer to a kind – a kind that is in the process of being established. The "referent function," therefore, must be supplemented by something else.

The satellites of Jupiter were a new object in the sense that no one had seen them before, but they were not a completely new kind of object because they could be potentially labeled as "errand stars" – a category accepted since antiquity. Moreover, astronomers had long developed protocols to sort out errand stars from fixed stars – protocols Galileo adopted in his representation of Jupiter's satellites in the *Sidereus nuncius*. Because of the combination of previous beliefs about "errand stars," protocols to identify them, and the fact that there was

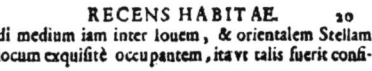

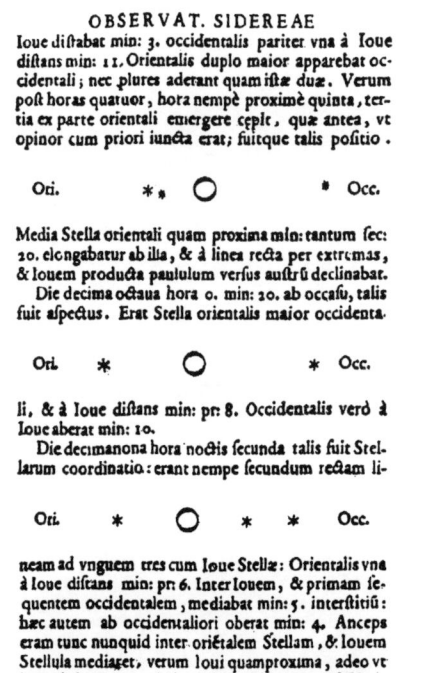

OBSERVAT. SIDEREAE

Ioue diſtabat min: 3. occidentalis pariter vna à Ioue diſtans min: 11. Orientalis duplo maior apparebat occidentali; nec plures aderant quam iſtæ duæ. Verum poſt horas quatuor, hora nempè proximè quinta, tertia ex parte orientali emergere cœpit, quæ antea, vt opinor cum priori iuncta erat; fuitque talis poſitio.

Ori.   * *   O   * Occ.

Media Stella orientali quam proxima min: tantum ſec: 20. elongabatur ab illa, & à linea recta per extremas, & Iouem producta paululum verſus auſtrũ declinabat.
Die decima octaua hora 0. min: 20. ab occaſu, talis fuit aſpectus. Erat Stella orientalis maior occidenta-

Ori.   *   O   * Occ.

li, & à Ioue diſtans min: pr: 8. Occidentalis verò à Ioue aberat min: 10.
Die decimanona hora noctis ſecunda talis fuit Stellarum coordinatio: erant nempe ſecundum rectam li-

Ori.   *   O   * * Occ.

neam ad vnguem tres cum Ioue Stellæ: Orientalis vna à Ioue diſtans min: pr: 6. Inter Iouem, & primam ſequentem occidentalem, mediabat min: 5. interſtitiũ: hæc autem ab occidentaliori oberat min: 4. Anceps eram tunc nunquid inter orientalem Stellam, & Iouem Stellula mediaret, verum Ioui quamproxima, adeo vt illum ferè tangeret; At hora quinta hanc manifeſtè vidi

RECENS HABITAE.   20

di medium iam inter Iouem, & orientalem Stellam locum exquiſitè occupantem, ita vt talis fuerit confi-

Ori.   * • O   * *   Occ.

guratio. Stella inſuper nouiſſimè conſpecta admodum exigua fuit; veruntamen hora ſexta reliquis magnitudine ferè fuit æqualis.
Die vigeſima hora 1. min: 15. conſtitutio conſimilis viſa eſt. Aderant tres Stellulæ adeo exiguæ, vt vix

Ori.   • O • •   Occ.

percipi poſſent; à Ioue, & inter ſe non magis diſtabant minuto vno: incertus eram nunquid ex occidente duæ, an tres adeſſent Stellulæ. Circa horam ſextam hoc pacto erant diſpoſitæ. Orientalis enim à Ioue

Ori.   • O ••   Occ.

duplo magis aberat quam antea, nempe min: 2. media occidentalis à Ioue diſtabat min: 0. ſec: 40. ab occidentaliori vero min: 0. ſec: 20. Tandem hora ſeptima tres ex occidente viſæ fuerunt Stellulæ. Ioui proxima abe-

Ori.   • O• ••   Occ.

rat ab eo min: 0. ſec: 20. inter hanc & occidentaliorem interuallũ erat minutorum ſecundorum 40. inter has vero alia ſpectabatur paululum ad meridiem deflectés; ab

Figure 3: An example of Galileo's maps of Jupiter and its satellites in *Sidereus nuncius* (1610), courtesy of Owen Gingerich.

no explicit authoritative objection to the possibility of finding errand stars orbiting other planets, Galileo could simply tabulate the satellites' periodical motions around Jupiter (no matter how schematic Jupiter or the satellites were presented in print) to convince his audience that these were real astronomical objects.[11]

In this case, then, the referent of the satellites' pictures was not the satellites "in and of themselves" (an object that was not established yet), but the features that made them classifiable within a category of traditionally accepted kind: the errand stars. The traditional features of errand stars informed also the parameters of accuracy according to which Galileo's pictures were to be evaluated. Galileo's pictures did not need to be as pictorially detailed as, say, photographs. That kind of detail would not have mattered in this case. What mattered, instead, was that his sequential pictures of the satellites represented their movements around Jupiter with sufficient accuracy to show that they behaved neither randomly nor like fixed

---

11 The discovery of satellites orbiting planets posed philosophical problems to Aristotelean philosophers who wished to reconcile the Aristotelian belief in crystalline spheres with the existence of these new bodies which, presumably, were carried around by their own spheres. It was not clear what kind of arrangements could make the two spheres coexist. This, however, did not appear to be a cosmological nightmare (Lattis (1994), p. 201). While the existence of satellites was not philosophically unproblematic, it was not a claim that had been ruled out by Aristotelian philosophy (like the existence of the vacuum, action at a distance, etc.).

stars, but that their orbits around Jupiter showed periodical patterns that could not be associated with any known optical artifact.

Assumptions about the possibility or impossibility of new kinds of objects played a direct role in Galileo's representation of the irregularity of the lunar surface too, but ended up producing a very different-looking kind of picture. One could say (as Galileo did) that the valleys and mountains the telescope detected on the Moon belonged to the same category as the topographical features of the Earth (the same way the satellites of Jupiter could be said to belong to the category of "errand stars.")[12] The crucial difference was that while there was no explicit theological or philosophical veto against the possibility of satellites orbiting Jupiter, in the case of the lunar surface Galileo had to confront the authoritative Aristotelian assertion of the incorruptibility of the heavens that ruled out the possibility of topographical irregularities outside the Earth. This constraint, however, came with a few loop-holes.

It may be puzzling that Galileo used pictorially "realistic" but substantially inaccurate pictures to make the particularly controversial claim about the lunar surface's irregularities. The inaccuracy of Galileo's pictures, however, was of a specific kind. Like the diagrams of the positions of the satellites of Jupiter, the pictures of the lunar surface worked as a sequence (albeit a short one). They did show that the changing patterns of bright and dark areas cast by the topographical features of the Moon through different phases displayed a distinct periodicity and that (like the changing patterns of bright dots one could see move around Jupiter) they could not be written off as artifacts of the telescope. Despite their very different pictorial look, both the "diagrams" of the satellites and the "realistic pictures" of the Moon worked as visual narratives. The pictures of the lunar surface were accurate only in the sense that they could sustain a visual narrative about the periodicity of the changing patterns of lights and shadows on the Moon. Galileo's pictures of the Moon (like those of the satellites of Jupiter) did not need to be mimetic maps of its surface. They just needed to convey what mattered to construe his object as non-artifactual: periodicity.[13]

After the existence of the new object was established, Galileo could then use the Aristotelian assertion of the incorruptibility of the superlunary sphere to his own advantage. Because the Aristotelians denied the possibility of the Moon having an irregular surface, all Galileo had to do was to show that, instead, it did display such irregularities. That did not hinge on how accurately Galileo depicted a specific crater. He was arguing for the existence of a kind, not for the particular manifestations of that kind. The "referent" of Galileo's pictures of the Moon, therefore, was both the irregularity of the lunar surface (as a kind) and the erroneousness of Aristotle's cosmological assumptions. And because Aristotle had been so categorical in ruling out celestial corruptability, one just needed to represent some degree of "corruption" to make the point that he was wrong. Paradoxically, the Aristotelians' denial of change, novelty, and corruptibility gave a headstart to critics like Galileo by freeing them from the need to produce "accurate" pictures.

---

12 Galilei (1989), p. 47. In fact Galileo applied optical methods for the computation of terrestrial mountains' height to the features of the lunar surface, the same way he had applied the method for sorting errand from fixed stars to the satellites of Jupiter (Galilei (1989), pp. 51-2). On the the measurement of the height of the lunar mountains see Cajori (1929); Adams (1932).

13 This argument is developed in Biagioli (2000), esp. 561-9.

This may also explain why Galileo's pictures of the Moon were very effective in 1610 but were harshly criticized only a few decades later. That criticism did not come from an Aristotelian, but from a fellow astronomer. In his monumental 1647 *Selenographia* – the most detailed and extensive 17th-century description of the lunar surface – Hevelius commented on the inaccuracy of Galileo's pictures and attributed it to the poor quality of either Galileo's telescope or of his drawing skills.[14] But because Galileo had substantial pictorial skills and telescopes improved only incrementally between 1610 and 1647, Hevelius' statement reflects not so much the evolution of the visual language of astronomy but the transition from a phase in which new kinds of objects were being introduced to one in which they could be simply mapped and catalogued.[15] Galileo's *Nuncius* belongs to the former phase, while Hevelius' *Selenographia* exemplifies the latter (Fig. 4).

Figure 4: Map R from Hevelius' *Selenographia* (1647)

---

14 Winkler/van Helden (1993), pp. 107-8.

15 In a private communication, Claus Zittel has suggested that Hevelius' remarks about the poor quality of Galileo's pictures may not be the result of the different epistemological status of lunar topography in 1610 and

We cannot say, therefore, that Galileo's pictures of the Moon were less realistic or less detailed than Hevelius'. What we find between 1610 and 1647 is not a simple trajectory toward "better" astronomical illustrations, but rather a shift in the definition of what counted as relevant detail. Once you accept that the moon is not a perfect body, you can then use your newly-constituted physical assumptions about the existence of lunar craters and mountains to argue that your pictures, telescopes, and drawing methods are better than those of somebody else because they can resolve more craters and mountains. Similarly, once you have accepted that there might be satellites orbiting planets (that is, after you have accepted the category "satellite"), you can then engage in the search of new satellites, and the telescope which can spot the most of them could be represented as the "best." Observations, of course, could still be easily contested, but the contestations would most likely be about the presence of a specific object, not about the category that object belonged to.

Pursuing novelty, controlling change

If the *Sidereus nuncius* marked astronomy's shift toward questions of a more physical nature, it also epitomized its transformation into a discovery-intensive discipline.[16] New astronomical objects that were almost unthinkable in 1609 had become commonplaces by 1611. As people awaited more discoveries, high-power telescopes were built and used throughout Europe by a range of people hunting for astronomical novelties. It was precisely this atmosphere of competition that provided the context for the debate on sunspots between Galileo and the Jesuit astronomer Christopher Scheiner.

The pursuit of new discoveries was not necessarily connected to a support of the new astronomy. New discoveries could provide evidence against the dominant Aristotelian cosmology by uncovering new objects and processes that may not be accomodated within that framework, but it was often possible to neutralize their disruptive potential – at least for a while. For instance, the Jesuits were able to reframe Galileo's discovery of the phases of Venus within the Tychonic system, thereby deflecting their challenge to geocentric cosmolo-

---

1647. I take Zittel's argument to be that Galileo's brand of empiricism (and the notion of discovery that went with it) was very different from Hevelius'. Hevelius' notion of discovery (which he applied to his selenography) was about placing objects in their place within a comprehensive framework, not about showing the non-artifactual nature of a new finding (as Galileo tended to do). Hevelius' framing of discovery placed him closer to Scheiner (or Bacon) than Galileo. It was this notion of discovery that led him to perceive Galileo's images as flawed. I take Zittel's point about the philosophical underpinnings of Hevelius' remarks, and about how Hevelius' and Galileo's different concepts of discovery would have informed different representational styles. At the same time, while I agree that the philosophical differences between Hevelius and Galileo may explain the specificity of Hevelius' critique, I do not believe they explain its conditions of possibility of Hevelius' criticism of Galileo. That is, I believe that it was because Galileo had succeeded at establishing the lunar topography as an object that, thirty-seven years later, Hevelius had empirical grounds for criticizing it (no matter what the nature of that disagreement may have been about).

16 The discovery of novas in 1572, 1600, and 1604, and of the superlunar trajectory of the comets of 1577 are the exception to the meager list of discoveries one can attribute to ancient and early modern astronomy before the introduction of the telescope in 1609.

gy.[17] Furthermore, some of the new findings were not directly threatening to traditional philosophy, while others that challenged Aristotelian cosmology did not necessarily challenge Ptolemaic astronomy. For instance, the topographical irregularities of the lunar surface were of no relevance to Ptolemaic astronomy, but contradicted the Aristotelian belief in the immutability of non-terrestrial bodies. But, even in this case, some philosophical footwork could suffice to control the anomaly. One could still acknowledge the rugged look of the Moon while denying that it reflected actual topographical features. Some Aristotelians and Jesuits argued that what one saw from the earth was not the Moon's actual surface but an inner surface – a messy layer encased within a perfectly smooth, transparent (and thus invisible) outer sphere. That transparent outer sphere was the real "lunar surface."[18] In sum, while telescopic discoveries had the potential to challenge the cosmological status quo, the actualization of that threat was by no means automatic.

If it was often possible to include some buffer between a discovery and its cosmological implications, it was also possible, to some extent, to separate the discoverer's credit from the philosophical and theological liability his findings may accrue down the line.[19] Most of the audience shared such an attitude about discoveries as it allowed them to be curious and feel pious at the same time. While not everyone shared Galileo's strategy of dedicating discoveries to patrons, discoveries did become standard material for patronage and were keenly pursued by astronomers of different cosmological commitments, Jesuits included.[20]

The main difference between Copernicans and non-Copernicans was not in their interest in new astronomical objects, but in the way they packaged their discoveries. All of them wanted to be first, but while Galileo hoped to find pro-Copernican evidence in forthcoming discoveries, some of the Jesuits and other non-Copernican astronomers hoped to be able to interpret their discoveries in ways that would allow them to gain priority credit without requiring anything more than minor readjustments to the established cosmological framework. The way a discovery was labeled, the way it was severed from or connected to cosmological claims was a subtle exercise in philosophical marketing that could have a great impact on how much credit or trouble a discoverer would incur. Early modern cosmology was such a

---

17 The Jesuits did not accept the Tychonic system until 1620, but their public acceptance of Galileo's discovery of the phases of Venus in the spring of 1611 suggests that they were aware of the possibility of placing those discoveries in a Tychonic framework to diffuse their anti-geocentric potential. It would have been most unlikely that Clavius and his students would endorse a discovery that could effectively refute the philosophical framework in which the Church operated. On the Jesuits' slow path toward Tycho see Baldini (1992b).

18 See footnote no. 45 below.

19 Sometimes one could receive credit for a discovery whose philosophical significance was still debatable. For instance, in the case of the changing appearance of Saturn (announced by Galileo in 1610 after the publication of the *Sidereus*), Galileo received credit for the discovery despite the fact that for a long time it was not clear what the object behind those observations may have been.

20 Although members of a religious institution, the Jesuits were keenly interested in patronage on behalf of their order. In their case, the production or dedication of scientific discoveries was not necessarily a means for developing personal patronage, but a way to open the door for other forms of institutional patronage on behalf of the Society. For instance, Scheiner's work on sunspots and his relationship with Welser was the first step in a long a patronage-intensive life that culminated in his position of confessor of Archduke Maximilian (brother of Emperor Rudolph II) and then of Archduke Karl (brother of Emperor Ferdinand). He developed patronage connections also with Archduke Leopold. Through these appointments, Scheiner managed to gain financial support for the building of churches (Innsbruck) and the establishment of new colleges (Neisse), Gorman (1996), p. 307.

philosophical and theological minefield that all discoverers, no matter what their party affiliation was, had to perform a fine balance between credit and risk. If Copernican advocacy put Galileo in a difficult position vis a vis the Church, the Jesuits' "philosophical laundering" of discoveries may have put them in an even more complex and slippery predicament.[21]

From realism vs. nominalism to credit vs. risk

As the discovery of artifacts carried no credit, both Galileo and non-Copernican astronomers like the Jesuits were eager to claim that what they discovered was something quite real (while occasionally arguing that it was not real if others had claimed priority over it).[22] Their registers of the "real", however, diverged substantially as they reflected differences between the Jesuits and Galileo concerning the philosophical status of mixed mathematics and the relationship between astronomy, philosophy, and theology. In that regard, Galileo has been cast in a realist position, while the Jesuits have been placed on the more nominalist side of the methodological spectrum – a position closely connected to their institutional and doctrinal predicament.[23] But Galileo's position and that of the Jesuits appear inverted when we look at how they tended to conceptualize astronomical discoveries. In that case the Jesuits were more eager than Galileo to determine the physical nature of newly discovered objects, not just the geometrical description of their motions. While Galileo claimed to know how sunspots moved but not what they were, Scheiner stated that they were solar satellites.

The Jesuits' apparently realist stance about discoveries was a direct consequence of their cosmological "bricolages", that is, their recombinations of elements of orthodox cosmological pictures into frameworks that could account for new discoveries or even planetary models. For example, the Jesuits acknowledged the existence of the phases of Venus while bypassing their anti-Ptolemaic implications by framing them within the Tychonic system (which al-

---

21 Much has been made of the tensions between Galileo and the Jesuits – interpretations that cast Galileo as the progressive and the Jesuits as the conservative element. But it may be worth considering that, in some cases, they may have implicitly played a symbiotic "good cop, bad cop" routine. For instance, it is very unlikely that the Jesuits would have been able to publish a text like the *Sidereus nuncius* had they wished to do so. At the same time, the discoveries listed in the *Nuncius* and in Galileo's later letters allowed the Jesuits to endorse claims they may have not been able to make on their own. The enthusiasm with which they endorsed Galileo in 1611 indicates that they were not just "corroborating" his claims, but using them as resources for their internal debates with or against the philosophers of their order. Similarly, while the Jesuits were probably unhappy when Galileo went public with the discovery of the phases of Venus just before they did, they may have also been relieved that they did not have to be the ones to refute Ptolemaic astronomy (Biagioli (2000), pp. 558-9).

22 In the dispute on comets of 1619-1623, Galileo argued that the comets observed by the Roman Jesuits may have been not real physical objects but optical artifacts (Biagioli (1993), pp. 273-80). Galileo's claim emerged in a context in which the Jesuits had been first to publish observations of the comets while Galileo had been sick and unable to produce a comparable body of observations.

23 Different views on the Jesuit mathematicians' (especially Clavius') complex attempt to give epistemological status to the mixed mathematics while avoiding mathematical realism – a stance that would have brought them into conflict with the philosophers and theologians of their order are in: Dear (1987); Jardine (1979); Baldini (1992), pp. 36-56; Cosentino (1970); Crombie (1974); Lattis (1994), pp. 30-8.

lowed them to maintain a geocentric cosmology). They later picked Kepler's elliptical planetary orbits out of their original Copernican framework and plugged them into the more pious Tychonic system. They also endorsed Galileo's discoveries of the satellites of Jupiter (which posed no direct threat to geocentrism) and later used them to explain more controversial discoveries (some of which – like the sunspots – were claimed by members of the Society of Jesus). Therefore, while they may have perceived Galileo's early discoveries as a potential "poison" to the philosophical status quo, the Jesuits quickly turned them into an antidote against more dangerous claims.[24]

The Jesuits' bricolages could be seen as a cosmological version of the astronomers' traditional "save the phenomena" tactics. Traditionally, the astronomers' task had been to break down complex planetary motions to combinations of eccentrics, epicycles, and equants. Although all planetary motions appeared to be philosophically anomalous (as they did not display the uniformity and circularity expected of proper celestial motions), the anomaly was then solved by showing that these motions could be simulated – "saved" – by a combination of circular and uniform motions made possible by eccentrics, epicycles, and equants. By following that practice, astronomers could gain credit for their geometrical descriptions of planetary motions without questioning the philosophers' cosmological postulates.

Similarly, the Jesuits tended to reduce new discoveries to a combination of unproblematic astronomical objects and gain credit for the discoveries (or for their domestication) without destabilizing the philosophico-theological status quo. Their initial tool-box was constituted by those discoveries of Galileo's that had been publicly endorsed by the Jesuit mathematicians of the Collegio Romano in the spring of 1611: the satellites of Jupiter, the phases of Venus, the existence of many more fixed stars, the rough appearance of the Moon, and the non-circular shape of Saturn. For instance, configurations of satellites could be invoked to account for the puzzling and apparently anomalous phenomena of sunspots and of the rings of Saturn.[25] The discovery of the phases of Venus was also enlisted to explain some features of the sunspots. This normalization game could not be played unless one defined what object a given appearance was being reduced to (e. g., that sunspots were satellites). The Jesuits' apparently "realist" stance about discoveries, therefore, had distinctly defensive roots. Not only did the categorization of a discovery allow for its processing into cosmological orthodoxy, but it also entailed fewer philosophical risks than leaving that discovery in epistemological limbo. Not reaching a prompt closure about a new object may have allowed it to develop into an anomaly to Aristotelian cosmology.

---

24 Interesting, Clavius used the term "monster" in relation to the new telescopic discoveries: "And thus I believe that gradually other *mostrosita'* about the planets will be discovered." (Clavius to Welser, 29 January 1611, *GO*, Vol. XX, pp. 600-601).

25 Another satellite-based model of the sunspots was published in *Austriaca Sidera heliocyclica ...* in 1633. Its author, the Belgian Jesuit Charles Malapert, had completed the text in 1628, and died in 1630. He had named the spots after the Habsburg family. Jean Tarde (not a Jesuit) published his *Borbonia sidera ...* in 1620 in an attempt to show (with more technical details than Scheiner had provided) how the phenomenon of sunspots could be saved by an arrangement of satellites. On Tarde and, to a lesser extent, Malapert see Baumgartner (1987). Concerning the peculiar appearance of Saturn, Scheiner himself hinted to its possible explanation through an arrangement of satellites in his early texts on sunspots, but had provided no details about his hypothesis. When Christiaan Huygens put forward the claim that Saturn was surrounded by a ring in his *Systema Saturnium* (The Hague, 1659), the Jesuit Honore' Fabri responded with a satellite-based model of the ring, publish under Eusta-

Galileo, instead, tended to limit himself to showing that the phenomena he had observed could not be dismissed as artifacts. While in some cases he attached physical labels to his discoveries (the satellites of Jupiter, the mountains of the Moon, the phases of Venus), in some others (the sunspots, the peculiar appearance of Saturn) he simply argued that what he had observed was "real" but that he did not quite know, physically speaking, what it was or what caused it. Galileo's apparent philosophical humility, however, was anything but that. First of all, casting his discoveries in phenomenological, non-essentialist terms did not diminish his credit as a discoverer. Although he did not always place his discoveries in specific object categories, he nevertheless managed to convince most of his readers that his discoveries were not observational artifacts by displaying their periodic features (as he did in the *Sidereus nuncius*). Once that step was achieved, Galileo had little to gain and possibly much to lose by rushing to conclusions about the nature of his discoveries. The open-endedness of questions about the physical nature of his discoveries may be also seen as a long-term philosophical investment. Processes that were not understood either by him or the Aristotelians at the time of their discovery were not likely to be explained in Aristotelian terms at a later time. An unexplained new phenomenon was not likely to become a problem for someone like Galileo who did not present himself as a system-builder. It could actually provide him with a double credit opportunity: Galileo could first claim credit for the discovery and then, after the object was recognized, he might receive further credit for explaining its nature or causes. The Aristotelians, instead, were in a much different position. Because of the comprehensiveness, authority, and interconnectness of their system, they may have perceived new unexplained phenomenona as philosophical time bombs.[26]

The Jesuits' and Galileo's epistemological stances about discoveries seem to contradict the former's generally nominalist and the latter's generally realist stance about the epistemological status of astronomy and mixed mathematics. But the contradiction disappears when we cease to treat "nominalism" and "realism" as elements of a philosophical taxonomy and see them, instead, as symptoms of contextual philosophical tactics. For instance, the Jesuits approached the reduction of new discoveries to known physical objects as analogous to the reduction of potentially anomalous celestial motions down to fictional astronomical devices. Satellites were physical objects, but the Jesuits deployed them the way they would have deployed epicycles, eccentrics, or equants, that is, as fictional devices useful for "saving the appearances." The apparent mixing of apples and oranges does not imply a contradiction. The Jesuits' mathematicians wanted (or needed) to defend cosmological orthodoxy or some acceptable variant of it. In one case that goal was achieved through an apparently "realist" stance (the reduction of potentially anomalous discoveries to known objects), in the other with an apparently "nominalist" one (the accounting of anomalous motions through fictional devices).

---

chio Divini's name as *Brevis Annotatio in Systema Saturnium Christiani Hugenii* (Rome 1660). On the debate and on Fabri's model see van Helden (1973).

26 Galileo made a closely related point in a June 16, 1612 to Paolo Gualdo: "These sunspots and my other discoveries are not things that will go away not to return anytime soon, like the new stars of [15]72 and [1]604, or the comets. [New stars and comets] eventually go away thus giving – through their disappearance – a chance to rest to those who, when these [phenomena] were present, experienced some anxiety. But these [discoveries of mine] will torment them forever because they will always be visible." (GO, Vol. XI, p. 326-7, translation mine). I would like to thank Claus Zittel for pointing this letter to me.

The instability of the distinction between "fictional" devices and "real" new objects is also demonstrated by the fact that geometrical devices could be moved from the category of fictions to that of physical objects without modifying their (low) epistemological status. It did not help the philosophical status of astronomy to say that epicycles were not geometrical models but were, instead, physically real. Philosophers could construe that statement as implying that astronomers could only describe (but not explain) planetary motions. Because any given planetary movement could be accounted for through different combinations of epicycles and eccentrics, it meant that those devices (no matter whether one thought of them in physical or abstract terms) could not qualify as the real causes of that planetary motion. "Realism" about devices, therefore, could be easily turned into "nominalism" about astronomy's philosophical status. One ended up in a nominalist position by saying that epicycles and eccentrics were *either* physically real *or* fictional geometrical devices.

This helps to resolve the apparent contradiction between Galileo's aggressive stance about the epistemological status of mixed mathematics and his apparently humbler stance about the physical nature of astronomical discoveries.[27] While in the Jesuits' case the contradiction was solved by finding a conservative tactical common denominator underneath their apparently realist and nominalist positions, a closer look at Galileo's positions shows that his tactics hinged on avoiding that very dichotomy. In one of his letters to Welser, Galileo criticized those who discussed geometrical devices as if they were physical, material objects. But this apparently "nominalist" view did not lead him to say that devices were fictions useful only for astronomical calculations. He argued instead that epicycles were not real the way mechanical gears were, but they were nevertheless very real in the sense that their steady, periodical effects (the orbits of planets) could be clearly observed.[28] Knowing that (for the arguments discussed above) both a strictly "nominalist" or "realist" position about the status of these devices would have undermined the high philosophical status he wanted to give astronomy, he simply bypassed the dichotomy altogether. He did not state that astronomical devices were physically real, but simply that they were not artifactual. It was only by rejecting the essentialist notion of the "real" that informed the dichotomy between real and fictional used by philosophers and theologians to marginalize astronomy that he could be a "realist" about astronomy's epistemological status.

This coincides with Galileo's stance about discoveries. Unlike the Jesuits who tended to put new discoveries in old "real" objects categories when it seemed necessary to control their cosmological implications (the same conservative philosophical politics that led them to treat astronomical devices as "fictional"), Galileo treated both astronomical devices and his discoveries as not artifactual, that is, as not unreal. As with the satellites of Jupiter, the irregularities of the Moon, and the sunspots, Galileo's principal concern was not to say what the "essence" of epicycles and eccentrics was, but to show that, based on their periodical patterns, those discoveries could not be dismissed as observational artifacts even though their physical nature was yet undetermined.

---

27 I am referring to the 1613 "Letter to Castelli" and the 1615 "Letter to the Grand Duchess Christina", in Finocchiaro (1989), pp. 47-54; 87-118.
28 Galileo to Monsignor Dini, 23 March 1615, in Finocchiaro (1989), pp. 60-7; and Galileo Galilei, *Istoria e dimostrazioni intorno alle macchie solari e loro accidenti...*, (Roma: Mascardi, 1613), in *GO*, Vol.V, pp.102-3.

While he did attach traditional nouns to some of his discoveries (satellites of Jupiter, phases of Venus, etc.) those were linguistic labels, not essentialistic definitions. He did not need to attach names to objects to show they existed. And it is very much for the same reasons that he did not need to produce "mimetic" pictures of the objects he had discovered (thus showing how the dichotomy between diagrammatic and identifying pictures is closely related to that between nominalism and realism). He did not need to picture their "essence" but only map their periodic motions to show that they existed. From astronomical devices, to discoveries, to pictorial evidence, to naming, Galileo's positions were structured around a non-essentialist working definition of the real as the non-artifactual.[29]

Change in the Sun?

The tension between a desire to gain credit for discoveries and the need to present them as cosmologically unthreatening framed much of Christopher Scheiner's attitude and behavior during the debate on sunspots. Scheiner had just become teacher of mathematics and Hebrew at the Jesuit college at Ingolstadt when, in the Spring of 1611, he briefly observed with a colleague "some rather blackish spots like dark specks" on the Sun.[30] When Scheiner returned to observe the Sun in October, he started to draw pictures of the spots to map their movements and changing appearances as they slowly moved across the solar disk. He was unaware that the sunspots had already been observed with telescopes by several other people in different European countries (and, without telescopes, by many people in ancient China and medieval Europe).[31] Thomas Harriot was probably the first to observe them with a telescope in December 1610.[32] Galileo showed them to a few friends in the spring of 1611 (including some Roman Jesuits), though he later claimed to have observed them before August 1610.[33] Gentlemen in Padua and painters in Rome had also been observing the spots since

---

29 It also allowed him to avoid unnecessary risks. Venturing into definitions could have brought him into conflict with philosophers and possibly even with theologians. For instance, he did not know what epicycles or sunspots were, but he also had nothing to gain (and much to lose) by saying what they were. Galileo's non-essentialism was also tactically astute. It allowed him to stay clear of the traps of the realist/nominalist dichotomy, and to introduce new astronomical objects without having to define them – something that, given the state of his knowledge, he could not do.

30 Christopher Scheiner, *Tres epistolae de maculis solaribus scriptae ad Marcum Welserum*, (Augsburg: Ad insigne pinus, 1612), in *GO*, Vol. V, p. 25.

31 A survey of pre-telescopic sunspots observations is in van Helden (1996), pp. 368-9. Van Helden argues that while sunspots were frequently observed in the West, they were usually read not as sunspots but as transits of Mercury or Venus across the solar disk. See also Yau/Stephenson (1988), Hosie (1879), p. 131-2; Schove (1948); Goldstein (1969); Sarton (1947).

32 North (1974), pp. 129-157.

33 It is well documented that Galileo showed the spots to people in Rome during his visit in the spring of 1610. If and when he observed them prior to his trip to Rome is a more open question. It is also puzzling that he never published the discovery or communicated it to friends through letters as he did with the phases of Venus or with the three-bodied Saturn. In the *Istoria* he claimed that he had been observing sunspots since November 1610 (*GO*, Vol. V, p. 95). In a letter to Barberini, however, he put that date at December 1610 (*GO*, Vol. XI, p. 305).

the fall of 1611.[34] For sure, the first to publish the discovery was the German Johannes Fabricius. His observations were not dated, but the book's dedication carried the June 13, 1611 date.[35] Ironically, while the discovery of sunspots has been identified with the priority dispute between Scheiner and Galileo, neither of them was first to publish that finding nor could prove to have been the first to observe them.[36]

Based on the observations he conducted in the Fall of 1611, Scheiner wrote three letters to Marc Welser, a politically influential patrician of Augsburg, and a patron of the Society of Jesus.[37] The two had corresponded before about other telescopic discoveries, and Scheiner was eager to pursue the connection. Described as ambitious by his own colleagues (and being once reproached by his superiors for that reason), Scheiner had already perfected his patronage skills with Duke William V of Bavaria, and knew how much curious instruments, objects, and discoveries were appreciated in aristocratic circles.[38] Since hearing about the new telescopic discoveries, Scheiner had been working at building telescopes to test the claims of Galileo and others. He was eager to impress both his patrons and the Jesuit mathematicians of the Collegio Romano.

The brevity and fragmentary nature of Scheiner's three letters on sunspots testify to his concern about establishing his priority over the discovery. One of the letters had been written in less than an hour, and the entire set did not take up ten printed pages.[39] The volume came off the press less than ten days after the completion of the third letter.[40] Scheiner seemed so concerned with priority that he wrote a fourth letter to Welser on January 16, 1612, that is, only about ten days after Welser had published the *Tres epistolae de maculis solaribus*, asking him to publish this fourth letter as soon as he could:

Later on, in the *Dialogue on the Two Chief World Systems*, he claimed to have observed the spots when he was still at Padua, that is, before September 1610. These various statements are discusses in Favaro (1882), pp. 730-41.

34 Righini Bonelli (1970).

35 Fabricius (1611). Substantial portions of Fabricius' small book – it is about 22 pages long – are reproduced in Favaro (1882), pp. 767-76. Favaro argues that Fabricius's text was written in the middle of June 1611 (p. 777). However, it does not seem to have been distributed till the Frankfurt book fair in the fall.

36 Clearly this was a case of independent discovery by a number of people, and for very good reasons. Telescopes had become quickly available since late 1610 or early 1611, and there had been so many discoveries about so many different planets (Moon, Jupiter, Venus, Saturn) that it would have occurred to many people to aim a telescope to the Sun. As Welser himself put it in March 1612, "One should not think it a novelty that in natural philosophy one can find various inventors, each of them ignorant about the other." (*GO*, Vol. V, p. 282). Slow means of communication and book distribution and especially the lack of a standardized concept of what counted as an acceptable priority claim fueled the practitioners' sense of priority.

37 On Welser see Evans (1984); Favaro (1884); Gabrieli (1938).

38 The standard but limited biography of Scheiner is von Braunmühl (1891), but see also Favaro (1919), and, for the later period of his life, Daxecker (1995). The reproach from the General can later in his life, when he was perceived as overextending his duties as archducal confessor to include those of political advisor (Gorman (1996), p. 308).

39 Scheiner's text occupies eight pages (illustrations excluded) in the version reissued together with Galileo's *Istoria e dimostrazioni* (Rome: Mascardi, 1613). I have not been able to access the original Augsburg edition of his work. Scheiner's mention of having written one of the three letters in less than one hour is at the very beginning of the *Accuratior* (*GO, Vol. V, p. 39*).

40 Scheiner's last letter to Welser is dated December 26, 1611. The book came off the press on January 5, 1612.

"I have sent this letter, which has matured for a long time [?], to you especially as a matter of priority, so that [...] you will preserve undiminished this glory of our Germany and your Augsburg, which I trust can be done if the publication is no ways delayed ... Hence I fear that, unless you will anticipate them, they will almost be forced from our hands."[41]

Scheiner did not write the *Tres epistolae* as a book but, quite literally, as three letters. These letters were not installments of a book-length argument, but time-sensitive periodical reports of his work and findings. Each was dated so as to register the time of the discovery, and was addressed to a well-known and internationally respected figure who could testify to having received those letters at that time. Welser was Scheiner's publisher, but he also functioned as the register of his discoveries. Between January and July 1612, Scheiner wrote three more letters on sunspots that Welser then published as the *De maculis solaribus et stellis circa Iovem errantibus accuratior disquisitio* in September 1612. He wrote one of these letters while conducting observations for the next, and sent all of them off to Welser in the hope he would publish them as soon as possible in whatever format and arrangement he wished.[42] The fragmentarity of Scheiner's arguments and the somewhat hurried nature of his sunspots illustrations was, therefore, anything but accidental.

His simultaneous concern for gaining priority credit while maintaing cosmological orthodoxy was inscribed in his argument. His first goal was to establish the sunspots as real objects, not artifacts produced by the telescope, the eyes, or by meteorological conditions. He reported having observed the spots with eight different telescopes to make sure they were not optical illusions. He tried "turning and shaking the tubes back and forth," but even these drastic interventions "never moved the spots along with the tubes, which ought to happen if the tube produced this phenomenon."[43] He also had several witnesses confirm his observations.[44]

Scheiner could have stopped there and say that he did not have sufficient evidence to determine what kind of objects the spots were. Instead, fearing perhaps that his discovery could be used by less pious minds as evidence for the existence of change and corruption in the Sun – a claim that was opposed by the Jesuit theologians and philosophers – he exorcised that possibility by stating that the spots were *not* in the Sun. Scheiner had already taken a stance against the possibility of corruptability in the heavens in 1610 when (with other Jesuits) he disputed Galileo's observation of the irregularities of the lunar surface.[45] It seemed

---

41 Christopher Scheiner, *De maculis solaribus et stellis circa Iovem errantibus, accuratior disquisitio ad Marcum Welserum ...*, Augsburg: Ad insigne pinus, 1612, in *GO*, Vol. V, pp. 53-4.

42 The observations in his fourth letter (the first of the *Accuratior*) was from December 10th, that is, from when he was still composing the third and last letter of the *Tres epistolae*, which he completed on December 26th.

43 *GO*, Vol. V, p. 26.

44 *GO*, Vol. V, p. 25.

45 When the mathematicians of the Collegio Romano were asked by Cardinal Bellarmine to assess Galileo's discoveries, they disagreed over how to interpret the irregularity of the lunar surface – the findings that was most directly related to the issue of celestial change. All of them agreed that the Moon looked rough, but while some of the younger mathematicians thought that its rough appearance reflected actual topographical features, Clavius did not draw that conclusion: "[...] non si puo' negare la grande inequalita' della Luna, ma pare al P. Clavio che piu' probabile che non sia la superficie inequale, ma piu' presto che il corpo lunare non sia denso uniformemente, et che habbia parti piu' dense, e piu' rare, come sono le macchie ordinarie, che si vedono con la vista naturale.

impossible to him to admit that the Sun could display even darker spots or, even worse, to provide ammunition for those who may have wanted to say that the existence of change in the Sun could justify, *a fortiori*, the presence of corruption on the lunar surface:

"It has always seemed to me unfitting and, in fact, unlikely, that on the most lucid body of the Sun there are spots and that these are far darker than any ever observed on the Moon [...] Moreover, if they were on the Sun, the Sun would necessarily rotate on its axis and cause them to move, and those seen first would at length return in the same arrangement and in the same place with respect to each other on the Sun. But so far they have never returned, yet successive new ones have run their course across the solar hemisphere visible to us. This proves that they are not on the Sun. Indeed, I would judge that they are not true spots but rather bodies partially eclipsing the Sun from us and are therefore stars ..."[46]

According to Scheiner, the spots were not dark stains on the solar surface but the shadows of opaque bodies close to the Sun's surface. What people observed as black spots were, in fact, partial eclipses of the Sun. Scheiner had to pull the spots out of the Sun to defend the Aristotelian belief in the incorruptability of the heavens, but taking that step significantly narrowed the range of acceptable objects with which he could identify the sunspots. Scheiner's identification of the spots with "stars," therefore, was almost a default move resulting from his decision to remove the spots from the Sun.[47]

The second letter did not deal with the spots but with what he presented as a new discovery concerning the orbit of Venus. Scheiner seemed strangely unaware of both Galileo's and the Roman Jesuits' observation of the phases of Venus at the end of 1610 – a discovery that was commonly taken to imply that Venus orbited the Sun. Scheiner reached Galileo's and the Roman Jesuits' same conclusion about Venus orbiting the Sun, but he claimed to have arrived there not by observing the phases of Venus but by trying – and failing – to observe that planet's transit across the solar disk. He read his failure to observe Venus' transit as unequivocal evidence that, at that time, Venus must have been above the Sun – a position it could have occupied only if its orbit was centered on the Sun, not the Earth.[48] However, it is virtu-

---

Altri pensano essere veramente inequale la superficie: ma infin hora noi non habbiamo intorno a questo tanta certezza, che lo possiamo affermare indubitatamente." (Christopher Clavius et. al. to Cardinal Bellarmino, 24 April 1611, *GO*, Vol. XI, pp. 92-3). The hypothesis about the transparent sphere was proposed by the Florentine Aristotelian Ludovico delle Colombe to Calvius in a May 27, 1611 letter (*GO*, Vol. XI, p. 118). We know that Scheiner shared Delle Colombe's opinion about the Moon. On January 7, 1611 (before the sunspots debate) Welser wrote Galileo that an unnamed friend of his (Scheiner) did not believe that the lunar surface was physically irregular (*GO*, Vol. XI, p. 13-4). Other debates concerning the Moon's surface involved Jesuits, like Giovanni Biancani at Mantua (*GO*, Vol. XI, pp. 126-7, 130-1; Vol. III, Part I, pp. 301-7). On the broader philosophical and religious implications of claims about the irregularities of the Moon, see Reeves (1997), pp. 138-83.
46 *GO*, Vol. V, p. 26.
47 Another implication Scheiner wanted to exorcise by denying the existence of spots on the solar surface was the rotation of the sun on its own axis. If the spots were found to be on the Sun (not just very close to it) then their motion across the solar disk would be strong evidence of the Sun's own rotation – another sign of celestial change. Scheiner's anxieties would have substantially increased had he realized that the Sun's velocity of rotation itself changed in time and it was going through a substantial accelleration in the years Scheiner was observing it (Eddy/Gilman/Trotter (1977); Herr (1978).

ally impossible that Scheiner had not heard of the discovery of the phases of Venus either from Welser (who was in continuous correspondence with Clavius, the chief Jesuit mathematician in Rome) or from the German Jesuits who had been receiving letters from the Collegio Romano about the progress in telescopic astronomy since early 1611.[49] Scheiner's silence about the phases of Venus in the *Tres epistolae* is all the more puzzling given that, in a January 12 letter to Welser (written less than three weeks after completing the *Tres epistolae*) he did mention both Galileo's and the Roman Jesuits' observations of the phases of Venus.[50]

I believe Scheiner remained silent about the observation of the phases of Venus by Galileo and the Roman Jesuits in order to defend or constitute a small priority claim. He was trying to create the impression that he had independently discovered the true orbit of Venus, and that he had reached that conclusion through a different kind of evidence that was different from theirs.[51] He could, in a sense, claim priority over the method and evidence, if not on the finding itself. Scheiner's competitiveness, then, seemed directed to his own senior colleagues in Rome as much as toward lay astronomers like Galileo.[52] Unfortunately, the alleged discovery turned quickly into an embarassment when Scheiner realized that he had misread the time of the conjunction of Venus and the Sun in the tables he was using. He tried to control

---

48 The orbit of Venus came up as a topic because Scheiner claimed that, if Venus went around the Earth it should have shown itself against the solar disk when at upper conjunction. But when Scheiner tried to observe what he expected to be a partial eclipse, he saw nothing. That led him to believe that Venus at that time was on the other side of the Sun.

49 A student of Clavius in Rome, Paul Guldin, sent a long letter to Lanz (Scheiner's former teacher) at the Jesuit college in Munich on February 13, 1611 detailing the corroboration of Galileo's discoveries by his group (Guldin to Lanz, 13 February 1611, Dillingen, Studienbibliothek, MSS 2, 247, pp. 220-222). Lanz reported on Guldin's letter to Tanner at the Jesuit college at Ingolstadt, adding "I would like to have these things also communicated to Father Scheiner and others who are interested in these things" (Lanz to Tanner, March 1, 1611, Graz, Universitatsbibliothek, MS 159, no. 17,2, Van Helden trans.).

50 "For at about the same time at which Galileo observed Venus horned in several Italian cities, in Rome too ather mathematicians admired them and indeed discovered her in the same shape, horned, bisected, and gibbous." This letter was to become the first section of the *Accuratior* (*GO*, Vol. V, p. 46).

51 This reading is confirmed by Scheiner's own discussion of the discovery of the orbit of Venus at the end of the *Accuratior*. There he lists his "discovery" of the heliocentric orbit of Venus first though it is clear from his own narrative that he did not mean to claim to have been the first to have discovered it: "For if Venus goes aroun the Sun, as was made known in the first painting by Apelles, and gradually established from its daily transformations, and as Tycho Brahe taught some time ago, and as the Roman mathematicians and Galileo observed at about the same time..." (*GO*, Vol. V, p. 69).

52 Scheiner's ambition to be treated on par with the Society's mathematical elite in Rome (and his displeasure with their criticism) comes through in his correspondence. He tried to walk a thin line. As a Jesuit, he was not supposed to let his private ambition override his complete allegiance to the order. Therefore, his discoveries were expected to support the Society's interests, not his personal ambitions as an astronomer. But there was always a slippage between the two ethoses. For instance, in a letter to Paul Guldin in March 1613, Scheiner distinguished priority credit within and without the Society. He cast his discoveries as his own within the boundaries of the order, but as belonging to the order when they were presented to people outside of the Society: "It never came to my mind to be afraid that you would appropriate mine for yourselves. But what I feared has happened, and therefore I was eager to guard against others rushing into the harvest – not mine (for it is mine inside the Society) but ours." The distinction between intra- and extra-Society credit reemerges later on in the same letter: "I write this, Paul, not out of ambition – I have none – but so that we will disclose our [discoveries] to others as *ours*, for in this way ten times more esteem and honor will accrue to the Society in the eyes of other, I beg that on this score you do not fail me, and that if you deem it worthy of light you rescue it from obscurity." (Scheiner to Guldin, 31 March 1613, Graz, Universitätsbibliothek, MS 159, fasc. 1, no. 3, van Helden trans.). The context of the letter,

the damage from this (rush-induced?) mistake by acknowledging that the second letter was "incomplete and not perfect" in a postscript which, however, could not be included in the original printing.[53] Additional logical flaws in his argument put him at risk of looking like an over-eager and impatient novice.[54]

In his third letter, Scheiner went back to sunspots reporting a new observation that "at most, they spend no more than fifteen days on the Sun." He argued, however, that they did not reappear after another fifteen days because "as is apparent from a course of observations of about two months, no spot has returned to the same place and arrangement." One could have argued that the spots' irregular motions and elusiveness contradicted Scheiner's hypothesis that they were hard and permanent objects like stars or satellites. Instead he took that evidence to confirm that "it is impossible that any spot is on the Sun."[55] The puzzling conclusions Scheiner drew from this evidence suggest that he could conceptualize the presence of spots in the Sun only by making an analogy to the topographical features of a rigid planet (like the mountains and valleys on the Moon).[56] Scheiner's "proof" would have collapsed if he had left open the possibility that the Sun was not a rigid body, or that the spots could change so much in fifteen days so as to be unecognizable when they came around again.[57] Scheiner seemed to conceptualize new objects in terms of old objects not only when he was tried to figure out what they were (sunspots as stars or satellites) but also when he was thought about what they could not be (sunspots as spots on the Moon).

Confident of having demonstrated the impossibility of the spots being on the surface of the Sun, Scheiner went on to prove to his satisfaction that they could be neither in the sublunary sphere, nor in the orbs of the Moon, Mercury, or Venus and concluded that: "What remains is that these shadows must revolve in the heaven of the Sun."[58] Finally, having ruled out that the spots could be either clouds ("who would assume clouds there?") or comets, Scheiner stated that they must be "solar stars."[59] But as soon as he selected a safe object category for the sunspots, he found it difficult to say what kind of stars the spots were, or what the cause of their motions might have been. Scheiner could not come up with anything more specific than that "they are moved with their proper motions and those either in a fixed or wandering fashion – which of the two I cannot yet say."[60] Similarly, while he tended to

---

however, shows that Scheiner was trying to enlist the Roman Jesuits to help him establish his priority claims. Scheiner's emphasis on "our discoveries", therefore, may have been something of a carrot to mobilize Clavius' students' support so that he could establish those discoveries as his own.

53 *GO*, Vol. V, p. 32. Welser had this short addendum printed and then mailed to the people he had previously sent the *Tres epistolae* to.

54 That Scheiner did not observe Venus going across the solar disk could not, in and of itself, prove that Venus was orbiting above the Sun. The tables used by Scheiner could have been wrong, Venus could have been too small to be observable against the Sun, etc.

55 *GO*, Vol. V, p. 29.

56 As shown by his correspondence with Welser in 1610 (prior to the sunspots debate), Scheiner did not believe there were mountains and valleys on the Moon. At the same time, he seemed to assume that the only way someone could think of the spots on the Sun would be by analogy to the irregularities of the lunar surface (a phenomenon he denied). In short, the very analogy between the Sun and the Moon implied its own dismissal.

57 Galileo remarked on the circularity of Scheiner's argument in his first letter to Welser, (*GO*, Vol. V, p. 101).

58 *GO*, Vol. V, p. 29.

59 *GO*, Vol. V, p. 30.

60 *GO*, Vol. V, p. 29.

compare these "solar stars" to the satellites of Jupiter, he remained vague when describing exactly what kind of stars produced the spots: "It remains therefore that they are either some denser part of the heaven – and according to the philosophers that is what stars are – or solid and opaque bodies existing in themselves."[61]

He did however address one aspect of the sunspots' behavior that could have been easily turned against his claim that the spots were not in the Sun. As the spots reached the limb of the Sun and moved out of the solar disk, one would have expected them to be observable as distinct bodies located right next to the Sun. Instead they did not appear round but very flat and extremely close to the solar surface (if not actually on it). That is, they looked exactly like spots would have looked, due to foreshortening, had they been on the Sun. To control this serious anomaly, Scheiner drew an analogy between the appearance of his "solar stars" and that of Venus as it went through its phases.[62] He argued that the intense light reflected by one side of the spots made it so bright that it became virtually indistinguishable from the nearby Sun. Under those circumstances all we can observe is the other darker half of the "solar stars."[63] The "thinning" of the spots toward the limb was, in fact, the result of the solar satellites going through phases. In sum, Scheiner was using two recent discoveries endorsed by the Collegio Romano to support his attempt to explain the sunspots in a cosmologically safe manner: the satellites of Jupiter (at least as a model of star-like bodies going around other planets), and the phases of Venus (as a way to explain the effacement of the solar stars). By doing so, he also constructed the sunspots as physically real, that is, as objects he should received credit for discovering.

Scheiner's vagueness was defensive, but had proactive dimensions too. For instance, while he took the satellites of Jupiter to be the solar stars' most likely kin ("It is also consistent that the companions of Jupiter are by no means of an unlike nature as far as motion and place is concerned") he kept open the possibility that the spots could be an example of a larger category of yet undiscovered objects.[64] Perhaps his attempted observation of the transit of Venus over the solar disk made him think that there were many unknown objects orbiting the Sun – objects that became visible only when they crossed over the solar surface.[65] This may explain Scheiner's quizzical quotation at the end of the *Tres epistolae* that "The Sun will also give signs: who would hear the Sun speak a falsehood?"[66] Far from being corrupted, the Sun functioned as a "detector" of new astronomical bodies that it made visible by projecting their shadows toward the observer. The Sun's apparent stains were actually projections of true knowledge. It literally "brought to light" previously "obscured" astronomical objects.

Similarly, Scheiner hinted at a possible use for his "effacing stars" model which could have explained the peculiar appearance of Saturn: "Nor am I altogether afraid to believe something similar about Saturn, namely that it appears at one time of an oblong shape and at

---

61 *GO*, Vol. V, p. 30.
62 Scheiner, however, did not acknowledge the analogy in this first text, probably to pretend he did not know about the phases of Venus.
63 *GO*, Vol. V, pp. 30-1.
64 *GO*, Vol. V, p. 31.
65 "It occurs to me to think that from the Sun all the way to Mercury and Venus, at proper distances and proportions, very many wandering stars turn, of which to us only those have become known which their motion fall in with the Sun." *GO*, Vol. V, p. 31.
66 *GO*, Vol. V, p. 32.

other times accompanied by two lateral stars touching it."[67] Discovered by Galileo in late 1610, the changing appearance of Saturn (later known as Saturn's rings) was still unexplained in 1612 and Scheiner thought he could use a cosmologically safe device to solve a well-known astronomical puzzle. A further example of Scheiner's *modus operandi* comes from his use of the satellites of Jupiter. After comparing the satellites of Jupiter to the sunspots, Scheiner added: "we consider it almost certain that of these [satellites of Jupiter] there are not just four but many, and not carried just in one circle but many."[68] That is, he first analogized the sunspots to the satellites of Jupiter to diffuse their challenge to cosmology, but right after this defensive analogy he moved forward to predict new discoveries by positing another analogy between the plurality of orbital planes of the sunspots and those of the satellites of Jupiter. As with his claims that the puzzling appearance of Saturn could be explained through "effacing stars," Scheiner's first analogy was cosmologically defensive, but the second was astronomically aggressive (if not plainly reckless). The same can be said about his more general conceptualization of the sunspots: the first move hinged on a cosmologically defensive analogy (spots as shadows of stars), but the second (many sunspots suggest the existence of many yet unknown bodies between Venus and the Sun) was an astronomically aggressive analogy.

This process is quite different from a conceptual scheme's ability to direct its supporters toward other discoveries (the way a Copernican would tend to look for the phases of Venus). The discoveries foretold by Scheiner did not fit any specific cosmological "paradigm," but could be made to fit each and all of them. That is why Scheiner was so eager to predict them. His tactic to get credit from philosophers and theologians for making new findings safe, and from astronomers and mathematicians for making or forecasting new safe discoveries matched Scheiner's disciplinary and institutional predicament as a Jesuit mathematician. The problem was that his analogy-based thought style had a tendency to multiply patterns of similarity while expanding their scope. While his initial analogical impulse was a conservative one, the play of analogies (coupled with his desire for recognition) could quickly lead him to make increasingly sweeping and therefore risky claims. For a very short and hastily written tract, the *Tres epistolae* contained enough hypotheses and predictions to keep a school of astronomers busy for decades.

Between one debate and two monologues

Scheiner needed to publish as quickly as possible to claim priority. But, as a Jesuit, he was expected to submit his book manuscript to an internal review committee composed of mathematicians, philosophers, and theologians.[69] That process could have taken months.[70] And because the Society was going through a period of doctrinal retrenchment, it is unlikely that the theologians and philosophers on the review committee would have approved Scheiner's

---

67 *GO*, Vol. V, p. 31.
68 *GO*, Vol. V, p. 31.
69 Baldini (1985).

manuscript.[71] Nor is it obvious that the mathematicians would have supported it. The sensitivity of the topic and the epistolary and fragmentary nature of Scheiner's texts would have not helped his case. It was in this context that his connection with Welser, and Welser's willingness to publish the letters, made all the difference.

Following the orders of his superiors at Ingolstadt who worried about the controversial nature of the topic, Scheiner published pseudonymously as "Apelles latens post tabulam" – a pun on the fact that, like Apelles who hid behind his paintings to hear what people had to say about his work, Scheiner hoped to elicit more candid comments from his readers by not revealing his identity.[72] By having Scheiner publish pseudonymously and without mentioning his institutional affiliation, his superiors, willing to please a powerful patron like Welser, allowed the *Tres epistolae* to be printed without undergoing the internal review and censoring process.[73]

The months Scheiner saved by bypassing the review process strengthened his priority claim over the discovery of sunspots, but it also deprived him of feedback from the mathematicians at the Collegio Romano – feedback that could have helped him avoid the transit of Venus blunder and frame his claims more carefully in the context of current telescopic astronomy. For instance, the *Tres epistolae* did not mention any astronomer (Jesuit or not) or publication concerning the first wave of telescopic discoveries – an absence that may have created the appearance of ungenerosity and excessive concern with credit.[74] More important, senior Jesuit mathematicians could have advised him on how to align his cosmological claims within the positions being discussed by his Roman colleagues around that time. This was a period of intense change for Clavius' students. The discovery of the phases of Venus just a few months earlier had undermined Clavius' defense of Ptolemaic astronomy, while the observations of astronomical "monsters" (as Clavius termed new phenomena like the irregular surface of the Moon, the non-spherical appearance of Saturn, the satellites of Jupiter, and most recently the sunspots) were challenging traditional Aristotelian cosmology.[75] These

---

70 The central censorship board was in Rome, with satellite boards in the various Provinces. Had it gone through a normal review, Scheiner's manuscript could have been sent to the censors of the German Province. But, given its controversial topic, it could also have been sent to Rome.

71 On the Society's General renewed call to doctrinal orthodoxy in these years, see Blackwell (1991), pp. 135-140. Additionally, I doubt that, given the preliminary and rushed nature of Scheiner's text, the mathematicians of the censoring board would have have gone to war with the philosophers and theologians in the committee to get the book published. Baldini's study of the censurae librorum shows that the mathematicians on the board were careful at picking their fights (Baldini (1992a), pp. 217-50).

72 Scheiner discussed the reasons behind his use of pseudonimity in his later *Rosa Ursina*, (1630), Book I, Ch. 2, pp. 6-7.

73 In a 17 January 1612 letter to Guldin in Rome, Scheiner relayed that "Mr. Welser has prevailed on the Father Provincial that he might publish it without the Society of the name of any of us mentioned [...] You in Rome may not reveal Apelles hiding under the painting for it would not please the superiors, but neither does Apelles himself desire it," Scheiner to Guldin, Graz, Universitätsbibliothek, MS 159, 1, 1 (cited in Van Helden, "Scheiner"). On this point see also van Helden (1996), p. 393, note 32.

74 The exception is Giovanni Antonio Magini who was mentioned only as the author of the tables (incorrectly) used by Scheiner to find the time of conjunction of Venus and the Sun.

75 "And thus I believe that gradually other *mostrosita* ‘ about the planets will be discovered." (Clavius to Welser, 29 January 1611, *GO*, Vol. XX, pp. 600-601). The topos of "monster" – fabulosa monstrorum prodigia – was also used in a public lecture at the Collegio in february 1612 to refer to the continuing emergence of new astronomical objects (*GO*, Vol. XI, p. 274).

developments reconfigured not only the Jesuit mathematicians' own positions, but also their relation to the philosophers and theologians within their order.[76] Furthermore, Clavius' terminal illness – he died in February 1612 right as Scheiner's *Tres epistolae* were arriving in Rome – made things more complicated by adding a generational shift to an already delicate scenario.

For instance, Scheiner's near-explicit endorsement of Tycho's planetary model and, later on, of the doctrine of the fluid skies shows that he vastly underestimated the amount of work and negotiations still required to get the Society to accept those positions.[77] And his cheerful prediction of the discovery of more new astronomical objects did not jibe with the Society's increasing conservatism on philosophical matters. While he seemed aware of the scope and stakes of his claims, he did not seem to understand how much the hurried nature of his text (not to mention his reluctance to cite anyone who was anyone in telescopic astronomy) could have turned him into an easy target for mathematicians and theologians alike.[78]

As soon as the *Tres epistolae* came off the press, Welser sent a copy of the letters to Galileo and many other astronomers and savants asking for their comments.[79] Welser's request initiated an exchange that produced three letters from Galileo and a further text from Scheiner. Galileo's letters were eventually collected and printed in 1613 in Rome by the Accademia dei Lincei as *Istoria e dimostrazioni intorno alle macchie solari*.[80] Part of this edition included a reprint of Scheiner's two previous publications. As the small Augsburg editions of the *Tres epistolae* and the *Accuratior* went immediately out of print, most people became familiar with Scheiner's arguments only by reading thems as an appendix to Galileo's book. This bit of publication history greatly contributed to constituting Galileo and Scheiner's texts as a dispute – an effect that has since been canonized by the historiography of science. The chronology of Scheiner's and Galileo's texts, however, suggest a different picture.

It took Galileo several months to respond to Welser. Welser's letter, however, made an impression on Galileo as he rushed to include a few lines about *his* discovery of the sunspots (but no mention of Scheiner's *Tres epistolae*) in the manuscript of his *Discourse on Floating*

---

76 In the *Nuntius sidereus Collegii Romani*, the oration given in Rome in honor of Galileo during his visit in the spring of 1611, Oto Maelcote (Clavius' student) subscribed to Galileo's conclusion that the phases of Venus showed that it went around the Sun. There is no evidence that Maelcote was deviating from the consensus position within Clavius' group. An observer, however, reported that the philosophers were less pleased, saying that Clavius' students had demonstrated "to the scandal of the philosophers, that Venus circles about the sun." (Lattis (1994), pp. 193-4, 197).

77 At the end of the *Accuratior*, Scheiner claimed that "And therefore Christopher Clavius, the choragus of mathematicians of his age, should rightly and deservedly be heeded. In the final edition of his works he warns astronomers that, on account of such new and hitherto invisible phenomena (although it is a very old problem), they must unhesitatingly provide themselves with another system of the world" (*GO*, Vol. V, p. 69). But in the 1611 edition of his textbook on Sacrobosco's *Sphaera*, after referring to Galileo's discoveries, Clavius made the much more modest and conservative claim that "Since things are thus, astronomers ought to consider how the celestial orbs may be arranged to save these phenomena." (Clavius, *Sphaera (1611)*, p. 75, cited in Lattis (1994), p. 198).

78 It is difficult to gauge the reaction of the mathematicians at the Collegio Romano to the Tres epistolae, except that they were probably critical in private and supportive in public. See note no. 127 below.

79 *GO*, Vol. V, p. 93.

80 Galilei (1613).

Bodies which was licensed on May 5, 1612.[81] When he eventually wrote back to Welser in May 1612, Galileo apologized saying he could say little about the spots, and certainly nothing definitive, without conducting more systematic observations.[82] He endorsed Scheiner's claim that the spots were neither metereological phenomena nor optical illusion produced by the telescope, and agreed with his observation of how complicated the motions of the spots appeared to be.[83] Galileo had no problem with all these claims because, as he told Welser, he had been observing the spots and showing them around for the last eighteen months.[84] He had no qualms about Scheiner's claims about the orbit of Venus either because, as he reminded Welser, he had reached the same conclusion "almost two years ago."[85] Galileo, however, was surprised by Scheiner's apparent ignorance of his discovery of the phases of Venus – a discovery that, he claimed, would have also provided Scheiner with a more direct and conclusive way to determine the heliocentric orbit of Venus.[86] Galileo must have been also surprised not to see his name or work ever mentioned in the *Tres epistolae* despite the fact that it discussed some of his other discoveries like the irregularities of the Moon, the satellites of Jupiter, and the peculiar appearance of Saturn.[87]

Therefore, even when Scheiner and Galileo agreed, their consensus was fraught with concerns about priority and credit – concerns that were to color most of their future interactions. Galileo grew more vocal about his priority claims in the 1623 *Assayer* and in the 1632 *Dialogue on the Two Chief System*.[88] That, and the disrespectful remarks that colored those passages, fueled Scheiner's increasing resentment and personal emnity that, eventually, may have played a role in Galileo's trial.[89]

---

81 In the *Discorso intorno alle cose che stanno in su l'acqua, o che in quella si muovono* (Florence: Giunti, 1612), p. 2, Galileo claimed that "Aggiungo a queste cose l'osservazione d'alcune macchiette oscure, che si scorgono nel corpo solare, le quali mutando positura in quello, porgono grand'argomento, o che 'l Sole si rivolga in se' stesso, o che forese altre stelle, nella guisa di Venere e di Mercurio se gli volgano intorno, invisibli in altri tempi, per le piccole digressioni e minori di quella di Mercurio, e solo visibili, quando s'interpongono tra 'l Sole e l'occhio nostro, o pur danno segno, che sia vero e questo e quello; la certezza delle quali cose non debbe disprezzarsi, o trascurarsi." This shows that although Galileo had already observed the sunspots since 1611 or perhaps 1610, he had no clear position about them. The quote from his *Discorso intorno alle cose che stanno sull'acqua...* shows that Galileo had read Scheiner's *Tres epistolae* and acknowledged, as an hypothesis, the possibility of the sunspots being solar satellites. He did not, however, mention Scheiner's publication in order not to support his priority claim.

82 Galileo started to observe the spots on a quasi-regular basis only after he got Welser letter. The first observation recorded is from 12 February 1612. If he did observe the spots regularly before then, the records are lost. The first set of observations stops on May 3. Very likely, this is when Galileo switched to Castelli's system. Between February 12 and May 3, Galileo recorded only 23 observations (*GO*, Vol. V, pp. 253-4.).

83 *GO*, Vol. V, p. 95. Although Galileo always refer to "Apelles" in his letters to Welser, I have decided to drop Scheiner's pseudonym throughout this essay to avoid confusion.

84 *GO*, Vol. V, p. 95.

85 *GO*, Vol. V, p. 98.

86 "I am a bit surprised that it has not come to his ears, or if it has that he has not relied on the msot esquisite and judicious means that can often be used, discovered by me about two years ago and communicated to so many that by now it has become well known; and this is that Venus changes its shapes in the same way as the Moon ..." *(GO*, Vol. V, p. 98).

87 Galileo's surprise is recorded in a letter to Cesi in *GO*, Vol. V, p. 426.

88 For a synopsis of the priority dispute between Galileo and Scheiner see Favaro (1882), pp. 739, 751. Favaro's text, however, is strongly biased in Galileo's favor. Righini Bonelli (1970), pp. 405-10, reviews what Galileo knew about other people's observations of sunspots as he was writing his *Istoria*.

The agreement between Scheiner and Galileo was short-lived anyway. It collapsed around the question of the sunspots' physical nature. Galileo argued that the question was so complex that it possibly exceeded human comprehension, and faulted Scheiner for claiming so confidently that sunspots were the effects of solar stars.[90] Although Scheiner had not taken a consistent position about what "solar stars" meant, he had nevertheless put the spots in an object category. Galileo, instead, wanted to limit himself to an apparently phenomenological discussion of their location, appearances, and motions. This was more than a disagreement about methodology. Galileo saw Scheiner's confident claims about the spots being solar stars as the result of an a priori assumption about the impossibility of change on the solar surface – a postulate Scheiner had openly stated at the beginning of his first letter to Welser.[91] It was that assumption that forced Scheiner to claim that the spots were not on the solar surface but orbited just above it.

By tracking the spots motions and location, Galileo tried to show that although he did not (and maybe could not) know what the spots were about, he could tell that they were not the solar stars Scheiner had claimed they were. By showing that the spots were not above the solar surface but on it, Galileo would have shown that they were not satellites. No matter what those spots may have turned out to be, their location on the solar surface meant that there was change in the Sun, and that meant that Aristotle was wrong in his fundamental assumptions about the property of the elements.[92] As he was sending off his first letter to Welser, Galileo wrote a friend that: "I believe this discovery will be the funeral or perhaps the last judgment of the pseudo-philosophy..."[93]

Scheiner's and Galileo's different stances about the issue of change in the heavens informed the logic and structure of their arguments down to their different use of pictorial evidence. While Scheiner claimed that the spots' uneven and changing appearance could be

---

89 For a critical reassessment of the claims about Scheiner's involvement in Galileo's trial see Gorman (1996).

90 "It remains to consider that which Apelles decides about the essences and substances of these spots, which is, that they are neither clouds nor comets, but rather stars that revolve about the Sun. About such a decision, I confess to Your Most Illustrious Lordship that I do not yet have enough certainty to dare to establish and affirm any conclusion as certain, for I am very certain that the substance of the spots can be a thousand things unknown and unimaginable to us, and the accidents which we observe in them, that is, the shape, the opacity, and the motion, which are very common, can provide us with no, or very little, or too general information. Therefore I do not believe that the philosopher who confesses that he does not know – and cannot know – the material nature of sunspots deserves any reproach." (*GO*, Vol. V, pp. 105-6).

91 "But that they are cannot be on the solar body does not appear to me to have been demonstrated with entire necessity. For it is not conclusive to say, as he does in the first letter, that because the solar body is very bright it is not credible that there are dark spots on it, because as long as no cloud or impurity whatsoever has been seen on it we have to give it the title of most pure and most bright, but when it reveals itself to be partly impure and spotted why should we not have to call it spotted and impure? Names and attributes must accommodate themselves to the essence of the things, and not the essence to the names because the things come first and the names afterwards." *(GO*, Vol. V, p. 97). Scheiner's remark that "It has always seemed to me unfitting and in fact unlikely that on the Sun, a very bright body, there are spots, and that these spots are far darker than any ever seen on the Moon..." (*GO*, Vol. V, p. 26).

92 The publication history of Galileo's *Istoria* shows that the issue of the corruptability of the heavens was a sensitive one even for the censors. As he was writing his text, Galileo asked a friendly theologian whether the corruptability of the heavens went against scriptural teachings. When he was told by cardinal Conti that not only the corruptability of the heavens did not contradict the Scripture, but that it was actually closer to biblical teachings than the Aristotelean veto on it (*GO*, Vol. XI, pp. 354-5, 376) Galileo went on the attack and tried to

reduced to satellites, Galileo argued that the sunspots displayed so many kinds of change that, if one grasped them, it would be unthinkable to try to save the phenomena through complicated arrangements of satellites. For Galileo, the sunspots behaved more like non-homogeneous, blurry, fickle, and fast-changing clouds:

> "The sunspots appear and vanish in variably short periods of time. Some shrink and draw apart greatly from one day to the next. Their shapes change, and most of them are very irregular and display varying degrees of darkness. Because they are on the solar body or very close to it, they must be very large. Because of their varying opacity, they impede the transmission of sunlight to different degrees. Sometimes many appear, other times only a few, and then again none. Now, very large and immense masses, which appear and disappear in brief times, which sometimes last longer and at other times shorter, which rarify and condense, which easily change their shapes, and which display varying degrees of density and opacity, nothing like that is found near us except for clouds…"[94]

The sunspot-cloud analogy was presented as a mere similarity and not as a hypothesis about the true nature of the spots. Galileo's emphasis on the exceptional mutability of the spots was also reflected in the tool he mentioned in the first letter, but did not employ until the second letter: the use of pictorial evidence. Galileo's approach to the sunspots was the same sequential mapping of their movements he had used in the *Sidereus nuncius* to map the periods of the satellites of Jupiter and of the irregularities of the lunar surface.

The ensuing debate, however, was so asymmetrical that "debate" may not be the right term for it. Because of the delays in the correspondence between Florence and Augsburg, the time it took to translate Galileo's letters (Scheiner did not read Italian but Galileo did not want to write in Latin, fearing he would lose his Italian audience), Galileo's own delay in responding to Scheiner's first set of letters, and the delay in the publication of Scheiner's second set of letters (the first letter was written in January, the last in July, but the set came out only in September), the two rarely addressed the other's current position. Galileo had Scheiner's first set of letters in front of him as he wrote his first and second responses to Welser, but did not know that, by the time his second letter reached Welser, Scheiner had already published the *Accuratior*. Similarly, Scheiner completed his second set of letters after having seen only Galileo's preliminary response in his first letter to Welser, but he had not see Galileo's main body of evidence – the extensive observations and illustrations that were included in the second letter. Scheiner also missed a detailed description of Galileo's observational apparatus that made observations easy (and painless) while greatly simplifying the production of printed illustrations of sunspots. The timing of Scheiner's second texts in rela-

---

deploy the Bible against the Aristotelians. This move, however, was blocked by the censors who required the eliminations of those passages (*GO*, Vol. XI, pp. 428, 437-8, 439, 446, 453, 460, 465). Welser also expressed the possibility of "various oppositions" against the publication of Galileo's text in Augsburg ,were he to decide to publish there (*GO*, Vol. XI, p. 361). It is not clear, however, whether Welser thought those "oppositions" were going to come from censors.

93 Galileo to Cesi, May 12, 1612, *GO*, Vol. XI, p. 296. An almost identical remark is in a letter to cardinal Barberini, *GO*, Vol. V, p. 311.

94 *GO*, Vol. V, p. 106.

tion to Galileo's may have sealed him into a position he might have regretted (and which, in fact, he would modify years later).[95]

In sum, throughout the debate Scheiner had access only to Galileo's briefest and most general reply. Galileo, instead, was not only better informed about the state of Scheiner's work, but he had the last voice as he wrote his third and last letter to Welser in response to Scheiner's second text, the *Accuratior*. This is not to say that the debate was unfair, but that it may have been something else – something that was rewritten as a debate between Galileo and Scheiner through the work of later commentators and historians who saw it mostly as an episode in Galileo's career. Perhaps the conceptualization of the discovery of sunspots as a priority dispute between Scheiner and Galileo has led historians first to reduce a multivocal debate to a bipartisan dispute, and then to assume that each text produced by one opponent must have been address to the other. Instead, it is important to realize that while Galileo's discussion of sunspots was framed as a response to Scheiner's work (and to his priority claim), Scheiner did not see Galileo as his main interlocutor – at least not during the first phase of the debate. During that phase, Galileo was only one of the many people Welser had sent copies of Apelles' work, asking for comments.[96] The content and style of his arguments, his uses of pictorial evidence, and his language choice indicate that Scheiner addressed communities that overlapped but did not coincide with Galileo's.[97]

Picturing sunspots

Scheiner had already used pictures in the *Tres epistolae* to map out the movement and appearance of the spots (Fig. 5, p. 67). These illustrations were small (about an inch in diameter), crammed in one fold-out page and, by Scheiner's own admission, not very accurate:

---

95 Scheiner had dropped the satellite hypothesis by the time he published his *Rosa Ursina* (1630). The book's printing started in 1626. For a discussion of Scheiner's much more sophisticated use of visual imagery in the *Rosa Ursina*, see van Helden (1996), pp. 384-9.

96 It is important to realize that Welser was almost in complete control of the first phase of the dispute, of who got a copy of the book, and who was asked to respond. Scheiner was given only two [!] copies of his books to share with his colleagues (copies which he sent to the Collegio Romano). In the *Accuratior*, Scheiner gives a long list of names of people whose opinions about sunspots he had heard through Welser: Cardinal Borromeo, Andrea Chiocco, Giovanni Antonio Magini, Angelo Grillo, Ottavio Brenzoni, Leonardus Canonicus. Scheiner then lists people he already knew by name, though they too may have been contacted by Welser: Reinhard Ziegler, Simon Stevin, Johannes Kepler, Johannes Pretorius, Johann Brengger. (*GO*, Vol.V, p.62). In other passages he also mentions Vincenzo Dotti, "Protogenes," and Galileo. Like Galileo, the other two were also contacted by Welser. The only people mentioned in the book whom we know to have been contacted by Scheiner directly were Guldin and Griemberger. Of course, the picture could have been very different if Scheiner had not published pseudonymously. In that case, he would not have needed to rely on Welser as a "front man" to protect his pseudonymity.

97 Galileo was not alone in responding to Scheiner's text as it was distributed throughout Europe by Welser. An anonymous *De maculis in sole animadversis et tamquam ab Apelle in tabula spectandum in publica luce expositi*. Batavi dissertatiuncula ad Amplissimus Nobilissimumque Virum Cornelium Vander-Millium, Academiae Lugodinensis Curatorem vigilantissimum. Ex officina Plantiniana Raphelengii, MDCXII. is mentioned

"About the observations shown, I have the following admonitions. They are not terribly exact, but rather hand-drawn on paper as they appeared to the eye without certain and precise measurement, which could not be done sometimes due to the inclement weather, sometimes due to lack of time, and at other times due to other impediments."[98]

Accuracy was only part of the problem. Scheiner's one-page map of the sunspots' motions was hard to read not only because of its size but because the various conventions he had adopted while drawing and assembling the pictures:

"The more notable spots that appeared unchanged are marked by the same letters [...]. If I added spots without letters they were either not seen constantly because of the turbulence of the air or, if they appeared constant, they did not need to be observed in comparison to the others because of their smallness. But this is to be noted as well: the proportion of the spots to the Sun should not be taken from the drawing, for I made them larger than they ought to be so that they would be more conspicuous [...]. Fre-

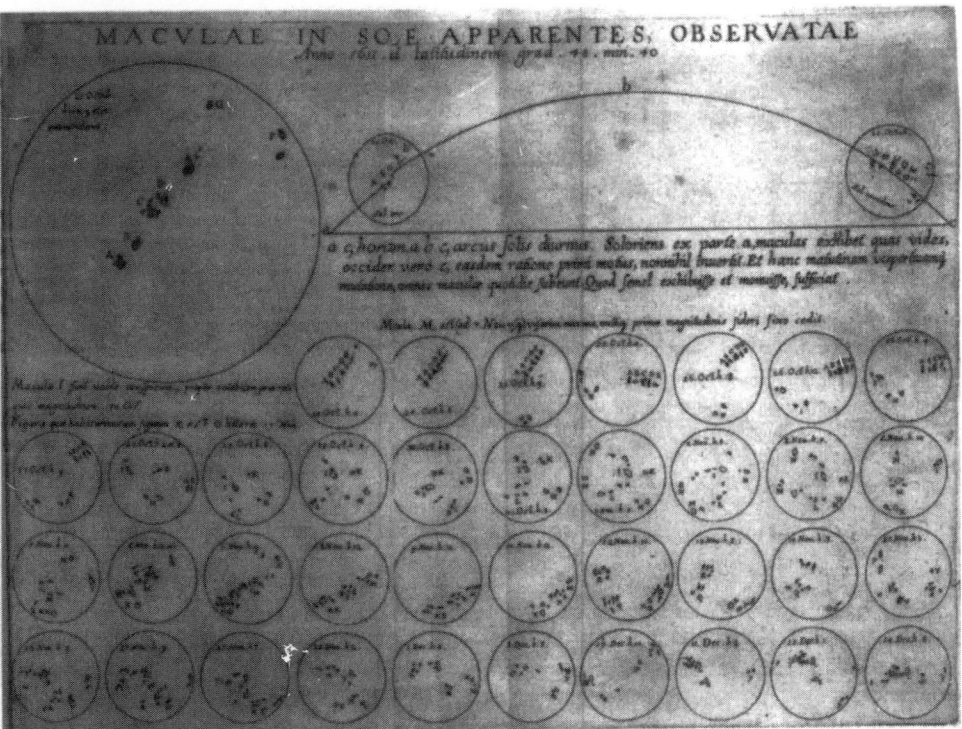

Figure 5: Foldout of sunspots illustrations from Scheiner *Tres epistolae* (1612). This reproduction is taken from the 1613 Roman edition, courtesy of the William Andrews Clark Memorial Library, UCLA.

(and dismissed) by Cesi in a September 29, 1612 letter to Galileo (*GO*, Vol. XI, p. 402). I have not been able to access a copy of this text.
98 *GO*, Vol. V, pp. 26-7. Only one of Scheiner's illustrations was larger (86 mm. diameter).

quently many small ones were conflated into one large one [...]. The spots which
always retain the same letters next to them are always the same, although they are
depicted as they appeared at the time they were drawn. When some spots and their
letters are no longer drawn, these had ceased at that time to appear on the Sun. But
when different spots are designated by different letters, these are different newly appearing spots. When, however, spots not designated by any letters are at times represented and at times not, these either have entirely set [disappeared from the edge of the
Sun] and so are not drawn, or (which happens often) they have not appeared due to
thick air, since these kinds of spots only offer themselves to view when the Sun is very
bright and the air very pure."[99]

Given the small size as well as the multiple codes inscribed on the spots (the lettering, scale
relative to the Sun, graphical rendition in relation to their size and relative permanence, etc.)
Scheiner's pictures were far from self-evident.[100] One had to study that page quite carefully to
figure out how to read it. The arrangement of the pictures on the foldout page did not help to
make patterns of change more visible either, especially because several of the pictures did not
show the spots aligned in a uniform direction so as to facilitate the detection of visual patterns.[101] This, however, was not the result of Scheiner's poor design skills or of the engraver's
ability (which was excellent), but rather of the constraints of his observation system and,
perhaps, of German winter weather.[102]

Scheiner pointed his telescope directly to the sun and, not to be blinded, he added heavily
colored blue or green glass filters (but not smoked glass) between the eye and the eyepiece.[103]
These filters, I believe, worsened the telescope's resolution and created more distortions.
Given that colored glass was not part of standard astronomical equipment and that Scheiner
did not seem to have access to glass-making facilities, he probably relied on glass made for
other commercial purposes.[104] If Scheiner used this kind of colored glass it is likely that his
filters were quite thick, as commercial colored glass is not nearly as opaque as the smoked
glass commonly used to observe solar eclipses.[105] These problems may have encouraged Scheiner to observe with filter-less telescopes, using metereological or athmospheric conditions
(not filters) to abate the Sun's luminosity. In fact, he often observed either through thin

---

99 *GO*, Vol. V, p. 27.

100 We do not know, however, if this was the original size of the pictures in Scheiner's manuscript. The foldout
engraving included in the printed text was commissioned by Welser, not Scheiner. Quite probably, the hand-
drawings and the engravings were the same size, but we do not know for sure.

101 Van Helden (1996), pp. 370-2.

102 On the effect of weather conditions, see van Helden (1996), p. 378. Alexander Mayr, the engraver selected
by Welser, was a well-known and accomplished engraver (*ibid.*, p. 372).

103 "The Sun can be observed everywhere through a tube equipped with a convex and concave lens and also a
dark blue or green glass, plane on both sides and of the appropriate thickness, at the end that is applied to the
eye. [A tube so equipped] will protect the eyes from injury even [when the Sun is] on the meridian." *GO*, Vol. V,
p. 27.

104 These considerations begin in the *Tres epistolae*, but are much expanded in Scheiner's second text, the
Accuratior, making up most of the second letter (*GO*, Vol. V, pp. 57-62).

105 Scheiner acknowledges that his colored filters could use some additional help from cloudy conditions: "[...]
and this even better if to this blue or green glass which is not sufficiently tempered [rendered opaque] a thin air
vapor or mist is added, the Sun being wrapped as it were in shadow." *GO*, Vol. V, p. 27.

clouds or at dusk or dawn. In the latter case, however, the observational window shrunk down to only a quarter of an hour.[106] It may have been difficult to draw detailed pictures of all the sunspots in such a short period of time.

All these constraints produced frequent gaps in Scheiner's observational record – gaps that further disrupted the flow of his visual narrative. Then, being produced at different times of the day, Scheiner's pictures reflected the observer's changing positions due to the earth's daily rotation. As a result, the spots appeared to be aligned in different directions in different pictures – a difference that was particularly confusing when comparing observations conducted at dusk with those made at dawn. The juxtaposition of pictures with different orientations added more confusion to an already complicated display.[107] The constraints of his system of observation imposed more constraints on the kind of visual narratives he could tell, which in turn hampered the intelligibility of those narratives.[108] The compounded effect was that he could not provide a viewer-friendly "movie" of the spots' motions, emergence, and disappearance over several consecutive days.

It is unclear whether Scheiner was simply unskilled with visual representations or whether, driven by the assumption that the spots could not be anything but satellites, he was not eager to produce the kind of images and visual narratives that could have provided evidence against his cosmological beliefs. What is clear is that Scheiner's small and viewer-unfriendly pictures did not damage but actually boosted his argument.[109] While Galileo relied on visual representations of the spots that could highlight their complicated motions and metamorphoses, Scheiner benefited from simplifying the visual complexity of the sunspots by making them look like dots or patterns that appeared to be reducible to planets. On the contrary, Galileo needed to make the spots look as mutable as possible so that no conceivable arrangement of satellites could seem able to simulate their appearance. Scheiner's "realist" claims was better served by "schematic" pictures while Galileo's non-essentialist stance about their nature had to be supported by "realistic" representations.

The most conspicuous conceptual and visual element of Galileo's second letter was a long, day-by-day sequence of 35 large-format illustrations of the sunspots' changing positions and appearances from June 2, 1612 till July 8 (Figs. 6, 7, 8, 9; pp. 70-1).[110] There were only two one-day gaps in the month-long sequence (compared to 24 gaps in Scheiner's illustrations between October 21 and December 14, 1611).[111] Each picture (almost five times larger than Scheiner's in diameter) occupied one full page in the book. In many ways, the

---

106 "Near the horizon, the morning and evening Sun can be observed for the fourth part of an hour without any danger whatsoever with a simple [filter-less] tube (but a good one) when [the sky] is cloudless and clear." (*GO*, Vol. V, p. 27). That he tried to avoid filters is also confirmed by the fact that most of his recorded observations were taken at dusk or dawn.

107 Scheiner's introduction, in his 1630 *Rosa Ursina*, of the "heliotropic telescope" (the first equatorially mounted telescope) was a response to this problem.

108 There were substantial gaps in Scheiner's visual record of the sunspots. The sequence included 31 days (sometimes with more than one illustration per day) from October 21 to December 14, with 24 days left uncovered.

109 Only one illustration (that of October 21, 1611) was published in a size sufficiently large to detect with clarity the irregularities of the spots.

110 Galileo added also a set of three drawings (August 19 to August 21) at the end of the sequence. Galileo added this separate set because it depicted spots that, due to the unusually large size, could have been observed without the telescope.

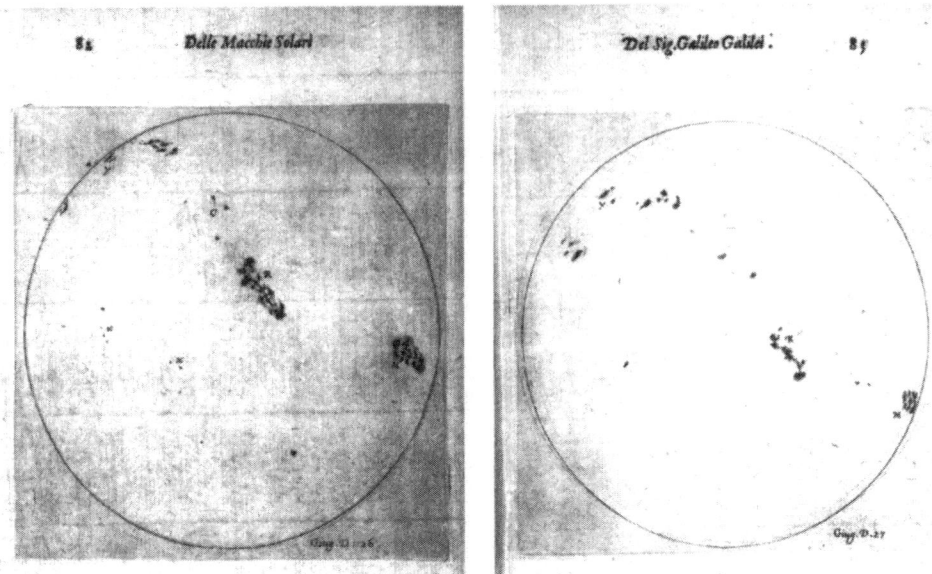

Figures 6, 7: Galileo's illustrations of sunspots for June 26-9 from his *Istoria e dimonstrationi* (1613), courtesy of the William Andrews Clark Memorial Library, UCLA.

sequence of pictures appended to Galileo's second letter was a book of its own. The size of the pictures, the virtual absence of gaps, the detail with which the spots' peculiarities were represented, coupled with the viewer-friendliness of the illustrations (which, unlike Scheiner's, required little or no decoding) turned that section of Galileo's letter into a virtual movie any viewer (not only astronomers) could watch.[112]

Galileo's pictures of the sunspots provided visual narratives that functioned much like those of his previous astronomical illustrations. Their look, however, was substantially different. In the *Sidereus nuncius* Galileo could use diagrams of the satellites, or exaggerated depictions of the topography of the Moon as evidence of his discoveries, but now he needed to show the details of the spots' irregular and ever-changing contours.[113] While he had previously used visual sequences to argue that satellites and lunar mountains and valley were real because of the periodical patterns they displayed, now he used visual sequences to argue that

---

111 Weather conditions certainly helped the tightness of Galileo's visual record. It could not hurt that he observed in the spring and summer in Italy while Scheiner had observed in the fall and winter and in Germany. Van Helden (1996), p. 378.

112 The claim that Galileo tried to develop an alternative audience for his discoveries and natural philosophy rather than convince the traditional philosophers had already been introduced by Feyerabend (1975), pp. 141-3.

113 However, this new difficulty came with new resources. In 1610 Galileo could attach physical meaning to pictures of the moon which (by later standards) were quite inaccurate both because he did not need anything better, and because pictures which (by later standards) would have been better may not have been read as such at that time. Instead, by this time "telescopic accuracy" had become a culturally recognizable category, as Scheiner's own claims were predicated on the acceptance of telescopic evidence. Consequently, Galileo could rely on these new culturally established parameters and on the fact that, since 1610, he had managed to represent himself as the producer of the best telescopes.

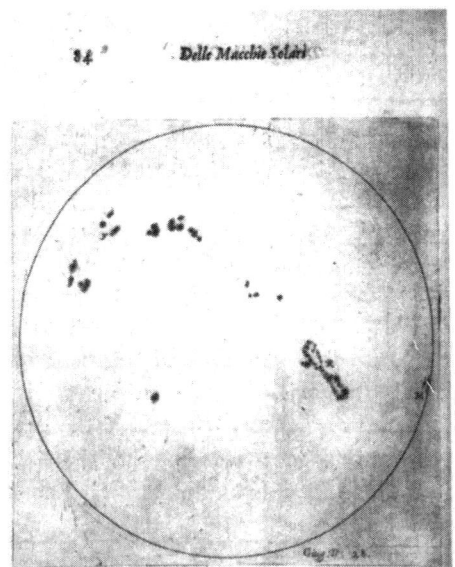 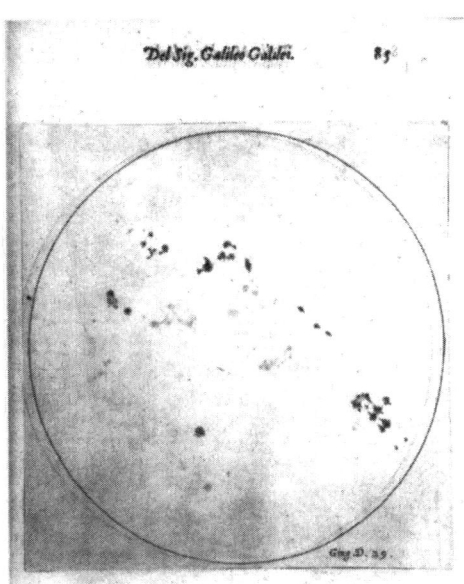

Figures 8, 9: Galileo's illustrations of sunspots for June 26-9 from his *Istoria e dimonstrationi* (1613), courtesy of the William Andrews Clark Memorial Library, UCLA.

the spots were *not* satellites by showing that they did *not* display the periodical patterns of satellites. His critique of Scheiner's claims had to be visually "finer" than his earlier dismissal of Aristotle' assumption of the perfect nature of the moon – a task he had been able to accomplish with coarse engravings.

This is not an indication that the visual language of astronomy was evolving toward more accurate depictions, nor that such a language existed as a coherent canon of pictorial representations. The different look of Galileo's pictures of sunspots reflected contextual factors. The first was the nature of Scheiner's claims and the timing of their publication. Galileo did not have to confront alternative telescopic representations of Jupiter, the Moon, or the Milky Way when he wrote the *Nuncius*, but he had to contend with Scheiner's depictions of sunspots (which predated Galileo's) when he wrote the *Istoria*. This does not mean that Galileo's pictures of sunspots were part of an ad hominem argument while those in the *Nuncius* were not. The *Nuncius* was ad hominem too, though in that case Galileo's adversary was a corporate persona – the Aristotelians. The more important difference is that while in 1610 Galileo crafted his illustrations of the Moon to refute the Aristotelians' *sweeping* and *negative* claim (an across-the-board veto) about celestial change, his 1612 pictures of sunspots were produced to refute Scheiner's *specific* and *positive* claim about the existence and nature of sunspots. If Galileo's pictures of sunspots seemed more "accurate" than those of the Moon it is because they were deployed against narrower claims.

The second factor was the sunspots themselves. New spots emerged, fell apart, came together, and disappeared and they drifted eastwardly over the solar disk. But if some reemerged from the backside of the Sun, they were so changed that they could not be conclusively identified as an "old" spot. The spots, therefore, were not virtually unchanging bodies like the satellites of Jupiter or the mountains of the Moon whose cyclical appearances could be observed ad infinitum. In the case of sunspots, Galileo needed to make two interrelated claims: (1) that the spots were real despite their irregular behavior and (2) that their irregular behavior showed they could not be the satellites Scheiner had claimed they were. This required a visual sequence able to show that the spots displayed certain kinds of periodical regularities (so as to show that they were real objects) but did not display either the kind of periodical regularities of satellites or the appearance of stars (so as to refute Scheiner). Luckily for Galileo, these two goals were not mutually exclusive.

The fact that the spots emerged and fell apart indicated that they could not be satellites – at least in the traditional, pre-Scheiner sense of the term. But because of the specificity of their cycle of emergence, growth, and eventual disappearance, the spots' lack of permanence actually testified to their non-artifactual nature. The lifecycle of sunspots was, I believe, the "period" that was most important to establishing them as objects. Furthermore, the spots were not observed just anywhere on the Sun, but tended to concentrate within a band along the solar equator. They also moved eastwardly, making the trip across the solar disk in about fifteen days. Although Galileo was unable to prove that some of the bigger spots came back after surviving the trip across the backside of the Sun, he could offer a strong argument (based on the cycle of the spots' birth and decay) about why that was the case. He could not track one specific spot during its many trips over the solar surface in the same way he had been able to track the revolutions of the satellites of Jupiter, but he could determine the spots' half period (about fifteen days) and argue that the fact that they did not come around did not mean that they were artifacts, but simply objects with a short lifecycle – a lifecycle he could map and share with his readers. The spots that did not come back around constituted an anomaly for Scheiner's position, but not Galileo's.

These intricate patterns of change needed to be represented with pictures detailed enough to display the stage of development or decay of a certain spot on a certain day, but also through a fast-paced narrative that could cover the spots' whole lifecycle. One detailed picture taken every fifth day would have not allowed a reader to observe the lifecycle of the spots and conclude that they were not artifacts. "Detail" was in both the pictures and in the *frequency of the sampling*.[114] Not surprisingly, there were only two one-day gaps in the sequence of observations from June 2 to July 8 that Galileo attached to his second letter to Welser.

If Galileo's sunspots pictures appear remarkably more detailed than those of the satellites of Jupiter or of the Moon, it is not because they represent a more mature stage of the visual language of astronomy, but because they were asked to map periodical patterns that were more complex than those needed to establish the satellites of Jupiter or the mountains of the Moon as physical objects. We cannot say that the pictures in the *Istoria* were more accurate than those in the *Nuncius* because we have no context-independent yardstick to measure accuracy. All we can say is that each set of pictorial sequences managed to convince their readers of the periodicities of their respective phenomena and that, in this sense, they were "accurate." That some of them may appear "schematic" while others "realistic" is quite irrelevant.

---

114 *GO*, Vol. V, pp. 132-3.

Galileo admitted that not all issues of the sunspots debate could be answered by pictures.[115] And yet, the impact of their narrative, of the miriads of details and changes they put in motion, created a "reality effect" that undermined the plausibility of Scheiner's claims (without, however, refuting them). As he put it:

"Their different densities and blackness, the changes in shape, and the mingling and separation, are in themselves manifest to vision without the need of further discussion, and therefore a few simple comparisons will suffice of those accidents in the drawings that I am sending you ..."[116]

As in Scheiner's case, the appearance, display, and kind of visual narratives produced by Galileo were closely connected to his observational system. His system, however, was radically different from Scheiner's:

"Direct the telescope upon the sun as if you were going to observe that body. Having focused and steadied it, expose a flat white sheet of paper about a foot from the concave lense [i. e. the eyepiece]; upon this will fall a circular image of the sun's disk, with all the spots that are on it arranged and disposed with exactly the same symmetry as in the sun. The more the paper is moved away from the tube, the larger this image will become, and the better the spots will be depicted. Thus they will all be seen without damage to the eye, even the smallest of them – which, when observed through the telescope, can scarcely be perceived, and only with fatigue and injury to the eyes. In order to picture them accurately, I first describe on the paper a circle of the size that best suits me, and then by moving the paper towards or away from the tube I find the exact place where the image of the sun is enlarged to the measure of the circle I have drawn. This also serves me as a norm and rule for getting the plane of the paper right, so that it will not be tilted to the luminous cone of sunlight that emerges from the telescope. For if the paper is oblique, the section will be oval and not circular, and therefore will not perfectly fit the circumference drawn on the paper. By tilting the paper the proper position is easily found, and then with a pen one may mark out the spots in their right sizes, shapes, and positions. But one must work dexterously, fol-

---

115 "[...] a few simple comparisons will suffice of those accidents in the drawings that I am sending you [...], but that they are contiguous to the Sun and that they are carried around by its revolution, this must be deduced and concluded by reasoning from certain particular accidents provided to us by sensory observations." *GO*, Vol. V, p. 117.

116 *GO*, Vol. V, p. 117. Also: "Convinced that it is a falsity to introduce such a sphere between the Sun and us [the sphere where Scheiner located the solar satellites] which alone could satisfy most of the phenomena, [...] it is not necessary to lose time in re-examining every other conceivable position, for each of them by itself will immediately encounter manifest impossibilities and contradictions, even if it were quite capable [of accomodating] all the phenomena I have recounted above and which continuously are truly observed in these spots. And so that Your Lordship [Welser] may have examples of all the particulars, I send You the drawings of 35 days [...]. In these Your Lordship will first of all have examples of these spots appearing shorter and thinner in the parts very near the circumference of the solar disk, comparing the spots marked A of the 2nd and 3rd day ... " *GO*, Vol. V, p. 130. And: "And from all these and other accidents which Your Lordship will be able to observe in the same drawings, it can be seen to what irregular changes these spots are subject ..." *GO*, Vol. V, p. 132.

lowing the movement of the sun and frequently adjusting the position the telescope which must be kept directly on the sun."[117]

The system adopted by Galileo (but developed by his friend, Benedetto Castelli) was as close as one could get to a mechanically produced image – to the "pencil of nature." The advantages were many: no filters, a much higher level of detail, no need to limit observations to certain times of the day, no need to go back and forth between observing the sun (in near-blinding conditions) and drawing from memory on a piece of paper (under very different lighting conditions), no problems with maintaing constant scale while changing the size of the image, better visibility of weak spots, and minimization of the impact of personal drawing skills. It probably saved a lot of time too, while greatly routinizing observation. This "blackboxing" and de-skilling of sunspots observation (and interpretation) also facilitated its spread among non-specialists and helped solidify Galileo's claim that the spots were not satellites. Galileo could encourage other people to observe and publicize the vagaries of sunspots, but he could also compare their drawings – produced in other parts of Europe – with his own.[118] As he put it at the end of the second letter to Welser:

---

117 *GO*, Vol. V, pp. 136-7.

118 A few of Galileo's friends, especially Domenico Cresti da Passignano (commonly known as Domenico Passignani), had been observing the spots since the fall of 1611 without Castelli's projection system (*GO*, Vol. XI, pp. 208-9, 212). Galileo gave Passignani advice and asked for drawings of his observations to compare them with his own (*GO*, Vol. XI, pp. 214, 229). Passignani sent the drawings on December 30, adding that he had shown them to Griemberger and Maelcote, who had asked about how his eyes could stand obsrving the Sun. Passignani replied that he used a blue filter attached to the eyepiece (*GO*, Vol. XI, p. 253). At the beginning of February, Cigoli writes to Galileo that Passignani, probably after having heard of Apelles' *Tres epistolae*, felt that he should be given priority credit for the discovery (*GO*, Vol. XI, p. 268). On February 17, 1612, Domenico Passignani wrote Galileo that he had been observing the spots since mid-September and that he disagreed with Scheiner's claim that they were not in the Sun (*GO*, Vol. XI, pp. 276-7). It is not clear whether Galileo ever told Passignani about Castelli's apparatus. If Passignani's observations appear to have started without Galileo's prodding, Daniello Antonini became interested in sunspots after receiving a letter from Galileo: "Cominciai, subito doppo hauta la lettera di V. S., a dipingere il sole ..." (*GO*, Vol. XI, p. 363). After that, Antonini wrote him about his observations in Bruxelless in July 1612, enclosing a number of drawings. Antonini requested Galileo's own drawings, which he received in October and found a close match to his own (*GO*, Vol. XI, p. 406). Antonini's early drawings were mentioned by Galileo in his *Istoria* (*GO*, Vol. V, p. 140). At first, Ludovico Cardi Cigoli acted as a trait d'union between Galileo and Passignani, but soon started his own observation, under Galileo's direct prodding. On March 23, 1612 he wrote sent Galileo a number of drawings of the spots which he had produced without Castelli's system (which was unknown to Galileo himself at that time) but with a telescope equipped with thick green filters (*GO*, Vol. XI, pp. 287-8). He's not too pleased with his illustrations because he cannot frame the whole Sun in the field of vision so that he needs to move it to observe and depict all the spots. He sent more drawings (which Galileo had requested) on June 30, saying that, due to time constraints, he had observed little, and that he had passed much of the observing and drawing task to "Cosimino" – most likely and assistant. A July 14 letter confirms that Cigoli is using the projection system (which he had heard of from another painter, Sigismondo Coccapani) and that he's still using "Cosimino" who is "being trained" (*GO*, Vol. XI, pp. 361-2). He adds that his pictures are now drawn to the diameter specified by Galileo. More projective drawings are sent on July 28 (*GO*, Vol. XI, p. 369). Cigoli sent more pictures to Galileo on August 31, still made to his specification and through the projections system. He also asked if Galileo needed more (*GO*, Vol. XI, pp. 386-7). Galileo, who by this time had concluded the second letter to Welser, must have ended his requests for drawings, as Cigoli stopped mentioning them in their letter. That Galileo told him to stop (and that, therefore, he was using him as part of an "extended observatory" made possible by the drawing technology) is confirmed by Cigoli on

"[...] we must recognize the divine kindness because the means needed for such understanding [of the spots' changing nature] are very easily and quickly learned. And he who is not capable of more [mathematical and philosophical arguments] may arrange to have drawings made in far-flung regions and compare them with the ones made by himself on the same days because he will find them absolutely to agree with his own. And I have just received some made in Brussels by Mr. Daniello Antonini [...] which fit exactly with mine, and with others sent to me from Rome ..."[119]

A few months later, Antonini (who in the meantime had received some drawings from Galileo) confirmed the evidentiary power of these pictures: "[Your] images of the Sun [...] match to the dot those I made on the same days in Bruxelles, so that I do not need your [mathematical] demonstration to be sure that the spots that appear on the surface of the Sun are contiguous to it."[120] With another correspondent, Galileo requested the drawings be made of the same size of his own, probably to allow for easy comparison through superimposition.[121] On top of gaining external confirmations for his observations, I believe Galileo was trying to make sure that, between his own drawing and those of his collaborators, he could secure an uninterrupted visual narrative spanning many weeks.

Not only did Castelli's apparatus improve observation, but it also greatly improved the *communication* of images of sunspots, both as printed images and as drawings. The projection system was the first step toward producing convincing visual sequences like the one included in Galileo's second letter and then printed in his 1613 *Istoria*, but those pictures could also be easily copied by putting another piece of paper over them, place the sandwich against the light, and trace the contours of the original image. This system allowed him to make "limited editions" of these drawings and circulate them before they were printed, like the sets he sent to Prince Cesi and Cardinal Barberini in May and June 1612 – sets that could be further copied and showed around without much quality loss.[122] Thanks to this picture-making system, Galileo could also send two "originals" of his second letter and its pictorial appendix to Welser in Augsburg and to Cesi in Rome at the same time. While Welser shared one with his northern European correspondents, Cesi readied the other for publication in Rome. Furthermore, the drawings attached to the second letter to Welser were copied once more during the letter's stopover in Venice.[123]

May 3: "I told him [Virginio Cesarini] that you had made me observe them, and that Your Lordship then told me not to observe them any longer, and how you had told me to observe them in that specific size, and told me how to do that ..." 1613 (*GO*, Vol. XI, p. 501). Benedetto Castelli, the inventor of Galileo's observational apparatus, also sent him drawings of sunspot observations on May 8, 1612 (*GO*, Vol. XI, pp. 294-5).

119 *GO*, Vol. V, p. 140.
120 *GO*, Vol. XI, p. 406.
121 Letters from Cigoli in Rome to Galileo in Florence refer to Galileo's instructions about the size of the drawings (*GO*, Vol. XI, pp. 362, 502).
122 *GO*, Vol. XI, p. 297 and pp. 307-11. The set sent to Cesi was then copied (and perhaps further distributed) by Cigoli (*GO*, Vol. XI, p. 302). I believe that Galileo's choice of Barberini and Cesi as recepients of these pictures reflects his perception that, given the large networks Barberini and Cesi were centers of, many other people would be able, so to speak, to "see" Galileo's argument before reading it.
123 In a September 22, 1612 letter to Galileo, Giovanfrancesco Sagredo reports that "I had the letter for Augsburg copied, illustrations included." (*GO*, Vol. XI, p. 398).

More important, Galileo's image-making system allowed for high-quality engravings, not just drawings. It was common practice among engravers to glue a drawing to a copper plate and then "burn" through it with the engraving tool. Unlike woocuts, engraving plates worked by capturing ink in the cavities cut into them and by then releasing it to the paper that was pressed against it. Therefore, if one put a drawing of sunspots on a copper plate and engraved through the dark parts of the drawing, that would produce a plate that yield positive prints of the original drawing. While we do not have the original drawings for the engravings that Galileo included in his *Istoria*, we do have a contemporary set of drawings that Galileo sent to Cardinal Barberini.[124] Because these drawings concerned observations that were not included in the *Istoria*, a comparison of their diameter (about 127 mm.) with that of the printed images (about 124 mm.) cannot tell us whether Galileo's drawings and prints were exactly the same size.[125] However, the match is close enough to support the hypothesis that the engravings in the *Istoria* were produced by glueing Galileo's drawings (or rather a direct copy of them) on the copper plate. If so, the quality of the printed illustrations was not just the result of a detailed, quasi-mechanically produced drawing, but also of an equally quasi-mechanical transfer from drawing to plate.[126]

Judging from the reception of the *Istoria*, Galileo's strategy paid off. Few people outside of the Society of Jesus seemed to believe that the complex patterns presented by his images could be accounted for by Scheiner's acrobatically-behaving satellites.[127] It did not help Schei-

---

124 The originals of the drawings sent to Cesi are lost, but those to Barberini are at Biblioteca Apostolica Vaticana, MSS Barberini Latini 7479.

125 I would like to thank Ill.mo Signor Christoph Lüthy for checking the size of the drawing at the Vaticana for me.

126 The production of the copper plates for Galileo's *Istoria* was a careful and expensive business. Prince Cesi recognized the importance of these illustrations and spared little money, time, and talent to produce and revise the plates according to Galileo's desires (*GO* Vol. XI, pp. 404, 409, 416, 418, 422, 424, 472, 475).

127 This does not mean that everyone gave Galileo credit for the discovery of sunspots (nor that they should have). In the decade after the debate, the views on the priority claims tended to be distributed along party lines, with Jesuit authors tending to cite Scheiner as the discoverer and the authority on sunspots. There is no question that Scheiner's later opus, the *Rosa Ursina* (1630, but the printing began in 1626) was the most detailed and comprehensive seventeenth-century text on sunspots. Similarly, in the period between the publication of Scheiner's *Tres epistolae* and Galileo's *Istoria*, the Jesuits of the Collegio Romano publicly sided with Scheiner although some expressed serious qualms about Scheiner's positions in private. On February 17, 1612, Passignani told Galileo that Griemberger "is of the same opinion of the writer [Apelles], that is, that the spots one sees are stars like those seen around Jupiter." (*GO*, Vol. XI, p. 276). A week earlier, in a letter to Galileo, Griemberger had taken a more ambiguous stance saying that Apelles' claims, which he had just read, were "not improbable" and that he had managed to keep the stars off the Sun. But, Griemberger continued, he could not, at that point, either certify or refute Apelles' claims (*GO*, Vol. XI, p. 273). In September 14, during a public disputation at the Collegio Romano, a Dominican friar defended the claim that the Sun was at the center of the cosmos and invoked the observation of sunspots to buttress his argument. The Jesuits replied that the sunspots were very minute stars that are visible only when they are grouped together but become invisible when they are isolated (*GO*, Vol. XI, p. 395). The Jesuits had assumed a pro-Scheiner position (though a more muted one) at a similar event in late January 1612 (*GO*, Vol. XI, p. 274). On October 19, 1612, Cigoli wrote Galileo that Griemberger still defended the position that the sunspots were stars (*GO*, Vol. XI, p. 418). However, on November 23, 1612 Johannes Faber wrote Galileo that, about a week earlier, Griemberger had visited him at home and that "he agrees more with you than with Apelles, as he finds very convincing the arguments Your Lordship uses to refute the assumption that they are not [sic] stars. However, as a child of holy obedience, he does not dare say it." It is not clear from the letter whether this is a quote from Griemberger or Faber's opinion (*GO*, Vol. XI, p. 434).

ner's case that most people became familiar with his argument and illustrations through the reprint of the *Tres epistolae* and *Accuratior* appended to Galileo's 1613 *Istoria*. While Prince Cesi – the sponsor of Galileo's publication – invested much time and money to produce the best plates for the *Istoria*'s illustrations, he explicitly cut costs on the reproduction of Scheiner's work and images.[128]

From public movies to private dark rooms

Scheiner had not seen either Galileo's images of the sunspots or the description of his apparatus by the time he wrote the *Accuratior*. All he had heard was Galileo's announcement, included in the very last paragraph of his first letter to Welser, that:

> "in a few days I will send him [Scheiner] some observations and drawings of the solar spots of absolute precision, indeed the shapes of these spots and of the places that change from day to day, without an error of the smallest hair, made by a most exquisite method discovered by one of my students."[129]

By the time he read this sometime in mid June, Scheiner had already completed all his observations as well as two of the three letters that were to make up the *Accuratior*. He decided to not to delay the publication and completed the last letter by July 25.[130] It is not clear, however, whether exposure to Galileo's pictures and method would have changed Scheiner's own use of pictorial evidence.

Scheiner accompanied his second much longer set of letters with new illustrations. His second generations pictures, however, were not very different in size, number, or organization from the earlier ones (Figs. 10, 11; pp. 78-9). Some of the illustrations he included in the first letter of the *Accuratior* overlapped with some from the *Tres epistolae* and were barely distinguishable from them.[131] The main improvement came from the frequency of observations, not their quality or technique. For his second publication, Scheiner (possibly helped by better weather) observed more regularly and was able to avoid many of the gaps that had marred his earlier visual narratives. He also drew lines across the solar circles to mark the orientation of the ecliptic and of the spots' path across the Sun. Despite these improvements, however, Scheiner did not seem impressed by his own pictures, nor did he seem concerned about their flaws:

---

128 Cesi decided to reprint Scheiner's two texts together with Galileo's because Scheiner's original editions were already rare in 1612. Without a reprint, Cesi worried that people would read Galileo's text without being familiar with its counterpart. Welser did not send the original plates to Rome, so Cesi had all the images reengraved. More precisely, he used engravings for Scheiner's pictures of sunspots (and kept to the same size), but used woodcuts for all other images (and made those images smaller), (*GO*, Vol. V, pp. 404, 472, 474, 482).
129 *GO*, Vol. V, p. 113.
130 Welser acknowledged receipt of Galileo's first letter on June 1, 1612 (*GO*, Vol. XI, pp. 303-4).
131 The overlap concerns December 10, 11, 12, 13, and 14 (*GO*, Vol. V pp. 33, 47). The slight difference between the two editions may be due to the engraver's different rendering of the same drawings on two different occasions.

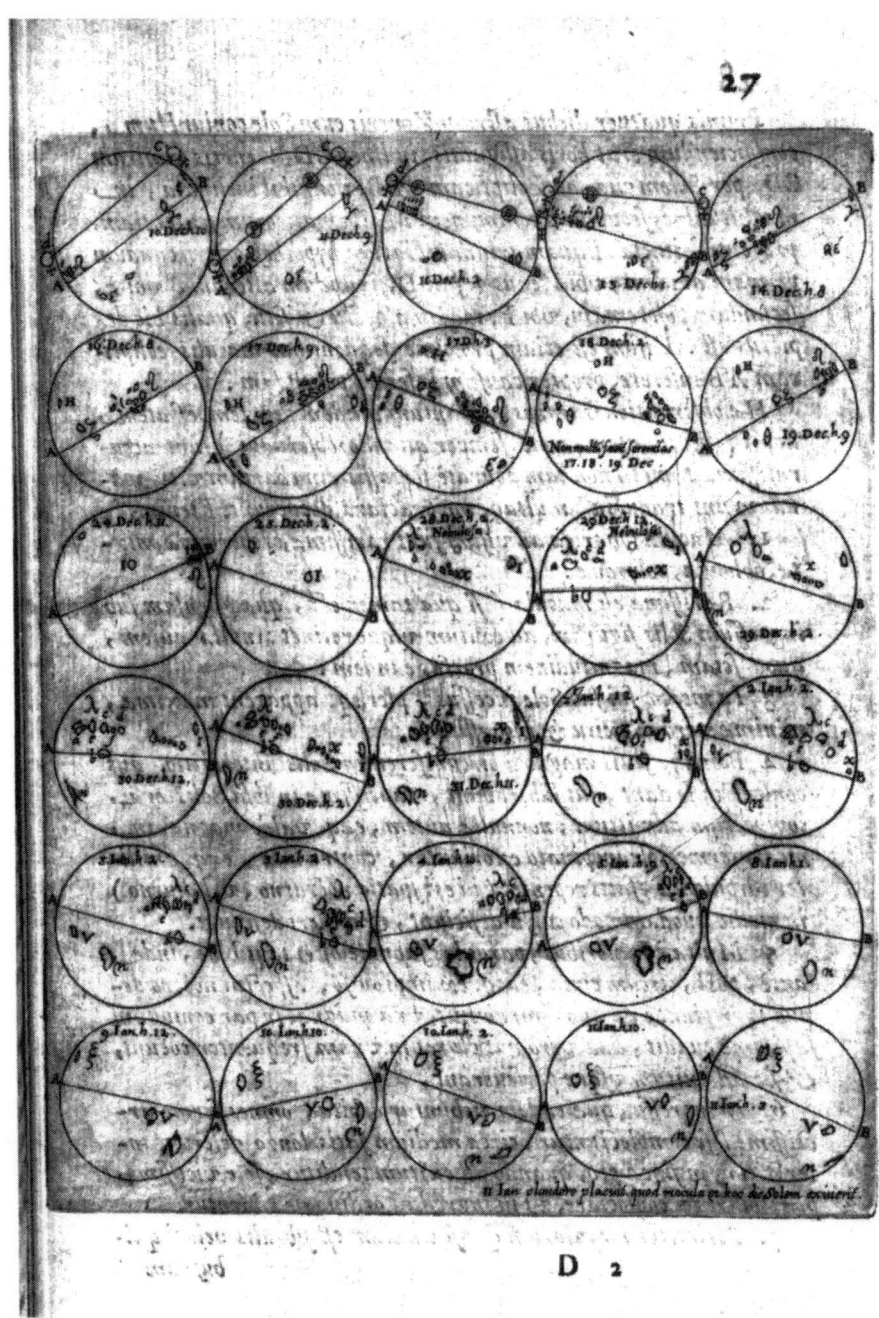

Figure 10: Sunspots illustrations from Scheiner's *Accuratior disquisitio* (1612) as reproduced in the Roman 1613 edition, courtesy of the William Andrews Clark Memorial Library, UCLA.

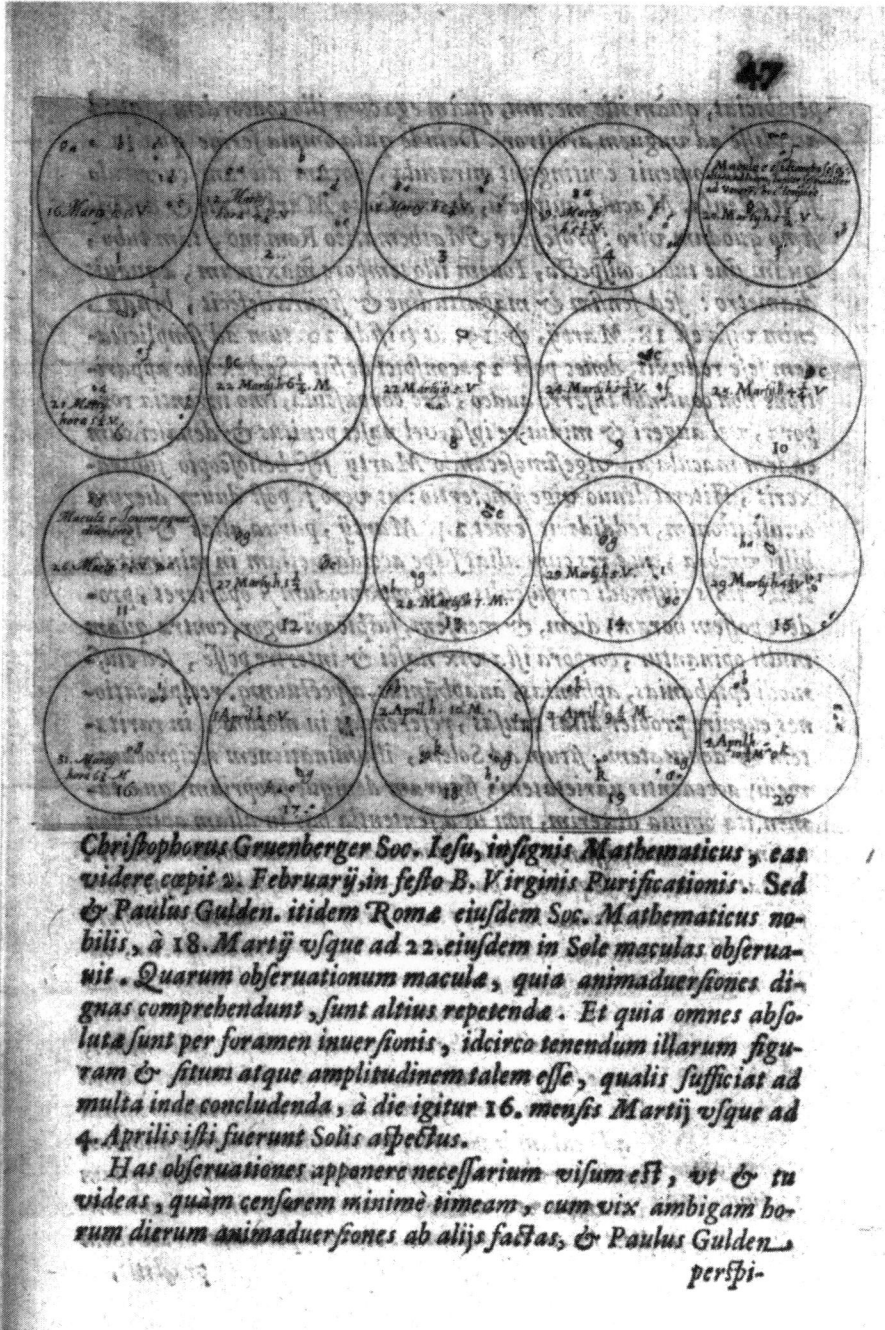

Christophorus Gruenberger Soc. Iesu, insignis Mathematicus, eas videre cœpit 2. Februarÿ, in festo B. Virginis Purificationis. Sed & Paulus Gulden. itidem Roma eiusdem Soc. Mathematicus nobilis, à 18. Martÿ vsque ad 22. eiusdem in Sole maculas obseruauit. Quarum obseruationum maculæ, quia animaduersiones dignas comprehendunt, sunt altius repetenda. Et quia omnes absolutæ sunt per foramen inuersionis, idcirco tenendum illarum figuram & situm atque amplitudinem talem esse, qualis sufficiat ad multa inde concludenda, à die igitur 16. mensis Martÿ vsque ad 4. Aprilis isti fuerunt Solis aspectus.

Has obseruationes apponere necessarium visum est, vt & tu videas, quàm censerem minimè timeam, cum vix ambigam horum dierum animaduersiones ab alijs factas, & Paulus Gulden perspi-

Figure 11: Sunspots illustrations from Scheiner's *Accuratior disquisitio* (1612) as reproduced in the Roman 1613 edition, courtesy of the William Andrews Clark Memorial Library, UCLA.

"All these observations, [made] as often as the weather allowed (and that was almost always when I observed) are the most accurate possible, though they are perhaps not so accurately drawn on the paper because of the failing of my hand."[132]

Similar cautionary claims occur elsewhere in the *Accuratior*.[133] While it might have been counterproductive for Scheiner to produce detailed pictures, there may have been other considerations behind his lukewarm interest in visual representations. Although he had not seen a description of Galileo's projection system, Scheiner too reported several experiments involving the projection of images from the telescope against a flat white surface:

"If during the day you place a tube [telescope], which is positioned before you in a window of your room, before a nearby white wall or hold a sheet of very white paper up to it, you will still observe all these appearances as before."[134]

And, in another passage:

"This [effect] is evident when you transmit the Sun through a similar lens onto a smooth wall or reflect it onto a wall from a similar lens, for the entire image of the Sun will undulate with these tracks."[135]

Strikingly, the subject matter of these projections is *not* sunspots. The images Scheiner was studying on walls or sheets of paper were of flaws in the lenses (bubbles in the first case, and swirls – "tracks" – in the second). He used the projection system not to make pictures of sunspots, but to map out how the optical artifacts produced by the telescope looked like, and then to demonstrate that sunspots are clearly distinct from those artifacts. In fact, what characterizes the *Accuratior* is not the detail visual mapping of sunspots (which Scheiner was technically equipped to produce), but a truly relentless analysis of optical effects in the atmosphere, in the eye, and in the telescope that could be used to dismiss the reality of his discovery. Evidently, the "more accurate" in the title of his second text did not refer to the pictures but to its philosophical arguments.

This points to the very different ecologies and markets in which Scheiner and Galileo operated. In both of his texts, Scheiner seemed much more concerned than Galileo with responding to possible philosophical objections to his use of the telescope, and described the painstaking procedures he went through to prove that the spots were not optical artifacts. The physical existence of the spots was as important to him as their categorization as solar satellites. Galileo, instead, appeared to take the reliability of the telescope for granted and did not seem overly concerned about people taking the sunspots to be optical artifacts. His own recent career was a tangible reminder of the epistemological reliability of that instrument. Scheiner, instead, was behaving as if he were in 1609, when the telescope had just been introduced and many were still skeptical about it. His apparently anachronistic behavior may have reflected

---

132 *GO*, Vol. V, p. 48.
133 "Because if the drawing on paper of their shadows does not agree to a hair, it is to be attributed to my eyes and hand," or "… I tried to transfer the shapes to paper faithfully" (*GO*, Vol. V, pp. 53, 49).
134 *GO*, Vol. V, p. 59.
135 *GO*, Vol. V, p. 58.

Scheiner's institutional affiliation: Jesuit mathematicians needed to be concerned about what their philosophers and theologians thought of their work – work they could censor. He seemed to be addressing his superiors (and other philosophically and theologically concerned people) much more directly than astronomers like Galileo or Kepler.

Scheiner's concerns with philosophers and theologians explain, I believe, the *Accuratior*'s frequent references to observations of sunspots not with the telescope but with the camera obscura:

> "These spots are neither delusions of the eye nor a mockery of the tube or its lenses, since without a tube they are seen on paper."[136]

Or:

> "If I now show that the solar spots are also seen without a tube, by the eye of any man, what will he oppose, whoever opposes, that this is not a fraud? Certainly neither the eye, nor the glasses, nor the air can be blamed."[137]

Scheiner's camera obscura (like the projection of telescopic images on walls or paper) was not part of a system to produce visual sequences about sunspots – sequences that could then be printed and distributed to a wide audience of non-specialists. It was, instead, a "natural instrument," one that could not lie.[138]

Then, because several people could be admitted in the camera obscura, it was also a place where perceptual biases could be collectively checked and corrected:

> "If through a round hole of about this size – O – or a bit larger, the Sun is admitted perpendicularly onto a clean sheet of paper or some other white plane, it shows itself and all the bodies below it in proportion to the distance, position and number which they retain among themselves and to the Sun. And I have made observations in this manner, *and to all willing I have shown*, whenever possible, spots so large, dense, and black, that they were quite apparent even through thin clouds."[139]

Scheiner's camera obscura was, quite literally, a darkened room, *a place for in-house demonstrations*. The camera obscura was where he brought the people he needed to convince, peo-

---

136 *GO*, Vol. V, p. 53. This point is repeated in the *Accuratior*: "For almost all of these observations were made not only with a tube but also with the Sun projected through an aperture onto a sheet of paper held perpendicularly, and thus the disk of the Sun, cast on the paper supplied the true location and motion of the spots, and the tube directed to the Sun supplied the shape." (*GO*, Vol. V, p. 64). And elsewhere: "This is confirmed by the fact that the Sun, projected through an aperture onto a sheet of paper, also distinctly represented the shadows of the spots." (*GO*, Vol. V, p. 67).
137 *GO*, Vol. V, p. 61.
138 Scheiner also mentions another lens-less method of observation: "If you hold a clean mirror to the Sun and reflect the species of the Sun from the mirror onto a clean wall or sheet of paper at the required distance, you will see spots on the Sun in number, arrangement, and size respect to each other and the Sun. And this method of observing, sought after in vain for a long time, I learned from a very good friend." (*GO*, Vol. V, p. 62).
139 *GO*, Vol.V, pp. 61-2 (emphasis mine).

ple who lived close to him, that is, his colleagues. Interestingly, Scheiner reported that the Jesuits at the Collegio Romano had also used the camera obscura for their observations.[140]

The same reasons that led Scheiner to see the camera obscura as an important "conversion site" may have also discouraged him from spending much time devising a better system for the pictorial display of sunspots. Given Scheiner's skills in instrument making and in drawing technology – he was the inventor of the pantograph – there is little doubt he could have developed a system similar to Galileo's.[141] That he aknowledged the shortcomings of his images while apparently doing little to correct them (when he had the technical capability to do so), suggests that Scheiner did not see those pictures as his best argument to convince his audience of the spots' existence.[142] And he may have had a point: philosophers and theologians were not known for their reliance on pictorial evidence.

What differentiated Scheiner's and Galileo's sunspot illustrations, therefore, was not not just their pictorial style or the narratives they told. They were deployed in two quite distinct economies and aimed at different consumers. Scheiner and Galileo were not competing in producing the "best" pictures of sunspots, nor where they targeting them primarily to each other. Rather, they made different tactical decisions about what claims to make, what kind of credit to seek, what audiences to target. Given where Scheiner stood institutionally and philosophically, he would have not necessarily benefited from a different kind of pictures. His audience may not have appreciated them and his permanence-saving arguments may have been weakened by more detailed mapping of the spots' changes. Galileo, instead, needed detailed sequential pictures because his argument hinged on narratives of change. And he needed to distribute his visual narratives because, unlike Scheiner, his audience was not primarily local or concentrated in a few institutional sites like Jesuit colleges.[143]

From fluid skies to blurry satellites

Scheiner had reasons to be apprehensive about what his superiors thought of his claims.[144] While the observations he had conducted since the *Tres epistolae* had strengthened his belief in the reality of the sunspots, those same observations had made him much less certain about their permanence. Scheiner kept referring to the sunspots as stars, but he also kept redefining, with relative clarity, what he meant by that. He kept stating his belief in the incorrupt-

---

140 "And because all were made with a pinhole camera, therefore the shape, place and magnitude of the spots is to be maintained so that they may be sufficient for concluding much from them." (*GO*, Vol. V, p. 63).

141 Besides building his own telescopes, Scheiner was an accomplished sundial-maker and the author of *Pantographice, su ars delineandi res quaslibet* ...(Rome: Grignani, 1631).

142 Scheiner, however, changed his mind quite drastically a few years later. By the time he published the *Rosa Ursina* in 1630 he had become a major user of pictorial evidence. But by that time he did not have to worry anymore about people not believeing the existence of sunspots.

143 The Jesuit College at Ingolstadt was not Scheiner's sole audience. My argument is that while Scheiner's most important audience was made up by Jesuits living and working in specific sites (colleges), Galileo's audience was not nearly as institution-based as Scheiner's. Scheiner's text, therefore, could be also read as a how-to manual providing instructions to Jesuits mathematicians (wherever they may be) about how to convince the philosophers and theologians in their respective colleges of the existence of sunspots.

ability of the Sun, but the evidence he shared with his readers pointed in the opposite direction. Unless he managed to keep the sunspots safely away from the solar surface, his new evidence about the extensive variability of sunspots could have ended up supporting to the very claim he had been trying to refute all along: that there was plenty of change and corruption in the Sun.

For example, he acknowledged that the spots' shape was hardly circular to begin with and that if further changed as they moved across the solar disk:

"Spherical spots appear rarely, while most spots are combined, oblong, and polygonal [...] Very rare is the spot (if it exists at all) which retains the shape that it shows at ingress of the Sun all the way to egress. Indeed there are none that I know of that displays exactly the same size."[145]

He also acknowledged other features that were hardly reducible to the behavior of satellites:

"The perimeter of almost all spots was roughened with, as it were, whitish and blackish fibers; and most spots, wherever they appear, were diluted by a greater whitenes around the edges than in the middle of their bodies. Indeed the shape of very many of the spots reminds the observer now of some blackish snowflake, now of some small piece of black bread, now of a balled-up mass of hair hidden in a large torch, and now of a blackish cloud."[146]

He finally remarked quite candidly on the ephemeral nature of the spots:

"[some large spots] suddenly spring up around the middle of the sun, but some almost equally large suddenly spring up around the middle of the Sun. Others just as large, on the contrary, suddenly decay in some way in the middle of their path and cease to be seen."[147]

But although these observations suggested the impossibility of reducing the remarkable variety of sunspot phenomena to configurations of satellites, Scheiner claimed that: "From this follows a new and evident argument: these spots are not in the Sun."[148] Paradoxically, the spots' extraordinary range of variability that made Galileo say they could not be satellites "proved" to Scheiner that they could not be on the Sun because:

"Since the Sun is a hard and unchanging body (according to the general opinion of all philosophers and mathematicians ...) it is impossible that such a great variation of dark shapes occurs anywhere except outside the Sun."[149]

---

144 Scheiner concludes this section by listing several authoritative people (some Jesuits, some not) who have endorsed the existence of sunspots (some by observing them, and some by declaring themselves convinced by other people's arguments and evidence – people he calls "ear witnesses"), (GO, Vol. V, pp. 62-3).
145 GO, Vol. V, p. 48.
146 GO, Vol. V, p. 48.
147 GO, Vol. V, p. 48.
148 GO, Vol. V, p. 49.

But while he was reaffirming the demarcation between the solar surface and its immediate surroundings, one of his observations made him realize that the distance between the spots and the Sun could be indefinitely small:

"[This spot] underwent ingress and egress almost on the circumference of the Sun itself, in the shape of some very thin and black little line, not making more of a white space between itself and the Sun than the thickness it shows to the eye, which is scarcely equal to the thinness of the italic letter 'l'."[150]

Aware that the evidence of radical change in the proximity of the Sun was growing as the space between the solar surface and the spots' orbit was shrinking, Scheiner seemed to realize that he could no longer save the permanence of both the Sun and the solar stars. Wanting to save the Sun, he attributed the change to the solar stars and started to redefine them accordingly.

In the last page of the *Accuratior*, Scheiner made the surprising remark that "about only one thing we are still at a loss: whether these bodies are generated and perish, or whether they are eternal."[151] Galileo himself was puzzled by the seeming inconsistency of Scheiner's claims. In a handwritten comment to this page of the *Accuratior*, Galileo remarked that: "He says a thousand times that they are stars, but now he doubts whether or not they are generated and perish."[152] Galileo was correct in his critique that Scheiner had modified his claims, but not in assuming that they were contradictory. Scheiner's tentative attribution of generation and dissolution to stars reflected the corner in which he had put himself. His prematurely categorical statement about the sunspots being shadows of solar stars (a claim he had made in the name of cosmological orthodoxy) had become increasingly less tenable as he conducted more observations, some of them just a few days after having rushed to publish the *Tres epistolae*. Additionally, in the second letter of the *Accuratior* (dated April 4, 1612), he claimed to have discovered yet another novelty in the heavens: a fifth satellite of Jupiter (which he dedicated to Welser and his family).[153] With typical concern for priority, Scheiner waited only five days from the first observation of the satellite to write up his discovery and send it to Welser. Unfortunately the new satellite seemed to disappear shortly after the letter was sent.[154]

In sum, I believe that Scheiner was in trouble on three fronts as he was writing the *Accuratior*: (1) he could have lost face with astronomers for having blundered the transit of Venus and for having probably mistaken sunspots for satellites; (2) he could have embarassed Welser both by dedicating an artifact to him and his family, and by backtracking on claims Welser had been instrumental in publishing and disseminating; and (3) he could have run afoul with Jesuit philosophers and theologians for declaring (without censors' approval) that the spots were real and satellite-like, and then to make things worse for the Society by backtracking on those claims. Were Scheiner to decide that the spots were no longer satellites, that would have

---

149 *GO*, Vol. V, p. 49.
150 *GO*, Vol. V, p. 51.
151 *GO*, Vol. V, pp. 69-70.
152 *GO*, Vol. V, p. 70.
153 *GO*, Vol. V, p. 56.
154 It is unclear what Scheiner had observed. A few modern astronomers have suggested that his "fifth satellite" was, in fact, a variable star (Ashbrook (1977), pp. 344-345).

not erased his well-supported claim that they were real. His rush to say that the spots were satellite-like had committed the Society to a position that could have later prevented them from explaining them in other ways. Since Scheiner had shown so convincingly that the spots were real and remarkably close to the Sun, there would have been few ways left to deny that these phenomena were signs of corruption in the Sun if in the end it turned out that they were not satellites.

As the costs of backtracking would have been too onerous, Scheiner charged ahead, determined to prove that *all* of his claims in the *Tres epistolae* were correct. He did so by introducing even more sweeping and potentially controversial cosmological claims.[155] For example, his remark about not knowing if the sunspots were eternal or generated and perishable does not mean that, as Galileo believed, Scheiner doubted whether the sunspots were satellites or not. Instead it suggests that Scheiner was attempting to redefine what satellites and, more generally, what stars and planets were about so as to save both his claims about sunspots and about his "fifth satellite" of Jupiter.

To Scheiner, the analogy between the sunspots and the satellites of Jupiter had become so strong that, toward the end of the *Accuratior*, he stated that: "I said not in vain [...] that the theory of the sunspots and of the stars of Jupiter appeared to be the same."[156] That is, he did not simply use the satellites of Jupiter to explain the sunspots, but he also used what he had observed about the behavior of sunspots to explain what he had observed (or not observed) about the satellites of Jupiter.[157] For instance, he used the observation that some sunspots did not appear to return to suggest that maybe his newly discovered fifth satellite of Jupiter would not return either: "since some [satellites] suddenly appear and others suddenly disappear in almost the same way as the shadows on the sun."[158] Vanishing sunspots helped him defend his discovery of the vanishing satellite of Jupiter while the vanishing satellite of Jupiter helped him defend that the sunspots could be both real and disappearing at the same time.

In Scheiner's hands, the phases of Venus became the paradigm of one of the ways in which real astronomical bodies appear to almost disappear, that is, of how spherical bodies like planets end up assuming very different, non-spherical (and less visible) appearances. In the *Tres epistolae* Scheiner had already used the phases of Venus to explain why sunspots seemed to become thinner as they approached the limb of the solar disk, that is, how a presumably hard and round object would end up looking less so.[159] He had also suggested that the peculiar appearance of Saturn may be explained through a combination of satellites and phases.[160] The implication was that all astronomical objects, due to how light strikes them, are bound to appear irregularly shaped like the sunspots.[161]

---

155 "And I have gladly communicated this completion of my earlier work to Your Excellence so that you would know how wrongly this great phenomenon is called into doubt by some and how wrongly torn to pieces by most. For all the other things which I have shown in the first painting are correct" (*GO*, Vol. V, p. 69).
156 *GO*, Vol. V, p. 57.
157 Van Helden argues that poor telescopes were at the roots of Scheiner's position. With the instrumenst he had available, Scheiner could not detect the periods of the satellites of Jupiter. Scheiner did not think there was a problem with his telescopes, but that his difficulty in tracking the satellites reflected the satellites' inconsistent behavior or appearance. Once he came to believe that even the satellites of Jupiter behaved somewhat erratically, it was easy for him to assume that sunspots coud behave that way too (van Helden: Scheiner).
158 *GO*, Vol. V, p. 56.
159 *GO*, Vol. V, pp. 29-31.
160 *GO*, Vol. V, p. 31.

Scheiner introduced another analogy to cover the opposite case – that of irregularly shaped bodies appearing circular. This time it was the humble candle, not the phases of Venus, to supply Scheiner with the paradigmatic case. He argued that although a candle's flame is far from circular, it appears like a luminous dot when observed from a distance. To Scheiner this meant that we cannot be sure that the stars are really round (implying that even the fixed stars – the emblems of permanence – may actually be as messy as the sunspots).[162] In a way reminiscent of the astronomers' devices aimed at accounting for different anomalies in planetary motions, Scheiner created a toolbox of analogies through which some of the apparent irregularities of sunspots (their tendency to emerge and disapper, their non-circular appearance, their thinness) could be either explained away as optical artifacts or could be claimed to be shared by other canonical astronomical objects. Either way, he could diffuse the claim that the sunspots complex appearance showed that they were not physical objects or, worse, that they were signs of corruption. That is, he was trying to decouple changing appearances from corruption.

A last "normalizing" analogy introduced by Scheiner involved the relative transparency of apparently opaque bodies. After he had openly admitted that some spots appeared quite frayed while other seemed denser and more compact, he tried to explain their less than starry appearance by attributing their apparently irregular contours to an effect of the spots' uneven opacity. The Moon was the body that, according to Scheiner, displayed a similarly uneven opacity. He described a partial eclipse of the Sun he had observed with several people, some of them using telescopes, other only their eyes. Scheiner drew two lessons from these observations: One was that sunspots were as opaque as the Moon (something he decided after observing sunspots in the section of the solar disk that was not obscured by the Moon and judging them as black as the lunar disk). The other was that the Sun shone *through* the Moon, though feebly. This was an astonishing claim, but Scheiner trusted the observation of one of his colleagues who "most firmly asserted that through a tube he saw the entire circumference of the Sun even through the Moon still occupied som portion of it."[163] Scheiner had much to gain by believing this observation. A partially transparent Moon allowed Scheiner to explain both the cosmologically thorny question of the Moon secondary (or ashen) light while lending support to his and other Jesuits' attempt to explain the apparent irregularities of the lunar surface by invoking differences in the Moon's density.[164] It was also an observation that could be used to argue that sunspots were no less real because they seemed to be sometimes "transparent" and some other times dense and black:

"... the Moon itself is transparent throughout, more and less according to the greater or lesser density (which is also the case with many spots, and because of which it is

---

161 His statement that the dramatic changes in the sunspots' appearance "are to be referred to motion: to rarity and density, position with respect to the Sun, mutual illumination, change of the accidental medium, and finally, particular shape" (*GO*, Vol.V, p. 64) was meant to apply, I believe, to all celestial bodies, not only sunspots. The mention of "mutual illumination" refers, I believe, to a generalized view of phases involving more than one light-emitting (or light reflecting) body.

162 *GO*, Vol. V, p. 53.

163 *GO*, Vol. V, p. 68.

164 *GO*, Vol. V, p. 68.

maintained that many have tears in them [i. e. that they appeare frayed at the edges]).[165]

That at times the sunspots appeared "transparent" was not evidence of them not being real. Rather, they were as opaque (or as "transparent") as the Moon.

Scheiner's extraordinary claims about the variable appearance of planets, stars, sunspots, and satellites were no sign of skepticism. On the contrary, by emphasizing that things were not what they seemed, Scheiner tried to explain that although sunspots became invisible or looked ephemeral, they were as real as other very real astronomical objects like stars or planets – objects which all shared, to different extents, the peculiar appearances of sunspots. This is also reflected in the long discussion of the many varieties of optical distortions involved in telescopic observations that fill the *Accuratior*'s third letter. Scheiner's point there was not so much to prove how unreliable the telescope was, but to analyze distortions and artifacts in order to show how the sunspots could *not* be reduced to them. The message he delivered over and over was that while the disappearance of the spots and their changes were artifacts, the spots themselves were real. I would frame Scheiner's candid assessment of the flaws in his illustrations in this context. He probably saw his poor pictures as just another "distortion" which he admitted like all other optical distortions involved in the spots' observations, but which was ultimately irrelevant because it did not take away from the necessary reality of the spots. In sum, he started with a cosmologically orthodox assumption about the permanence of astronomical objects (not unlike those astronomers who assumed circularity and uniformity as the paradigm for celestial motions) and then showed how much that permanence could appear to disappear:

> "I am forced to suspect, against what many believe, that these bodies can hardly be born and perish, but rather that such appearances, disappearances, and changes back and forth of appearances result from other causes, which are to be referred to motion: to rarity and density, position with respect to the Sun, mutual illumination, change of the accidental medium, and finally, particular shape."[166]

Scheiner's attempt to show that there was something very real (and cosmologically orthodox) behind the appearances of his observations and that, among other things, he deserved credit for the discovery of real but tricky objects (sunspots and additional satellites of Jupiter) involved a remarkable redefinition of the features of most astronomical objects. Scheiner did take substantial additional risks to salvage his claims and the credit he believed he deserved for them, but he did so by articulating what he imagined to be the cosmological directions being adopted by his senior colleagues in Rome.

At the end of the *Accuratior*, he stated that "according to the opinion of the astronomers [i. e. Clavius' group], hardness and this constitution of the heavens cannot endure, especially in the heaven of the Sun and Jupiter."[167] This passage suggests Scheiner's endorsement of the

---

165 *GO*, Vol. V, p. 68. The mention to sunspots "transparent" in the middle due to sunlight shining through them is at p. 50.
166 *GO*, Vol. V, p. 64.
167 *GO*, Vol. V, p. 69.

doctrine of the "fluid skies" ("hardness cannot endure"), and of Tycho's planetary model ("this constitution of the heavens cannot endure"). Cardinal Bellarmine, a leading Jesuit theologian, had been a supporter of the "fluid skies" since 1572 – a support he confirmed in a 1618 letter to one of Galileo's closest supporters, Prince Federico Cesi.[168] (Not surprisingly, Scheiner reprinted Bellarmine's letter in his final text on sunspots, the 1630 *Rosa Ursina*.[169]) According to Bellarmine, the cosmos above the Moon was not made of rigid crystalline spheres, nor was it composed of Aristotle's fifth incorruptable element, the aether. More likely, it was composed of a fire-like substance that was *not* incorruptable, could undergo substantial change, and could be easily penetrated. Because of these features, the fluid skies could account for the novas observed in 1572 and 1604, the superlunary trajectory of comets (that otherwise would have had to plow through various crystalline spheres), and the satellites of Jupiter (whose orbits would have been hard to accomodate within the crystalline sphere of their 'host' planet). The changeability of the fluid skies could also explain the existence of sunspots. The mathematicians at the Collegio Romano were fully familiar with this doctrine and it is most likely that Scheiner had read some of the letters that circulated between Rome and Germany on this topic.[170]

Scheiner's simultaneous affirmation of the reality of astronomical objects and of their changing appearances reflected his belief that the heavens were not made of rigid aether but of a more fluid substance. The changes affecting the material that made up the fluid skies could account for many of the changes Scheiner observed in the sunspots and in the satellites of Jupiter. That same changing substance could have easily produced optical distorsions. Then, if one accepted that the cosmos was not made up of a rigid and incorruptible element, it would not have taken a big step to extend the same assumption to celestial bodies and to entertain the possibility that they too may not be perfectly spherical, hard, opaque, unchanging, uniformly dense, etc. If the fluid skies could explaing the vanishing act of novas, they probably could explain also the generation and disappearance of both sunspots and of Jupiter's fifth satellite.

More important, the doctrine of the fluid skies allowed Scheiner to switch from defense to offense. He did not have to struggle to explain the anomalous behavior of sunspots and of the vanishing satellite of Jupiter, but could simply declare them paradigmatic. The new bodies he had discovered were not anomalous but epitomized the features of *all* known astronomical bodies once those were redefined within the fluid skies doctrine. He could hope to keep the credit for the discovery of the sunspots and the fifth satellite of Jupiter, gain further credit for having explained the secondary light of the Moon, all the while helping his Jesuit superiors develop a new theologically safe cosmology.

---

168 Bellarmine's own cosmological views are in Baldini and Coyne (1984). See also Baldini (1984) and (1992c), pp. 305-344.

169 Scheiner (1630), pp. 783-4.

170 Paul Guldin (one of Clavius' students at the Collegio Romano) wrote to his collegues in Munich on February 1611 reporting on the Roman Jesuits' corroboration of Galileo's discoveries. This letter, which Scheiner is very likely to have read, asked the question: "Where shall we locate these new planets? Which and how many orbs and epicycles shall we attribute to them? Do not Tycho's views encompass them? Shall we have it that *the stars to move freely like the fish in the sea*?" (Guldin to Lanz, 13 February 1611, Dillingen, Studienbibliothek, MSS 2, 247, pp. 220-2, cited in Lattis (1994), p. 210. The last line is a clear reference to Bellarmine's statement that the planets "move by themselves like the birds in the air and the fish in the water" (Baldini and Coyne(1984),

It was not unreasonable for Scheiner to believe that the discoveries of 1609-1611 could force the Jesuits to consider replacing Ptolemy's astronomy and his crystalline spheres with a cosmology that would combine the fluid skies with Tycho's planetary model. Those issues were discussed (as questions) in an informal letter sent from Guldin in Rome to Johann Lanz in Munich in the spring of 1611 – a letter Scheiner was most likely to have seen.[171] However, it is far from clear that these options were being seriously discussed in Rome by mid 1612 when Scheiner was writing the *Accuratior*.[172] For all we know, the mathematicians of the Collegio Romano implicitly adopted Tycho and the fluid skies only in 1620 with the publication of Giovanni Biancani's *Cosmographia*.[173] If Scheiner misjudged the general state of the debate in Rome in 1612, he was also surprisingly ignorant that the head mathematician in Rome, Clavius, was a staunch opponent of the fluid skies. As Clavius' closest associate, Christoph Griemberger, put it, "I know, as does anyone who was familiar with Clavius, that up to the end of his life he abhorred the fluidity of the heavens, and that he constantly sought arguments to explain the phenomena by ordinary means."[174]

Clavius was not the highest-ranking Jesuit to condemn the fluid skies. In May 1611 and again in December 1613, the General of the Jesuits had written detailed letters to all heads of Jesuit colleges stressing the necessity to follow Aristotle (in philosophy) and Aquinas (in theology) in all teachings and publications.[175] The doctrine of the fluid skies had been specifically singled out as non-Aristotelian (which may explain why Bellarmine never published anything about it).[176] The opposition continued well after 1612. In July 1616, the General congratulated a provincial censor in Sicily for doing "very well to oppose Father Blandino's teaching on the fluidity of the heavens," and in 1631 he ordered the rector of the Jesuit College of Avignon to "make sure that this opinion [the fluid skies] is not proposed or defended in any way in the theses of our pupils."[177] Scheiner was caught in the same net. Toward the end of 1614, Scheiner's superiors in Germany complained to General Aquaviva about his philosophical orthodoxy. The General admonished Scheiner not to uphold the doctrine of the fluid skies and instructed the head of the Society's Upper German Province to "take care that Father Scheiner not put forward any new opinions about the fluidity of the heavens and the movements of the stars on the basis of some still uncertain observations."[178]

p. 20). On the Jesuit mathematicians trouble with their fellow theologians concerning the publication of claims about the "fluid skies" doctrine see Blackwell (1991), pp. 148-53, and Lattis (1994), pp. 94-102, 211-6.

171 Guldin to Lanz, March 1, 1611, quoted in Lattis (1994), p. 210.

172 On the ambiguous cosmological position of Clavius right before his death, see Lattis (1994), pp. 180-202.

173 Baldini (1992b).

174 Griemberger to Biancani, 1618, quoted in Lattis (1994), p. 201.

175 Blackwell (1991), pp. 137-141.

176 That Bellarmino did not publish on the topic because of the Society's ban on it is reported in a letter by Cesi in *GO*, Vol. XIII, pp. 429-30.

177 Gorman (1996), pp. 294-5.

178 Duhr (1913), Vol. 2, Part II , p. 438. The full quote reads: "This man [Adam Tanner] turned, at the end of 1614 to the General with a complaint. Aquaviva [the General] answered him on 23 december 1614 that he had admonished Father Scheiner to abandon the new opinions about the heavens. This admonition really did happen on 13 December 1614. After a decided encouragement of his studies, which Scheiner related in a letter of 11 November 1614, Aquaviva, before anything else, answers the question of Scheiner if in a new edition of his writings on sunspots he should add his name: on good grounds it would be better to let them appear under the old name of Apelles. In refuting Galileo, Aquaviva elaborates, it would seem to me more fitting to put forward the evidence for the truth, then to refute the opposing evidence without mentioning the author, and finally to darw

There was, therefore, a substantial gap between the state of the debate in Rome (both among mathematicians and between mathematicians and philosophers) and Scheiner's representation of it. It is hard to gauge whether this was the effect of ignorance, ambition, or both. Being a junior mathematician without direct contacts with Rome, Scheiner may have mistaken Guldin's letter – a student's excited report about the new discoveries and the problems and possibilities they brought up – as the cosmological manifesto of Clavius' group. The enthusiasm and pride he must have experienced for moving, in a few months, from being a rookie teacher of mathematics to being the discoverer (and domesticizer) of new important and controversial objects may have made him feel like a major participant in the Society's growing involvement with cutting-edge astronomical research. As a result, he may have felt entitled to be a participant in the development of the new Jesuit cosmology rather than the mere recepient of the decisions of Clavius' group. Or perhaps he had invested so much in his claims about the spots being stars and about Jupiter having an extra satellite that, in order to save his cultural capital, he ignored what he knew about the cosmological debates in Rome and charged ahead with claims that effectively committed the Society to positions it was not ready to endorse at that time. Scheiner's ability to bypass the internal review expected of all Jesuit manuscripts has deprived us of the kind of material that could have answered some of these questions. At the same time, had those censorship rules been followed, it is almost certain that there would have been no *Tres epistolae*, no *Accuratior*, and no sunspots dispute as we know it.

Conclusion: Excremental realism

Only two years after the publication of the *Istoria e dimostrazioni intorno alle macchie solari*, Galileo wrote a series of letters on Copernicanism and the epistemological status of astronomy. They were never published during his lifetime. One of these letters contains a startling claim about sunspots:

> "I have already discovered the constant generation on the solar body of some dark substances, which appear to the eye as very black spots and then are consumed and dissolved; and I have discussed how they could perhaps be regarded as part of the nourishment (or perhaps its excrements) which some ancient philosophers thought the sun needed for it sustenance. By constantly observing these dark substances, I have

---

the necessary conclusions. In this way the entire refutation wil proceed with more benevolence and humility. This one thing I would like to counsel you, that, basing yourself on the solid doctrine of the ancients, you avoid new opinions about certain novelties. Please be convinced that such things displease us very much and that we will not allow their publication by our members. In particular, against the universal doctrine of the Fathers and the Schoolmen no new hypothesis about the fluidity of the heavens will be presented, or about the stars that move like fish in the sea or birds through the air. Over the same date and instruction from Aquaviva to Hartel, the Upper German Provincial, was issued: he should take care that Father Scheiner not put forward any new opinions about the fluidity of the heavens ..." (Duhr, pp. 436-8, van Helden trans.).

demonstrated how the solar body necessarily turns on itself, and I have also speculated how reasonable it is to believe that the motion of the planets around the sun depends on such thing."[179]

It is difficult to gauge how literally one should take Galileo's organicistic picture of the cosmos, but it is clear from this letter that he believed that light and heat were life-giving substances:

"So I should say it seems to me that there is in nature a very spirited, tenuous and fast substance which spreads throughout the universe, penetrates everything without difficulty, and warms up, gives life, and renders fertile all living creatures. It also seems to me that the senses themselves show us the body of the Sun to be by far the principal receptacle of this spirit, and that from there an immense amount of light spreads throughout the universe and, together with such a calorific and penetrating spirit, gives life and fertility to all vegetable bodies."[180]

The Sun is not just the physical center of the cosmos, but also its source of movement and life. The Sun moves all the other planets through its own rotation, and gives life to the cosmos through light and heat. Believing the Sun to be the vital center of the cosmos, Galileo was not surprised to find that it also displayed some of the attributes of life such as the production of excrements.

This, I believe, points to the most fundamental difference between Scheiner's and Galileo's positions – a difference that can be traced to their epistemological practices and ontological assumptions. Like the Aristotelians, Scheiner saw change in negative terms, that is, as a sign of corruption or of "reduced being." Galileo instead treated change as a positive notion because he linked it to movement, life, and generation. He made that very claim in the last of his sunspots letters to Welser, where he criticized the Aristoteleans' fear of celestial change. Fearing their own mortality, Galileo argued, humans idolize what they do not and cannot have (permanence) and hate what they fear (fragility).[181] But their fears of mortality get in the way of conceptualizing the world, leading them to confuse change for death:

"If that which is called corruption were annihilation, the Peripatetics would have some reason for being such staunch enemies of it. But if it is nothing else than a mutation, it does not merit so much hatred. Nor does it seem reasonable to me that anyone would complain about the corruption of the egg if what results from it is a chick."[182]

The remark about the mistake of saying that an egg is corrupted when in fact its "corruption" leads to the production of a chick is, I believe, closely aligned with his later claim about the sunspots being signs of the Sun's role as the "power generator" of the cosmos, and not of its

---

179 Galileo to Dini, 23 March 1615, in Finocchiaro (1989), p. 66.
180 Galileo to Dini, 23 March 1615, in Finocchiaro (1989), p. 63.
181 "I suspect that our wanting to measure the universe by our own inadequate yardstick, makes us fall into strange fantasies, and that our particular hatred of death makes us hate fragility." *GO*, Vol. V, p. 235.
182 *GO*, Vol. V, pp. 234-5.

corruption. An incorruptable egg is a dead egg. An incorruptable Sun does not belong in a live cosmos.

This suggests that Galileo's notion of the real was closely linked to his notion of life (which appeared to encompass both physical motion and generation of living beings). In this sense, Galileo's "real" was inherently and positively tied to change, not permanence. He spelled this out in a letter he sent to Paolo Gualdo in the midst of the sunspots debate:

"Now she [nature] finally shows us with indelible characters [the sunspots] who she is and how much she dislikes otium. Instead, she likes to work, generate, produce, and dissolve always and everywhere. These are her highest achievements."[183]

This adds yet another twist to the distinction between realism and nominalism by indicating that Galileo could indeed be seen as a closet realist in the sense that he ontologized change. Compared to Scheiner, whom he criticized for conceptualizing the sunspots according to his ontological boxes, Galileo did not appear to structure his argument around ontological assumptions, but that was only because his assumptions were so unorthodox that it would have been very unwise for him to publish them. If we look beyond his published texts, Galileo was as dogmatic about change as Scheiner was dogmatic about its impossibility. Similarly, Galileo's ontological assumptions involving solar excrements were no less extravagant than Scheiner's comprehensive "fluidification" of things astronomical.

Be as it may, Galileo's non-essentialist conceptualization of new objects was closely related to his ontology of change. His claim that the periodic features of his discoveries proved them to be non-artifactual dovetailed with his organicistic view of change and of the real. Because he ultimately linked the real to life, and because life is change, what made something real was how it "moved," not what linguistic labels one may attach to it. For the same reasons, his pictorial evidence about discoveries was not made up of mimetic snapshots but of visual narratives about the "life" of the new object. They were "movies" because only the movie (not the individual frames) could capture the real.[184]

References

Ackerman, James (1985): "Early Renaissance 'Naturalism' and Scientific Illustration," in A. Ellenius (ed), *The Natural Sciences and the Arts*. Uppsala, 1-17.

Adams, C. W. (1932): "A Note on Galileo's Determination of the Height of Lunar Mountains," *Isis* 17 (1932), 427-429.

Ashbrook, Joseph (1977): "Christopher Scheiner's Observations of an object near Jupiter," *Sky and Telescope* 42 (1977), 344-345.

---

183 *GO*, Vol. XI, p. 327 (translation mine).

184 It would be interesting to compare these uses of illustrations in astronomy (based on narratives of change and periodicity) to the function of illustrations in other disciplines like medicine or natural history (where periodicity may not be an important epistemological issue, and where the objects of study are physically – not only visually – accessible).

Ashworth, William (1991): *The Scientific Revolution: The Problem of Visual Authority*. Paper presented at the conference on "Critical Problems and Research Frontiers." Madison, Wisconsin.

Baldini, Ugo and Coyne, George (1984): *The Louvain Lectures of Bellarmine and the Autograph Copy of his 1616 Declaration to Galileo*. Vatican City: Specola Vaticana.

Baldini,Ugo (1984): "L'astronomia del Cardinale Bellarmino," in P. Galluzzi (ed), *Novita' celesti e crisi del sapere*. Florence: Giunti, 293-305.

Baldini, Ugo (1985): "Una fonte poco utilizzata per la storia intellettuale: le 'censurae librorum' e 'opinionum' nell'antica Compagnia di Gesu'," *Annali dell'Istituto storico italo-germanico in Trento*, 11(1985), 19-67, reprinted in Ugo Baldini (1992a), 75-119.

Baldini, Ugo (1992a): *Legem impone subactis, Studi su filosofia e scienza dei gesuiti in Italia: 1540-1632*. Rome: Bulzoni.

Baldini, Ugo (1992b): "Dal geocentrismo alfonsino al modello di Brahe: La discussione Griemberger-Biancani," in Baldini (1992a), 217-250.

Baldini, Ugo (1992c): "Bellarmino tra vecchia e nuova scienza," in Baldini (1992a), 305-344.

Baumgartner, Frederic (1987): "Sunspots or Sun's Planets: Jean Tarde and the Sunspots Controversy of the Early Seventeenth Century," *Journal for the History of Astronomy* 18 (1987), 44-54.

Biagioli, Mario (1989): "The Anthropology of Incommensurability," *Studies in History and Philosophy of Science* 21 (1990), 183-209.

Biagioli, Mario (1993): *Galileo Courtier*. Chicago: University of Chicago Press.

Biagioli, Mario (2000): "Replication or Monopoly: The Economies of Invention and Discovery in Galileo's Observations of 1610," *Science in Context* 13 (2000), 547-590.

Blackwell, Richard (1991): *Galileo, Bellarmine, and the Bible*. Notre Dame: University of Notre Dame Press.

Braunmühl, Anton von (1891): *Christoph Scheiner als Mathematiker, Physiker, und Astronom*. Bamberg: Buchnersche Verlagsbuchhandlung.

Bredekamp, Horst (2000): "Gazing Hands and Blind Spots: Galileo as Draftsman," *Science in Context* 13 (2000), 423-462.

Cajori, Florian (1929): "History of Determinations of the Heights of Mountains," *Isis* 12 (1929), 482-514.

Cosentino, G. (1970): "Le matematiche nella Ratio Studiorum della Compagnia di Gesu'," *Miscellanea storica ligure*, nuova serie, 2 (1970), 171-213.

Crombie, A. C. (1974): "Mathematics and Platonism in the Sixteenth-Century Italian Universities and in Jesuit Educational Policy," in Y. Maeyama, W. G. Saltzer (eds), *Prismata*. Wiesbaden: Franz Steiner Verlag, 63-94.

Daston, Lorraine (1991): "Marvelous Facts and Miraculous Evidence in Early Modern Europe," *Critical Inquiry* 18 (1991), 243-274.

Daston, Lorraine and Park, Katharine (1998): *Wonders and the Order of Nature, 1150-1750*. New York: Zone Books.

Daxecker, Franz (1995): *Briefe des Naturwissenschaftlers Christoph Scheiner SJ an Erzherzog Leopold V von Österreich Tirol 1620-1632*. Innsbruck: Publikationsstelle der Universitat Innsbruck.

Dear, Peter (1987): "Jesuit Mathematical Science and the Reconstitution of Experience in the Early Seventeenth Century," *Studies in History and Philosophy of Science* 18 (1987), 133-175.

Duhem, Pierre (1969): *To Save the Phenomena*. Chicago: University of Chicago Press.

Duhr, Bernhard (1913): *Geschichte der Jesuiten in den Ländern Deutscher Zunge*. Freiburg: Herdersche Verlagshandlung.

Eddy, J., Gilman, P. and Trotter, D. (1977): "Anomalous Solar Rotation in the Early 17th Century," *Science* 198 (1977), 824-829.

Edgerton, Samuel (1984): "Galileo, Florentine 'Disegno' and the 'Strange Spottedness' of the Moon," *Art Journal* 44 (1984), 225-232.

Edgerton, Samuel (1991): *The Heritage of Giotto's Geometry*. Ithaca: Cornell University Press.

Evans, R. J. W. (1984): "Rantzau and Welser: Aspects of Later German Humanism," *History of European Ideas* 5 (1984), 257-272.

Fabricius, Johannes (1611): *De maculis in sole observatis et apparente earum cum Sole conversione* .... Wittemberg: Typis Laurentii Seuberlichii.

Favaro, Antonio (1882): "Sulla priorita' della scoperta e della osservazione delle macchie solari," *Memorie del Reale Istituto Veneto di Scienze, Lettere, ed Arti*, 22 (1882), 730-741.

Favaro, Antonio (1884): "Sulla morte di Marco Velsero e sopra alcuni particolari della vita di Galileo," *Bullettino di bibliografia e storia delle scienze matematiche e fisiche* 17 (1884), 252-270.

Favaro, Antonio (1919): *Oppositori di Galileo, III. Cristoforo Scheiner*. Venice.

Feyerabend, Paul (1975): *Against Method*. London: Verso.

Finocchiaro, Maurice (ed) (1989): *The Galileo Affair*. Berkeley: University of California Press.

Gabrieli, Giuseppe (1938): "Marco Welser Linceo augustano," *Rendiconti della Reale Accademia Nazionale dei Lincei, Classe di Scienze Morali, Storiche e Filologiche*, serie VI, 14 (1938), 74-99.

Galilei, Galileo (1890-1909): *Opere*, Antonio Favaro (ed). Florence: Giunti Barbera 1890-1909, cited as *GO*.

Galilei, Galileo (1613): *Istoria e dimostrazioni intorno alle macchie solari e loro accidenti* .... Roma: Mascardi, reprinted in *GO*, Vol. V., 72-249.

Galilei, Galileo (1610): *Sidereus nuncius*. Frankfurt: Paltheniano.

Galilei, Galileo (1989): *Sidereus nuncius*. Albert van Helden (trans). Chicago: University of Chicago Press.

Gingerich, Owen (1975): "Dissertatio cum Profesor Righini and Sidereo Nuncio", in: Righini Bonelli/Shea (eds) (1975), 77-88.

Goldstein, Bernard (1969): "Some Medieval Reports of Venus and Mercury Transits," *Centaurus* 14 (1969), 49-59.

Gorman, Michael John (1996): "A Matter of Faith? Christoph Scheiner, Jesuit Censorship, and the Trial of Galileo," *Perspectives on Science* 4 (1996).

Harcourt, Glenn (1987): "Andreas Vesalius and the Anatomy of Antique Sculpture," *Representations* 17 (1987), 28-61.

Herr, Richard (1978): "Solar Rotation Determined from Thomas Harriot's Sunspots Observations of 1611 to 1613," *Science* 202 (1978), 1079-1081.

Hosie, Alexander (1879): "The First Observations of Sun-Spots," *Nature* 20 (1879), 131-132.

Ivins, William (1953): *Prints and Visual Communication*. Cambridge, MA: Harvard University Press.

Ivins, William (1952): "What About the Fabrica of Vesalius?," in S. Lambert (ed), *Three Vesalian Essays*. New York: McMillan, 45-99.

Jardine, Nicholas (1979): "The Forging of Modern Realism: Clavius and Kepler against the Sceptics," *Studies in History and Philosophy of Science* 10 (1979), 141-173.

Jardine, Nicholas (1984): *The Birth of the History and Philosophy of Science*. Cambridge: Cambridge University Press.

Jones, Matthew (2001): "Writing and Sentiment: Blaise Pascal, the Vacuum, and the Pensees," *Studies in History and Philosophy of Science* 32 (2001), 139-181.

Lattis, James (1994): *Between Copernicus and Galileo: Christoph Clavius and the Collapse of Ptolemaic Astronomy*. Chicago: University of Chicago Press.

Licoppe, Christian (1996): *La formation de la pratique scientifique: le discours de l'experimente en France et en Angleterre (1630-1820)*. Paris: Editions La Decouverte.

Lynch, Michael (1985): "Discipline and the Material Form of Images: An Analysis of Scientific Visibility," *Social Studies of Science* 15 (1985), 37-66.

North, John (1974): "Thomas Harriot and the First Telescopic Observations of Sunspots," in John Shirley (ed), *Thomas Harriot: Renaissance Scientist*. Oxford: Clarendon Press, 129-157.

Reeves, Eileen (1997): *Painting the Heavens: Art and Science in the Age of Galileo*. Princeton: Princeton University Press.

Righini, Guglielmo (1975): "New Light on Galileo's Lunar Observations," in Righini Bonelli/Shea (1975), 59-76.

Righini Bonelli, Maria L. (1970): "Le posizioni relative di Galileo e dello Scheiner nelle scoperte delle macchie solari nelle pubblicazioni edite entro il 1612," *Physis* 12 (1970), 405-410.

Righini Bonelli, Maria L. and Shea, William (eds) (1975): *Reason, Experiment, and Mysticism*. New York: Science History Publications.

Rudwick, Martin (1976): "The Emergence of a Visual Language for Geological Science, 1760-1840," *History of Science* 14 (1976), 149-195.

Sarton, George (1947): "Early Observations of the Sunspots?," *Isis* 37 (1947), 69-71.

Scheiner, Christopher (1612): *Tres epistolae de maculis solaribus scriptae ad Marcum Welserum*. Augsburg: Ad insigne pinus, in *GO*, Vol.V.

Scheiner, Christopher (1630): *Rosa Ursina*. Bracciano: Apud Andream Phaeum.

Schove, D. J. (1948): "Sunspots and Aurorae," *Journal of the British Astronomical Association* 58 (1948), 178-190.

Shapin, Steven and Schaffer, Simon (1984): *Leviathan and the Air Pump*. Princeton: Princeton University Press.

Van Helden, Albert: *Scheiner*. Unpublished manuscript.

Van Helden, Albert (1973): "The Accademia del Cimento and Saturn's Ring," *Physis* 15 (1973), 237-259.

Van Helden, Albert (1996): "Galileo and Scheiner on Sunspots: A Case Study in the Visual Language of Astronomy," *Proceedings of the American Philosophical Society*, Vol. 140. no. 3, September 1995, 357-395.

Van Helden, Albert and Reeves, Eileen (forthcoming): *Galileo vs. Scheiner: The Sunspots Controversy of 1611-1613*. Chicago: University of Chicago Press.

Westman, Robert (1989): "The Astronomer's Role in the Sixteenth Century," *History of Science* 18 (1980), 105-147.

Whitaker, Bruce (1978): "Galileo's Lunar Observations and the Dating of the Composition of the 'Sidereus nuncius'," *Journal of the History of Astronomy* 9 (1978), 155-169.

Winkler, Mary and van Helden, Albert (1992): "Representing the Heavens: Galileo and Visual Astronomy," *Isis* 83 (1992), 195-217.

Winkler, Mary and van Helden, Albert (1993): "Johannes Hevelius and the Visual Language of Astronomy," in Judith Field and Frank James (eds): *Renaissance and Revolution: Humanists, Scholars, Craftsmen, and Natural Philosophers in Early Modern Europe.* Cambridge: Cambridge University Press, 97-116.

Eileen Reeves

Occult Sympathies and Antipathies:
The Case of Early Modern Magnetism

"I have been involved, for about three months now, in marvelous handiwork," Galileo Galilei confided in a letter of June 1626,

"[t]hat is, in greatly multiplying, through artifice, a magnet's power to hold up iron. I have put aside all other matters so as to devote myself to this, and becoming ever more avid and greedy, I can no longer be satisfied, for while at first it seemed to me a matter of great profit if I could make the magnet bear forty times its weight, now the exorbitant increase or usury of 150 times its weight cannot content me, and I go struggling away for each new gain, even the smallest one, and at the same time I am learning just what it is that moves and torments a miser."[1]

Galileo's suggestion that the gains one might make with the magnet in its armature were somehow economic – a usurious bit of earning wrung from the original deal – are part of a larger argument, one in which early moderns from Thomas More to William Gilbert sought to present ugly, useful, and ubiquitous iron ore as a practical alternative to precious metals.[2] What I wish to examine in this essay is the convergence of epistemological, cultural, and scientific questions in the treatment of magnetism in the Third Day of the *Dialogue concerning the Two Chief World Systems,* and I will suggest that Galileo's peculiarly overdrawn portrait of the Aristotelian view of the lodestone is the obverse face of his own uncertainties about the way in which this new discipline might contribute to his stature as a court scientist.[3]

There was surely a certain ambiguity in Galileo's posture in June 1626, for while he presented himself as a toiling mechanic eager to profit from the magnetic device, it was difficult to maintain that either this lodestone or its costly predecessor, displayed in the gallery of the Medici Grand Dukes since 1608, was other than a curious trinket for the idle amateur experimentalist and the highest bidders among his patrons.[4] Whereas More and Gilbert had argued for the inherent value of iron in general and the lodestone in particular on the basis of their utility, and in disregard for the fact that they were neither beautiful nor especially rare, the substances they prized had little to do with the fetishized object of Galileo's investigations.

---

1 Galilei (1968), XIII. p. 328.
2 See Freudenthal (1983), especially pp. 28-30, and Eileen Reeves (1999).
3 On Galileo as a court scientist, see the pioneering study of Biagioli (1993).
4 See Allan-Olney (1870), cited and slightly emended by Drake (1967), pp. 488-489.

The recognition that magnets were, in theory, closely connected to the commercial interests of any sea-going nation was commonplace by the mid-sixteenth century, and it followed that those men who maintained these instruments naturally laid claim to such profits. There was considerable litigation in Seville and Madrid in the 1560s over a certain instrument maker's assertion of his monopolistic rights to "feed" or remagnetize the compasses of ships venturing to the East for trade; the fact that these rights were recognized by the courts suggests, paradoxically, both an appreciation of the technological and economic importance of the device, and a presumption that the power in this particular lodestone was extraordinary, passably occult, and neither present nor producible in all such tools.[5] The publication of William Gilbert's On the Loadstone in 1600, and its enthusiastic reception in Galileo's circle, signaled a certain willingness to depart from the occult associations of that nascent discipline, and yet at least one of Galileo's colleagues, Mario Guiducci, shared his resistance to the Englishman's mercantile presentation of the lodestone. In 1626, Guiducci, Galileo's confederate in the controversy over the comets of 1618, offered a series of orations to the Florentine Academy on the poetry of Michelangelo Buonarroti.[6] Under consideration in the second of these meditations was imagery involving the lodestone, and Guiducci offered a brief summary of Gilbert's arguments for a magnetic earth and for verticity.[7] Much more remarkable, however, was Guiducci's apparent reluctance to acknowledge the lodestone's alleged connection to economic profit, as the following observations will suggest:

> "While the magnet is a little earth, it is worth noting that it guides and leads sailors across the vast oceans in every part of this wider world. Through this device every province has news of those goods that it needs, and opportunities for trade and for very rich treasures and most ample kingdoms thus are made available to industrious and warlike nations; indeed they have new worlds to acquire. In truth, one might say that through the use of the compass man is made a citizen of every part of the world, and it is in this way and no other that human beauty invests in and offers us the benefits and privileges of that great and supreme Beauty, as if it were a tiny colony of it, and reveals to us through the doubtful sea of this life the conquest of those precious and rich mines of Divine Love, showing us the way to become happy owners of the Celestial Kingdom."[8]

Suggesting that the economic advantages seemingly provided by the lodestone existed solely as an index of that other and better conquest, Galileo's disciple turned to the question of spiritual reorientation. The direction provided by the magnet, predictably, was the physical analogue of the sluggish soul's dawning sense of its true goals, and here Guiducci developed one of his most peculiar comparisons. Drawing on a tale well-known to his audience, the first story of the fifth day of Giovanni Boccaccio's Decameron, Guiducci recalled that the handsome and rich Cimone was the despair of his parents, being "coarsely inarticulate and with the manners rather of a wild beast than of a man," and was thus banished to a country estate.[9]

---

5 Lamb (1987).
6 Guiducci (1817). The first two of these lezioni are by Guiducci, while the third is the work of the Florentine Humanist Benedetto Varchi.
7 Guiducci (1817), "Lezione seconda," pp. 115-120.
8 Guiducci (1817), "Lezione seconda," p. 122.

While living the life that best suited him, that of a rustic bumpkin, Cimone chanced upon a flimsily attired maiden sleeping in the woods beside a fountain; inspired by love for her, he moved back to the city, acquired a sumptuous wardrobe, adopted "manners befitting a gentleman," "became a paragon of intelligence and wit," and distinguished himself in singing, music-making, horse-riding and martial prowess.[10] Guiducci wisely omitted the rest of Cimone's ventures – his two attempts to kidnap the maiden from her betrothed, and his ungentlemanly slaughter of his rivals – for these exploits would with difficulty illustrate the soul's magnet-like reorientation from bestial to celestial pursuits.[11] I will return shortly to Guiducci's strange analogy, showing both the cultural logic it would have had for the members of the Florentine Academy, and its affinity with the more familiar arguments about magnetism aired by Galileo. For now I would like merely to insist upon the most obvious aspects of Guiducci's meditation on the magnet: a pronounced refusal to rehearse even the slightest detail about the occult qualities of the stone, a genuine resistance to its mercantile associations, and a resultant uncertainty about the nature and purpose of this discipline.

Consider, then, the discussion of magnetism in Day Three of the *Dialogue concerning the Two Chief World Systems*. Galileo's intention in his rather brief presentation of this field was to use Gilbert's suggestions about the magnetic core of the earth as evidence for Copernicanism. In this he was successful; whereas Gilbert's Copernicanism, relegated to the last book of *On the Loadstone,* appeared a tentative afterthought to some of his earliest readers, Galileo's spokesmen Salviati and Sagredo related little about the magnet that did not support the burden of a larger cosmological argument.[12] Where they digressed from that issue, it was to present Galileo's study of the lodestone in its iron armature, for here his weight-bearing experiments far surpassed those reported by his English predecessor.[13] And it was in this connection that Salviati described, as if to underscore the Humanist assertion of the intrinsic value of iron and the lodestone, the way in which he used gem-cutting tools in the grand ducal Workshop for Precious Stones to reveal the porous nature of a large magnet once owned by Sagredo and gallantly ceded to Cosimo de' Medici.[14] His conclusion – that the presence within the lodestone of hard and lustrous gemstone, the very substance so prized by the Grand Duke and his artisans, diminished its attractive force and thus its worth – is generally consonant with Gilbert's insistence on the superiority of the magnet to more familiar icons of wealth. On the other hand, the setting and tenor of the experimentation, and Salviati's absolute indifference to the topos of navigation, exploration, and commerce, make the entire study more the self-indulgent pursuit of the privileged amateur than the useful undertaking of the practical scientist.

This apparent ambivalence about the purpose of the discipline reemerges in Galileo's presentation of the Aristotelian view of the lodestone in this part of the *Dialogue*. Though the first chapter of the first book of Gilbert's *On the Loadstone* offered an abundant catalogue of folkloric beliefs about the magnet – that it works only by day, that its vapor allows thieves to enter a locked house, that it might be foiled by a diamond and restored by the blood of a buck,

---

9 Boccaccio (1995), V 1, p. 367; Guiducci (1817), "Lezione seconda," p. 123.
10 Boccaccio (1995), p. 370; Guiducci (1817), "Lezione seconda," 123-124.
11 On the utility of the "dichotomous structure" of this tale, see Marcus.
12 Galilei (1967), pp. 399-415.
13 Galilei (1967), pp. 405-409.
14 Galilei (1967), pp. 408-409.

that it is useful in the detection of adulterous women, that shoals composed of this substance draw out all the nails of passing ships, and so on[15] – the Aristotelian Simplicio appears to hold no such beliefs, and his sole affirmation about magnetic phenomena – that they involve sympathy and antipathy – is one of surpassing banality. That this anti-Copernican thinker is resistant to and largely ignorant of the claims made by William Gilbert is, of course, no surprise, but it is the way in which Sagredo characterizes his outlook that is of particular interest here. When Simplicio admits that what little he has heard of magnetism makes it seem "very laborious and difficult to master,"[16] Sagredo assures him that understanding the issue at hand is no great task, and that

> "[i]f you like, I can prove to you that you are creating the darkness for yourself, and feeling a horror of things which are not in themselves dreadful, like a little boy that is afraid of the *tregenda* without knowing anything about it except its name, since nothing else exists beyond the name."[17]

The *tregenda* to which Sagredo refers, from the Low Latin *transienda,* a place of passage, was the supposed gathering point of sorcerers and demons, and to this image I will return. That Gilbert's work, with its stout defense of all that is sturdy and utilitarian, its impatient rejection of centuries of folkloric beliefs, its straightforward Latin prose and workmanlike engravings of bustling artisans and lumpen lodestones, could provoke terror in any reader is openly absurd, and part of a pattern of ridicule to which the bumbling Simplicio is routinely subjected. On the other hand, it is worth noting that early modern Europeans *were* aware of a context in which magnetic phenomena figured at once as fearful and as the object of hearsay, little being known of the lodestone "except its name." This was the necromantic *Clavicula Salomonis,* a handbook of invocations and diagrams, allegedly of Jewish origin, variously translated into Latin and most European languages, and on the *Index of Forbidden Books* since the mid-sixteenth century.[18] There one finds a working definition of the *tregenda,*

> "[t]he places best fitted for exercising and accomplishing Magical Arts and Operations are those that are concealed, removed, and separated from the habitations of men. Wherefore desolate and uninhabited regions are most appropriate, such as the borders of lakes, forests, dark and obscure places, old and deserted houses, whither

---

15 Gilbert (1952).

16 Galilei (1967), p. 401.

17 Galilei (1967), p. 401. I have somewhat altered Drake's translation of this passage.

18 See MacGregor Mathers (1974); Gollancz (1903); Scholem (1965). For recent work on the provenance of certain versions of the *Clavicula,* see Rohrbacher-Sticker (1993/1994) and (1995). Rohrbacher-Sticker, confining herself to a meticulous examination of MSS British Library Or. 14759 and 6360, follows Scholem's suggestions that *Clavicula* is a compilation of Christian, Jewish, and Arabic elements, and that the Sephardic manuscript version discovered by Gollancz in 1903, copied in Amsterdam around 1700, is a "very late Jewish adaptation of a Latin (or rather Italian) *Clavicula* text of the Renaissance period." As Rohrbacher-Sticker points out, there is nonetheless some evidence that an old (if not original) Hebrew version of the *Clavicula* existed alongside of, or even predated, some of the many Latin and vernacular versions of the text. On the *Clavicula*'s appearance on the Index, and for its cultural importance in the Veneto in the early modern period, see the illuminating study of Barbierato (1998), p. 254.

rarely and scarce ever men do come, mountains, caves, caverns, grottos, gardens, orchards; but best of all are cross-roads, and where four roads meet, during the depth and silence of night."[19]

In both his *Teatro dei vari e diversi cervelli mondani* of 1583 and his very popular *Piazza universale di tutte le professioni del mondo* of 1585, Tomaso Garzoni mentioned in passing what was for many the most salient aspect of the *Clavicula Salomonis*:

"And it is well demonstrated that these perverse sorcerers do everything by diabolic arts, inducing insane love and extravagant hatred in men through incantations, using the so-called and profane *Clavicula Salomonis*, wickedly and sacrilegiously baptizing magnets for this effect, and using images of melted wax, and hair-raising, unspeakable imprecations. Thus they make madmen of men, and turn them frantic, as if they were trapped, or rapt by a higher power, and lifted by force from their true selves."[20]

Garzoni's impression of the *Clavicula*, while not entirely accurate, derived from his close friendship with the noted Mantuan Jew Abramo Colorni, to whom the *Piazza universale* is dedicated, and whose engineering feats were popularly associated with black magic.[21] Colorni's Italian translation of a Hebrew version of the *Clavicula*, made at the request of the Duke of Mantua, was subsequently translated to Latin and to French, and is well represented among extant pseudo-Solomonic manuscripts today.[22] Colorni's version includes chapters on how to use waxen images and "a needle or other iron instrument" – which Garzoni and his contemporaries evidently understood to be a magnetic needle – to produce unseemly love and wild

---

19 "Of Places Wherein We May Conveniently Execute the Experiments and Operations of the Art," MacGregor Mathers (1974), II vii, p. 94). Mathers drew mostly on the Colorni class of Solomonic manuscripts, where the seventh chapter of the second of the two books treats the places appropriate for these practices. See the chapter headings for British Library MS. Sloane 3091, British Library MS. Harley 3981, British Library MS. Kings 288, Wellcome Institute MS. 4658, Wellcome Insitute MS. 4666, and Alnwick Castle MS. 584; all but this last are described in "Manuscripts of the 'Key of Solomon' in French: Abraham Colorni versions," www.levity.com/alchemy/clav_fr1.html. The Alnwick manuscript is described in "Manuscripts of the 'Key of Solomon' in Latin," www.levity.com/alchemy/clav_lat.html.
20 Garzoni (1585b), p. 515. In Garzoni (1583) Garzoni observes that those who are "mad with love go in search of the *Clavicula Salomonis* in order to get the *calamita* [magnet] that inevitably fills them with more *calamità* [calamities] than happiness."
21 On Colorni see Garzoni (1585c) and (1613); Gaddi (1636); Jarè (1891); Roth (1934); Colorni (1983), esp. pp. 644-646; Colombero.
22 The Italian version, allegedly translated from the Hebrew original, appears less common than its French derivatives. Kristeller, however, lists manuscripts of Colorni's Italian version in the Biblioteca Zayas of Seville, Spain (C V 1; C XIV 1), in the Stadtbibliothek of Zittau, Germany (B 107), among the Mss. Hamiltoniana of the Deutsche Staatsbibliothek of Berlin, Germany (589), and in the Universitätsbibliothek of Münster, Germany (Ms. Nordkirchen 169). Before its dispersal, the Royal Library of Hannover, Germany, held *La clavicola di Salomone di Abramo Colorni;* see Lewinsky (1904). Finally, there is some suggestion in two manuscripts that the Duke of Mantua ordered a Latin translation as well, though not necessarily by Colorni. For the Devonshire Collections at Chatsworth, Great Britain, Kristeller lists a sixteenth century manuscript entitled *Clavicula Salomonis ... traducta in Latinum idioma ex Hebraeo per F ... L ... C ... ex mandato ... Mantuæ Ducis* (Shelf 73 D); the Evangelische Kirchenbibliothek of Neustadt an der Aisch, Germany likewise has an eighteenth-century *Clavicula Salomonis ... translata in latinum idioma ex Hebræo mandato ... Mantuæ Ducis* (Cod. 31).

hatred in men and women, and the method does indeed depend upon "wicked and sacrilegious baptisms" of the requisite equipment and the recitation of "unspeakable imprecations."[23] As is natural in the case of translations of magical formulae, these incantations were literally "unspeakable" in the sense that they were perforce an admixture of familiar prayers and *nomina barbara* of doubtful meaning and transcription.[24] These foreign words were sometimes submitted to a chain of translations designed to preserve sound rather than sense, other times retained as lines of apparent gibberish whose magical efficacy was believed ultimately to depend upon an unknowable, unspeakable name, and still elsewhere altogether omitted by fearful or disapproving scribes.[25] It is not surprising to find that Colorni's Italian translation, or rather his particular rendering of "Hebrew and Chaldean names, as well as those in Arabic and the like" was explicitly criticized in an undated letter written by a "G. G. I. E. of Antwerp, Philosopher and Professor of Astrology," who proposed to emend and amplify the work himself.[26]

The inference that the "needle or other iron instrument" was necessarily magnetized is erroneous, but not implausible, in view of the fact that the initial conjurations and relevant diagrams emphasized the necromancer's orientation within the four cardinal points of the compass. Something of the sort was in fact reported about the magnet, albeit in more critical fashion, in a late work of Giovanni Battista della Porta (d. 1615). The great defender of Natural Magic related in his *Criptologia* that demons had induced men "to blend with the names of Saints and of God unknown words [*incognita verba*], which are blasphemies and curses," in order that such practitioners attribute legitimate effects to these "words and characters," and he affirmed that he himself had investigated "the works of doctors, some printed and others in manuscript, which are full of magic rituals and useless words."[27] The white magnet, he found, "was ill famed for being used to engender love by foolish women and seductresses, who complement it with infamous rituals and dire imprecations."[28] An argument of this sort – one which insisted both on the inefficacy of these "unknown words having no meaning in any language" and the evil intent of those who uttered them – also appeared, around 1620, in the more menacing context of an influential Inquisitorial manual attributed to Desiderio Scaglia, an eventual signatory to the sentence against Galileo in 1633.[29] In its discussion of charms the manual noted that the dwelling places of those under suspicion were to be searched for books such as the *Clavicula* and for items such as "magnets, which they often baptize."[30]

---

23 Chapters 8-11 of the first book concern ways of arousing love and inducing hatred in others; the appropriate conjurations are given here, but the methods of preparing the necessary equipment (wax images, ink, parchment, and needles) are not described until Chapters 13-17 of the second book. These involve conjurations, the recitation of certain psalms, and sprinkling with "exorcized" water; three masses must also be said over the needle and the parchment.
24 Barbierato (1998), pp. 268-269, offers the instance of a manuscript version of the *Clavicula* examined by the Holy Office in Venice in 1636 featuring prayers in slightly garbled early modern English.
25 For an informative discussion of these processes, see Rohrbacher-Sticker (1996).
26 See Jarè (1891), pp. 286-287.
27 Della Porta (1982), pp. 158-159.
28 Della Porta (1982), pp. 188-189.
29 Scaglia (1633), fol. 16. On Scaglia's treatise, its circulation, its influence, and the difficult question of its author's private views, see in particular Tedeschi (1991b) and (1991c), both reprinted in Tedeschi (1991a), and Slawinski (1999), especially pp. 53-69. For one version of the *Prattica,* see Mirto (1986).

In his *Treatise on Divination,* composed before 1602 and posthumously published in 1616, Jean-Jacques Boissard noted that men were moved to "impure and wicked love" through the magic of the *Clavicula,* which involved the baptism of magnets and subsequent incantations.[31] When the German alchemist Michael Maier warned against the demonic magic of the *Clavicula Salomonis* in 1617, he defined it as a work "to be greatly detested by all Christians," and stated that in order to bring men to excessive love or hatred "a lodestone must be sacrilegiously baptized" and certain incantations be undertaken.[32] Both Boissard and Maier wrote that the *Clavicula Salomonis* had been used by "Aaron the Jew," a twelfth-century magician who served as interpreter and adviser to a Byzantine emperor, to command legions of demons for nefarious purposes.[33] This impression of the *Clavicula* as a Jewish specialty, morally repulsive and linguistically inaccessible to good Christians, was a common one, and designed to minimize the cultural impact of translations of the sort that the very well-assimilated courtier Abramo Colorni had made.

Galileo was, in fact, probably familiar with other aspects of Colorni's work, though not necessarily with the *Clavicula* itself. The Mantuan engineer's unpublished treatise on measurement had aroused great curiosity in the decade in which Galileo was perfecting his geometrical sector,[34] and when the astronomer subsequently encountered suggestions that his instrument had predecessors elsewhere, it is significant that he portrayed the source of such rumors as satanic, and a practitioner of occult arts. "My ancient adversary," he wrote in 1607, "an envious enemy not just to me, but to all mankind, ... is always busy consulting diabolical treatises."[35] His probable target was Giacomo Antonio Gromis, a minor nobleman and former minister to the Duke of Mantua, who upon retiring to Padua enthusiastically devoted himself to alchemy, if not to necromancy.[36] Galileo referred to Gromis by name but once in this

---

30 "... ordinariamente hanno scritture di caratteri et esperimenti magici, carte vergini, clavicole, Alma del Centum Regum, Arte Notoria, Paolina, Cornelio Agrippa, Pietro d'Abano, l'Opus mathematicum, instrumenti magici, come spade caratterizzate, specchi, anelli, pentacoli, verghe, calamita che sogliono battezzare et altre cose ...", Scaglia (1633), fol. 17 r-v. While the text cited by Barbierato ((1998), p. 258, n. 76) is substantially the same as the version to which I had access, apart from a reference to "spade catechizzate" and "calamita battezzata," that used by Tedeschi ((1991c), pp. 248-249, n. 32) is closer still. That published by Mirto ((1986), pp. 114-115) omits the allusion to mirrors, rings, pentacles, wands, and baptized lodestones. All versions, however, refer to the use of baptized lodestones later in the same chapter.

31 Boissard (1616?), p. 44.

32 Michael Maier (1617), p. 69.

33 Boissard (1616?), p. 59, and Maier (1617), p. 69. Both authors draw on Niceta Choniates' *Chronicle* V: 7: 9.

34 Colorni's *Euthimetria,* though licensed for publication in 1580, remains unpublished today in the Herzog-August Bibliothek of Wolfenbüttel, Germany; see Lewinsky (1904), p. 237. Rafaele Mirami noted in his *Compendiosa introduttione alla prima parte della specularia* that mirrors were useful in "measuring from afar heights, depths, and distances, as Abram Colorni the Jew, the most ingenious engineer for the Most Serene Duke of Ferrara, has amply shown in a treatise of his." See Mirami (1582). In the prefatory letter that accompanies the 1592 and 1601 editions of the *Piazza universale* – a letter which was suppressed in 1617 – Tomaso Garzoni alluded to the catoptrical nature of one of Colorni's measuring devices, calling the still unpublished treatise itself a "bright mirror exposed to the world." (Garzoni (1585a), p. 21.) Colorni's treatise describes an odometer and a device composed of three calibrated pivoting sighting rods designed to form triangles whose congruency to actual objectives allowed for altimetry. The altimetric device could be further modified by the attachment of a plane mirror.

35 Galileo, *Difesa contro alle calunnie ed imposture di Baldessar Capra,* in Galilei (1968), 2: 519. On the events leading to this publication, see Drake (1978), pp. 44-48, 83 and 120-123.

diatribe; the fact that he called him "Gromo" may have misled some readers into associating him, like Abramo Colorni, with the Italian Jewry.[37]

There is no need to suppose, however, that Galileo knew more of the *Clavicula Salomonis* than what Garzoni, della Porta, Scaglia, Boissard, or Mayer reported about the baptized lodestones and unspeakable formulae. These passages alone would suggest that in the view of some, an obscure combination of magnetic phenomena and strange words in the context of what is known as "sympathetic magic" resulted in the onset of otherwise inexplicable love and hatred.

Thus while part of the humor of Sagredo's remark to Simplicio lies in the fact that nothing in Gilbert's work could possibly evoke the fearful specter of the *tregenda,* such an impression of the lodestone nevertheless had a place within the cultural imagination. The fact that Salviati claimed that he came by his copy of *On the Loadstone* via a Peripatetic friend of his, who, fearing that the treatise would infect his personal library, made a gift of it, is also noteworthy, if not especially believable, in this connection.[38] It was not uncommon for scholars to segregate their necromantic books from the rest of their collection, consigning works concerned with magic to a separate vault or room; the suggestion is that Simplicio is not the sole Aristotelian to associate the study of magnetism with the practice of the occult.[39] Finally, Sagredo's effort to liberate Simplicio of his fears of the *tregenda* is tellingly preceded by his promise to prove to him that he is creating such darkness for himself: *vi farò toccar con mano come voi da per voi stesso vi fate ombra.* Galileo used the resonant phrase *far toccar con mano* elsewhere, but its original context is crucial here: it is the way in which the resurrected Christ converted the doubting Jew Thomas into an exemplary early Christian.[40]

Sagredo's second reference to whatever it is that Simplicio understands about magnetism involves the Aristotelian's unremarkable reliance on the terms "sympathy" and "antipathy." After introducing Gilbert's theory by insisting that the earth itself was an immense lodestone, and had like that substance two poles of attraction, Salviati and Sagredo discuss at some length the experiments performed with the magnet in its iron armature, their conclusions about the porous nature of the former, and the importance of the concave shape of the latter.[41]

---

36 Gromis was portrayed as the hero of Ingegneri's *Argonautica* of 1601, an allegorical meditation on alchemy. Unfortunately for Gromis, Ingegneri eventually changed his mind about the value of the discipline, publishing *Contra l'alchimia e gli alchimisti. Palinodia dell'Argonautica* in 1606. See Ingegneri (1989), xxvii. Gromis evidently owned several alchemical manuscripts: his name appears, for example, in Wellcome Institute MS. 310 ff. 120-156, *Secreta secretissima,* and in an alchemical miscellany in King's Library in Copenhagen, MS. 1722, *Secreta.*

37 Galileo, *Difesa,* in Galilei (1968), 2: 594. In northern and central Italy, some Jewish families, as for example the Colorni, took place names as surnames. See Roth (1946), p. 359. It is just possible that Galileo's other targets, the Catholic Milanese Baldessar Capra and the Protestant Simon Mayr of Guntzenhausen, Germany, would have appeared to be Jewish or Judaizers to readers of the *Difesa.* Animal names such as "Capra" were common surnames among Spanish Jewry and *conversos;* Francisco Quevedo's antisemitic *Buscón* (1626), for example, features the *converso* schoolmaster Cabra. See on this point Ettinghausen (1987), esp. p. 246, and Villanueva (1960), p. 47 n. 9. Simon Mayr's name recalls that of Abramo Colorni's ancestor, "Simone de Maya," this last being perhaps an Italianized form of the German name "Simon Meyer;" see Colorni (1983), pp. 640-641.

38 Galilei (1967), p. 400.

39 Scaglia's presentation of the books and magical instruments described above in Scaglia (1633) of the *Prattica* as "*materia contagiosa*" is typical.

40 *John* 20: 24-29.

When called upon to express admiration for their explanations of the way in which the Medici magnet surpassed that of the Englishman, Simplicio has recourse to little other than traditional terminology for the phenomena at hand.

"Truly, I think that Salviati's eloquence has so clearly explained the cause of this effect that the most mediocre mind, however unscientific, would be persuaded. But we who restrict ourselves to philosophical terminology reduce the cause of this and other similar effects to *sympathy*, which is a certain agreement and mutual desire that arise between things which are similar in quality among themselves, just as on the other hand that hatred and enmity through which other things naturally fly apart and abhor each other is called by us *antipathy*."[42]

Simplicio's position, and his implicit resistance to the neologisms scientific developments would seemingly require, is an unsurprising, if naïve, subscription to Pliny's first century adoption of the Greek terms *sympathy* and *antipathy* to describe the accord and discord of all natural things.[43] Such postures were obliquely criticized by William Gilbert in the preface to his work: rather than relying on the two traditional terms, he would use "words new and unheard of not (as alchemists are wont to do) in order to veil things with a pedantic terminology," but "in order that hidden things with no name and up to this time unnoticed [might] be plainly and fully published."[44] Mario Guiducci had likewise ridiculed those who supposed that *sympathy* and *antipathy* adequately explained magnetism in his *Oration* of 1626, associating such reductive language with the "infancy of Philosophy."[45] Given that Simplicio's reliance on the outworn and not especially useful doctrine of sympathy and antipathy is no more than what one would expect, and that it comes very near the end of the Third Day of discussions, the sheer extravagance of Sagredo's reaction demands attention.

"And thus, by means of two words, causes are given for a large number of events and effects which we behold with amazement when they occur in nature. Now this method of philosophizing seems to me to have great sympathy with a certain manner of painting used by a friend of mine. He would write on the canvas with chalk, 'This is where I'll have the fountain, with Diana and her nymphs; here, some greyhounds; there, a hunter with a stag's head. The rest is a field, a forest, and hillocks.' He left everything else to be filled in with color by a painter, and with this he was satisfied that he himself had painted the story of Acteon – not having contributed anything of his own except the title.[46]

This characterization of Simplicio's approach to magnetism and numberless other phenomena – more colorful by far than anything the Aristotelian has enunciated – may have an actual historical referent. It is not inconceivable, after all, that Galileo had in mind a particular

---

41 Galilei (1967), pp. 405-410.
42 Galilei (1967), p. 410.
43 Pliny, *Natural History* XXXVII: 15; see on this point French (1994), pp. 239-240.
44 Gilbert (1952), p. 2.
45 Guiducci (1817), p. 117.
46 Galilei (1967), p. 410.

artist whose studio assistants produced a painting bearing little but the crucial elements of his name and the work's title, and that the alleged technique of marking the canvas with *chalk* and *words* is his approximate description of a man with some prestige as a painter who routinely abandoned the task to others after the initial phases of the *gesso* and *imprimatura*.[47]

But what is more striking, of course, is the suitability of the painting in this discussion, the tale of Diana and Acteon being the most apposite example of states that are named, but in no way explained, by the words *sympathy* and *antipathy*. As Ovid tells the story, when the young Acteon, obsessed with hunting, happened upon a cave where Diana and her nymphs were bathing, the enraged goddess splashed her unwilling observer with water, and turned him to a stag, the inarticulate and inevitable victim of his own dogs. It is easy to see that this version of the tale would lend itself, as did much of the Ovidian corpus, to a kind of scientific allegory, as if the focal point of the episode were not its divine, human, and canine protagonists, but rather the various chemical combinations so obscurely described by the poet. The fact that the title of the *Metamorphoses* was sometimes rendered *Trasformationi,* a word of strong alchemical associations, merely reinforced such tendencies. To choose one of the several such readings of this Ovidian myth in early modern Italy, we learn in Vincenzo Percolla's *Auriloquio* that

> "[t]he story of Acteon involves the division of elements. It is said that when Diana was washing herself in a fountain, Acteon passed that way and having stopped to contemplate her, the goddess threw water on him and changed him into a stag, such that he was torn to bits by his own dogs. Diana bathing in a fountain is quicksilver, which dissolves in the bath of refined and mixed mercury. Acteon, superheated gold, is nearby. He is a hunter, being one that catches every quarry. He contemplates Diana, when he approaches the refined and animated quicksilver to dissolve himself. Diana throws water on him; ... transformed into a stag, he is dissolved and becomes a light animal, such that in this dissolved form he leaps and passes through the alembic. And he is torn to bits by its dogs, that is, by its fires, which wears down and catches everything it desires. With the first heating he is torn apart by putrefaction, then with the second, water comes forth, and in the third, air is produced. In the fourth and final heating the element of fire is produced and thus all the elements are separated."[48]

One can be certain of Galileo's contempt for those readings of the *Metamorphoses* designed to explain "the secrets of Nature." In Day Two of the *Dialogue,* Sagredo scorns the alchemists' gold-mongering approach to Ovid, especially their interpretations of

> "what the loves of the moon [goddess Diana] mean, and her descent to the earth for Endymion; her displeasure with Acteon; the significance of Jupiter's turning himself into a rain of gold, or a fiery flame; what great secrets of the [alchemical] art there are in Mercury the interpreter, in Pluto's kidnappings, and in golden boughs."[49]

---

47 While it would be hazardous to interpret this remark as a reference to a particular painter's practice, it is the case that Titian, much admired by Galileo, did leave a number of *abbozzi* of just this subject. See in this connection Pignatti (1994).

48 Vincenzo Percolla (1996), pp. 127-128.

But we can also see the rhetorical usefulness of this very reductive reading of the Acteon episode, and of the suggestion that the alchemists' habits of mind, while not identical to those of Simplicio, were somehow related. Recall that Sagredo first mockingly characterized Simplicio's fearful impression of the science of magnetism in terms evoking the sympathetic magic of the *Clavicula Salomonis,* where the combination of "unspeakable imprecations" and baptized lodestones was thought sufficient to explain otherwise unfathomable bouts of love and hatred. His second and rather more accurate description of the man who understands all phenomena in terms of the meaningless rubrics *sympathy* and *antipathy* as a painter who contributes nothing but a few words shorn of context to a canvas is clearly a variant upon this structure. Whereas the necromancer would insist that it is the sacred nature of the *nomina barbara* he uses that has an effect on the lodestone and so incites men to love and hatred, it is the seemingly scientific aspect of the Greek words that misleadingly serves not to induce, but rather to explain, all forms of attraction and repulsion. And there is a third portrait here: it satirizes anyone who misinterprets the subject of the painting, Diana and Acteon, as an alchemical process, and implicitly criticizes allegoresis, the process of "speaking of something other," as a means of "explaining" the complicated narrative of the goddess and the hunter.

While it is clear that the general mechanisms of sympathy and antipathy subtend both black magic and alchemy, little or nothing in the behavior of the bland and upright Simplicio can be interpreted as evidence for his practice of either art. It is as if some inevitable collapse of distinct social groups must occur under the force of a world view based solely on these two terms: or so Sagredo implies when he remarks that the Aristotelian's "way of philosophizing seems to have some sympathy" with the gesture of a man who contributed nothing but a few words to the *istoria* of *Diana and Acteon,* itself, inevitably, an inarticulate story of human attraction and repulsion, misread in its turn by alchemists as the tale of quicksilver and superheated gold. And Sagredo's observation anticipates that of the skeptical François la Mothe le Vayer, who noted in 1640 that "those who know most of natural sympathies and antipathies, or of these occult and specific properties of which philosophers speak, are the greatest magicians of all, in the view of the people."[50]

Galileo's effort to associate Simplicio's unthinking preference for the terms "sympathy" and "antipathy" with the *Clavicula* on the one hand, and with alchemy on the other, was clearly of some cultural moment in early modern Europe. Though many versions of the *Clavicula,* including Colorni's, offered no explicit advice about the transformation of metals, the treatise was and remains of considerable interest to alchemists. In the early seventeenth century, those who sought to obtain licensing for works of alchemical content sometimes distinguished their efforts from the Solomonic text; Antoine Domayron, a Frenchman resident in Venice around 1605-1606, for example, rejected in his discussion of alchemy any common ground with "all those keys or clavicles of the Rabbis, which some falsely attribute to the wise Solomon."[51] Michael Maier, as we have seen, made much the same gesture in 1617, Saint Louis-Marie Grignion de Montfort criticized alchemists' reliance on that "secret but false and pernicious book, which they call the *Clavicula Salomonis*" in the early eighteenth centu-

---

49 Galilei (1967), p. 110.
50 Mothe le Vayer (1761), 1: 1: p. 364.
51 Secret (1973), on p. 112.

ry,[52] and in his *Dictionnaire philosophique* Voltaire would likewise make the work the very emblem of the vain hopes of gold-seekers: "'If we haven't yet found [the philosopher's stone],' they say, 'it is because we are not advanced enough yet, but it is certain that it is in the *Clavicula Salomonis.*' And with this fine certitude, more than two hundred families have been ruined in Germany and France."[53]

Yet more pertinent for our inquiry, however, is the fact that another early modern version of the *Clavicula Salomonis,* an Italian rival to Colorni's translation and one that insists at length on astral influences, explicitly connects "sympathy" with magnetic attraction, and argues for the likeness of phenomena involving action at a distance to those produced by incantation. Thus, in an effort to present the *Clavicula* as something other than a work "to be greatly detested by all Christians," the unknown editor tells us,

> "[n]o ignorant person, condemning [this Solomonic art] would calmly expose himself to ridicule by maintaining that the surprising effects that take place every day before our eyes actually involve magic, or sortilege, or enchantment or devilry. Thus for example is sympathy, which is found in the natural operation of the stars, whose occult virtues are known in no other way than through the quotidian experience we have of them; for the stars remain hidden in their very essence, are impenetrable to reason itself, ... and not even the most capable philosophers can at present establish solid and evident laws about them. Likewise are the occult powers of the lodestone, which is able to draw iron from afar, and of amber when it attracts straw, and of a cock's eyes when they render someone ardent and intrepid, and of a dog's tongue and heart, which, if carried behind one's back, prevent bites from this animal, and of infinite other things, some of which we will examine through most curious and useful experiments in the following chapters of [the *Clavicula.*]"[54]

We cannot, of course, allege Galileo's familiarity with this passage, but neither can we overlook its resemblance to his portrait of Simplicio as one who obscurely connects the lodestone with the *tregenda.* As the two texts stand, in other words, we may see them as part of a general dialogue in which the presumed association of magnetism with the occult is at once ridiculed and rendered suspect, as in the *Dialogue,* or normalized through the promise of "curious and useful experiments," as happens in this version of the *Clavicula.*

It is also worth noting that any popular impression of magnetic phenomena as a quotidian version of the diabolic practices described in the *Clavicula* found confirmation in Milan in 1630, during the infamous trials of suspected plague-spreaders. One of the accused confessed under torture that he had been given a toxic ointment that "drew men more than the lodestone does iron" by a man named Baruello;[55] the unfortunate Baruello, likewise subjected to torture

---

52 De Montfort (1966), p. 138.

53 Voltaire: "Astronomie, et quelques réflexions sur l'astrologie," in *Dictionnaire philosophique.*

54 *La clavicola magica e cabalistica del re Salamone, tradota dal testo original ebraico in latino dal virtuoso negromante Cornelio Agrippa, e posta in francese dal Rabino Nazar,* fol. 17. I have used the incomplete eighteenth-century manuscript in possession of the Rare Book Room of Van Pelt Library at the University of Pennsylvania (Ms. Codex 515); I am especially grateful to Michael Ryan and Dan Traister for making this work available to me. Codex 515 is related to the French Rabbi Abognazar versions; on the latter see http://www.levity.com/alchemy/clav_fr3.html.

and urged to renounce all prior pacts with the devil, at length stated that he had been recruited for plague-spreading by a French priest who bore a long black wand and the *Clavicula Salomonis*, repeated the unspeakable "*Gola Gibla* and other Hebrew words," and drew circles upon the ground.[56] Baruello's other confederates were also said to have written incantations "with circles and diabolic characters" – conventional necromantic practices described in detail in the *Clavicula* – to ensure that their enemies would endure lifelong torment.[57]

Let me return, however, to what Galileo had had to say about the lodestone just *prior* to the plague of 1630. Here I would like to emphasize a certain interestedness in the astronomer's entire treatment of the Aristotelian's alleged view of magnetic phenomena, and one that may explain his willingness to shift the focus of his rhetorical energy to whatever it was that he perceived behind Simplicio's bland remarks. Consider, in this connection, the story of Diana and Acteon. That the alchemical interpretation of the Ovidian tale failed to explain, in Galileo's view, its obscure motives of attraction and repulsion is clear enough, but what I want to insist upon is the rival reading of the legend. A number of mythographers from the fourth century B. C. through the eighteenth century C. E., unable to fathom the disastrous dynamic between the goddess and the hunter, and like Ovid, unwilling to depict Acteon as a simple sexual predator, tended to scrutinize the other and seemingly less central events of the tale. In some of these readings, Acteon's death bore no particular connection to Diana herself; the fact that he had spent all his substance on his dogs meant that he would inevitably be consumed by them. [58] This interpretation, surprising in that it renders the dramatic encounter at the cave largely superfluous, makes some sense of the extraordinarily long catalog of dogs in the accounts of Ovid and his predecessors, where thirty-six hounds are named and described before they devour their master. It also proved remarkably adaptable as a cautionary tale: though Acteon's flaw was sometimes narrowly interpreted as an obsession with hunting, it was more generally read as any socially irresponsible form of profligacy.

Early modern Englishmen, relatively fond of hunting and ruled by those for whom the activity was a true passion, tended therefore to conflate that sport with other excesses. Arthur Golding's *Metamorphoses,* for example, states in 1567 that

"[a]ll such as doo in flattring freaks, and hawks, and hownds delight,
And dyce, and cards, and for too spend the tyme both day and nyght
In foule excesse of chamberworke, or too much meate and drink:
Uppon the piteous storie of Acteon ought too think.
For theis and theyr adherents used excessive are in deede
The dogs that dayly doo devour theyr followers on with speede."[59]

Robert Burton, moving yet further from Ovid's lonely and abstemious sportsman, noted in 1621 that "some men [who] live like the rich glutton, consuming themselves and their sub-

---

55 Farinelli/Paccagnini (1988), p. 250.
56 Farinelli/Paccagnini (1988), pp. 374-375, 540, 570.
57 Cusani (1841), p. 136; Farinelli/Paccagnini (1988), pp. 570-571. On the related and more diffuse representation of the plague as a diabolic process, and of the supposed *untori* as *negromanti,* see Farinelli/Paccagnini (1988), pp. 78-80, 83, 131-133, and Calvi (1984), p. 198.
58 See Schlam (1984), pp. 82-110, and Barkan (1980), especially pp. 326-329.
59 Ovid (1567), vv. 97-100.

stance by continuall feasting & invitations, ... keeping a table beyond their meanes, and a company of idle servants ... are blowne up on a sudden, and as *Actæon* was by his hounds, devoured by their kinsmen, friends, and multitude of followers."[60] And Alexander Ross bluntly advised his audience in 1642,

> "[t]hink you on this, who spend your dayes, and strength,
> And means, on Whores, Dogs, Parasites; at length
> They'l worry you: before you feel their wounds,
> Look to their teeth, shun these Actæons hounds."[61]

More pertinent to Galileo's own experience, however, was the interpretation of the tale reiterated in numerous Venetian editions of Giovanni dell'Anguillara's translation of the *Metamorphoses* and published from the 1560s through 1624. "The allegory is the following," readers were told,

> "[h]e who diligently gives himself over to studying the mysterious arrangement of the heavens, and the variation of the moon, which is symbolized here by Diana, is transformed into a wild stag, staying in the wilderness, and in solitary places, drawn by curiosity about that science, such that when his own familial concerns, which are symbolized by the dogs, catch up with him, he is consumed by them, since such concerns never permit a man to live unto himself."[62]

It is hard to imagine a more fitting description of Galileo's difficult management of the evident conflict between familial burdens and intellectual endeavor. His sometime success as a courtier – a social role that would have at once fed his family and made whatever he accomplished "in the wilderness and in solitary places" more accessible to patrons and wealthy dilettanti – did not prevent a friend from portraying him, late in life, as a more unfortunate version of the hunter. Thus in July 1637, when the sequestered astronomer was blind in one eye, Vincenzo Renieri wrote, "my dear Galileo, the sun is competing with his sister; while she punished Acteon for seeing her unclothed, he wants to cloud the eye that spied him *al vivo*."[63]

What I would suggest, then, is that Galileo's apparent readiness to associate Simplicio's reliance on the terms *sympathy* and *antipathy* with the black magic of the *Clavicula*, with profitable workshop productions, and with alchemical investigations is motivated, at least in part, by his sense of the embarrassingly apposite application of the story of the mute Acteon to his own life. Put differently, Galileo's depiction of the Aristotelian's view of magnetism as somehow contiguous to practices whose primary focus was the pursuit of wealth cannot be unrelated to his own quite real anxieties about his precarious financial situation. Given that his domestic burdens increased dramatically between January 1627 and the summer of 1628

---

60 Burton (2001), p. 108.
61 Ross (1642).
62 I have used the following version, dell'Anguillara (1584), p. 98. For the importance of this vernacular version of the *Metamorphoses* for Titian in particular, see Ginzburg (1986), esp. pp. 144-146. For an early modern Spanish version of the dell'Anguillara reading, see dell'Anguillara (1609), fol. 81 v.
63 Galileo (1968), 17, p. 133.

when his younger brother Michelangelo, a musician in Munich, left to him the education of his intractable eldest son, and sent his wife and four of his other children to stay with the scientist in Florence, Galileo may indeed have felt, to use the terms of the Ovidian allegory, hounded, dogged, and consumed by his own extended family.[64] That he had chosen, in "studying the mysterious arrangement of the heavens, and the variation of the moon," and in pursuing the ancillary investigations of the ugly magnet, a path either more profitable or more acceptable to a princely patron than those of the practitioner of the occult or the master painter would have been by no means obvious. Michelangelo, for his part, prefaced his request of his brother's aid with the admonition that given the wartime conditions in Germany, there was little hope of interesting rulers such as Maximilian of Bavaria in trifles such as the magnet:

> "Here we skimp on everything, and the constraints all are under is unspeakable, the cause being these wars, and his Highness is therefore full of worries and grave concerns. Though brief mention was made of your business with the magnet, he doesn't seem to have any inclination for it, such that I think that it would be profitable for you to leave the matter for now ..."[65]

Let me conclude, then, by returning to Mario Guiducci's peculiar discussion of the magnet, one which avoided Gilbert's straightforward insistence on the economic value of the stone, and sought instead to present it as an emblem of the soul's inevitable attraction to loftier goals. As I have noted, Guiducci's analogy involved a tale from the *Decameron,* one which featured the metamorphosis of the uncouth Cimone from sullen rustic to accomplished courtier. That such a transformation came about because he chanced upon a beautiful and partially clad maiden beside a fountain is plainly a gratuitous appropriation of *dolce stil nuovo* doctrines; as Guiducci's audience would have known, however, the first part of the tale recast, in more felicitous and Florentine fashion, the equally implausible events of the Ovidian tale of Diana and Acteon. Rather than pursuing scientific studies to his detriment in some remote setting, the boorish Cimone bolts from the woods and converts his ardor into a kind of courtly capital; far from being the silent victim of those he once fed, he speaks with wit and eloquence, and consumes all that he desires. He stands as the apparent corrective to that too-available portrait of the astronomer-as-Acteon, and as a fantastic rival to that motley gang of necromancers, alchemists, idle artists, and simple Aristotelians whose professional success depended upon their various manipulations of formulaic language.

---

64 Much of the correspondence in the latter part of the 13th volume of Galileo's *Opere* is devoted to these unhappy episodes. See in particular pp. 346-348, 352-355, 417-418, 438-439 and Galileo (1968), 14, pp. 177-178 and 256-257.
65 Galileo (1968), 13, p. 347. Three and a half years earlier, Galileo had interested Duke Maximilian of Bavaria in a microscope; see Galileo (1968), 13, pp. 177-178.

References

Allan-Olney, Mary (1870): *The Private Life of Galileo*. London: Macmillan.

Barbierato, Federico (1998): "Il Testo Impossibile: La *Clavicula Salomonis* a Venezia (Secoli XVII-XVIII)," *Annali della Fondazione Luigi Einaudi* 32 (1) (1998), 235-284.

Barkan, Leonard (1980): "Diana and Actæon: The Myth as Synthesis," *English Literary Renaissance* 10 (1980), 317-359.

Biagioli, Mario (1993): *Galileo, Courtier: The Practice of Science in the Culture of Absolutism*. Chicago/London: University of Chicago Press.

Boccaccio, Giovanni (1995): *The Decameron*. Translated by G. H. McWilliam. London: Penguin Books.

Boissard, Jean J. (1616?): *Tractatus posthumus ... de divinatione & magicis præstigiis*. Oppenheim: Typis Hieronymi Galleri.

Burton, Robert (2001): "Democritus to the Reader," in Holbrook Jackson and William H. Gass (eds.), *The Anatomy of Melancholy*. New York: New York Review of Books.

Calvi, Giulia (1984): *Storie di un anno di peste*. Milan: Bompiani.

Colombero, C. : "Colorni, Abramo," *Dizionario biografico degli italiani* [   ] 27, 466-469.

Colorni, Vittore (1983): "Genealogia della famiglia Colorni (1477-1977)," in *Judaica minora: Saggi sulla storia dell'ebraismo italiano dall'antichità all'età moderna*. Milan: A. Giuffrè Editore, 637-660.

Cusani, Francesco (ed. and trans.) (1841): "Appendice al libro secondo", in *La Peste di Milano del 1630. Libri cinque cavati dagli annali della città e scritto per ordine dei LX decurioni dal Canonico della Scala Giuseppe Ripamonti istoriografo Milanese*. Milan: Tipografia e Libreria Pirotta.

De Montfort, Saint Louis-Marie G. (1966): "L'Amour de la Sagesse Éternelle," in *Œuvres completes*. Paris: Éditions de Seuil.

Dell'Anguillara, Giovanni A. (1584): *Le Metamorfosi d'Ovidio in ottava rima, con le Annotationi di M. Gioseppe Horologgi, & gli Argomenti, & Postille di M. Francesco Turchi*. Venice/Bern: Giunti.

Dell'Anguillara, Giovanni A. (1609): *Metamorphoseos del excelente poeta Ovidio Nasson. Traduzidos en verso svelto y octava rima' con sus allegories al fin de cada libro por el Doctor Antonio Perez Sigler natural de Salamanca*. Burgos: I. Baptista Varesio.

Della Porta, Giovanni Battista (1982): *Criptologia*. Edited by Gabriella Belloni. Rome: Centro Internazionale di Studi Umanisitic.

Drake, Stillman (1978): *Galileo at Work*. Chicago and London: University of Chicago Press.

Ettinghausen, Henry (1987): "Quevedo's Converso Pícaro," *MLN* 102 (2) (1987), 241-254.

Farinelli, Giuseppe and Paccagnini, Ermanno (ed.) (1988): *Processo agli untori. Milano 1630: cronaca e atti giudiziari in edizione integrale*. Milan: Garzanti.

French, Roger (1994): *Ancient Natural History: Histories of Nature*. London/New York: Routledge.

Freudenthal, Gad (1983): "Theory of Matter and Cosmology in William Gilbert's *De magnete*," *Isis* 74 (1983), 22-37.

Gaddi, Iacopo (1636): "Elogio Poetico Oratorio di Abramo Colorni," in *Adlocutiones et elogia exemplaria, cabalistica, oratoria, mixta, sepulcralia*. Florence: Typis Petri Nestei, 167-169.

Galilei, Galileo (1967): *Dialogue concerning the Two Chief World Systems*. Translated by Stillman Drake. Berkeley/Los Angeles/London: University of California Press.

Galilei, Galileo (1968): *Le Opere*. Edited by Antonio Favaro (20 vols.). Florence: Giunti Barbèra.

Garzoni, Tomaso (1583): "Il Teatro de' vari e diversi cervelli mondani," in Paolo Cerchi (ed.), *Opere*. Ravenna: Longo Editore 1993, 223-224.

Garzoni, Tomaso (1585a): *La piazza universale di tutte le professioni del mondo*, 2. vol. Edited by Giovanni Battista Bronzini, Pina De Meo, and Luciano Carcereri. Florence: Leo S. Olschki 1996.

Garzoni, Tomaso (1585b): "De' Maghi incantatori," in Garzoni (1585a).

Garzoni, Tomaso (1585c): "Lettera scritta al Sopradetto m[aestro] Abramo con occasione del sonetto, et d'alcune annotationi antecedenti nella Piazza," in Garzoni (1585a), 17-22.

Garzoni, Tomaso (1613): *Il serraglio de gli stupori del mondo ... opera ... arrichita di varie annotationi dal M. R. P. D. Bartolomeo Garzoni suo fratello*. Venice: Apresso Ambrosio, et Bartolomeo Dei, 225-226.

Gilbert, William (1952): "On the Loadstone" (translated by P. Fleury Mottelay), in Robert M. Hutchins (ed. in chief), *Great Books of the Western World*, vol. 28. Chicago/London/Toronto: William Benton, 3-7.

Ginzburg, Carlo (1986): "Tiziano, Ovidio e I codici della figurazione erotica del Cinquecento," in *Miti Emblemi Spie. Morfologia e storia*. Turin: Einaudi, 133-157.

Gollancz, Hermann (ed.) (1903): *Mafteah Shelomoh. Clavicula Salomonis. A Hebrew Manuscript*. Frankfurt/London: J. Kauffmann and D. Nutt.

Guiducci, Mario (1817): "Tre Lezioni sopra le Rime di Michelagnolo Buonarroti," in *Le Rime di Michelagnolo Buonarroti pittore, scultore, architetto e poeta fiorentino*. Vicentius Columna.

Ingegneri, Angelo (1601): *Argonautica*.

Ingegneri, Angelo (1989): *Della poesia rappresentativa e del modo di rappresentare le favole sceniche*. Edited by Maria L. Doglio. Ferrara: Edizioni Panini

Jarè, Giuseppe (1891): "Abramo Colorni, Ingegnere del secolo XVI," *Deputazione proviciale ferrarese di storia patria: Atti e memorie* 3 (1891), 257-312.

La Mothe le Vayer, François de (1761): "De la magie," in *Oeuvres de François de la Mothe le Vayer*. Dresden: Michel Groell. (Originally published in *De l'Instruction de m. le Dauphin*. Paris 1640: Sebastien Cramoisy.)

Lamb, Ursula (1987): "The Sevillian Lodestone: Science and Circumstance," *Terræ Incognitæ* 19 (1987), 29-39.

Lewinsky, A. (1904): "Alcuni manoscritti italiani nella biblioteca ducale di Wolfenbüttel," *Rivista israelitica* 1 (1904), 150, 237.

MacGregor Mathers, S. Liddell (ed. and transl.) (1974): *The Key of Solomon the King (Clavicula Salomonis)*. York Beach, Maine: Samuel Weiser.

Maier, Michael (1617): *Symbola aureæ mensæ duodecim nationum*. Frankfurt: Anton Humm.

Marcus, Millicent (1980): "The Sweet New Style Reconsidered: A Gloss on the Tale of Cimone (*Decameron* V, 1)," *Italian Quarterly* 81 (1980), 5-16.

Mirami, Rafaele (1582): "Ai benigni, & giudiciosi Lettori," in *Compendiosa introduttione alla prima parte della specularia*. Ferrara: Apresso gli Heredi di Francesco Rossi, & Paolo Tortorino.

Mirto, Alfonso (1986): "Un inedito del seicento sull'Inquisizione," *Nouvelles de la République des Lettres* 1 (1986), 99-138.

Ovid (1567): *Metamorphoses*. Translated by Arthur Golding.

Percolla, Vincenzo (1996): *Auriloquio*. Paris: SEHA; Milan, Archè.

Pignatti, Terisio (1994): "*Abbozzi* and *Ricordi*: New Observations on Titian's Technique," *Studies in the History of Art* 45 (1994), 72-83.

Reeves, Eileen (1999): "As Good as Gold: The Mobile Earth and Early Modern Economics," *Journal of the Warburg and Courtauld Institutes* 62 (1999), 126-166.

Rohrbacher-Sticker, Claudia (1993/1994): "*Mafteah Shelomoh:* A New Acquisition of the British Library," *Jewish Studies Quarterly* 1 (1993/1994), 263-270.

Rohrbacher-Sticker, Claudia (1995): "A Hebrew Manuscript of *Clavicula Salomonis,* Part II," *British Library Journal* 21 (1995), 128-136.

Rohrbacher-Sticker, Claudia (1996): "From Sense to Nonsense, From Incantation Prayer to Magical Spell," *Jewish Studies Quarterly* 3 (1996), 24-46.

Ross, Alexander (1642): "Actæon," *Mel Heliconium*.

Roth, Cecil (1934): "Abramo Colorni geniale inventore mantovano," *La Rassegna Mensile di Israele* 9 (3-4) (1934), 147-158.

Roth, Cecil (1946): *The History of the Jews of Italy*. Philadelphia: The Jewish Publication Society.

Scaglia, Desiderio (1633): "Dei Sortileghi," in *Prattica,* ms. 602. Van Pelt Library, University of Pennsylvania.

Schlam, Carl C. (1984): "Diana and Actæon: Metamorphoses of a Myth," *Classical Antiquity* 3 (1984), 82-110.

Scholem, Gershom (1965): "Some Sources of Jewish-Arabic Demonology," *Journal of Jewish Studies* 16 (1965), 1-13.

Secret, François (1973): "Littérature et Alchimie à la fin du XVIe et au début du XVIIe siècle," *Bibliothèque de l'Humanisme et de la Renaissance* 36 (1973), 103-116.

Slawinski, Maurizio (1999): "Marino, le streghe, il cardinale," *Italian Studies* 54 (1999), 52-84.

Tedeschi, John (1991a): *Prosecution of Heresy: Collected Studies on the Inquisition in Early Modern Italy*. Binghamton, NY: Medieval & Renaissance Texts & Studies.

Tedeschi, John (1991b): "The Roman Inquisition and Witchcraft: An Early Seventeenth-Century 'Instruction' on Correct Trial Procedure," in Tedeschi (1991a), 205-227.

Tedeschi, John (1991c): "The Question of Magic and Witchcraft in Two Inquisitorial Manuals of the Seventeenth Century," in Tedeschi (1991a), 229-258.

Villanueva, Francisco M. (1960): *Investigaciones sobre Juan Alvarez Gato*. Madrid: Academia Española.

Laurence Brockliss

Harvey, Torricelli and the Institutionalization of New Ideas in the 17ᵗʰ Century France

1.

Much to the pleasure of many learned physicians, on 17 February 1673, the actor-dramatist Jean-Baptiste Poquelin, commonly known as Molière, dropped down dead in the middle of the fourth performance of *Le Malade imaginaire*. In this play, as in a number of others, Moliere presented the learned physicians of Paris as antediluvian ignoramuses, who *inter alia* refused to accept the circulation of the blood. In Act II, scene V, the physician, Diafoirus, introduces his son, Thomas, to the hypochondriac, Argan, with the ringing endorsement that he is a medical conservative:

> "Mais sur toute chose ce qui me plaît en lui, en quoi il suit mon exemple, c'est qu'il s'attache aveuglément aux opinions de nos anciens, et que jamais il n'a voulu comprendre ni écouter les raisons et les expériences des prétendues découvertes de notre siècle, touchant la circulation du sang, et autres opinions de même farine."

> ["Above all, what pleases me about him – and in this he follows my example – is that he is blindly attached to the opinions of the ancients and that he has never wanted to understand or listen to the arguments and the experiments underlying the pretended discoveries of this century concerning the circulation of the blood and other opinions of a like kind."]

The young Thomas then confirmed his father's confidence in his conservatism by offering Angélique, Argan's daughter, a copy of a thesis, which he had sustained "contre les circulateurs," as "un hommage que je lui dois des prémices de mon esprit."[1]

Molière was not alone in his low opinion of the Paris medical faculty. Two years before, his court colleague, the poet, Nicolas Boileau, had published a mock *arrêt de parlement*, in which the University of Paris (loosely disguised as the University of the Stagyrite) was judged to have won its case to outlaw Reason and her modern defenders from its classrooms. In particular, Reason was accused of attributing to the heart:

> "... la charge de recevoir le chile appartenante cy-devant au Foye; comme aussi de faire voiturer le Sang par tout le corps avec plein pouvoir audit Sang d'y vaguer, errer et circuler impunément par les veines et les arteres, n'ayant droit ni titre pour faire

---

1 Molière (1900), p. 620 (lines 144-51, 154-6). For the presentation of medical practitioners in Molière's comedies, see Calder (1993), ch. 12; Brockliss (2000).

lesdites vexations que la seule Experience, dont le témoignage n'a jamais esté recu dans lesdites Ecôles."

["... the task of receiving the chyle which had formerly been the prerogative of the liver', and also causing 'the blood to travel around the body, wandering and circulating impudently through the veins and the arteries, with no right to perform such vexatious acts except by the right of experience, whose testimony had never been accepted in the said University."][2]

On other occasions, I have argued that this picture was unjust on the grounds that the very last time any Paris student of medicine attacked the circulation of the blood was in a dissertation, sustained by one Jean Cordelle in 1670. Indeed, I have argued frequently over the last twenty-five years that the Paris faculty in the 17th century was far more open to new ideas than contemporary satirists and later historians ever admitted.[3] On the other hand, it is always possible to take arguments too far. It may be the case that by the early 1670s there were very view opponents of circulation in the Faculty, but it also true that no student can be found who clearly supported the discovery before 1665.[4] In other words, it took nearly forty years for the account of the circulation of the blood presented in Harvey's *De motu cordis* (1628) to be accepted in one of the leading medical faculties in Europe. In fact, it was 1642 before there is any sign that the Faculty even discussed the new idea in public.[5]

The slowness with which the Faculty took on board the circulation of the blood is all the more interesting in that the University frequently had little problem in absorbing or at least bringing to public notice other discoveries virtually over night. As I have again emphasized on numerous occasions,.the Aristotelian allegiance of the University of Paris (and other universities) was no bar to the reception of many aspects of the new science. 17th-century Aristotelianism was a plastic and flexible natural philosophy that could easily incorporate new material provided certain fundamentals – such as hylomorphism and the essential division between the sub and superlunary universe – remained unquestioned.[6] A good case in point would be the speedy reception of Torricelli's work on air-pressure. The Italian mathematician first announced to the world his experiments undermining the traditional Aristotelian view

---

2 Boileau (1978), p. 328. There were four separate printings of the satire in 1671-2. Boileau admitted his authorship when he published his collected works in 1701, where the text was very slightly altered. See Magne (1929), i. 191-4. It is the 1701 text which is cited here.
3 For the first time in Brockliss (1978). The Cordelle thesis can be found in Bibliothèque de la Faculté de Médecine de Paris, Collection *Theses medicae Parisiensis*, in folio, volume viii, no. 1235: "An sanguis per omnes corporis venas et arterias iugiter circumfertur?"
4 Paul Mattot: "An motus cordis a sanguinis fermentatione?" (idem, no. 1160). The Faculty's reception of new medical theories and remedies is best studied through the printed abstracts of the oral theses sustained by graduands. The record is virtually complete from the mid-sixteenth century: see Legrand (1913).
5 The first student to discuss the theory, only to dismiss it, was Simon Boullot: "An motus sanguis circularis?," *Theses medicae Parisiensis*, in folio, volume vi, no. 901. The thesis does not name Harvey personally. In the mid-sixteenth century the Paris Faculty had played a major role in the development of anatomical dissection: see Cunningham (1997).
6 E. g. Brockliss (1981). The plasticity of sixteenth- and seventeenth-century Aristotelianism was first recognized by Charles Schmitt: see Schmitt (1973). This is now generally accepted. For a good summary of the existing literature, see Mercer (1997); Ariew/Gabbey (1998).

that nature abhorred a vacuum in June 1644 in two letters to Ricci. Three and a half years years later his experiments were being cited by a professor of philosophy at the University of Paris, called Jacques Desperiers, in the first part of his course of physics, devoted to the discussion of physical principles, in the section on the locus.[7] A physics course given at the Jesuits' Paris college, the Collège de Clermont, in 1650 by one Garnier was even more *au fait* with current knowledge. In addition to telling his students about Torricelli's work, he also informed them of the crucial experiment performed by Pascal's brother-in-law at the foot and the summit of the Puy-de-Dôme, which had only been carried out in September 1648.[8] A Jesuit textbook published at Caen in 1656 by Pierre Gautruche, who had taught at Descartes's college, La Flèche, in the 1640s, was just as informative, as was a second, anonymous course given at the Collège de Clermont the following year.[9]

The fact that some aspects of the new science were taken up more quickly than others in institutions of higher education (be it at Paris or elsewhere in France and Europe) is not something to which I or other historians have paid much attention. We have been so busy demonstrating that universities did have a positive disseminatory role in the Scientific Revolution that we have not looked closely at the pattern of appropriation.[10] It is time, though, that our gaze was turned in this direction. To all intents and purposes the secular role played by universities and colleges in the institutionalization of the new science has been mapped. We know that in most parts of Europe an eclectic Aristotelianism or Aristotelo-Galenism dominated the natural philosophy and medical curriculum until the final quarter of the 17ᵗʰ century. It was then gradually replaced (though in some places not until 1750) by an eclectic form of Cartesianism, which in turn gave way to a multiple-force Newtonian physics and, in many faculties, a vitalist medicine.[11] We are now faced with two principal questions. Why was Aristotle replaced by Descartes, and why within both traditions were some discoveries more easily assimilable than others? Both these questions, it should be said, will have to be pursued synchronically as well as diachronically. We need not only to identify the general factors which encouraged or discouraged the reception of new ideas and discoveries across the higher educational world, but we also need to explore the factors that may have influenced a differential take-up between individual institutions, between sets of institutions and between countries. One thing that can be said for certain is that there is no single history of the reception of the new science. Oxford was not Paris; Paris was not Louvain.[12] It is not my

---

7 Desperiers (1648-49), tractatus iii, on the properties of natural bodies. Desperiers was a bachelor of theology who taught philosophy at the Collège de Lisieux. In the 17ᵗʰ century courses in the humanities and philosophy at the University of Paris were given in ten of the some forty residential colleges founded for poor scholars in the late middle ages. Desperiers began the course on 29 December 1648.

8 Bibliothèque Nationale, Paris, MS Latin 11257, pt. 1, Jean Garnier SJ: *Tractatus de mundo*, fos. 98-128. The Jesuit college in Paris was not part of the University and provided independent tuition in philosophy.

9 Gautruche (1656), ii. 287-93. Bibliothèque Nationale,.MS Latin 10214, fos. 362-82. Gautruche's textbook first appeared in 1653, but I have not seen the edition. Another version was published in 1665. For an account of his teaching, see Brockliss (1995), ch. 9.

10 For the best general account of the universities' role, see Porter (1996), ch. 13.

11 Brockliss, *Curriculum*, ch. 14.

12 At Louvain, for instance, there is little sign that Torricelli swiftly entered the physics curriculum. Transcriptions exist of three physics courses in the 1650s but none mentions Torricelli, his experiments with the column of mercury or air-pressure: On the other hand, a philosophy thesis was defended in the Louvain School of Arts on 31 July 1651under the presidency of the professor Willem van Gutschoven (brother of the more famous Carte-

intention in this paper to get very far down the road to an answer to either question. All I want to do is to use the history of the Paris reception of Torricelli and Harvey as a case study through which to examine some of the ideological and material factors, both general and particular, which could shape the appropriation of new ideas within institutions of higher education.

The first point to be made is that Steven Shapin's work seems of limited value here. Although he more than anyone in recent years has forced historians of science think about the way in which new discoveries and ideas are assimilated, he has primarily been interested in their spread within the community of experimental philosophers.[13] Although many experimental philosophers did hold teaching posts for some of their adult lives, most did not, and those who did – such as Galileo or Newton – were under no obligation to pursue scientific research. Institutions of higher education were not in the business of creating knowledge. In Paris, in the second quarter of the 17th century, for instance, only a handful of experimental philosophers dwelt in the Latin Quarter, where the University was to be found. The newly created, botanical garden, the *Jardin du Roi*; sheltered one or two botanists and chemists, while Gassendi and Roberval held posts in mathematics at the Collège Royal.[14] The heartland of Parisian experimental philosophy was on the Ile de Cité, where Renaudot's *Bureau de l'adresse* was located, and, on other side of the Seine, around Richelieu's Palais Cardinal and in the Marais, at the Minims' convent which sheltered Mersenne.[15] Pascal ascended the Tour Saint-Jacques on the edge of the Marais to replicate the crucial experiment which proved the existence of air-pressure; he did not take his makeshift barometer to the dome of Lemercier's recently finished Eglise de la Sorbonne.[16]

Most professors of philosophy and medicine did not come into contact with new scientific ideas and discoveries at the moment when experimental philosophers first placed their results before their peers. Most, one suspects, only heard about them once they had been digested by the *virtuosi*; most, too, probably learnt about the developments second hand, perhaps even by word of mouth. There is no reason to believe that they read much at all of the new scientific literature. In fact, from the garbled way in which Descartes's theories, notably his theory of the formation of the world, were initially discussed at Paris, it is hard to imagine that our Paris professors even read the most seminal and provocative texts.[17] Given, too, their woeful preparation in mathematics, much of the work in certain areas of the new science would have been beyond their comprehension.[18] It matters little therefore whether Torricelli

---

sian, Gerard) which endorsed Torricelli's work. Unfortunately Gutschoven's course does not survive. Personal communication from Geert Vanpaemel: letter, 26 July 1987. Vanpaemel is the author of Vanpaemel (1986).
13 Shapin (1994).
14 The Jardin du Roi was a royal foundation distinct from the University established in 1635 through the efforts of its first director, the Montpellier-trained physician, Gui de la Brosse: see Rio C. Howard (1974), ch. 2. The Collège Royal was also an independent institution, set up by Francis 1 in 1530 to provide lectures in subjects (like mathematics) thought to be undervalued in the University. It occupied one of the University's residential colleges, the Collège de Cambrai.
15 In the 1630s Renaudot organized twice-weekly *conférences* which debated scientific questions: see Solomon (1972), ch. 3. Renaudot also ran the government's gazette, a city employment agency and health practice offering free medical advice.
16 Jouy (1928). As it is difficult to imagine that the barometer readings at the bottom and top of the tower were significantly different, historians have wondered whether the experiment ever took place. Pascal never attended a university.

or Harvey displayed a gentlemanly persona in presenting their discoveries and presented the material in a non-dogmatic way, which Shapin believes was the key to Boyle's success as an experimental philosopher. The Paris professors were not involved in assaying the reliability of the testimony, merely deciding when and in what way to incorporate a new, publicly accepted truth into their curriculum. To use Peter Dear's classic distinction, they were not dealing with individually-constructed experimental data but traditional experience.[19]

At the same time, theirs was a relatively free act. Throughout the 17th century the community of experimental philosophers in Paris, as in other major cities, was expanding. At no point, though, for most of the period between the publication of *De motu cordis* and the first performance of *Le Malade imaginaire* could it be said to have had any cultural authority. The appearance of the first unofficial academies made little difference in this regard: they were merely private clubs where consulting adults met behind closed doors. It was only in 1666 with the foundation of the French Academy of Sciences that the new science received a royal warranty. Before then, despite Mersenne's valiant efforts, the community of Parisian experimental philosophers was split into several, mutually hostile groups – absolutist clients of Cardinals Richelieu and Mazarin; *frondeur* friends of the Duc d'Orléans (Louis XIII's detested brother); Gallican academics, clerics, lawyers and physicians, such as the high-court judge and *salon* host, Habert de Montmor; and ultramontane Jesuits and their allies in the regular clergy. All were caught up in factions with distinctive political and religious agendas. Together, they scarcely had the moral power to force the university's professors to take up a new discovery.[20] Indeed, even after the establishment of the Academy, the community of experimental philosophy still lacked the power to command consent. Too many of the members were Protestants and foreigners for them to be able to instantly legitimize a new scientific fact, especially as the king took little interest in their activities. It would be the turn of the century and the reconstruction of the Academy under Fontenelle's control, which would give the body an authoritative voice.[21]

On the other hand, neither were the professors under any compulsion to turn their back on the discoveries of the *virtuosi*. In 1624 the chemist Etienne de Clave and two friends had attempted to hold a public debate in the University in defence of atomist principles. As a result, the Parlement of Paris (the local high court) had forbade in the future "a toutes personnes à peine de vie tenir ni enseigner aucunes maximes contre les Auteurs anciens et approvez" throughout its jurisdiction (which embraced a third of France)[22]. This, though, was suitably vague, even if its Aristotelian bias was clear. As neither the Church, the state

---

17 E. g. Du Hamel (1705), iv. 29. Du Hamel had taught philosophy at Paris in the 1670s and 1680s. His course was belatedly published under the auspices of the Paris Faculty of Theology in a vain attempt to counter the Cartesian influence in the University.

18 According to the 1601 statutes, philosophy students at Paris were expected to have studied Euclid and the *Sphere* of the 13th-century John of Holywood, but mathematics had a low priority in the course: see Brockliss (1987), pp. 381-2. In other universities the level of mathematics tuition available was much more impressive: see Feingold (1984).

19 Dear (1985).

20 The era of the Cardinals was a period of great political instability as the representatives of competing political and religious traditions attempted to gain the ear of the crown. For an introduction, see Le Roy Ladurie (1991), chs. 1-3. The only general account of the first French private academies remains Brown (1934). Gaston d'Orléans was a keen botanist. For a recent account of Montmor's academy, see Perkins (1985).

21 Stroup (1990), esp. chs. 1, pp. 14-5; Hahn (1971): chiefly on the 18th century.

nor the University moved to outlaw specific aspects of the new science in the forty years following the publication of *De motu cordis*, the professors were under no constraint to toe a particular Aristotelian party line. Even the judgement against Galileo did not operate in France where there was no Inquisition. The fact then that the University was a confessionally closed Catholic institution and that most professors of philosophy were secular clerics from relatively disadvantaged backgrounds who looked to be placed in a lucrative ecclesiastical position after a period of teaching can have had little effect on their attitude to particular aspects of the new science: the Church did not give a clear lead.[23] Professors of philosophy may have been to some degree under the thumb of the College principals who hired them, and professors of medicine could be censored by the assembled corporation of the Faculty's doctors. In both cases, though, the manner in which control was operated is unclear. Only student dissertations in the higher faculties were definitely subject to censorship prior to their delivery, as they were supposed to be shown to a faculty official before they were printed. It was only after 1670, and especially in the period 1690 to 1710, that the king, Louis XIV, through the Archbishop of Paris tried, albeit unsuccessfully, to force the University's professors of philosophy to toe a particular, in this case anti-Cartesian party line.[24]

2.

The factors that governed the professors' appropriation of any particular fact of the new science were arguably primarily three: philosophical commitment, religious prejudice and peer-group pressure. Throughout the forty year period which interests us the professors were all unreconstructed Thomist Aristotelians, secure in their belief in the unimpeachability of Aristotelian principles. The one professor of philosophy, Janus Cecilius Frey, who showed some interest in the chemical philosophy at the beginning of the period never strayed from Aristotelian fundamentals.[25] The professors were Aristotelians by acculturation, not force. Alternative physical theories were incomprehensible. It was only right at the end of the period that there was even felt the need to discuss them.[26] They were also fervent and orthodox Catholic Christians. There was no sign before 1670 that any had even drifted into the Jansenist camp, which Pascal so warmly supported.[27] Provided therefore that a new scientific fact

---

22 D'Argentré (1728-31), ch. ii, pp.146-7. For further details, see the paper in this collection by Daniel Garber.
23 I have unpublished biographical information on some seventy professors of philosophy at the University of Paris in the 17th century. The overwhelming majority were clerics.
24 Brockliss (1976), pp. 159-63; Du Hamel (1705), v. appendix, pp.33ff; Jourdain (1862-66), part i., pp. 269-70; part. ii, pp. 19-53.
25 Blair (1993). Frey, a physician not a cleric, taught at the Collège de Montaigu.
26 The first professor indubitably to preside over a thesis discussing Cartesian physics was Jean Du Hamel in 1672. See René Pucelle: "Conclusiones philosphicae": theses sustained in the Collège du Sorbonne-Plessis, 3 July 1672.: Bibliothèque Nationale,Départment des Imprimés, Rp 108.
27 Jansenists were austere Augustinians who followed the teaching on grace of Jansen, Bishop of Ypres. They technically became heretics when the pope denounced Jansen's views in 1653. Although Jansenists were a powerful force in the Paris Faculty of Theology in the mid-seventeenth century, they do not seem to have had much support among philosophy professors. The Jansenist theologians were attacked, then expelled, in 1656,

could be incorporated into their Aristotelian (or Aristotelo-Galenic) world view without causing problems, provided it was not a threat to Counter-Reformation Catholicism, and provided its harmlessness was generally accepted within the University world, then it could be taken on board without difficulty.

This immediately helps to explain the relative fortunes of Torricelli's and Harvey's discovery. At first sight, the Englishman might seem the easier to accommodate. Harvey presented himself as an Aristotelian and claimed to be restoring an Aristotelian physiology. If he were being deceitful, he certainly couched his theory of the heart in the familiar language of forms and faculties.[28] Torricelli's work, however, undermined one of the fundamentals of Aristotle's theory of the elements. Air-pressure was an Aristotelian nonsense. Airy and fiery particles were always attempting to rise, while terrestrial and aqueous ones attempted to fall. The acceptance of Torricelli's experiments therefore would seem to have been an immediate threat to one of the pillars of Aristotelian physics.

In fact, this was not the case because the Aristotelian professors at Paris and elsewhere in the third quarter of the 17ᵗʰ century carefully distinguished between facts and the conclusions drawn from them. Desperiers, Garnier and their successors accepted the reality of the barometrical experiments, even, starting with Garnier, Pascal's crucial experiment. They did not, though, accept that these experiments demonstrated that the atmosphere had weight (or spring, as Boyle later emphasized). Rather, they virtually ignored the existence of this conclusion, usually not even mentioning it. Instead, they continued to treat the phenomena as good Aristotelians. The experiments were further evidence of something known since the time of the Stagyrite: that nature abhorred a vacuum. For them the subject of interest in the experiments was thus the gap above the inverted mercury column. If the vacuum was impossible, how could the gap be explained. This was done in one of three ways. Either it was assumed that the glass of the tube was porous and particles of air had penetrated the apparent space, or particles of air in the mercury had floated upwards, or the more subtle and invisible particles of mercury had precipitated out to fill the gap. The Jesuit, Pierre Gautruche went for the second explanation in the edition of his textbook which appeared in 1665.

"Superiorem illam tubi partem non esse omni corpore vacuam. Manifestumque est, quia experimur per eam traijici lumen et qualitatem magneticam, tum inde exaudiri sonum, si quod parvulum tintinnabulum incluseris. Autem qualitates debent per eiusmodi medium diffundi, quod reale sit. Item, si in ea tubi parte fuerit inclusa carpionis vesicula intumescat. Ac denique si parti illi apponantur aqua calidior, hydrargyrum descendet tantisper, quod signum est rarefactionis alicuius factae intra illud spatiae, dum e contra, fiat condensatio ascendente hydrargyro, si effundatur aqua frigidior. Quae quidem postremae duae experientiae satis manifeste indicant contineri illo spatio aerem, cuius aliqua particula semper cum hydrargyro insinuat, quaeque metu vacui in hunc modum rarefiat ad descensum hydrargyri."

---

leading Pascal to pen the *Lettres provinciales*, and the movement did not regain its foothold in the university until the turn of the 18ᵗʰ century. By then the professors of philosophy must have been infected, for, after 1690, the king and the archbishop moved against Jansenism as well as Cartesianism: Adam (1968); Brockliss (1987), ch. 5, section iii.

28 See especially Pagel (1967).

["The upper part of the tube is not empty. This is clear because light passes through it, magnets can attract across it, and a sound can be heard if a small bell is shut up inside it. These, though, are qualities which must be transmitted across a medium. Similarly if you shut up in the empty part of the tube a carp's bladder, it will swell up. And finally, if you apply hot water to the side of the tube, the mercury will gradually descend, which is a sign of rarefaction occurring within the space, while, on the contrary, the mercury column will ascend if cold water is poured over the tube. Now these last two experiences definitely and clearly indicate that the space contains air particles, which are always found inside the mercury and which by fear of the vacuum will rarefy according to the level of the mercury."][29]

Whatever the method, Aristotle's elemental theory was saved. Air pressure did not exist. The phenomenon that occurred when a tube of mercury was inverted over a bath could be explained by the earthy and aqueous particles in the atmosphere pushing the mercury down, but, lacking the weight of elemental earth or water, only doing so to a limited degree. The higher up you went, the fewer the earthy and aqueous particles in the air: hence the lower the column. Nature then ensured that a airy or subtle particles could enter the space to forestall a vacuum.[30]

Harvey's work, on the other hand, could not be so easily disaggregated. Although the circulation of the blood could be called a theory in that it was a conclusion drawn from a series of observations, just like air pressure, Harvey's experiments were of a different order from Torricelli's. The Italian's observations confirmed an already known natural effect – he merely used his experimental work to establish a novel explanation of the phenomenon. Harvey's experiments in contrast claimed to establish a new effect – that venal blood must flow from the extremities to the vena cava, not the other way round as had always been assumed. If the reality of his discoveries was once granted, then traditional Galenic physiology was completely undone: the parts could not be fed directly by venal blood; all blood in the vena cava must pass through the heart; arterial blood must somehow be passed to the veins; and the blood must continually circulate rather than be continually used up and replenished.

Harvey's experimental work therefore could not be detached from his new physiology. And this was extremely difficult for a traditional physician to swallow. Although Harvey eased the pain by trying to uncouple Aristotle from Galen, the two were so closely wedded together and Harvey's conception of the circulation of the blood so far removed from Aristotle's albeit simplistic physiology that few can ever have believed that the new theory was merely perfecting traditional teaching. As it was, the Faculty physicians must have been aware that the theory had been early championed by anti-Aristotelians, such as the hermeticist, Fludd (in 1629 and 1633), and the mechanist, Descartes, in the *Discours de la Méthode* (1637).[31] In addition, there were strong practical reasons for keeping the theory at arms length,

---

29 Gautruche (1665), ch. ii, pp. 354-55. Gautruche's publication of 1656 (and presumably 1653) does not explain what the apparently empty space contains. Desperiers, Garnier and the anonymous Jesuit course of 1657, on the other hand, all offer some form of particulate explanation.

30 This was the account given in the anonymous Jesuit course.

31 Fludd's immediate support for Harvey seems to have led Gassendi to attack the circulation of the blood in his 1631 critique of the hermeticist's works: see Bloch (1971), ch. viii. Descartes, of course, offered a mechanical, not a qualitative, explanation of the motion of the heart: Gilson (1930), ch. 2.

because its acceptance seemed to have profound therapeutic consequences. As the first Paris dissertation to discuss the circulation of the blood clearly recognized, it made the favoured therapy of bleeding a nonsense.

"Ars alioquin medica quae et salutifera est et fallere nescia, inani remedii nomine nugaretur, ubi ophthalmiae, pleuritidi, anginae, phlebotomia succurrendum putat, et inflammationem prohiberi curat humormem avocando aut revellendo, sedat dolores, menstrua cohibita revocat, male affecti genitalis arvi putrescentem sanguinem secta in malleoloa saphena vacuari iubet, si a capillaribus venis in maiores …, alius confestim isque copioior adventaret."

["The medical art, which is safe and unknown to deceive, would be deservedly called an empty remedy, when it advocates phlebotomy in eye diseases, pleurisy and angina, looks to stop humoral inflammation by summoning or recalling blood, sedates pain and promotes suppressed periods by bleeding, or orders putrescent blood in the genital region to be drawn through severing the hammer-like saphene vein. For if blood were really to flow from the capillaries to the major veins, then it would only flow more quickly and copiously thereby."][32]

The therapeutic consequences were all the greater in the case of the Paris faculty in that many of its members had a particular penchant for bleeding. Although all Galenic physicians used phlebotomy frequently as part of an arsenal of remedies that were deployed to fight virtually every disease, Paris faculty physicians were particularly prone to gild the lily. The lawyer and diarist, Pierre L'Estoile, at the beginning of the century revealed the case of an eighty year old woman suffering from the stone, one Madame de Douaile, who died after being bled thirty times in ten weeks and losing ninety palettes of blood! Paris physicians treated members of their own family no less rigorously than other clients. In 1633 the Faculty physician, Jacques Cousinot (died 1646), in later life briefly premier physician of Louis XIV, went down with "a rude and violent" rheumatism. On the orders of his father and his father-in-law Charles Bouvard, premier physician to Louis XIII, he was bled sixty-four times in eight months.[33]

According to Richelieu's physician, Citois, a graduate of Poitiers and Montpellier, the logic for such intensive phlebotomizing lay in the Parisian character and the geographical location of the city.

"En une region temperée, comme est celle de la moitié du sixiesme climat, & sous le septiesme tout entier, ... qui s'estend depuis le 45. Degré iusqu'à 49. ou 50. les habitans sont à leur aise, oisifs, grands carnassiers, abondans en sang, il leur faut tirer du sang abondamment, c'est pourquoy à Paris, qui a pour hauteur du ciel 48. degrez & 50. Minutes, & en toute cette contree qui est comprise entre les rivieres de la Seine & Loire, où ils vivent à leur aise, & font bonne chere, ils engendrent beaucoup de sang, leur en faut tirer bonne quantité, le pouvant mieux supporter, sans affoiblissement,

---

32 See note 5 above.
33 L'Estoile (1958), p. 505; Gui Patin (1846),ch. i, pp. 353-4: letter to Spon, 16 April 1645.

que ceux de Narbonne & Toulouze, & de toute la contrée, qui confine la mer Mediter-
ranée, où ils vivent plus sobrement."

["In a temperate region, as is the half of the sixth climatic zone which lies under the
seventh – that is, the region from the 45th to the 49th or 50th degree of latitude – the
inhabitants are comfortable, great meat eaters and overflow with blood. It is thus
necessary to phlebotomize them abundantly. That is why at Paris (whose latitude is 48
degrees 50 minutes) and in all the country between the Seine and the Loire, where
people live at their ease and make merry, lots of blood is created which needs to be
drawn off in great quantities. This is something the northerners can support without
weakening, much more than the inhabitants of Narbonne and Toulouse and all the
land around the Mediterranean, where they live more soberly."][34]

Admittedly, not all the Parisian doctors in the 17th century were ardent phlebotomizers.
When L'Estoile needed a physician in the reign of Henri IV, he was confronted with a number
of possible alternatives in his part of the city whose use of the practice ranged from the
moderate to the heroic.[35] None the less, the Faculty's therapeutic identity was closely bound
up with the frequent use of the therapy, while an acceptance of Harvey's theory seemed to
demand a change of therapeutic style. To a group of people extremely anxious to present
themselves as consistent, rationalist practitioners whose therapeutic decisions were taken on
their intimate knowledge of the internal structure of the body in health and disease, there
were clear, professional reasons to hold fast to traditionalist physiology. Furthermore in the
middle of the seventeenth century maintaining the phlebotomizing tradition was a matter of
honour. In the 1640s and early 1650s the practice came under attack from the Montpellier
Faculty (the only other medical faculty of any significance in France), whose dean, Siméon
Courtaud was anxious to promote the right of Montpellier graduates, such as Renaudot, to ply
their profession in the capital regardless of the legal monopoly possessed by Paris graduates.
The fact that other learned physicians were calling them bloodsuckers and therapeutic dino-
saurs who had turned their back on the potential of Paracelsian chemical remedies and tra-
duced the good name of Galen must have only encouraged the Paris faculty to dig its heels
in.[36]

A variety of ideological and material factors therefore evidently militated against the
rapid assimilation of all or part of Harvey's work on the circulation of the blood by the Paris.
In fact, so destabilizing was his work that it could not even be properly discussed in student

34 Cited in Riolan (1651), pp. 240-1. A similar explanation is given by Patin in Patin (1846), ch. i, pp. 353-4:
to Spon, 1645. Patin was a member of the Paris medical faculty. Although often considered a freethinker because
of his hostility to witchcraft and his association with Gassendi, he was in fact a hardline Aristotelo-Galenist: see
Brockliss (1992).
35 L'Estoile was chary of phlebotomy. When his own doctor , Albert Le Fevbre, died, most of his neighbours
became clients of Alphonse Le Moyne, who had a reputation for bleeding copiously. L'Estoile therefore took up
with another Faculty physician, Hélin, who treated him more gently: see L'Estoile (1958), pp. 464, 532.
36 Courtaud fired the opening salvo in a speech to mark the beginning of the academic year, 1644-5: see
Courtaud (1645), esp. pp. 24-9, where he accuses his Parisian colleagues of littering France with dead bodies
from excessive bleeding. In most large towns in France physicians were members of a corporation whose members alone could practise locally. Members of the Paris corporation all had to be Paris graduates. The two
faculties were long-standing rivals: see Brockliss (1993).

dissertations until somebody in the Faculty had found a way of sanitizing the theory. The first dissertation to talk about Harvey, only did so tangentially. The thesis title "An motus sanguis circularis?" seemed to promise a detailed assessment of the new discovery but in fact it only contained a restatement of the traditional Galenic theory and the assertion that the circulation of the blood was unnecessary and, as was said above, therapeutically subversive. Harvey himself was never mentioned.

Three years later – and seventeen after the publication of *De motu cordis* – a much more interesting thesis was sustained in the Faculty by one Jean Maurin under the presidency of Jean Riolan II. Riolan was no ordinary member of the Faculty corporation of physicians. In the early seventeenth century he was recognized to be one of, if not the leading, anatomist in Europe, who believed that only through hands-on dissection could the science of anatomy by learnt. In the 1626 edition of his anatomy textbook, *Anthropographia*, he claimed to have personally dissected one hundred corpses in twenty-four winters. Moreover, he had actually met Harvey in England in the 1630s and discussed the theory of circulation with him.[37] If there was one man, then, who could guide the Faculty in its response to the Englishman's work, it was clearly Riolan. Indeed, it seems reasonable to believe that the Faculty would take its cue from its leading anatomist and make no public statements about the theory until the maestro had spoken. This would presumably account for the silence of the 1630s. For the whole of the decade Riolan was absent from France in the entourage of the exiled Queen-Mother, Marie de Medici, acting ostensibly as her personal physician, but actually spying for Richelieu[38] It was only with Marie's death in July 1642 that he was free to return to France and the Faculty could fashion a response to this new, disturbing physiological theory which was already gaining support from many anatomists in other parts of Europe, such as the young Dutch anatomist, Jan de Wales (Wallaeus).[39]

Riolan's riposte eventually came in the thesis of 1645 and was then developed in his *Encheiridium anatomicum et pathologicum*, an abridged version of his anatomical textbook, published in 1648, and in a lengthy treatise which appeared in the following year as part of a larger work, containing a final edition of the *Anthropographia* and a number of critiques of other contemporary anatomists.[40] Riolan's argument, which eventually received two stinging replies from Harvey[41] and has generally been dismissed as of little interest, was a masterpiece of accommodation. In the initial student dissertation, Maurin made no attempt to discuss Harvey's experiments. He simply accepted the circulation of the blood as a fact not a theory. His description of the phenomenon, however, was very different from Harvey's. The blood passed from the right to the left ventricle through the septum not via the lungs; it circulated

---

37 Mani (1967).
38 du Plessis (1835-76), vii. 911-14: Riolan's letter to Richelieu after Marie's death.
39 Wallaeus had the theory sustained in student dissertations in 1639 and 1640: see French (1989), p. 55.
40 Riolan (1648), *sub* "De corde;" Riolan (1649), pp. 553-604. This contains a copy of the Maurin thesis, plus objections supposedly raised against it by Faculty conservatives (pp. 543-52). The title reflected Riolan's belief that his own circulation theory both safeguarded and perfected traditional medical practice. The treatise and the other critiques were republished in London the following year under the title *Opuscula nova anatomica*. The *Encheiridium* was dedicated to Gui Patin, one of the most enthusiastic phlebotomists in the capital: I have used the 1649 Lyons edition, and have no reason to believe it was any different. Patin's support for phlebotomy was balanced by a deep suspicion of the metallic remedies, which were increasingly in vogue: see Patin (1846), passim.
41 Harvey (1649a); Harvey (1649b). Harvey had been sent a copy of Riolan's *Encheiridium*.

extremely slowly, normally only once or twice a day; and it was continually used up in feeding and enlivening the parts, thus needing constant replenishing. In consequence in answer to the question the student had set himself – "An propter motum sanguinis in corde circulatorium, mutanda Galeni methodus medendi?" – there was no need to change traditional therapeutics. The sick, it was implied, would still have blood which was too copious or corrupted, and which could be cleansed by phlebotomy.[42] Riolan's treatise took the revisionism further by stressing that Harvey was wrong to give the heart such primacy. It was the liver which turned the chyle into both nutritive and vital spirit, and the heart merely pushed the blood round the body. Strictly-speaking then the circulation of the blood via the heart was not essential for life. Moreover, not all the blood was continually pumped round the body, but only the blood in the major veins and arteries of the second and third region (i. e. above the liver). Blood in the portal vein, the coeliac artery and their respective vascular networks simply flowed backwards and forwards. Blood in the minor veins and arteries branching from the vena cava and the aorta was equally stationary, though in this case it would sometimes be drawn back into the major vessels, if they needed topping up.. Finally, if too much blood poured into the right ventricle from the vena cava, it would be rerouted through the lungs to ease pressure across the septum.[43]

Riolan thereby showed the Faculty how it could embrace Harvey's discovery with the minimum inconvenience to physiological and therapeutic tradition. To do this, though, he had to rule out of court much of Harvey's experimental work from which the Englishman had drawn his conclusions about the speed of blood flow and the volume of blood in the body. He did this by questioning their value. Many of these results had been obtained by vivisectional experiments on animals. Riolan argued that such experiments were both useless and illegitimate. The physiology of animals and man was not necessarily the same, ligating vessels damaged the parts leading to doubtful results, while the suffering vivisection caused was unworthy of the Christian philosopher.[44] On the other hand, Riolan was anxious not to seem to be hostile to physiological research and be dismissed as a diehard conservative. If he opposed vivisection, he was quite happy to encourage the force-feeding of criminals about to be executed, so that when their bodies were subsequently dissected, it would be possible to explore the path of the chyle through the lacteals.[45] In the preface to his abridged anatomical textbook of 1648, he pointedly stressed his own "modern" experimental credentials.

---

42 Bibliothèque de la Faculté de Médecine de Paris, Collection Theses medicae Parisiensis, in folio, vol. vi, no. 939. The value of phlebotomy is asserted rather than explained.

43 Riolan (1649), esp. ch. 15. Riolan (1648), pp. 218-24. It should be pointed out that neither Harvey nor Riolan had seen the capillaries. The vessels connecting the arteries to the veins would first be seen by the microscopist, Malpighi, in 1661. Riolan developed his opposition to Harvey's ideas in a number of works published in Paris and London between 1648 and 1652, especially Riolan (1649).

44 Mani (1967), pp. 127-8, 140. For the development of vivisection in the 16th and early 17th centuries, see French (1999), ch. 6, and pp. 236-8 (on Riolan). For opposition to the practice, see Moehle (1992).

45 Riolan (1649), pp. 608, 610. This experiment tried was by Peiresc and Gassendi in 1634: see Gassendi (1992), p. 230.

Non mihi res, sed me rebus subiicere conor [Riolan's italics]

"Nec quae meditatus sum in rebus Anatomicis extare puto, nisi sedula indagatione saepius spectaverim & compromaverim."

["I do not attempt to accommodate things to my mind, but rather to submit my mind to the nature of things. And I never believe that things that I have anticipated in anatomy are so, until I have seen them confirmed several times through careful research."][46]

It need hardly be said that Riolan's scientific ethics were cleverly framed to stop his fellow anatomists or himself from ever confirming or pursuing Harvey's experiments.

3.

The history of the assimilation of Torricelli and Harvey by the Paris Aristotelians in the middle decades of the 17ᵗʰ century was thus very different. It was not just that one was accepted more quickly than the other, which was our point of departure. Rather, to the extent that they represented different sorts of challenge, so they were absorbed in different ways. Torricelli's experiments were taken on board, but not his discovery of air-pressure. Harvey's discovery of the circulation of the blood proved acceptable when suitably adjusted but not the experiments on which his conclusions were based.

Admittedly, the relative speed with which the two discoveries were integrated into the university curriculum is not simply to be explained in terms of the very different challenges they mounted to Aristotelian orthodoxy. Another significant factor was clearly the absence or presence of a figure or figures within the immediate Aristotelian camp who had the imagination to sanitize the new ideas and the authority to command assent. Just as the the absence of Riolan in the 1630s helps to explain the Faculty of Medicine's prolonged neglect of Harvey, so the fact that in October 1647 an Aristotelian experimental philosopher at the Collège de Clermont – one, Etienne Noël – had published an account of Torricelli's experiments in which the results were confirmed but their vacuist interpretation denied can only have encouraged their swift assimilation into the curriculum.[47] Arguably, potentially troublesome new ideas and discoveries, could only be incorporated into courses of physics and medicine once a creative mediator with a foot in both the new science and the university world had waved his magic wand. The professors and students did not take the initiative themselves.[48]

---

46 Riolan (1648), "Praemonitio ad lectorem et auditorem."

47 Noël (1648). Noël was a senior member of the Order who had taught Descartes philosophy at La Flèche over forty years before. Jesuit experimental philosophers could not be anything else but Aristotelians and proved extremely adept at neutralizing potentially harmful discoveries. French Jesuits, for instance, had earlier drawn the sting from Gilbert's work on magnetism. See Pumfrey (1990), ch. 12.

48 Although the courses were set out in a scholastic format which allowed room for many different approaches to a problem to be discussed, there is no sign that the professors (especially in philosophy) created their own

It might be wondered further how far the absence or presence of such a mediator also reflected the level of interest among local experimental philosophers, whatever their philosophical allegiance. Riolan was absent in the 1630s, but the Faculty's disregard for Harvey may well reflect a wider lack of concern for the Englishman's work in the French capital. There seems to have been very little interest in investigatory anatomizing in the French capital in this decade outside or inside the University. The situation would be very different in the late 1650s when Pecquet was demonstrating his discovery of the canal that bears his name to the Montmor academy.[49] In the second half of the 1640s, on the other hand, Paris was buzzing with interest in Torricelli's work. Mersenne had been quickly informed of the Italians experiments and tried, but failed to replicate them due to the difficulty he had finding glass which could bear the weight of mercury. By the spring of 1647, however, he was aware of the success the Pascals had had in studying the behaviour of inverted columns of liquid in nearby Rouen, and he and Roberval began their own experiments, as did Adrien Auzoult. Noël, therefore, was part of a regional experimental community deeply engaged by the puzzling question of the gap above the mercury column. When Pascal junior in his *Expériences nouvelles touchant la vide* of 1647 publicly asserted that a vacuum was now hypothetically possible, the Jesuit inevitably had to promote an orthodox Aristotelian interpretation of the phenomenon as soon as possible.[50]

Whatever the exact mix of these particular external factors in accounting for the distinctictly different receptions the University accorded Harvey and Torricelli, the assimilation of the two discoveries must also to be placed in the context of broader changes in the curriculum. Both "accommodations" were the product of the 1640s. Both, too, survived quite happily for some twenty years and then were abruptly jettisoned. Riolan died in 1657 at a ripe old age, having twice survived being lithotomized.[51] His *Enchiridium*, however, continued to be reprinted, reaching its fourth edition in 1658 and being translated into French, for the benefits of surgeons, in 1661.[52] The first sign that his control over the Faculty was finally waning came two years later when a student called Gui-Crescent Fagon, later first physician to Louis XIV, sustained a dissertation that gave the heart pride of place in the maintenance of native heat or life, insisted it was the first part of the body to develop in the womb, and stressed its close connection with the soul's emotions. "Inde in animi permotionibus diffunditur, eoque refugiat sanguis, non ad hepar aut cerebrum" ["Thence is diffused the perturbations of the soul and to it flows the blood, not to the liver or the brain."].[53] The Torricelli compromise

---

arguments. Probably only a handful of Aristotelians across Europe (many not even professors) provided the material that allowed university courses to absorb the new science.

49 Perkins (1986): an account of letters to Cardinal Mazarin on Pecquet's lectures.

50 Lenoble (1971), pp. 431-7; Davidson (1983), ch. 2; Garber (1992), pp. 136-43. Noël preceded his treatise with three letters attacking Pascal in October and December 1647. Noël must have been all the more confident of his ability to take on the vacuists in that Descartes and Mersenne also refused to accept that the upper part of the tube was empty: there was no common anti-Aristotelian front among the Parisian community of experimental philosophers. Mersenne's work on the phenomenon was published in the summer of 1648.

51 des Réaux (1960-61).

52 Riolin (1661): note the title.

53 Bibliothèque de la Faculté de Médecine de Paris, Collection *Theses medicae Parisiensis*, in folio, vol. viii. no. 1143: "An a sanguine impulsum cor salit?" The thesis connects the heart-beat with the expression of blood from the left ventricle and describes the pulmonary circulation, but it does not provide an account of the blood's circulation *tout court*.

lasted another decade. The first extant thesis in which a student rejected the traditional belief that nature abhorred a vacuum dates from 1672, when one René Pucelle declared that the spaces above inverted tubes of mercury "non sunt alteri causae quam incumbentis aëris gravitati tribuenda" ["were to be attributed to no other cause but the heaviness of the incumbent air"].[54]

Neither fact is too surprising. The 1640s was a creative decade. Not just Torricelli and Harvey, but Galileo's astronomical work, too, was being incorporated into the Paris curriculum for the first time. In previous decades, the new astronomy had been sometimes discussed but largely ignored. By 1650, however, professors had accepted Galileo's telescopic discoveries – Jupiter's satellites, sunspots, new stars, superlunary comets –, jettisoned planetary spheres, and begun to question the superiority of the Ptolemaic system. There then followed twenty years relative calm, to which succeeded a further creative outburst in the late 1660s and 1670s, which culminated in the establishment of the Tychonic system as the favoured theory of the world.[55] These few years, too, seem also to have seen the first discussion of Cartesian mechanist ideas in the University, which led to a growing belief in some quarters that philosophical orthodoxy had to be reinforced.[56]

The actual sequence of events surrounding the defense of Aristotle in these years is very difficult to unravel, but it would seem that in the summer of 1671, the Archbishop of Paris appeared before the Faculty of Theology, apparently on the behalf of the king,[57] to denounce certain ideas, "se répandant présentement non seulement dans l'université mais aussi dans le reste de cette ville et quelques autres du royaume. qui pourroit apporter quelque confusion dans l'explication de nos mystères" ["which were spreading at that time not only in the University but also in the rest of the town, and others in the kingdom, and which, if allowed to continue would bring confusion to the explanation of our holy mysteries"]. The Faculty of Theology in consequence issued a decree ordering nothing to be taught "contra statuta universitatis, censures Facultatis et Senatus Parisiensis decretum" ["against the statutes of the University, the censures of the Faculty and the Parlement de Paris' decree [of 1624]"].[58] The University's response, however, was more limited – it merely agreed to deprive anyone who taught anything contrary to the existing statutes (which dated from 1601 and simply demanded that professors lectured on Aristotle's philosophical works)[59] to be deprived of their degrees and privileges. In consequence, two years later – after another Cartesian thesis had been apparently sustained in the University – the Dean of the Faculty of Theology, Morel,

---

54 See above, note 26.
55 Brockliss (1981), pp. 43-6. Two textbooks, divided by thirty years, graphically chart the changes: Le Rées (1642); Barbay (1675). Both works were published after the professors had finished teaching.
56 Unfortunately there is no definite evidence that Descartes was being discussed in the University classroom before Pucelle's thesis of 1672: see above note 26. In the city Jacques Rohault had been giving private conférences on Cartesian science since the mid-1650s. Clair (1978), pp. 42-56. There were also members of the University anxious to stress that Aristotle was not an essential prop of Catholic orthodoxy, and therefore by implication could be safely jettisoned: see Launoy (1653). This work by a theologian emphasized that Aristotle's status as the philosopher in the University dated only from the 16ᵗʰ century!
57 Possibly a response to the appearance of Rohault's Cartesian textbook: Rohault (1671).
58 Archives de l'Université de Paris, Register 30, fos. 68, 75: University minute book; Bibliothèque de la Faculté de Médecine de Paris, MS 15, fos. 235-6: Faculty of Medicine, minute book; Jourdain (1862-66), pt. 1, pp. 234-5; Du Hamel (1705), vol. v, appendix, p. 19.
59 Réformation de l'univesité de Paris (Paris, 1601), pp. 30-3.

attempted to get the Parlement of Paris to put its weight behind philosophical conservatism.[60] Since Boileau by this time had produced his mock *arrêt* – possibly in response to the shenanigans of the summer of 1671 – the legal establishment had no wish to look fools, so did nothing.[61]

Given the fact that the conservatives had little effect on this new round of accommodation, the details of these moves are of a little importance. What deserves greater attention are the forces underlying these bursts of creativity. Were there general factors (as opposed to the particular ones we have already explored) encouraging a more dynamic engagement with the new science in the 1640s and the late 1660s and early 1670s, or do the developments simply reflect generational shifts in the professoriate? Unfortunately neither hypothesis is open to easy verification. Clearly, the foundation of the Académie des Sciences in 1666 with its mixed membership of Aristotelians, chemists and mechanists must have at the very least encouraged Aristotelian professors to pay some attention to alternative physical theories.[62] In the 1640s, on the other hand, there was no such potential catalyst. Renaudot's *conférences*, which might equally have stimulated an interest in new ideas and discoveries (albeit to a lesser degree) were a phenomenon of the 1630s and came to an end in September 1642, shortly before the death of his protector, Cardinal Richelieu.[63] Generation change is impossible to chart. Too little information survives about the professors of philosophy in the University of Paris in the 17th century to allow even a list of names to be compiled before the 1690s, let alone decide if a group of new, young professors took up their posts at a critical moment.[64] The strategies of accommodation are much easier to understand than the forces impelling curricular change, and this paper has been primarily about the former

What seems worth emphasizing in conclusion is that the theme of assimilation in stages can be pursued across the century. Certainly in the history of philosophy teaching at the University of Paris the 1610s and the 1680s were equally vital decades. At each stage, the professors opened up their teaching a little more to the new science, ultimately finding more and more ingenious ways of accommodating the old and new philosophy. In the 1680s they were even incorporating mechanical explanations into the course as "secondary causes". Assimilation was neither straightforward nor linear, but with each readjustment significantly more ground was always given to discoveries and eventually explanations which did not sit at all comfortably with traditional Aristotelian natural philosophy.[65] It is possible to see, as

---

60 Bibliothèque de la Faculté de Médecine de Paris, MS 15, fo. 33. Bibliothèque Nationale, MS Francais 14699, Arnauld (1673). Arnauld was a Cartesian as well as a Jansenist.

61 According to the court wit, Ménage, Boileau's satire had definitely stopped the Parlement from acting: see Desmarais (1777), vol i., p. 27.

62 The first members of the Academy were chosen for their non-dogmatic philosophical allegiance and were supposed to concentrate on practical rather than theoretical issues. The leading Aristotelian among them was the Secretary, Jean-Baptise Duhamel. See Hahn (1971), ch. 1.

63 The *conférenciers* were encouraged to discuss all possible opinions on a particular topic and no conclusion was deliberately reached. The University of Paris succeeded in getting the Bureau d'adresse closed on the grounds that the free medical advice given there was in breach of their monopoly of practice: see Solomon (1972), ch. vi.

64 Although over the years I have retrieved the names of a respectable number of Paris professors of philosophy (see above note 23) and a great deal of information has survived about their teaching, establishing detailed biographies is very difficult. It is seldom known when a professor began or finished teaching, although it is clear that some professors taught for a few years before obtaining a lucrative benefice, while others remained in the classroom for all their adult life.

many of us have done, 17ᵗʰ-century Aristotelianism as a dynamic and plastic philosophy enthusiastically embracing the new science and constructing a plausible alternative to the chemical or mechanical philosophies. The evidence of the University of Paris' encounter with the new science suggests this is stretching a point. The professors of philosophy at least were on the back foot from the beginning of the century. They did not obviously welcome the new science, but, for reasons that still await full elucidation, they periodically felt the need to address its discoveries. As a result, they were increasingly forced to abandon all but their most fundamental Aristotelian principles. By the 1690s Paris Aristotelian natural philosophy had become little more than a weak shell from which Cartesianism would easily hatch fully formed. The Archbishop of Paris finally tried to shore up the old philosophy in the early 1690s and again in the early 1700s but he shut the stable door after the horse had bolted. Parisian Aristotelianism was not cut off in its prime; it was not even superseded; it merely imploded under the weight of foreign bodies.

References

Adam, A. (1968): *Du mysticisme à la révolte: les Jansénistes du XVIIe siècle.* Paris.

Ariew, Roger and Gabbey, Alan (1998): "The Scholastic Background," in Daniel Garber and Michael Ayers (ed.), *The Cambridge History of Seventeenth-Century Philosophy.* Cambridge.

Arnauld, Antoine (1673): *Mémoire sur les sollicitations que fait M. Morel avec quelques autres docteurs pour obtenir du Parlement un arrêt qui condamne toute autre philosophie que celle d'Aristote.*

Barbay, Pierre (1675): *Commentarius in Aristotelis physicam,* 2 vols. Paris.

Blair, Ann (1993): "The Teaching of Natural Philosophy in Early Seventeenth Century Paris: The Case of Jean Cécile Frey," *History of Universities* xii (1993), 95-158.

Bloch, O. R. (1971): *La Philosophie de Gassendi.* Paris.

Boileau, Nicolas (1978): "Arrest donné en la 'Grand'Chambre du Parnasse, en faveur des maîtres-es-arts, medecins et professeurs de l'Université de Stagyre au pays des chimeres: pour le maintien de la doctrine d'Aristote" (1671), in *Oeuvres.* Paris.

Brockliss, Laurence W. B.: "Curriculum," in *History of the University in Europe,* vol. 2.

Brockliss, Laurence W. B. (1976): *The University of Paris in the Sixteenth and Seventeenth Centuries.* Ph. D. dissertation, Cambridge.

Brockliss, Laurence W. B. (1978): "Medical Teaching at the University of Paris 1600-1720," *Annals of Science* 35 (1978), 503-44.

Brockliss, Laurence W. B. (1981): "Aristotle, Descartes and the New Science: Natural Philosophy at the University of Paris, 1600-1740," *Annals of Science* 38 (1981), 33-69.

Brockliss, Laurence W. B. (1987): *French Higher Education in the Seventeenth and Eighteenth Centuries: A Cultural History.* Oxford.

Brockliss, Laurence W. B. (1992): "Seeing and Believing. Contrasting Attitudes to Observational Autonomy among French Galenists in the First Half of the Seventeenth

---

65 Brockliss (1981).

Century," in W. Bynum and R. S. Porter (ed.), *Medicine and the Five Senses*. Cambridge, pp. 69-84.

Brockliss, Laurence W. B. (1993): "La Querelle entre les facultés de médecine de Montpellier et de Paris au XVIIe siècle," in G. Cholvy (ed.), *Procès verbaux du colloque historique pour le VIIe centenaire de l'université de Montpellier, 1289-1989*. Montpellier.

Brockliss, Laurence W. B. (1995): "Pierre Gautruche et l'enseignement de la philosophie de la nature dans les collèges jésuites français vers 1650," in Luce Giard (ed.), *Les Jésuites à la Renaissance. Sysème éducatif et production du savoir*. Paris.

Brockliss, Laurence W. B. (2000): "Molière et le discours médical de son temps," in Jean-Louis Cabanès (ed.), *Littérature et médecine*, vol. II: Eidôlon: Cahiers du Laboratoire pluridisciplinaire de Recherches sur l'Imaginaire appliquées à la Littérature. Talence: Université Michel de Montaigne, Bordeaux III, 157-68.

Brown, Harcourt (1934): *Scientific Organizations in Seventeenth-Century France, 1620-1680*. Baltimore.

Calder, Andrew (1993): *Molière. The Theory and Practice of Comedy*. London/Atlantic Highlands, N. J.: Athlone Press.

Clair, Pierre (1978): *Jacques Rohault, 1618-72*. Paris.

Courtaud (1645): *Monspeliensis medicorum universitas, oratio pronunicata die vigesima prima mensa octobris anni MDCXLIV ...* . Montpellier.

Cunningham, Andrew (1997): *The Anatomical Renaissnce. The Resurrection of the Anatomical Project of the Ancients*. Aldershot.

D'Argentré, Charles D. (1728-31): *Collectio judiciorum de novis erroribus qui ab initio duodecim saeculi ... in Ecclesia proscripta sunt*, 3 vols. Paris

Davidson, H. M. (1983): *Blaise Pascal*. Boston.

Dear, Peter (1985): "*Totius in verba*: Rhetoric and Authority in the Early Royal Society," *Isis* 76 (1985), 145-61.

Des Réaux, Tallemant (1960-61): *Historiettes*, 2 vols. Edited by A. Adam. Paris.

Desmarais (1777): "La vie de Monsieur Boileau Despreaux," in *Oeuvres de Nicolas Boileau-Despréaux*, 4 vols. Dresden.

Desperiers, Jacques (1648-49): *Commentarius in physicam Aristotelis*. Paris: Bibliothèque Mazarine (MS 3536).

Du Hamel, Jean (1705): *Philosophia universalis sive commentarius in universam Aristotelis philosophia ad usum scholarum comparata*, 5 vols. Paris.

Du Plessis, Arrnand-Jean (cardinal-duc de Richelieu) (1835-76): *Lettres, instructions diplomatiques et papiers d'état*, 8 vols. Edited by D. L. M. Avenel. Paris.

Feingold, Mordechai (1984): *The Mathematicians' Apprenticeship: Science, Universities and Society in England, 1560-1640*. Cambridge.

French, Roger (1989): "Harvey in Holland: Circulation and the Calvinists," in Roger French and Andrew Wear (eds.), *The Medical Revolution of the Seventeenth Century*. Cambridge.

French, Roger (1999): *Dissection and Vivisection in the European Renaissance*. Aldershot.

Garber, Daniel (1992): *Descartes's Metaphysical Physics*. Chicago

Gassendi, Pierre (1992): *Vie de l'illustre Nicolas-Claude Fabri de Peiresc*. Translated by Roger Lassalle and Agnès Bresson. Paris.

Gautruche, Pierre (1656): *Philosophiae ac mathematicae totius institutio ... Ad usum studiosae iuventutis*, 5 vols. Caen.

Gilson, Etienne (1930): *Etudes sur le role de la pensée médiévale dans la formation du système cartésien*. Paris.

Hahn, Roger (1971): *The Anatomy of a Scientific Institution: The Paris Academy of Sciences, 1666-1803*. Berkeley.

Harvey, William (1649a): "*The First Anatomical Essay to Jean Riolan on the Circulation of the Blood*," in Harvey (1963).

Harvey, William (1649b): "A Second Essay to Jean Riolan," in Harvey (1963), pp. 121-82.

Harvey, William (1963): *The Circulation of the Blood*. Translated by Kenneth J. Franklin. London.

Howard, Rio C. (1974): *Gui de la Brosse: The Founder of the Jardin des Plantes in Paris*. Ph. D. dissertation, Cornell University.

Jourdain, Charles B. (1862-66): *Histoire de l'université de Paris au XVIIe et XVIIIe siècles*, 2 parts in 1 volume. Paris.

Jouy, E. (1928): "Pascal et la Tour Saint-Jacques (1648). A propos des expériences parisiennes de Pascal pour vérifier l'expérience du Puy-de-Dôme," in E. Jouy (ed.), *Etudes pascaliennes*, vol. iii. Paris, pp. 7-22

L'Estoile, Pierre de (1958): *Journal pour le règne de Henri IV*, vol. ii: 1601-1609. Edited by A. Martin. Paris.

Launoy, J. de (1653): *De varia Aristotelis in academia Parisiensis fortuna*. Paris.

Le Rées, Francois (1642): *Summa philosophia*, 4 vols. Paris.

Le Roy Ladurie, E. (1991): *The Ancien Régime. A History of France, 1610-1774*. Oxford.

Legrand, N. (1913): *La Collection de thèses de l'ancienne faculté de médecine depuis 1559 et son catalogue inédit jusqu'au 1793*. Paris.

Lenoble, Robert (1971): *Mersenne ou la naissance du mécanisme*. Paris.

Magne, E. (1929): *Bibliographie générale des oeuvres de Nicolas Boileau-Despréaux et de Gilles et Jacques Boileau*, 2 vols. Paris.

Mani, H. (1967): "Jean Riolan II (1580-1657) and Medical Research," *Bulletin of the History of Medicine* 42 (1967), 121-44.

Mercer, Christia (1997): "The Vitality and Importance of Early-Modern Aristotelianism," in Tom Sorrell (ed.), *The Rise of Modern Philosophy. The Tension Between the New and Traditional Philosophies from Machiavelli to Leibniz*. Oxford.

Moehle, Andreas H. (1992): *Kritik und Verteidigung des Tierversuchs im 17. und 18. Jahrhundert*. Stuttgart.

Molière (1900): *Oeuvres complètes de Molière*. Oxford.

Noël, Etienne (1648): *Plenum experimentis novis confirmatum*. Paris.

Pagel, Walter (1967): *William Pagel's Biological Ideas*. Basel.

Patin, Gui (1846): *Lettres*, 3 vol. Edited by J.-H. Reveillé-Parise. Paris.

Perkins, Wendy (1985): "The Uses of Science: The Montmor Academy, Samuel Sorbière and Francis Bacon," *Seventeenth-Century French Studies* vii (1985), 155-62.

Perkins, Wendy (1986): "Samuel Sorbière: Writings on Medicine," *Seventeenth-Century French Studies* viii (1986), 217-28.

Porter, Roy S. (1996): "The Universities and the Scientific Revolution," in H. de Ridder-Symoens (ed.), *History of the University in Europe*, vol. 2: Universities in the Early Modern Period (1500-1800). Cambridge.

Pumfrey, Stephen (1990): "Neo-Aristotelianism and the Magnetic Philosophy," in John Henry and Sarah Hutton (eds.), *New Perspectives on Renaissance Thought*. London.

Riolan, Jean (1648): *Encheiridium anatomicum et pathologicum*. Paris.

Riolan, Jean (1649): "Instauratio magna physicae &medicinae, per novam doctrinam de motu circulatoris sanguinis in corde," in *Opera anatomica vera, recognita, &auctoiora quam plura nova*. Paris, 539-622.

Riolan, Jean (1651): *Curieuses Recherches sur les escholes en médecine de Paris et de Montpellier*. Paris.

Riolin, Jean (1661): *Manuel anatomique et pathologique, où abrégé de toute l'anatomie, et des usages que l'on en peut tirer pour la connoissance, et pour la guerison des maladies*. Translated by F. Sauvin. Paris.

Rohault (1671): *Traité de physique*, 2 vols. in 1. Paris.

Schmitt, Charles (1973): "Towards a Reassessment of Renaissance Aristotelianism," *History of Science* xi (1973), 159-93.

Shapin, Steven (1994): *A Social History of Truth: Civility and Science in Seventeenth-Century England*. Chicago.

Solomon, Howard M. (1972): *Public Welfare, Science and Propaganda in Seventeenth-Century France. The Innovations of Théophraste Renaudot*. Princeton.

Stroup, Alice (1990): *A Company of Scientists. Botany, Patronage, and Community at the Seventeenth-Century Parisian Royal Academy of Sciences*. Berkeley.

Vanpaemel, Geert (1986): "Echo's van een wetenschappelijke revolutie. De mechanistische natuurwetenschap aan de Leuvense Artesfaculteit (1650-1797)," *Verhandelingen van de Koninklijke Academie voor Wetenschappen, Letteren en schone Kunsten van België: Klasse der Wetenschappen* 173 (1986). Brussels.

Daniel Garber

Defending Aristotle /
Defending Society in Early 17th Century Paris

It is commonplace to observe that the profound scientific changes that took place in the 16th and 17th centuries were deeply intertwined with issues and institutions in the larger society and culture, and that resistance to the new ideas then transforming the world picture went far beyond any purely rational doubts about their validity. At the one end of the historiographical spectrum we have the 19th-century caricatures of the far-sighted and noble seekers after truth, such as Copernicus, Galileo, and Bruno, facing the benighted forces of religious conservatism, who vainly sought to maintain an outdated picture of an anthropocentric universe and the authority of the oppressive religious institutions that they represented against the rising tide of truth.[1] Recent studies have gone a long way towards producing a more subtle and nuanced picture of the process of change during the so-called Scientific Revolution, and understanding the complex dance between tradition and innovation during this period. We now understand in much more detail the rise of Copernicanism and complex social factors that surrounded such events as the condemnation of Galileo. But there is still a great deal of work to be done.

In the literature of the last fifty years or so, perhaps from Kuhn's *Copernican Revolution*,[2] the emphasis has been on the astronomical revolution and the social factors surrounding the introduction and eventual acceptance of heliocentrism. However, there was another important scientific transformation during the same period: the rejection of Aristotelianism in natural philosophy and its eventual replacement by a new mathematical and corpuscular physics. Certainly there have been many studies of late Aristotelian natural philosophy and the development of early-modern alternatives.[3] But there has not been much attention to the social dynamics of the process of change in this case. The transformation between Aristotelian natural philosophy and mechanist was every bit as revolutionary and unsettling as was the transformation between the Ptolemaic universe and the Copernican, and while there is no event associated with that change quite as visible as the condemnation of Galileo, it occasioned, in its way, as much anxiety and social discord as heliocentrism did.

That is what I would like to explore in this essay, some of the social factors that surrounded the rise of an anti-Aristotelian natural philosophy. In particular, I would like to look at the specific case of the debate over Aristotelianism that took place in Paris in the mid-1620s. In 1624, a group of three scholars announced a public meeting in which they proposed to refute Aristotelian natural philosophy and defend a variety of atomism. This meeting was forbidden

---

1 The *locus classicus* for these views are Draper (1874) and White (1896). Both books were reprinted often well into the 20th Century.
2 Kuhn (1959).
3 Of particular note here are Ariew (1999), Des Chene (1996), and Dear (1995).

by the authorities, the theses condemned for opposing officially approved doctrines and authors, and the unfortunate three were exiled from Paris. In examining the reaction to this event in some detail, we shall confirm some generally accepted commonplaces about the opposition to the new anti-Aristotelian philosophies, that the opposition derived from a kind of blind conservatism, that medieval scholastic thinkers had so linked Aristotle with Christian theology that an attack on the one was regarded as an attack on the other, and so on. But we will also find something else: in the context of early 17th-century France, defending Aristotle against the new scientific ideas was seen as necessary for the stability of society itself. This, in any case, is what I shall argue.

The condemnation of 1624

My story begins in the summer of 1624, in Paris. There, three young scholars who opposed the peripatetic (Aristotelian) philosophy then taught at the schools announced a disputation to refute Aristotle. The three young scholars (whom I will call the Gang of Three for short) were Jean Bitaud, "Anthoine Villon, called the Soldier Philosopher," and Éstienne de Clave, identified in one document as "Physician and Chemist." The disputation, scheduled for Saturday and Sunday the 24th and 25th of August, in "the Palace of Queen Marguerite," was announced on Friday, the 23rd of August, by posting a broadside in Latin containing fourteen anti-Aristotelian theses on the street corners of Paris.[4]

The theses were something of a grab-bag of doctrines and criticisms. Theses I-III were directed against Aristotle's three principles in physics: matter, substantial form, and privation. What was being denied here, apparently, was the Aristotelian conception of body and change, the idea that bodies have innate natures and tendencies (forms), that inhere in undifferentiated matter, and that bodies can only change their characteristic behavior by shedding one form and acquiring another. However, the authors of the theses were careful enough in their broadside to follow the official Catholic doctrine that the human soul is the substantial form of the human body; in eliminating substantial forms, rational forms, human souls were excepted.[5] The remaining theses were somewhat more technical, and dealt with a theory of

---

4 The details of this event are reported in Morin (1624), pp. 5ff. According to his anonymous biographer, Morin was present for the affair; see Anonymous (1660), p. 39; Morin implies as much in Morin (1624), p. 18. Another shorter account, perhaps derived from Morin's, can be found in the *Mercure François*, vol. X (1625), 503-504. There are numerous short references to the affair in the literature, and a few articles that focus on it, but the disputation of August 1624 has received its most extensive treatment in an excellent recent article by Didier Kahn (Kahn (2001a)), pp. 241-286. In this article, and in his thesis (Kahn (1998)), forthcoming from Librarie Droz, Kahn exhaustively discusses the documents, people, and events connected with this affair. His very careful and thorough work renders all other accounts superfluous. Despite the impressive account he offers, though, I disagree with some aspects of his interpretation of the significance of the event. Kahn is mainly interested in showing that the condemnation of the theses was not a condemnation either of alchemy or of atomism, but was connected with some of the religious consequences of the theses. He also sees the objections to the theses as arising from disputes between the University of Paris and the Jesuit College de Clermont, and more generally, between the French universities and the Jesuit colleges. (See Kahn (2001a), pp. 243-244.) My claim is somewhat different, as will be seen.

elements and mixtures. Thesis IV dealt with the number of elements, which, according to V and VI, are the basic elements of everything, and are ungenerable and incorruptible. All diversity among things derives from different proportions among these elements in a given mixture (VII) and all action proceeds from the mixture of these elements. (VIII). Theses IX and XI dealt with the specific makeup of the elements found in our regions, earth and water, air and fire. Theses X, XII and XIII denied the Aristotelian account of change, and argued that all change happens by the rearrangement of parts, which themselves cannot be transmuted. And finally, the last thesis asserted that "everything is in everything" (that is, that every thing contains at least some amount of each of the elements), and that everything is composed of atoms or indivisibles (XIV).[6]

But things did not work out for the Gang of Three. On Thursday, the premier President de la Cour de Parlement saw copies of the theses, and he forbid them to sustain the theses on pain of death. The next day, Villon had the daring to seek him out at his home, and ask for the ban to be lifted. Though that was not granted, Villon and de Clave, without mentioning the ban, were able to gather eight or nine hundred people at their chosen site. The crowd waited in the late August heat until three o'clock for the disputation to begin. When people complained about the length of the wait, two large packets of the theses were handed out. After this, Villon went into the hall twice, and told the crowd that it was necessary to carry the seats to the courtyard, because the room couldn't hold the crowd. When it was pointed out that there was no problem in the hall, Villon went out a third time, and finally told the crowd that he was forbidden to sustain the theses, and so he would not sustain them. After much booing and whistling, the crowd left.

But the affair wasn't over. The Parlement sent the theses to the Faculty of Theology of the University of Paris (the Sorbonne) for their examination on 29 August. A few days later, the Sorbonne replied with a censure of some of the theses,[7] and through an Arrest of 4 September 1624, the "Cour de Parlement" ordered Villon, de Clave, and Bitaud to leave Paris, never to teach again within their jurisdiction, on pain of corporal punishment.[8] But the Parlement did not limit their attention to the unfortunate three. They wrote:

> "Be it forbidden to all persons, on pain of death, to hold or to teach any maxims that go against the ancient and approved authors, nor to conduct any disputations but those which are approved by the Doctors of the aforementioned Faculty of Theology."[9]

---

5 For the official statement of the Catholic dogma on this question, see Denziger (1952), pp. 222-223 (§ 481) and pp. 272-273 (§ 738). The former is from the Council of Vienne (1311-12) and the latter is from the Fifth Lateran Council (1512-17).

6 The theses are given in their entirety in translation in Appendix 1 below. In the notes I discuss the sources of the text.

7 The report of the Faculty of Theology can be found in Launoy (1653), pp. 125-134, and in d'Argentré (1728-31), vol. 3, pp. 215-216. It is interesting to note that the Launoy text was subsequently republished as an appendix to Bernier (1653), part of a defense of Gassendi to an attack made against him by Jean-Baptiste Morin, who was very much involved in the affair of 1624. (See Martinet (1992), pp. 47-64, esp. p. 62.) Launoy's point, and Bernier's point in including the text, was that Aristotle had been controversial since his reintroduction into the Latin West in the 13ᵗʰ century, and that contrary to Morin's claims, rejecting Aristotle is not contrary to religion. The comments of the Faculty of Theology on the theses can be found in translation, along with the theses, below in Appendix 1. The main body of the text of their report is translated below in Appendix 2.

8 The document can be found in a number of places, including in the *Mercure* X, pp. 504-506.

Among the "ancient and approved authors," Aristotle clearly stood at the head. While this event may not have had the visibility of the condemnation of Galileo in Rome eight years later, it did have its reverberations. Copies of the documents relating to the event are found both in published and unpublished sources outside of France, and as a legal precedent, it was prominently cited for many years to come.[10]

There are many fascinating questions that one might ask about this incident. But there is one big question I would like to address now: *Why did anyone care?* Why was a list of theses, posted in Paris in Latin, claiming to refute the thought of someone dead for 2,000 years and under serious attack for more than 200 years, sufficiently alarming to bring in the big guns of the civil government, the Church, and the University, and result in the exile of three scholars? Who objected and why?

Two Defenders of Aristotle in 1624

Among the many who were involved in this event and who argued against the Gang of Three, I would like to examine two, the young Marin Mersenne, author of a recently published commentary on Genesis, and a young physician and astrologer named Jean-Baptiste Morin.

Morin was later to become rather a visible person in the scientific life of early seventeenth-century Paris.[11] But at the time of the condemnation, he had done rather little of note, and had rather more in the way of ambition than success. At that point he was the physician to the Duke of Luxembourg, having passed through a number of other patrons earlier. But he was interested in much more than medical practice. Some years earlier, with the help of one of his patrons, he had undertaken a trip to Hungary and Transylvania to inspect the mines there. On the basis of his experiences there, he wrote a short treatise, published in Paris in 1619, *Nova Mundi sublunaris anatomia*, in which he argued for an astrological theory of the heat he found in the middle-European mines. In 1623 he had published a little book in Paris, *Astrologicarum domorum cabala detecta*, an argument for the twelve houses of the Zodiac largely on Cabalistic and numerological grounds, a work bizarre even for Morin; also, he was still basking in the glow of a correct astrological prediction he had made of a former patron's fall from power and influence, something perhaps not so surprising given that particular person's deep involvement in court politics. But then in late 1624 (the dedication is dated 29 November), Morin injected himself in another kind of debate by publishing a curious pamphlet entitled: *Refutation des theses erronees d'Anthoine Villon ... & Estienne de Claues...*, apparently self-published (on the title page it is said to be available "chez l'Autheur"). According to his contemporary biographer, he had attended the disputation, prepared to refute the group on the spot, and disappointed that he could not, he put his objections into print.[12]

---

9 *Mercure* X, p. 505.
10 On this see Kahn (2001a), pp. 279-284.
11 On Morin's life and career, see Anonymous (1660) and Martinet (1986), pp. 69-87. My account of Morin is taken from these two sources.
12 Anonymous (1660), p. 39.

Morin offers a long and complex attack against the posted theses, an attack too long even to summarize here; each one gets careful scrutiny and refutation. (In this, Morin goes even farther than the Fathers of the Sorbonne, who were willing to allow that *some* of the propositions were unproblematic.) But the most basic argument seems to be the following. Morin seems to take as basic and beyond serious question the Aristotelian view that "matter ... and form united are the essence of body as such."[13] He thus argues that without matter and form, there can be no bodies. And so, he argues, since they deny matter and form, for the Gang of Three the human being isn't a body. This leads to the denial of God. For if man is not a body, then neither is Jesus Christ. So, if there is no matter and form, Christ must have been lying when he declared that "this is my body." And if God can lie, then there is no God.[14] Thus heresy, blasphemy, and atheism follow "tres-euidemment" from the doctrines of these philosophers. A curious argument, but there it is.

We shall return to Morin shortly, but for the moment, I would like to turn to another contemporary writer who took an interest in this event, Marin Mersenne.[15] In 1624 Mersenne was a relatively young man (36 years old at the time), a graduate of the prestigious Jesuit College of La Flèche, who had recently established himself in Paris; a member of the order of Minims, he had already published a number of religious tracts by the time our three disputants had announced their refutation of Aristotle. Among Mersenne's works were two fat tomes, the *Quaestiones celeberrimae in Genesim* (Paris, 1623), and *L'impiété des deistes* (Paris, 1624). The *Quaestiones ... in Genesim* is in the form of a commentary on the book of Genesis (though the commentary only extends to the first six chapters). But in fact, the book is much, much more. One central theme concerns natural philosophy. Interspersed with the discussions of purely exegetical matters are many discussions of natural philosophy; God says "let there be light," and Mersenne responds with a treatise on light; God creates the earth, and Mersenne responds with a treatise on the earth and its place in the cosmos, including a detailed examination of the Copernican hypothesis (he was against).[16] There is method in all of this, though. In addition to the intrinsic interest there is in these discussions, Mersenne is out to demonstrate that a good Catholic can be a good natural philosopher too; Mersenne wants to undermine attacks from the new philosophers, not to mention the atheists and heretics against the Catholic church by showing that a rational person with an open mind should accept the doctrines there taught. Catholics had gained the reputation of being doctrinaire Aristotelians, Mersenne thinks; the new philosophers, the followers of Campanella, Bruno, Telesius, Kepler, Galileo, Gilbert, among others have spread the view that "that the Catholic

---

13 Morin (1624), p. 36.

14 *Ibid.*, pp. 48-49.

15 On Mersenne's life and career, see Lenoble (1971), chaps. 1 and 2, and Beaulieu (1995).

16 *Genisis*, chapter I verse 2 ("Dixitque Deus; fiat lux, et facta est lux") is finally given in column 731, followed in col. 735, after some philological and textual commentaries, by verse 3 ("et vidit Deus lucem, quod esset bona"), Mersenne launches into an extended treatise on light and topics in optics (col. 737ff). Chapter I verse 10 ("Et vocavit Deus aridam, terram ...") is given on col. 861, followed in col. 867ff by a lengthy treatise on the earth, focusing on the question as to whether or not the earth moves. Note that the references to the *Quaestiones ... in Genesim* are given by column number. The copy in the French Bibliothèque Nationale (cote A 952 (1)), which I am using, contains two versions of cols. 669-674, the famous Colophon against the Atheists. It seems that Mersenne pulled the first version because it was too violent, and substituted a somewhat milder version. Following normal practice, I shall refer to the first version by adding an asterisk to the column number (i. e., cols. 669*-674*).

Doctors and Theologians clearly follow only Aristotle and swear by his word." Mersenne continues:

> "These idlers try to persuade the world that Catholics ... are in the highest ignorance of philosophy; or that Catholics do not want to admit opinions that are true or probable, but instead pressure Christian souls, as if by ancient tyrannical persuasion, to accept false or less true opinions. But this is completely false. If indeed there are any to whom truth has ever been a friend, it is most friendly to Catholics ..."[17]

This leads to a second important theme of the work. Much of the commentary is directed squarely against atheism and heresy. When the word "God" first appears in Genesis, Mersenne follows it with 670 columns of attacks against the atheists. This attack is continued, in French, in *L'impiété des deistes*, which appeared in June of 1624, just a couple of months before the disputation.

Mersenne must have taken great interest in the disputation; like Morin, he may well have been in attendance. Mersenne's discussion of the event occupies a prominent place in the work that he was probably working on at the time, *La vérité des sciences*, published in Paris in August 1625. *La vérité des sciences* is a dialogue among three characters, the Christian Philosopher, the Sceptic, and the Alchemist. Needless to say, the Christian Philosopher wins the argument. But the treatment of the event in 1624 is interesting. None of Mersenne's three discussants has anything nice to say about the Gang of Three; both the Christian Philosopher and the Sceptic defend Aristotle against their attacks, and though the Gang of Three are identified as alchemists of sorts, even the Alchemist in the dialogue dismisses them as charlatans.[18] What is interesting, though, is how little argument there is against them. The Sceptic tells the other two about the events, and, in fact, goes through all fourteen theses that were posted.[19] But while all three express disapproval, the only real argument that any present is perhaps an abbreviated version of Morin's main argument, that the Christian Philosopher offers: "... if there is no matter and no form, then man has neither body nor soul, something contrary to the Catholic faith."[20]

Both Mersenne and Morin are interested in the content of the proposals put forward by the Gang of Three, and in refuting their position. One cannot doubt that both of them are deeply committed to defending Aristotle and Aristotelian at least in part because they genuinely believe that it is true, or, at least, the best that we have at that moment. But there is something else going on in their discussions of the issues, another theme that is at least as important as the discussion of the philosophical and theological issues. It is evident to anyone who reads the two polemics, that of Morin and that of Mersenne, that under the words and

---

17 Mersenne (1623), unpaginated. (Translation by Paul Mueller.) Though Mersenne stands up for reason, there still is a very conservative streak in the science of the *Quaestiones ... in Genesim*. Never does he suggest, for example, that Aristotle should be abandoned. However, Mersenne is clearly attracted by certain elements of the newly emerging mathematical sciences. This emphasis, together with the clear focus on the connection between science, mathematics, and piety make Mersenne appear very much the *ancien élève* of the Jesuits that he was. On the importance of mathematical science among the Jesuits, see Dear (1995).
18 See Mersenne (1625), pp. 100-101.
19 *Ibid.*, pp. 79f.
20 *Ibid.*, p. 81.

behind what arguments there are, there lies another, deeper reason for opposing the disputation. In the end, I think, it is not entirely a debate about Aristotle and certain abstract philosophical positions. What engages the polemics in Morin and Mersenne, and, I suspect, what draws the attention of the Parlement of Paris, the Sorbonne, is something else, issues concerning truth and debate in general and with respect to Aristotelianism in particular that underlies the whole incident.

Morin expresses an important sentiment in the very opening sentence of his letter of dedication:

"There is nothing more seditious and pernicious than a new doctrine. I speak not only in theology, but even in philosophy."[21]

Morin's sentiment here is by no means novel itself; his wording here, in fact, echoes the opening of the main body of the report of the Doctors of the Sorbonne.[22] Morin goes on to argue that since knowledge of natural philosophy leads us to knowledge of God, false principles lead us to heresy and atheism. It is obvious, then, why the Church should be interested in what philosophers teach. But, Morin thinks, so should the State. For false philosophical views, and the heresies they lead to, might cause sects to be formed, sects "from which follow division and the ruin of provinces and whole kingdoms."[23] Belief is not a matter of individual choice; it is a matter of politics, and well-ordered states, "tous Estats bien policez" have an obligation to prevent such intellectual novelties from arising, and to punish severely those who try to spread them. Thus, he argues, the Church, the University, and the State must oppose the Gang of Three. Aristotle, it must be remembered, was at the core of the educational system in the early seventeenth century, just as he had been since the fourteenth century. It was Aristotelian philosophy that you learned in college, and what was the background for any advanced work in law, medicine, and theology.[24] In denying Aristotle, these disputants, arrogant and proud, set themselves against the established authorities and best minds of the Church and the University. As Morin writes, the Gang of Three has presented "a public challenge to all the schools, sects, and great minds" of Paris.[25] They must be answered

"... to defend the truth, which is here impugned with great licentiousness, ... for the sake of the honor of the sect of Aristotle, which is here reviled, ... and for the honor of the celebrated city of Paris, and to prevent Villon from bragging here or elsewhere that in Paris there isn't any man who has the boldness or capacity to refute these theses, and that he [Villon] can overturn the doctrine of Aristotle."[26]

---

21 Morin (1624), Dedicatory letter to Monseigneur Halligre, Chancelier de France, p. 3. Note that the dedicatory letter is paginated separately from the rest of the text.
22 See Appendix 2 below.
23 Morin (1624), Dedicatory letter, pp. 4-5.
24 On the centrality of Aristotle in the philosophy curriculum in France in this period, see Brockliss (1987), chapt. 7.
25 Morin (1624), p. 6.
26 *Ibid.*, pp. 19-20. Morin seems particularly upset that the Gang of Three would dare to hold their disputation "not in a village, but in the city of Paris, opposite the Sorbonne, the entire University, and the most famous senate in the world."

Morin thus approves completely the official condemnation of the disputation. But, he feels, one should go further than that; one should also refute the theses, and show how they are wrong, something, strangely enough, that neither the Parlement nor the Sorbonne saw fit to do in issuing their condemnations.[27] This is what he tries to do in his pamphlet.[28]

For Morin, it is important to establish the falsity of the claims made by the disputants and the truth of the Aristotelian system. For Mersenne, on the other hand, truth seems irrelevant, in a strange way; true or false, Aristotle must be followed, and his competitors must be rejected. Like Morin, Mersenne holds that it is the responsibility of the authorities to determine what it is permissible to publish and teach. In *La vérité des sciences*, Mersenne's Christian Philosopher declares that it is not up to just anyone to decide what is true and false, orthodox or heretical; that is up to the Doctors and the Doctors alone to decide. Otherwise, just anyone could post whatever they liked, and through their sophisms, confuse the populace into believing things dangerous and heretical, he claims. As with Morin, belief is a matter of politics and not personal conscience. Just as the king can ban card games, dice, and the like, if he judges that such bans are needed to maintain order in his kingdom, he can ban any books heretics use to attack the faith, whether they are true or false.[29] Indeed, Mersenne argues, if "heretics or other enemies of God's Church" were to make use of Euclid's *Elements* or Aristotle's logic to undermine our faith, then the authorities would be right to ban them.[30] Under such circumstances, Mersenne argues, the authorities could even ban the publication of the Bible![31] And so, despite the fact that Mersenne doesn't hold that Aristotle's philosophy is completely true,[32] he still recommends it as worthy of our belief, in part because of the fact that it is generally accepted, tried and true if not true,[33] and the fact that alternative ways of thinking are likely to lead to heresy, atheism, and social chaos. Mersenne's Sceptic frankly admits that "ie n'estime pas la doctrine d'Aristote veritable;" but yet he still recommends it above all others, since it seems to be "better for human commerce, for the order [of society] ["*la police*"] and for common usage" than the main competitor Mersenne there considers, alchemy.[34] Even Mersenne's Sceptic is an Aristotelian in his own (sceptical) way.

For Morin and Mersenne, the problem is not merely that the theses advanced by the Gang of Three are false and heretical (though they are both); indeed, for Mersenne questions of truth and falsity seem to be almost irrelevant. The Gang of Three are not seen as disinterested seekers after truth. They are using their arguments to challenge legitimate authority, and there is a real danger that they will succeed in convincing many of the impressionable youth. What comes up again and again is their arrogance, the presumption that such know-nothings have in challenging the consensus of the learned world on the basis of nothing but their own flimsy intellects and their arrogant pride.

---

27 *Ibid.*, p. 19.
28 *Ibid.*, Dedicatory letter, p. 6, and main text, pp. 18-20.
29 Mersenne (1625), p. 111.
30 *Ibid.*, p. 112.
31 *Ibid.*, p. 113.
32 For Mersenne's critique of Aristotle, see Mersenne (1625), pp. 119-26.
33 I owe this phrase to Roger Ariew.
34 Mersenne (1625), p. 84.

Attacking Heterodoxy

The Gang of Three are perceived as troublemakers, people of bad character, arrogant opponents of established authority, intellectual, theological, and political and they are attacked as such. In this respect, the attack on the Gang of Three echoes a broader attack against heterodoxy in general, and religious heterodoxy in particular.

The period in which the Gang of Three announced their disputation was a period of intense nervousness about heterodox opinion in France. Starting in the mid and late 1620s, the so-called "libertins érudites" began to flourish, a loose movement of humanist free-thinkers in France that looked to the pagan past for a model for living largely outside of the constraints of traditional religion.[35] However, in the preceding years, for those of heterodox opinion who were not erudite, or not discrete enough or skilled enough at hiding their views, or unlucky enough to lack the proper connections, there could be big trouble and hell to pay. There are a number of instances of people brought to trial and in some cases burned at the stake, all punished for the supposed danger of their opinions. Often cited by contemporaries was Julius Caesar Vanini, condemned first in Paris, then burned as an atheist in Toulouse in 1619.[36] Two years after the execution of Vanini in Toulouse, in December 1621, there was another celebrated execution for spreading heresy and atheism, this time in Paris. The candidate this time was a young man of thirty-three named Jean Fontanier.[37] But the most celebrated attempt to suppress heterodoxy in this period is probably the famous trial of the poet Théophile de Viau, a trial that was going on at just the moment that our unlucky three disputants had announced their disputation in August 1624.[38] A sign of the nervousness of the times is the Rosicrucian scare of 1623. Though now thought to have been a student prank, posters announcing the arrival of members of this mysterious and secret alchemical sect in Paris from Protestant Germany produced considerable anxiety, and generated a considerable reaction.[39]

One of the more visible opponents of atheism and heterodoxy was Marin Mersenne. Writing in his *Quaestiones ... in Genesim* (1623), Mersenne reported 50,000 atheists in Paris alone, with as many as twelve to a house.[40] If true, that would mean that better than one in ten residing in Paris in 1623 was an atheist![41] One suspects that this must be something of an exaggeration, even if we remember that "atheist" was a term of derision that applied not only to non-believers, but to those whose beliefs were unorthodox or heretical, such as protestants would appear to a Catholic such as Mersenne.[42] But it certainly does convey a sense of alarm.

---

35 Françoise Charles-Daubert talks of "la crise des années 1623-1625;" see Charles-Daubert (1998), pp. 21ff. The central work on the movement is still Pintard (1983).

36 On Vanini, see Namer (1965); Namer (1980); and Spink (1960), pp. 28-42.

37 On Fontanier, see Lachèvre (1920), pp. 60-81; and Garasse (1623), pp. 147-153.

38 On Théophile and his trial, see Adam (1966); Lachèvre (1909); and Spink (1960), pp. 42-45.

39 On this incident and the Rosicrucian movement more generally, see Yates (1986), chapt. 8. For a more recent treatment, see Kahn (2001b), pp. 235-344. Kahn discusses the very convincing evidence that it was a student prank on pp. 244-252.

40 Mersenne (1623), col. 671*: "At non est quòd totam Galliam percurramus, nisi siquidem non semel dictum fuit unicam Lutetiam 50 saltem Atheorum millibus onustam esse, quae si luto plurimùm, multo verò magis Atheismo foeteat, adeo ut in unica domo possis aliquando reperire 12, qui hanc impietatem vomant."

41 Paris had fewer than half a million inhabitants at the time; see Chartier (1980/1998), pp. 31, 295. If one were to exclude young children from the count, then the proportion of atheists would be considerably more!

Mersenne explicitly linked the problem of heterodoxy with the problem of the stability of the state. According to Mersenne, heterodoxy often derives from a weak character, or is a consequence of having been seduced by bad arguments. But sometimes Mersenne sees something more nefarious, a desire to overthrow religion, Church, and even State. In one place he claims that "they appear to have had no other great plan in publishing their books but to make us give up the truth of Religion, and to make us suck the venom of their wretched opinions, and their fantastic and bizarre imaginations."[43] Similarly, the Deists have taken their name only "to abuse simpler and more credulous souls with the impression they give of recognizing a God." But their real aim is "silently to undermine the columns and foundations of the Catholic truth."[44] But even if the heterodox were honest and honestly mistaken, he and his writings are dangerous, and must be carefully controlled, indeed, eliminated. Some of these writings are so dangerous, Mersenne thinks, that "even a single page of these books damages the soul of curious men."[45] Even with books that are less dangerous, we must proceed with caution. In his discussion of Charron's *De la sagesse*, Mersenne concedes that even if someone "with a strong mind, well formed, and who has well beforehand had the fear of God imprinted in his soul, can profit from it," nevertheless "his book doesn't fail to be dangerous for weak minds, such as those of libertines and deists." For this reason he recommends that

"... it not be permitted to publish his view when one judges or when one ought probably to judge that it will harm and will be the cause of the loss of numerous minds. Now, I maintain that *De la sagesse* has caused more harm than good, and has led more people away from the true Religion than it has kept from error."[46]

More generally, Mersenne exhorts the booksellers to stop thinking about profits alone, and writers to teach true virtue and true knowledge, rather than writing the wretched books that they do. Had they done this, Mersenne thinks, "we wouldn't see as many restless, lost, and almost idiotic youth." He continues:

"It is to be desired that the authorities bring order to this, since it is very important for the public peace of mind, for the conservation of the state, and to maintain the respect that one ought to give Princes, Legislators, and the Law."[47]

More generally still, Mersenne addresses what should be done to combat those with heterodox opinions. One should, of course, address their arguments and refute their sophisms. This is an important part of the programs of the *Quaestiones ... in Genesim* and the *Impiété des déistes*. But not all of Mersenne's opponents are so reasonable: referring specifically to the atheists, Mersenne asserts that "... many don't obey reason, and live like beasts ... ."[48] Since

---

42 While Mersenne does distinguish atheists, strictly speaking, from others of doubtful character (Mersenne (1623), col. 225, e. g.), in at least one place he does identify the two. See the marginal note on Mersenne (1623), col. 235: "Haeretici sunt Athei." On the meaning of atheism in the period, see Kors (1990), chapt. 1.

43 Mersenne (1624), p. 236.

44 Mersenne (1624), letter of dedication, ãij v-ãiij r.

45 Mersenne (1623), col. 670*. The specific author under discussion in this passage is Lucian.

46 Mersenne (1624), pp. 197-198.

47 Mersenne (1624), pp. 94-95.

lack of appropriate faith is in part due to a weak intellect or a perverse will, rational arguments, however strong, will not always do to combat atheism. And so, Mersenne suggests, we must consider stronger methods. In the suppressed "Colophon against the Atheists" in *Quaestiones ... in Genesim*, Mersenne recommends that the rulers not allow books that promote atheism to be written or published in the state; indeed, he suggests that such books be cast into the flames.[49] But he is also capable of suggesting action even stronger than casting books into the flames. For example, in the *Quaestiones ... in Genesim* he writes that the victory over atheism

"... will easily be brought about if the magistrate, or the King, or another of their officials treat the atheists and the impious with the same punishment that the Athenians gave Diagoras Melius, indeed, who, because of the enormous monstrosity of the atheist, annihilated his homeland, one of the Cycladic Islands called Melos."[50]

He does, however, stop short of suggesting that the Prince cast the author into the flames along with his books, the fate that Bruno met in 1600 in Rome, and Vanini in 1619 in Toulouse. In the *Preface au lecteur* of the *Impiété des déistes*, discussing the quatrains of the Deist that are the focus of much of the book, Mersenne warns their author that he will be burned by the authorities, if he is found. Mersenne claims that he himself seeks only to convert the atheist or deist, and does not recommend that the Prince burn the unfortunate author. However, he does not dissuade the authorities from undertaking such a punishment either.[51]

Mersenne's reflections on religious heterodoxy extend directly to philosophical heterodoxy. What Mersenne said in the *Impiété* about the atomist Nicholas Hill and the Epicureans holds, for him, more generally about those who reject Aristotle: "... au bout du conte ils sont tous Heretiques ... ."[52] "... when you come right down to it, they are all heretics." More generally, in the *Quaestiones ... in Genesim*, Mersenne warns his readers against novelty:

"... I warn everyone that they should beware of new opinions, which in [these] days, the wickedly idle heads of men give birth to, which lead to the subversion of the true

---

48 Mersenne (1623), col. 669*.
49 Mersenne (1623), cols. 671*-672*. Cf. cols. 1829-30.
50 Mersenne (1623), col. 670*; cf. col. 1830. A short account of Diagoras (fl. 466 BC) can be found in Bayle's *Dictionaire Historique et Critique*, article "Diagoras." While Bayle wrote, of course, later than Mersenne, the article gives a good idea of the view taken of Diagoras in the 17ᵗʰ century. Bayle identifies Diagoras primarily as an atheist. In note E, Bayle relates the legend, which he attributes to a commentator on Aristophanes' *Frogs*, that to punish Diagoras for his atheism, the Athenians "did a great deal of mischief to Melos, the birthplace of that atheist." It is not clear what exactly Mersenne is proposing in this passage. Perhaps he thinks that France should invade Geneva? I would like to thank Shadi Bartsch for helping me track Diagoras down.
51 Mersenne (1624), *Preface au lecteur*, p. [ix]. (The pages are unnumbered, and I have numbered them in lower-case roman starting with the first page of the *Preface*.) Note also the following passage from Mersenne (1624): "Or s'il se retrouvoit quelque Mathematicien qui fust si étourdy, & si insensé que de s['] oublier de Dieu, & de sa providence, ie serois d'advis qu'on le bannist, & qu'on luy fist perdre la vie de laquelle il seroit tout à fait indigne. Mais on ne sera s'il plaist à Dieu en ceste peine, car ie ne croy pas qu'il y en ait aucun qui se laisse emporter à ceste extreme impieté ... ." [p. 138]
52 Mersenne (1624), p. 239.

philosophy under the shadow of some sophisms, and which carry in atheism. If you consider them more accurately, they will reveal their feebleness, and you will immediately judge that they were produced by a feeble mind, as I will show below in the proper place."[53]

It is no wonder, then, that Mersenne often treats the new philosophers in the same breath as atheists, heretics, and deists: they are all of a piece.[54] And like the atheists and their band, Mersenne recommends that we keep the new philosophers like Villon and de Clave on a short leash, that we refute their views and prohibit them from publishing, that we burn their books and broadsheets, that we exile them and forbid them to teach.

The attack on the Gang of Three in 1624 was less a genuine defense of Aristotle than it was an attack on those who challenge authority and received opinion; it was, in essence, an attack on the idea of open debate and the toleration of heterodox opinions on certain central issues. In this respect, it is very similar to the kinds of attacks that one finds in the same period against other kinds of heterodoxy. But to understand the extreme vehemence of these attacks, to understand why heresy and heterodoxy was such a pressing issue, we must turn to the larger historical context.

Civil War and Uncivil Dissent

In a way, the subtext that we find behind the critique of the theses of August 1624 is something common to much Christian literature, from the Bible on down. The emphasis on authority and on orthodoxy and the deep suspicion of novelty is a common theme running through a great deal of intellectual history. In their report to the Parlement on the disputation of August 1624, the Doctors of the Sorbonne supported their injunction against novelty with a long string of biblical quotations. It didn't matter much that many of them were taken out of context; the point, in a way, didn't need to be argued carefully.[55] The *Oxford Latin Dictionary*[56] lists „subversive, seditious" as one of the meanings of the word "nouus", literally "new" in classical Latin, supported with quotations from Cicero, Suetonius, Tacitus, and Hirtius. This view was reflected in the Jesuit's *Ratio Studiorum*, the rules governing their wide-flung network of schools. In the 1599 version of that document, teachers are warned that they must "flee novel opinions," "even in matters which don't endanger faith and piety."[57] Unlike today, novelty was not always considered an obvious good.

---

53 Mersenne (1623), col. 714.
54 Note here the encyclopedia of error that Mersenne claims that he is planning; see Mersenne (1624), pp. 237-241. There he mixes natural philosophers such as Gorleus, Basso, Hill, and Campanella, with Deists such as Bruno, Vanini, Charron, and Cardano. The association of the new natural philosophy with atheism and heresy is also found in England. See Briggs (1996), pp. 172-199, esp. p. 174 and p. 197n5.]
55 See the notes to the translation of that document in Appendix 2.
56 Oxford 1982: Oxford University Press.
57 Demoustier/Julia/Albrieux et. al (1997), p. 104.

But there are some very special reasons why people were suspicious of new ideas and hetero-dox opinions in France in the 1620s, on the eve of the Scientific Revolution. It must be remembered that the shadow of the religious wars of 16ᵗʰ-century France still darkened Paris when the three bold disputants announced their program in August 1624.[58] Though there were skirmishes between Huguenots and Catholics from the early 16ᵗʰ century, an outright civil war between the two parties began in early 1562. Bloody wars and civic violence contin-ued for more than thirty years, as armies led by royalty and nobility loyal to the Catholic Church fought those who had adopted the Protestant faith. The reign of Henri IV, starting in the early 1590s, was a respite from the violence and instability of the earlier part of the century. The religious wars were officially ended with the Edict of Nantes in April 1598, which established Catholicism as the official religion in France, while guaranteeing the Hu-guenots certain rights. Henri then set about rebuilding Paris and a country that had been torn apart by war. But on May 14, 1610, François Ravaillac, a fanatic Catholic, assassinated Henri in Paris, and political instability returned when the throne went to his nine-year-old son, Louis XIII. Stability had still not entirely returned in August 1624, the moment when Ar-mand-Jean du Plessis, the Cardinal de Richelieu, and later the architect of the absolute mon-archy, was appointed head minister in Louis' Royal Council.

It is hard to overestimate the violence of these religious wars, and the viciousness of the hostility between the different sides, between the Catholics and the Protestants, and on the Catholic side, between the fanatics who looked to Rome, and the more independent Gallican Catholics. The war was not one fought merely by professional soldiers and their commanders on the field of battle: it was fought in the streets, houses, and gutters of the cities and towns all over France, by farmer and townsman, housewife and maid. An extreme, but not atypical example of the kind of violence the dispute provoked can be found in the infamous St. Bar-tholomew's Day massacres of August 24-26, 1572. At the end of the three days of horror, roughly three thousand people lay dead, the Seine red with their blood and clogged with their bodies, and countless houses were burned and looted in Paris alone, as the violence spread to other parts of France.[59]

In this way, we can understand that heterodox belief was much more than an intellectual issue in theology for the French in the 16ᵗʰ and early 17ᵗʰ century. It was also very much a social and political issue. It was a political issue insofar as it led to the formation of armies that could and did oppose the legitimate power of the government. It was, in addition, a social issue insofar as the dispute elicited powerful hostilities between Catholics and Protestants, leading to massive civil disorder and brutal violence. In these circumstances, it is no wonder that novelty and heterodoxy was problematic. Writing in the 1588 version of his *Essais*, Montaigne expressed his genuine revulsion at the idea of intellectual innovation:

"I am disgusted with innovation [*nouvelleté*], in whatever guise, and with reason, for I have seen very harmful effects of it. The one that has been oppressing us for so many

---

58 For a general history of the wars of religion in France, see Holt (1995). The relations between Catholics and Protestants in Paris is treated in Diefendorf (1991). For an account of the city of Paris in the years following the accession of Henri IV, see Ranum (1968). The period following the death of Henri IV is treated in Tapié (1967).
59 On the St. Bartholomew's Day massacre, see especially (Diefendorf 1991), chapt. 6 and (Holt 1995),chapt. 3. The extreme violence of the conflicts is the focus of a justly famous essay by Davis (1987).

years [i. e. the Reformation] is not the sole author of our troubles, but one may say with good reason that it has accidentally produced and engendered everything, even the troubles and ruins that have been happening since without it, and against it; it has itself to blame. ... The unity and contexture of this monarchy, this great structure, having been dislocated and dissolved, especially in its old age, by this innovation, as wide an entry as one could wish is opened to similar attacks. ... But if the inventors have done more harm, the imitators are more vicious in that they wholeheartedly follow examples whose horror and evil they have felt and punished. ... Thus it seems to me, to speak frankly, that it takes a lot of self-love and presumption to have such esteem for one's own opinions that to establish them one must overthrow the public peace and introduce so many inevitable evils and such a horrible corruption of morals, as civil wars and political changes bring with them in a matter of such weight – and introduce them into one's own country."[60]

The Reformation in France gave heterodoxy a very bad name. In an age in which intellectual innovation had led to such disastrous consequences, intellectual conservatism must have looked enormously attractive.

It is unsurprising, then, that there was a feeling of crisis in France in the late 1610s and early 1620s. In August 1624, older residents of Paris must still have remembered the violence of the religious wars of the 16th century, and even the St. Bartholomew's Day massacres, which occurred almost fifty-two years to the day before the scheduled disputation. While there was no fighting in the streets, the political situation in Paris was unstable enough that a renewed outbreak of the earlier uncivil hostilities was reasonably to be feared.

The historical experience of the wars of religion in the sixteenth century and the events that followed in the early seventeenth century led the members of the Parlement, the doctors of the Faculty of Theology, and thinkers like Mersenne and Morin to the inescapable conclusion that difference in belief breeds violence. It is this specific historical experience that is behind Morin's observation in his pamphlet that false philosophical views, and the heresies they lead to might cause sects to be formed, sects "from which follow division and the ruin of provinces and whole kingdoms."[61] In this context, the new anti-Aristotelian philosophies seemed every bit as dangerous to the public welfare as the heresies of Luther and Calvin, and those who proposed them as dangerous to the state as Vanini, Fontanier, and Théophile, and of a piece with them. In the account of the August 1624 event in the *Mercure François*, the anonymous author made explicit reference to the trial of Théophile, then in progress, saying that Villon fled because "he didn't want to keep Théophile company in prison, which was threatened him."[62] Similarly, in his account of the affair, Morin makes explicit reference to one of the chief heretics of his day. Referring to the fact that Villon was a "professeur peripatetique en l'université de Paris," he notes: "He reminds me of Luther, haughty and seditious heretic as he always was, who after having crossed over into the error of heresy, did not fail to wear his Augustinian habit, although he preached against all of the orders of the Church, including his own."[63]

---

60 Montaigne (1962), vol. I, pp. 126-7, translated in Montaigne (1965), pp. 86-7.
61 Morin (1624), Dedicatory letter, p. 5.
62 *Mercure* X, p. 504.
63 Morin (1624), p. 7.

The response to the three unlucky disputants of August 1624 is in many ways a direct translation of the general arguments concerning *religious* heterodoxy and the proper treatment of heretics into the domain of *philosophical* heterodoxy. Or better, it represents the assimilation of philosophical heresy as a species of religious heresy. The danger the authorities saw was social disorder, the kind of civil war that France experienced in the 16ᵗʰ century. Their solution: forbid intellectual innovation.[64] In the Chapelle du Sépulcre in the Church of St. Étienne du Mont (Morin's parish church, in later and more prosperous times), there now hangs an anonymous 17ᵗʰ-century painting of Christ on the cross, with the Virgin, St. John, St. Louis and Louis XIII. Lurking in the shadows of the picture, on the left side is yet another figure, one that might seem extremely strange to us to include in such a scene of religious devotion. It is Aristotle himself, once a pagan, now symbolically present at the Crucifixion. United together in this one picture are the Church, the State, and Aristotle. So closely tied is Aristotle to the workings of both religious and secular society, that any attack on Aristotle is read as an attack on the whole enterprise.

Concluding Remarks

Let me end with a few brief remarks to pull the case together. The question posed was why there was such passionate opposition to those who would reject Aristotle and replace him with a different natural philosophy. Why was it so problematic? The answer I offered focused on the French context in the early seventeenth century. I argued that the immediate reason (at least among the figures I was examining) had to do with the challenge to legitimate authority: for Morin and Mersenne, at least, the challenge to Aristotle was a challenge to the university, to the learned professions, and to the state. At a deeper level, though, I argue, it was associated with other kinds of heterodoxy, in particular with the Protestant heterodoxy that lead to wars of religion of the 16ᵗʰ century. For them, and for their contemporaries, the very idea of novelty, the very idea of disagreement was deeply threatening, and could lead to violence and bloodshed and to the dissolution of civil society. In this respect, the defense of Aristotle was really, for them, part and parcel of the fight against heterodoxy in general, a defense of society as they knew it.

Let me add some final caveats and clarifications to my argument. First of all, my claim is made for France and France alone. I suspect that the argument generalizes, but I am reluctant to make that claim now. Second, I would claim that there is nothing special about the fact that it was Aristotle who was being defended, that Aristotle *qua* Aristotle had no particular internal link with the status quo. Aristotelianism was important not in and of itself, but only because it was the reigning orthodoxy. And challenging orthodoxy itself was what was dangerous. Here I diverge from the kind of sociological analysis offered, for example, by Shapin

---

64 Didier Kahn notes that the issue of the Wars of Religion also come up explicity in Naudé's reaction to the Rosicrucian scare of 1623; see "The Rosicrucian Scare," *loc. cit.*, pp. 191-94. Michael Hunter has made a similar argument for the "new philosophy" in England in the second half of the 17ᵗʰ century. See Hunter (1990), pp. 437-460.

and Schaffer, for whom there seems to be a kind of internal link between Boyle and Hobbes and their respective political views, if I understand them correctly.[65] I make no such claims. In the third place, I don't mean to suggest that this is the *only* reason why people supported Aristotle and Aristotelianism and resisted those who would replace him with one or another alternatives. Within the universities, for example, there are other reasons to be an Aristotelian. Given the extent to which Aristotelianism had permeated the curriculum, and given the investment that teachers had in their lecture notes and class preparations, it is not surprising that university teachers, for example, would strenuously resist change. (Think here of the conservatism of some of your own colleagues.) Nor do I want to deny that there must have been many who defended Aristotle and Aristotelianism because they genuinely *believed* that it was true and better fit the world than the alternatives then under consideration. In a celebrated line in *La vérité des sciences*, Mersenne wrote: "Aristotle is an eagle in philosophy, and the others are like chicks, who wish to fly before they have wings."[66] Mersenne may well have been justified in this remark, given the particular alternatives then available in 1625. And finally, in focusing on a broadly sociological explanation for those who defended Aristotle, I don't mean to suggest that we don't need to explain why the opponent to Aristotelianism opposes Aristotle every bit as much as we need to explain why the supporter supports him. Saying that Aristotelian physics is wrong is not good enough. But that is a different task than the one that I attempted here, one for another day.

All of this leaves the advocate of an alternative to Aristotelian natural philosophy with a double challenge. He must, of course, answer the Aristotelian arguments and make a convincing case for the philosophical and scientific superiority of the alternative that he supports. But it is at least as important for the opponent of Aristotelianism to show that giving up Aristotle and adopting his alternative will not undermine the stability of society. This, I think, was one of the central tasks for those who advocated a new philosophy.[67]

---

65 See Shapin/Schaffer (1985).

66 Mersenne (1625), pp. 109-110. Cf. here Galileo's comparison of the philosopher with an eagle in the *Assayer* (1623), in Drake (1957), p. 239.

67 This essay is a much abbreviated version of a chapter in my book, *How Aristotle Was Refuted: the Pre-History of the Mechanical Philosophy*, currently in progress. I have been working on this material for much too long, and have too many audiences to thank for their help. But I would like to thank especially Roger Ariew, who was with me when the larger project was first dreamed up, the participants in my NEH Summer Seminar, which I co-directed with Roger in the summer of 2000 at Virginia Tech, where I worked through the material. I would also like to thank the participants in the conference, "Ideal and Culture of Knowledge in Early Modern Europe: Concepts, Methods, Historical Conditions and Social Impact," at the Johann Wolfgang Goethe-Universität, December 1-3, 2000, especially Mario Biagioli and Eileen Reeves, both of whom gave me the hard time I so richly deserved. Finally, I am deeply indebted to Didier Kahn, who read an earlier version of this paper, and saved me from a number of embarassing mistakes. Though we still disagree about some of the larger questions, his archival research has reavealed a wealth of new insights into this and other related areas.

APPENDIX 1

*PUBLIC THESES*⁶⁸
AGAINST THE ARISTOTELIAN
Paracelsian and Cabalist⁶⁹ Dogmas

[DEDICATED TO] HOLY IMMORTALITY⁷⁰

1.

Primary matter, which the Peripatetics constitute as the subjective principle in transmutation, whether it has an existence in itself, or from form, is utterly fictitious and clearly has not been provided with any foundation by Aristotle. For in his [conception of] generation, which he believed to happen in these lower regions, he was mistaken, and with him, the others who embraced his position.⁷¹ [FACULTY OF THEOLOGY⁷² : *This first proposition is temerarious and erroneous in faith.*]

---

68 The text is taken from an original copy of the broadside conserved in the Paris Bibliothèque Nationale, Paris, ms. Dupuy 630 fol. 72. There were at least two versions of the text, as implied in Morin (1624), p. 6. Morin notes that the broadside was distributed twice, suggesting the origin of the two different versions. Kahn has recently discovered a copy of this second version, in the Bibliothèque Nationale, ms. Cinq Cents Colbert 163, fol. 168ro; see Kahn (2001a), p. 246. The text of this second version can also be found in Launoy (1653), pp. 128-135, as part of the text of the report of the Faculty of Theology. (The report consists of a text, given below in Appendix 2, followed by a copy of the broadside, noting their reaction to selected theses. That reaction is given below in translation.) The second version of the broadside, with further varients, can also be found in Sennert (1635), pp. 86-91. For a discussion of the two versions, see Kahn (2001a), pp. 245f. In this appendix I make no attempt at a genuine edition of the text. I will follow the text in the version in the ms. Dupuy, except for the one paragraph, indicated below, which is not found in that version.
69 This may be a specific reference to Morin's *Astrologicarum Domorum Cabala Detecta* of 1623, published, incidentally, by Jean Moreau, the same bookseller who published Villon's *L'usage des Ephemerides* the following year. See the following note. The Cabalists are mentioned only in thesis VIII, and then only in the earlier of the two versions of the broadside.
70 Note Morin's remark on this phrase this in Morin (1624), p. 22: "He wanted to imitate me in these words … ." Morin's *Astrologicarum Domorum Cabala Detecta* also had on the title page "Immortalitati Sacra." As Morin had intended it, the dedication was supposed to ensure the survival of the doctrine he was setting forth in the book: "... Lest it perish through the injustice of our times, we dedicate our Cabala to IMMORTALITY, with noble boldness." ["Cabalam nostram, ne iniuria temporum pereat, IMMORTALITATI generosa consecramus audacia." p. 7] This, together with the reference to "cabalist dogmas" in the title of the broadside suggest to me that the theses may have been intended to mock Morin, who, from all indications, took himself very seriously. (A careful reading of the text of Morin's *Refutation* suggests that he and Villon knew one another personally.) It would have been interesting to see the look on Morin's face when he first saw a copy of the theses posted on the Pont Neuf.
71 Primary matter is what is supposed to remain the same in substantial change. It is subjective in the sense that primary matter is the subject of change. For the Aristotelians, generation and corruption only happen in the sublunar world.
72 After each thesis on which the Faculty of Theology commented is given their remarks.

2.

Also, all substantial forms (except for the rational ones) are defended by the Aristotelians no less absurdly than matter is defended, since they understand by them certain incomplete substances constituting with matter a substantial composite that is one in itself. For if the matter is eliminated from the natural composite, then the forms, at very least the material forms, must be eliminated as well.[73] [FACULTY OF THEOLOGY: *This proposition is temerarious and near heresy.*]

3.

In natural transmutations (however they are understood to happen), since they are motions, the Aristotelians posit privation as the principle or terminus à quo. But this is incorrect, since even in the opinion of those who accept matter and form as the two other principles, generation is possible without any preexistent privation in matter, as can easily be noted by those who inquire carefully.

4.

The Peripatetics incorrectly determine the number of the elements, which they understand either as the integral parts of the sublunar world, or as the bodies from which mixtures are composed and into which they are resolved, for this world is composed of fewer than four [elements] and mixtures are composed of more. Both of these agree with experience, reason, and the anatomy of all mixtures.

5.

For a mixture is constituted of five simple bodies or elements, actually and formally existing in it, namely Earth, Water, Salt, Sulfur or Oil, and Mercury or Acidic Spirit.[74] These should be considered the only true natural principles, which cannot be brought about from one another or from anything else, but from which all physical composites are brought about.

6.

These principles are ingenerable and incorruptible, and in all mixtures, they are of the same lowest species, whatever the ignorant crowd of chemists might object, along with Paracelsus. For the diversity of sulfurs, salts, and mercuries, if they appear so in various resolutions of mixtures, in the end are reduced to homogeneity through purification and separation by those who are skillful. [FACULTY OF THEOLOGY: *Concerning the previous four theses the Faculty said nothing, since they are purely physical or chemical.*]

---

73 As noted above, De Clave and Villon are very careful to exempt rational forms from their claim here.
74 Earth and Water are two of the four Aristotelian elements. Of the other two, Air is here identified with Water, and Fire is banished from the sublunar world and identified with the Empyrian Heaven. (see thesis IX). Salt, Sulfur, and Mercury are the three basic principles (*tria prima*) of the Paracelsian alchemical philosophy.

7.

Nevertheless, from the different mixture and proportion of these five principles, in accordance with their quantities, arises all variety which is found in purely material composites, be they generic, specific, or individual, since all composites (with the exception of humans) consist of their union and mixture, without the production of any new entity.

8.

Also, all action and motion, at least the corporeal action and motion found in any sensible subject [*suppositum*] whatsoever, arises from differences in the mixture and proportion of these principles and not from that universal agent and spirit, namely fire, which the Cabalists contrive as the World Soul and the latent principle of all actions, I suppose, and which have been treated by their followers, quite ridiculously, as the greatest secret.[75]

9.

Moreover, this sublunar world consists of exactly two elements, Earth and Water, as its integral parts, for Air does not differ essentially from Water, nor is elemental Fire found within the concavity of the moon, since it is not different from the Empyrean Heaven. Even if this seems to differ from the commonly accepted philosophy, all of these things would not be very difficult to prove through demonstrations. [FACULTY OF THEOLOGY: *These two propositions should be sent to the Philosophers, as they may lead to the following theses, which are of some importance.*]

10.

The Peripatetics were dreaming, when, having spoken at odds with the nature of things, they said that true and physical alterations happen through the introduction or destruction of some new and unique accidental entity, while the subject remains invariant with respect to substance, since no [alteration] can happen naturally without the addition or subtraction of Principles, or their different mixtures. [FACULTY OF THEOLOGY: *This proposition is false, temerarious, scandalous, and in some way attacks the sacrosanct sacrament of the Eucharist.*][76]

---

75 In the version sent to the Faculty of Theology, this last phrase reads: "... which some contrive as the World Soul and the latent principle of all actions, and which have been treated as the greatest secret by some prominent persons in this city." [Launoy (1653), p. 131]

76 Thesis X claims that change does not happen through "the introduction or destruction of some new and unique accidental entity." This is precisely what it is that happens in transubstantiation, when the bread becomes the body of Christ and the wine becomes his blood through the introduction of a new form. Rather, the thesis claims, change must happen through the addition and subtraction of one or another of their five principles, earth, water, salt, sulfur, and mercury: bread can become body, and wine blood only through an alteration of their compositions. This would appear to be in clear contradiction to the official church doctrine of the Eucharist.

11.

Moreover, the Aristotelians err when they attribute to Fire the highest degree of dryness, since it is the wettest of all bodies, and the dryness, which the common among the philosophers think pertains to it, is fictitious. It is the same with the maximal heaviness of Earth, which in the true philosophy is lighter than Water. Though at first glance, Earth is observed hide itself under Water, this can be ascribed to mixture and heterogeneity.[77] [FACULTY OF THEOLOGY: *This proposition is false, and inconsistent with common sense.*]

12.

The Aristotelians admit without foundation virtual qualities productive of the primary [qualities], since all experiments which they toss about in favor of their opinion easily can be dispelled through substances actually and formally existing in such bodies, producing such actions, as a more subtle investigation of such effects can inform anyone even a little knowledgeable in natural things.[78] [FACULTY OF THEOLOGY: *This proposition is false, and temerarious.*]

13.

Nothing is more absurd, and nothing more inconsistent with experience than the transmutation that the Peripatetics acknowledge among the elements. For Earth is always Earth, nor can it be transmuted in any way into Water or any of the other Elements, nor can Water be transmuted into Earth, or Air into Fire. We strongly assert that the same should be said about the other principles, Salt, Oil, and Spirit. [FACULTY OF THEOLOGY: *This proposition is false, temerarious, and erroneous in faith.*]

14.

From all of these things, it is most obvious that everything is in everything, and that everything is composed of atoms or indivisibles, two statements of the Ancients that were mocked and insulted by Aristotle, either ignorantly or, rather, maliciously. Since they are in conformity with reason, the true philosophy, and the anatomy of bodies, we defend both tenaciously and sustain them intrepidly. [FACULTY OF THEOLOGY: *This proposition is false, temerarious, and erroneous in faith.*]

---

77 On standard Aristotelian doctrine, fire is supposed to be hot and dry, and earth heavier than water. This is a position shared with Campanella (see Morin (1624), pp. 98-99) and Telesio (see Frey (1646), pp. 29-89, esp. chapt. XI, p. 58).

78 For the Aristotelians, the primary qualities were hot, cold, wet, and dry, those in terms of which the four elements can be defined. The idea against which Villon and De Clave are arguing seems to be that a thing can have heat, say, virtually, in the sense that it can produce heat in other things without it being actually hot.

****

We reserve for the next set of theses to be published some other things which Aristotle, Paracelsus, and the entire Cabala of the ancients teach us on the qualities and mixture of the elements, on the organization and animation of living things, on the generation and alteration of meteors [i. e., meteorological phenomena], on the nature and properties of the heavens, all against the true way [*ratio*] of philosophizing.

****

The impregnable truth of these theses will be examined, God willing, by JEAN BITAUD of Saintonge, with ANTOINE DE VILLON, Soldier-Philosopher, and otherwise Peripatetic Professor at the University of Paris sitting as judge and president.

****

And so that everything might be even more entertaining for the listeners, Éstienne de Clave, Doctor in Medicine, will be present. De Clave, on the basis of his own experience, and supported by careful and profound meditation on nature, will offer some words, where necessary, in support of the truth which he has long since discovered against those who favor the elements of Aristotle and Paracelsus, this so that from next Monday on, by performing experiments and rehearsing the arguments, he can prove his point to such an extent, and open everyone's eyes so widely that all present will admit all of these truths together, in one single voice.[79]

****

Next Saturday and Sunday, the 24ᵗʰ and 25ᵗʰ of August, 1624, for the entire afternoon. In Paris, at the Palace of Queen Margaret.

---

79 This paragraph is not found in the copy preserved in the Dupuy version.

APPENDIX 2

Report of the Faculty of Theology of the University of Paris to the Parlement[80]

Here follows the censure of certain published theses put forth against the Aristotelian, Paracelsian, and Cabalist principles by Jean Bitaud of Saintonge, Antoine de Villon, and Éstienne de Clave, to which the said Bitaud was to have responded under the direction of de Villon, with the third using the chemical art to demonstrate their truths through experiments, were it not for the fact that the Supreme Senate had silenced them in order to make this request to the Faculty of Theology of the University of Paris.

Clearly, nothing is more dangerous in a Christian Republic and, according to the common judgment of the Fathers, nothing should be guarded against with more care than novelty, especially novelty which is known to be obviously opposed to true knowledge and sacred doctrine. Thus, from the earliest times of the Church, the Holy Spirit, the very author of truth, warned about this, through the witness of the Holy Apostle, who often recommended that new and foreign opinions in doctrinal matters should be avoided with all zeal and diligence. As is said in I Timothy 6 [20]: "O Timothy, keep safe that which has been entrusted to you, avoiding the profane novelties of the babble [*profanas vocum novitates*], and the contradictions of what is falsely called knowledge, expecting much from it with regard to faith, how certain have been disappointed!"[81] Ephesians 4 [14]: "We are no longer to be children, tossed by the waves and whirled about by every fresh gust of teaching, dupes of crafty rogues and their deceitful schemes."[82] And I Timothy 4 [1-2] "The Spirit says expressly that in after times some will desert from the faith and give their minds to subversive doctrines inspired by devils, through the specious falsehoods of men whose own conscience is branded with the devil's sign."[83] And Hebrews 13 [9]: "Do not be seduced by fickle and foreign [*peregrinis*] doctrines."[84] Therefore, we should take all care and zeal that what leads to the corruption of morals, the snaring of the people, the shaking of religion, and ruin of souls should not be published, or, if it happens to be brought to light, that it should immediately be taken out of public view, and, condemned with harsh criticism, crushed far and wide. Therefore, since recently, in these our wretched times, in which there is open access with impunity to all sorts of inquisitiveness and novelties, and to depraved and free thinkers [*liberioribus ingeniis*], certain printed propositions have come to the attention of the Faculty of Sacred Theology of the University of Paris which are to be discussed openly and publicly by certain people (driven by arrogance, doubtlessly), propositions which, as they acknowledge, oppose the Aristote-

---

80 The text is as given in Launoy (1653), pp. 126-128. I omit the introductory paragraphs, which describe the procedure by which the report was commissioned and approved.

81 In the text I have translated the Latin. The *New English Bible* (1961) reads a bit differently: "Timothy, keep safe that which has been entrusted to you. Turn a deaf ear to empty and worldly chatter, and the contradictions of so-called 'knowledge', for many who lay claim to it have shot far wide of the faith."

82 Translation from the *New English Bible*.

83 *New English Bible*. The examples that Paul goes on to give include forbidding marriage and dietary restrictions. That is to say, he isn't talking about natural philosophy.

84 "Do not be swept off your course by all sorts of outlandish teachings." *New English Bible*. Again, Paul is talking here of the dietary laws.

lian doctrine of all philosophers, without challenging the principles, and which also oppose the common and custom of all academies; several of them also seem to involve dangers against the principles of faith. These authors are, perhaps, among the number of those about whom the Apostle once wrote (Colossians 2 [8]) that we should be warned: "Beware that someone not deceive you by philosophy and empty fallacy."[85] Therefore, after they were diligently examined by several doctors of that same Faculty, specially deputized, and discussed in the private chambers of that same Faculty, at last, on September 2, 1624, after celebrating Mass of the Holy Spirit in the usual manner, having gathered the opinions of the individual masters in the general assembly, [the Faculty] expressed the opinion that these theses are extremely dangerous, not only from the point of view of the true philosophy, which has been received by the common consensus of all schools for many centuries, but also they were observed to oppose not a little the principles of faith and religion. Several of them, standing in for others, are noted for particular censor.

[What follows in the text are the notes of the Faculty on the individual theses. These notes are given above in Appendix 1.]

---

85 "Be on your guard; do not let your minds be captured by hollow and delusive speculations ... ." *New English Bible*. "Beware lest any man spoil you through philosophy and vain deceit ... ." Authorized King James version.

References

Adam, Antoine (1966): *Théophile de Viau et la libre pensée francaise in 1620.* Geneva: Slatkine Reprints.

Anonymous (1660): *La vie de Maistre Jean Baptiste Morin.* Paris: Chez Iean Henault.

Ariew, Roger (1999): *Descartes and the Last Scholastics.* Ithaca: Cornell University Press.

Bayle: *Dictionaire Historique et Critique.*

Beaulieu, Armand (1995): *Mersenne: Le Grand Minime.* Brussels: Fondation Nicolas-Claude Fabri de Pieresc.

Bernier, François (1653): *Favilla ridiculi muris.* Paris.

Briggs, John Channing (1996): "Bacon's science and religion," in Markku Peltonen (ed.), *The Cambridge Companion to Bacon.* Cambridge: Cambridge University Press, 172-199.

Brockliss, L. W. B. (1987): *French Higher Education in the Seventeenth and Eighteenth Centuries: A Cultural History.* Oxford: Oxford University Press.

Charles-Daubert, Françoise (1998): *Les libertins érudits en France au XVIIe siècle.* Paris: Presses Universitaires de France.

Chartier, Roger et al (1980/1998): *La ville des temps modernes.* Paris: Éditions du Seuil.

Duplessis d'Argentré, Charles (1728-31): *Collectio judiciorum de novis erroribus qui ab initio duodecim saeculi ... in Ecclesia proscripta sunt,* Vol. 3. Paris.

Davis, Natalie Zemon (1987): "The Rites of Violence," in Natalie Zemon Davies (1987), *Society and Culture in Early Modern France.* Cambridge: Polity Press.

Dear, Peter (1995): *Discipline and Experience: the mathematical way in the scientific revolution.* Chicago: University of Chicago Press.

Demoustier, A., Julia, D., Albrieux et. al (eds. & trans.) (1997): *Ratio Studiorum. Plan raisonné et institution des études dans la Compagnie de Jésus.* Paris: Belin.

Denziger, H. (1952): *Enchiridion symbolorum.* Barcelona: Herder.

Des Chene, Dennis (1996): *Physiologia: Natural Philosophy in Late Aristotelian and Cartesian Thought.* Ithaca: Cornell University Press.

Diefendorf, Barbara B. (1991): *Beneath the Cross: Catholics and Huguenots in Sixteenth-Century Paris.* Oxford: Oxford University Press.

Drake, Stillman (1957): *Discoveries and Opinions of Galileo.* Garden City, New York: Doubleday.

Draper, John William (1874): *History of the Conflict between Religion and Science.* New York: Appleton.

Frey, J. C. (1646): *Opuscula vairia nusquam edita, philosophis, medicis et curiosis omnibus utilissima.* Paris: Petrus Davis.

Garasse, François (1623): *La doctrine curieuse des beaux esprits de ce temps.* Paris.

Garber, Daniel (in progress): *How Aristotle Was Refuted: the Pre-History of the Mechanical Philosophy.*

Holt, Mack P. (1995): *The French Wars of Religion, 1562-1629.* Cambridge: Cambridge University Press.

Hunter, Michael (1990): "Science and heterodoxy: An early modern problem reconsidered," in David C. Lindberg and Robert S. Westman (eds.), *Reappraisals of the Scientific Revolution.* Cambridge: Cambridge University Press, 437-460.

Kahn, Didier (1998): *Paracelsisme et alchimie en France à la fin de la Renaissance (1567-1625).* Université de Paris IV: Librarie Droz.

Kahn, Didier (2001a): "Entre atomisme, alchimie et théologie: La réception des thèses d'Antoine de Villon et Éstienne de Clave contre Aristote, Paracelse et les 'cabalistes' (24-25 août 1624)," *Annals of Science* 58: 241-286.

Kahn, Didier (2001b): "The Rosicrucian Hoax in France," in William R. Newman & Anthony Grafton (eds.), *Secrets of Nature: Astrology and Alchemy in Early Modern Europe.* Cambridge, MA: MIT Press, 235-344.

Kors, Alan Charles (1990): *Atheism in France 1650-1729. Vol. I: The Orthodox Sources of Disbelief.* Princeton: Princeton University Press.

Kuhn, Thomas (1959): *Copernican Revolution.* New York, Random House.

Lachèvre, Frédéric (1909): *Le procès du poète Théophile de Viau* (2 vols.). Paris: Champion.

Lachèvre, Frédéric (1920): *Mélanges.* Paris: Champion.

Launoy, Jean de (1653): *De varia Aristotelis in academia Parisiensi fortuna.* Paris: Edmundi Martini.

Lenoble, Robert (1971): *Mersenne ou La naissance du mécanisme,* 2nd ed. Paris: Vrin.

Martinet, Monette (1986): "Jean-Baptiste Morin (1583-1656)," in Pierre Costabel & Monette Martinet (ed.), *Quelques savants et amateurs de science au XVIIe siècle: Sept notices biobibliographiques caractéristiques* (Cahiers d'histoire et de philosophie des sciences, NS no. 14). Paris: Société Française d'Histoire des Sciences et des Techniques et Éditions Belin.

Martinet, Monette (1992): "Chronique des relations orageuses de Gassendi et de ses satellites avec Jean-Baptiste Morin," *Corpus* 20/21.

Mercure François, Vol. X (1625).

Mersenne, Marin (1623): *Quaestiones celeberrimae in Genesim.* Paris.

Mersenne, Marin (1624): *L'impiété des deistes.* Paris.

Mersenne, Marin (1625): *La vérité des sciences.* Paris.

Montaigne, Michel de (1962): *Essais,* ed. M. Rat (2 vols.). Paris: Ganier.

Montaigne, Michel de (1965): *Complete Essays,* trans. Donald Frame. Stanford: Standford University Press.

Morin, Jean-Baptiste (1619): *Nova Mundi sublunaris anatomia.* Paris.

Morin, Jean-Baptiste (1623): *Astrologicarum domorum cabala detecta.* Paris.

Morin, Jean-Baptiste (1624): *Refutation des theses erronées d'Anthoine Villon dit le soldat Philosophe, & Estienne de Claves Medecin Chumiste ... ou sont doctement traictez les vrays principes des corps & plusieurs autres beaux poincts de la Nature; & prouvee la solidité de la Doctrine d'Aristote.* Paris: Chez l'Autheur.

Namer, Emil (1965): *Documents sur la vie de Jules-César Vanini de Taurisano.* Bari: Adriatica Editrice.

Namer, Emil (1980): *La vie et l'oeuvre de J.C. Vanini, Prince des Libertins, mort à Toulouse sur le bûcher en 1619.* Paris: Vrin.

New English Bible (1961), Cambridge University Press.

Oxford Latin Dictionary (1982), Oxford: Oxford University Press.

Pintard, René (1983): *Le libertinage érudit dans la première moitié du XVIIe siècle.* Geneva: Slatkine Reprints.

Ranum, Orest (1968): *Paris in the Age of Absolutism.* New York: John Wiley.

Sennert, Daniel (1635): *Auctarium Epitomes Physicae ... .* Hamburg: Jacob Rebenlinius.

Shapin, Steven and Schaffer, Simon (1985): *Leviathan and the Air-Pump.* Princeton: Princeton University Press.

Spink, J. S. (1960): *French Free-Thought from Gassendi to Voltaire*. New York: Greenwood Press.

Tapié, Victor L. (1967): *La France de Louis XIII et de Richelieu*. Paris: Flammarion.

White, Andrew Dickson (1896): *A History of the Warfare of Science with Theology in Christendom*. New York: Appleton.

Yates, Frances (1986): *The Rosicrucian Enlightenment*. London: Ark Paperbacks/Routledge and Kegan Paul.

Catherine Wilson

Corpuscular Effluvia:
Between Imagination and Experiment

Introduction

The metaphysics and natural philosophy of the mid-17th century constitute a largely favourable response to the atomism of Democritus, Epicurus, and Lucretius. Whether the recovery and dissemination of ancient atomism in the Renaissance contributed to the disaffection with Scholastic Aristotelianism that is so marked a feature of 17th century metaphysics and natural philosophy, or whether it merely supplied an alternative to a system that was crumbling for a variety of reasons, Christianized Epicureanism was clearly the preferred system of the moderns, with a few notable exceptions. Descartes, Gassendi, Boyle, Locke, Newton, and Leibniz each responded to the challenge to justify and develop the scheme of a latent corpuscular reality projecting into a manifest image, whilst modifying the remaining features of Epicureanism that were distasteful to them as sincere Christians, or that they correctly perceived would impede the reception of this fundamental doctrine.

My concern in this paper is a small part of the story of the 17th century reworking of Epicurean doctrine. Atoms, to paraphrase Claude Levi-Strauss, are good to think with, but this does not make them elements of a scientific view – fruitful or sterile – of the world. Elves, fairies, and extraterrestrials are also good to think with, but none of these entities has ever been an element of a scientific theory. Suppose it is possible to make a distinction between a hypothesized scientific entity, an element of philosophical dogma, and a poetic fiction. What is the status of the subvisible material corpuscle at the end of the 17th century? I propose that the answer has to be sought in three areas: general philosophy, the analysis of discourse, and the history of experiment. To become a scientific entity, the subvisible corpuscle had first to be detached from its locus in antitheology and moral theory. Second, it had to become an element of sober, restrained, discourse; and third, it had to be subjected to observational and experimental regimens. The process of conversion from poetic fiction and philosophical dogma to scientific hypothesis is well underway, but by no means complete by 1700. Epicurus's hedonism and Lucretius's mortalism and his scorn for religion continue to foster opposition to the material atom. The discourse of atoms retains much of its inspired character, and the process of subjecting the atom to experimental regimen has barely begun. I illustrate this thesis by focusing on the role of corporeal effluvia in Titus Carus Lucretius, (a pre-modern) Robert Boyle, (a modern) and John Mayow (a controversial modern).

Past historians exaggerated, either for pedagogical effect, or because they genuinely believed in epistemological ruptures and paradigm shifts, the contrast between Renaissance ontology and the corpuscularian-mechanical philosophy. The rejection of eclectic systems of form, matter, principles, and elements, and the enunciation of strict mechanism in place of sympathy, antipathy, correspondence, and "occult" action at a distance were seen as severe, clarifying moves that are characteristic of modernity along with the rejection of Aristotle's vocabulary of generation and corruption, formal and final causes. In *The Construction of Modern Science* Richard S. Westfall describes the second half of the 17th century as a "conversion" to corpuscles and the mechanical philosophy.[1] And it is often said that post-Newtonian chemistry and physiology in England and on the Continent involved a second conversion, this time from strict corpuscularianism to an enriched ontology of forces and active principles, and from the search for mechanical explanations to the search for descriptive mathematical formulas.

The theory that corpuscularianism and mechanism were first embraced, then rejected as experimentally inadequate in favour of a liberalized ontology and a new account of the aim of science I will refer to as the "two conversions" theory. The broad historiography of the "two conversions" is increasingly questioned,[2] and I shall try in what follows to show how Boyle's own case lend some support to the idea but also discredits it. Boyle was well known for maintaining, in his programmatic writings, a severe and restrictive position that I shall call "reductive atomism" (henceforth RA), that proved uninteresting to later generations of researchers. Yet his "historical"-experimental philosophy depended on a "qualitative corpuscularianism" (henceforth QC) that is continuous both with Lucretian atomism and with 18th century chemistry. Comparing 17th century RA with the eclectic systems of the Renaissance induces the impression of a major rupture. Comparing 17th century QC with the eclectic systems of the Renaissance does not. The impression of a rupture, in other words, arises from a selective and somewhat misleading comparative exercise. Nevertheless, Boyle's programmatic commitment to RA appears to have blinded him to the promise of some genuinely innovative trends in 17th century physiology.

Reductive atomism and mechanism might seem to offer themselves as a package, an all-or-nothing alternative to the ontology of matter-and-form and the normative theory that permits and even encourages intentional explanations, or to the ontology of elements and the normative theory that permits and encourages vitalistic explanations. Boyle, for example, describes a "mechanical" production of an effect as one performed "by such corporeal agents

---

1 "Alchemy expressed the organic conception of nature in its most vivid terms. Its vocabulary was filled with words unmistakable in their connotation – fermentation, vegetation, digestion, generation, maturation ... every major aspect of the chemical tradition expressed a view profoundly opposed to that which was becoming dominant elsewhere in physical science ... [B]ut chemistry was immediately and inescapably relevant to the mechanical philosophy of nature. If the properties of bodies are appearances caused by the particles of which they are composed, chemistry had much to say that a mechanical philosophy of nature could not ignore. The story of chemistry in the second half of the century is the story of its conversion to the mechanical philosophy. Perhaps one should say rather its subjection to the mechanical philosophy, since the growing role of mechanisms in chemical literature appears less to have sprung from the phenomena than to have been imposed upon them by external considerations." Westfall (1971), p. 68-69.

2 John Henry remarks in this connection, "As every historian of science 'knows,' the esssentially unworkable mechanical philosophy was transformed by the genius of Newton who re-introduced 'occult qualities' into natural philosophy ... The historiographical emphasis on the radical revisionism or innovatory nature of Newton's account of occult active principles and the concentration on non-mechanist sources of influence like alchemy tends to perpetuate the view that English mechanical philosophy before Newton followed the essentially unworkable Cartesian programme in which matter was completely inert and could only act by virtue of its 'force of motion' in collision with other parts of matter.." Henry (1986), p. 336f.

as do not appear to work either to work otherwise than by virtue of the motion, size, figure, and contrivance of their own parts, or ... by changing the texture, or motion, or some other mechanical affection, of the body wrought upon."[3] Because Boyle's corpuscles are corporeal and possess only motion, size, and figure, they might appear on first consideration to be the only entities that can produce effects mechanically, and to be specially disposed to produce them. But matters are not so simple.

1. The Christian Atom

As my focus in this paper is on English science, I will pass over Descartes's deeply-plotted effort to introduce the main features of the Epicurean system into textbook natural philosophy. Descartes's strategy was to frame this reintroduction by proving the existence of God, and by portraying God as the source of the atoms' motion and as the author of the simple and uniform laws of nature governing their motions. Further, Descartes wrote in an important exception to the Epicurean doctrine of universal dissolution: the human soul was, he argued, demonstrably immaterial and immortal. Within this framework, Descartes retained the key elements of Epicureanism, including the self-creation of the cosmos, and its plants, animals, and human bodies from an initial chaos, the material basis of sensation, and the moral adaptation to inevitability that naturalism entails. Moreover, he claimed all his results to be certain. Boyle's revisions and exceptions went much further. As Yung Sik Kim observes, his initial attraction to atomism as he encountered it in Diogenes Laertius's account of Epicurus was guarded; he was worried about its atheistic implications.[4] The chance play of atoms, he determined, cannot accomplish very much; God's action is needed to "dispose that *chaos* or confused heap of numberless atoms into the world, to establish the universal and conspiring harmony of things, and especially to connect those atoms into those various seminal contextures upon which most of the more abstruse operations and elaborate productions of nature appear to depend."[5] Unlike Descartes, Boyle did not imagine or boast that all phenomena except human volition could be explained by mechanical principles, and he regarded none of his hypotheses – including the atomical hypothesis, which he renamed corpuscularianism – as certain.

English scientists by and large rejected Cartesianism – its professions of certainty in natural philosophy, its rigorous dualism, and its universal mechanism – in favour of looser versions of Epicureanism propounded by Bacon and Gassendi. Joseph Glanville, writing in 1665, seems to accept some form of corpuscular hypothesis. But what he is prepared to deem a corpuscular-mechanical explanation is surprising, and he goes on to express doubt that the production of all effects is by mechanical means. He finds the notion that "all bodies, both

---

3 Boyle (1744b), p. 459. Reprinted in Stewart (1991) (hereafter SPP), p. 17.
4 Kim has argued that, given his profound reservations about atomism and mechanism, Boyle's propagation of them can only be understood as elements of a plan to elevate chemistry to the status of a dignified subject by linking it with natural philosophy. Corpuscularians were "ingenious persons" "found amongst [men of] nobility and genius" whilst chemists were seen as "illiterate operators" and "whimsical fanaticks." (Kim (1991), p. 1.)
5 Boyle (1744c), p. 452; SPP, p. 173.

Animal, Vegetable, and *Inanimate* are form'd out of such particles of matter, which by reason of their figures, will not cohaere or lye together, but in such an order as is necessary to such a specifical formation" a "pretty conceit."[6] He even cites several palingentical experiments in its support. A decoction of herbs that is frozen overnight forms from "*plantal emissions*" images of the original plant in ice that disappear when its crust is broken.[7] However,

> "[t]hough *blind matter* might reach some *elegancies* in individual effects; yet *specifick conformities* can be no *unadvised* productions, but in greatest likelyhood, are regulated by the immediate efficiency of some *knowing* agent; which whether it be *seminal Formes* ... or whatever else we please to suppose; the manner of its working is to us *unknown*: or if these effects are meerly *Mechanical*, yet to learn the method of such operations may, and indeed hath been ingeniously attempted; but I think cannot be performed to the satisfaction of the severer examination."[8]

The extent to which Boyle follows the sceptical pattern in *denying* that the production of nonchemical forms and qualities can be explained mechanically is surprising. In fact, one of his arguments runs as follows. Though we know that the forms of animals and plants *cannot* arise mechanically, but only through the operation of seminal principles that work upon matter, we should not therefore be led to conclude that the forms of crystals cannot arise mechanically. The experimental control one has over the production of such salts indicates that their production is mechanical. At the same time, we should resist that inference that if crystals can arise mechanically, so can plants, animals, and even humans, "It cannot be in the least inferred that, because such slight figurations need not be ascribed to the plastick power of seeds, it is not necessary, that the stupendous and incomparably more elaborate fabrick and structure of animals themselves should be so."[9]

It is true that Boyle is critical of "those who think the Mechanical principles may serve indeed to give an account of this or that particular part of natural philosophy, as statics, hydrostatics, the theory of planetary motions & c., but can never be applied to all the phenomena of things corporeal." He accuses them of being like those persons who think that alphabetic principles can account for English books, but not books in general.[10] But he is not a universal mechanist; his piety keeps his mechanism well in line. He rejects "Nature," substantial forms, real qualities, the archaeus, astral beings, gas, blas, the soul of the world, the universal spirit, and plastic powers.[11] But his central claim is that mechanism applies comprehensively to chemical and pharmacological phenomena – that is to elements of the microworld, not just to macrophenomena – and perhaps also to some elements of physiology. It is inadequate to the explanation of many effects observed in nature. It does not explain vitality, maturation or reproduction, which depend on the providential planning of God, special divine decrees, or on "seminal principles," endowed by God with developmental potential and scattered about at the creation.[12]

---

6 Glanville (1665), p. 34.
7 Glanville (1665), p. 34.
8 Glanville (1665), p. 34.
9 Boyle (1744b), p. 488; SPP, p. 80
10 Boyle (1744d), p. 452; SPP, p. 142.
11 Ibid., p. 450ff; SPP, p.139ff.

Boyle states, for example, that the onset of menstruation in young females can't be explained mechanically, no more than the causes and cures of some diseases, all of which "are performed under the superintendance and guidance of a wise and intelligent Author of things."[13] Unlike Glanville, he resorts to divine agency to explain the results of a palingentical experiment which a colleague

"took some ashes of a plant, just like our English red poppy: and having sowed these alksalizate ashes ... they did, sooner than was expected, produce certain plants larger and fairer than any of that kind, that had been seen in those parts. Which seems to argue that, in the saline and earthy, *i. e.* the fixed particles of a vegetable, that has been dissipated and destroyed by the violence of the fire, there may remain a plastic power, enabling them to contrive disposed matter so as to reproduce such a body, as was formerly destroyed. But to this plastic power ... it will not perhaps be necessary to have recourse, since an external and omnipotent Agent can without it perform all that I need contend for."[14]

Boyle's mechanism is not an entailment of an *a priori* commitment to a strict form of corpuscularianism. It is an structure imposed on atoms – including qualitative atoms – that confirms their status as supervised by the Christian providential God. It marks the distinction between pagan corpuscularianism – in which effects occur "by chance" and the corpsucularianism appropriate to creationism. As far as Boyle is concerned, there could have been a Lucretian universe in which the atoms mixed and mingled in the void. But such a world would not have resembled ours, for we live in a patterned and predictable world, in which law and order obtain. His theism encourages him to welcome nonmechanical explanations that Lucretius, as a mechanist *ab infra,* would have rejected as superstititious.

Occasionally, Boyle does try to explicate a number of particular phenomena in terms of the size, shape, and motion of the corpuscular constituents of matter and their interactions with one another. Consider his hypothesis about the springiness of air: namely, it is produced by tiny springiform particles. But this is not his usual mode of writing, which consists rather of "historical" observations and accounts of experiments which refer surprising or interesting effects to the agency of various *types* of corpuscles that are credited with highly specific modes of action. Where Descartes, with much hand-waving, claims to derive the patterned and speciated world from a Lucretian chaos of undifferentiated matter over which certain laws of motion have been superimposed, Boyle is a QC creationist, who thinks of corpuscles as issuing from the hand of God already formed into various *species*, all of whose members are identical, and each of which has its own particular, lawful mode of operation. In short: Boyle inherits an older – "Lucretian" manner of thinking about particles and invisible corporeal effluvia but employs it in a theistic context.

---

12 Such principles had to be distinguished from the apparently similar seminal principles of Epicurus and Lucretius that were intended to obviate the need for theistic accounts of the origins of life. See Clericuzio (1990), pp. 584-5.

13 Boyle (1744c), pp. 445-6; SPP, p. 159.

14 Boyle (1744e), p. 540; SPP, p. 196.

2. The Premodern and Modern Discourse of Atoms: QC vs. RA

QC is the teaching that, contrary to what Aristotle maintained, many ordinary substances are not continuous, and many substances – such as blood – that appear to be one substance are not homogeneous. They are composed of tiny, subvisible particles or mixture of different kinds of particles stuck or held together. These tiny particles compose fluids, solids, and also airs, vapours, mists, smokes, and so forth that emanate insensibly from terrestrial and celestial bodies. RA goes beyond QC in placing special restrictions on the corpuscles that are held to be truly basic units, assigning them primary properties only and treating secondary properties as effects of the interaction between sentient beings and figured, moving corpuscles.

Both QC and RA figure in Lucretius's *De Rerum Natura*. QC was an inference drawn from such observations as dancing motes in a sunbeam, the ease with which most hard substances such as horn can be reduced to a fine powder by grinding them, by the gradual wearing away of steps, cliffs, rings and tools, one invisible fragment at a time, and by the phenomena of erosion and diffusion. Nutrition, the penetration of noise through walls, sweat through the skin, light through glass, and the conduction of heat by metals, indicate that seemingly solid bodies are porous. Further, the particulate nature of matter was suggested by phenomena such as the following: a metal may be roasted until it becomes a fine powder; fire may penetrate a substance, indicating porosity, or separate a mixture, perhaps by forcing upward the lighter particles of which it is composed, leaving the heavier behind; the heaviness of substances like gold suggests that its parts are tightly packed together.[15]

All these observations suggest that there are corporeal particles that exist and act below the threshold of human perception, that all bodies are made up of them, and that they can be present and operative, invisibly in air, water, and even fire. While Aristotle would not have denied the existence of motes or powder particles, or of *minima naturalia* – the smallest quantities of a substance that can be taken – he did not consider liquids to be made up of particles separated by void; there is no reason in Aristotle's scheme to suppose tin, or water, or blood to be made up of discrete particles. Hence QC is a rival to matter-and form theories. Aristotle could acknowledge that "mixtures" like honey and water had the accidents of both, but this only confirms his thoroughly phenomenological approach. Notoriously, he maintained that a little wine poured into the ocean turns to sea water,[16] a claim which would be seen to jibe poorly with the experience of chemists in the recovery of dissolved substances or oxydized metals which had disappeared in the interim. Aristotelian theory deals very poorly with the phenomena of dissolution-and-recovery, while QC explains it easily: wine-atoms are diffused into the ocean where they continue to exist unchanged but are so sparsely distributed and so tiny that they remain below the threshold of perception. Recovery experiments, as W. R. Newman has argued, were critical in bringing QC into modern chemistry.[17]

---

15 Newman (1996), pp. 571-3.

16 Aristotle, *Generation and Corruption*, 1 10 328 a 19-31.

17 According to Newman, corpuscles had entered Renaissance chemical texts and theorizing through the qualitative atomism of "Geber's" 13th C. *Summa Perfectionis* and Daniel Sennert's *Hypomnemata physica* (1637), "Robert Boyle's Corpsucular Philosophy." A recent treatment of Sebastien Basson's atomism and its critical history can be found in Luethy (1997), esp. pp. 24-34.

The notion that there existed minimally-sized particles of an extreme smallness involved in chemical reactions and physical alterations was independent of the notion that these particles had some minimum number of characteristics. Christoph Meinel has argued that, by the standards of any era, 17th century arguments for and observations cited in favor of corpuscularianism were inconclusive and that its re-appearance and persistence in early modern science had as much to do with the charm of Lucretius's presentation in *On the Nature of Things* and its appeal to the senses and imagination as it did with arguments, observation, and evidence.[18] In favour of Meinel's contention, it can be observed that Lucretius the poet repeatedly evokes images of flocks, pastures, seashells, rivers, wild beasts, leaves, insects, storms, the ocean, and other beauties of nature to inculcate the theory of atoms. Consider one of his "arguments" for the claim that the atoms are colourless and odourless: "Just as when you are about to prepare the balmy tincture of marjoram and of myrrh, and the flower of spikenard which breathes nectar to the nostrils, amongst the first things you have to seek is olive oil, as scentless as may be … so that it may be as little as possible with any pungency of its own touch and destroy the scents that will be mixed and boiled up in it: for the same reason, the first-beginnings of things must not contribute any odour of their own to the making of things, nor any sound …"[19] But Meinel's criticism of its evidential basis is more pertinent to RA than to QC, which was in fact rather well supported. Because QC was not a very demanding or counterintuitive doctrine, it was welcome wherever Aristotelian orthodoxy did not hold absolute and exclusive sway. And QC was not an especially "modern" doctrine; its links to mechanism were more tenuous than RA's would be.

As Lynn Thorndyke informs us, eclectic systems, that embraced occult qualities, Paracelsian chemical principles, and corpuscles too were freely promulgated in the late 16th and early 17th centuries.[20] De Castro's four influential books (1571) on the microcosm and macrocosm treat the blood and its spirits, fevers, correspondences between meteors and diseases, epilepsy and catarrh, antidotes made from gold and pearls. De Castro defends the world soul and asserts the existence of atoms against Aristotelian elements.[21] Jean d'Espagnet's *Handbook of Restored Physics* (1623) deals with: celestial virtues, salt, sulphur, and mercury as secondary mixtures of the four elements, the world soul in the sun, rational forms in vegetation – and atoms.[22] Rosetti of Pisa in 1667 has a heart at the earth's center whose dilation and contraction explain the tides: the universe consists of atoms that attract or repel one another.[23]

RA, by contrast, is a restrictive and not very intuitive doctrine. It comprises the teachings that everything that exists (with the possible exception of God, and human and angelic souls) is corporeal and composed of indivisible particles possessing only magnitude, figure and motion that are differentiated by their size, their shape, and their movement. Qualities such

---

18 Meinel (1988); repr. in Dear (1997): "There is little doubt … that these phenomena were not adequate for definitely deciding the question of matter … Yet the frequent occurrence and repetition of these observations, the persuasive idea that truth should be visible or could be thought of in a pictorial way, infiltrated scholarly discourse and the very language of science. Atomism was an enticingly pictorial image of reality." (p. 193)
19 Lucretius (1992), 2:850ff., p. 161-3.
20 Thorndyke (1931), vol. 7, p. 124.
21 Thorndyke (1931), vol. 7, p. 124.
22 Thorndyke (1931), vol. 7, pp. 388-9.
23 Thorndyke (1931), vol. 7, p. 583.

as sweetness, coldness, redness, are "conventional," as Democritus had originally argued. No atom taken in isolation can have them. Moreover, according to RA, atoms have no powers except that by their motion they can strike one another, dislodge one another, filter through porous structures, abrade macroscopic structures by acting *en masse* and, by coming into contact with sentient beings, produce experiences. But they are devoid of qualities, since all qualities are perishable, and if the atoms are to be foundations, they must have only imperishable qualities. Arguments in both Epicurus and Lucretius purport to show the *a priori* necessity of RA:[24] The following, less picturesque argument succeeds Lucretius's "olive oil argument" quoted earlier:

> "The first-beginnings of things must not contribute any odour of their own to the making of things, nor any sound, since they can emit nothing from themselves, and similarly, no taste at all, nor cold, nor heat again and moderate warmth and rest; all these, since their nature is such that after all they are perishable ... all these must be kept apart from the first-beginnings, if we wish to lay an imperishable foundation for things upon which the sum of existence may rest: or else you will find all things passing back utterly to nothing."[25]

This is admittedly an elegant argument. But Aristotle had some similarly elegant arguments against the pretended emergence of qualities from the antecedently qualityless. Not only were the arguments for it technically inconclusive, but, unlike QC, RA is not an easily picturable doctrine. Meinel's view that corpuscularianism was largely inspired by the charm of the Lucretian narrative fits QC, while his view that atomism was radically underdetermined by available evidence and argument fits RA.

RA was suggested by the fact that manipulations that seem to alter only the size or relative orientation of the parts of a complex can result in qualitative changes in appearances. Mashing a pear, as Bacon noted, causes it to become sweet, a black horn can be shaved into a white powder, sea water whipped up by wind turns white, and the mere mixture of substances in the chemist's glass can produce dramatic changes in colour and odour. But such observations hardly establish that the atoms themselves are colourless. Locke says, as though this is obvious, that if a grain of wheat is divided and subdivided over and over, eventually we will reach, not just ever tinier particles of wheat, but particles that have no colour, odor or taste at all – only solidity, extension, figure, and motion.[26] Locke does not say whether these particles will be the *same* in every respect as the particles we will reach if we subdivide a particle of rye, or a nugget of gold until we – or God – can divide no more, and one might well find RA an intellectually difficult as well as confusing doctrine. While it is conceptually easy to decompose macroscopic substances into qualitative corpuscles, it is not conceptually easy to consti-

---

24 Cf. Epicurus "... [O]ne must believe that the atoms bring with them none of the qualities of things which appear except shape, weight and size and the [properties which] necessarily accompany shape. For every quality changes, while the atoms do not change in any respect; for it is necessary that during the dissolution of compounds something should remain solid and undissolved, which will guarantee that the changes are not into what is not nor from what is not, but come about by rearrangements ..." *(Letter to Herodotus* in Diogenes (1938), vol. 10, pp. 34-83, in Inwood/Gerson (1994), p. 10).

25 Lucretius (1992), Bk 2:855ff., p. 163.

26 Locke (1975), Bk 2, ch. 8 (§ 9), p. 135.

tute macroscopic substances from reductive atoms. RA allows only three interpretations (or combinations of them): the true indivisible corpuscles must all be identical; or they must come in species; or each atom must be a unique individual. (Lucretius seems to hold the intermediate position that *most* atoms are unique individuals: Just as each ear of corn and each seashell is slightly different from every other. The atoms, "since they exist by nature and are not made by hand after the fixed model of one single atom, must necessarily have some of them different shapes as they fly about."[27]) But on any alternative except the second, which Boyle professed to reject, it is unclear how the atoms project into natural kinds. How can a collection of perfectly heterogeneous, or perfectly identical particles, project into distinct substances, like tomatoes, or ink?

RA, in short, is a hard doctrine to think through, where QC is reasonably natural and intuitive, so long as one does not suffer from Aristotelian prejudices about the void or the authority of human sense-perception. It is not surprising that many 17th century philosophers took an agnostic position on the existence of reductive atoms and preferred not to propound definite theses concerning their forms or their variety. But even Berkeley (eventually) and Leibniz, both of whom vehemently rejected RA, accepted QC, for the evidence for QC is overwhelming. The phenomena of diffusion are qualitatively and quantitatively especially impressive. A few drops of scent can evidently perfume a whole room, or a drop of dye, a barrel of water, and when Boyle turns to consider *a posteriori* evidence for corpuscularianism[28] he observers that a portion of matter can impart a conspicuous colour to above 256,806 times its bulk of water and a faint but discernible tinge to a larger bulk above 513,620 times its size.[29]

Effluvia – smokes, steams, fumes, vapours, scents – everything airborne and perceptible by the senses or by its workings, had traditionally been imagined by QC atomists as definite *types* of particle, with specific effects. Here for example, is Lucretius:

"[I]n the earth are elements of every kind of thing: many (which belong to food) being useful to life, and many such as can strike us with disease and make death come quickly ... Many pernicious elements pass through the ears, many make their way into the very nostrils noxious and rough to the touch ..."[30]

---

27 Lucretius (1992), Bk 2:371, p. 124f.
28 Boyle (1744f), pp. 312ff.
29 Ibid., p. 315.
30 "Certain trees have a shade so dangerous that they often cause headache, if one has lain beneath stretched out on the herbiage. There is also in the great mountains of the Helicon a tree, which is accustomed to kill men by the vile stench of its flower. You may be sure that the reason why all these things rise from the soil is that the earth has many seeds of things which she holds mixed up in many ways and separates apart before passing them on. Again, when a night- light newly extinguished meets the nostrils with a sharp smell, it stupefies on the spot one who is accustomed to fall and foam at the mouth through disease. The heavy scent of castor makes a woman fall back asleep ... And how easily the strong fumes of droop in slumber ... How easily the strong heavy fumes of charcoal creep into the brain ..." (Lucretius (1992), Bk 6:780 ff., p. 552 ff).
[W]hen [men] follow veins of gold and silver, rummaging with their tools the innermost secret places of the earth, what smells Scaptensula exhales from below? Or what mischief do gold mines breathe out, what do they make men look like, what colours? Do you not see or hear in how short a time they are accustomed to perish ...?
All these streams therefore the earth streams out and breathes forth into the open and ready space of the sky." (Lucretius (1992), Bk 6: 805ff., p. 554ff.)

Lucretius took the visible species that mediated perception and dreams, as well as seeds, fumes, and other exhalations to be atomic, and many action-at-a distance or mysterious communication or transmission phenomena were explained by minute corpuscles. The 16th century corpuscularian Daniel Sennert regarded fascination, plague, and poisoning as proceeding from effluvia or minute corpuscles.[31] In the 17th century, the Gassendist Walter Charleton explained sympathies and antipathies in terms of a flow of atoms between the passional parties. Charleton illustrates the well-known contention of Keith Hutchison that 17th century corpuscularians did not reject a whole range of action-at-a-distance or otherwise mysterious effects as nonhappenings, but interpreted them in corpuscular terms.[32] In doing so, they were not however applying a modern scheme to old phenomena. They were applying a traditional scheme, doing precisely what Lucretius had done when he explained dreams, ghosts, plagues, and poisoning by the action of corporeal effluvia.

It would be convenient for proponents of the "two conversions" interpretation of 17th century science if the qualitative corpuscle, considered as a corporeal, airborne, effluvial particle was consistently distinguished from an active spiritual substance, and if it entered only into mechanical interactions. But the dividing line between invisible, light, tenuous particles, and active spirits is not very sharp: there are constant efforts by the English to undermine the Cartesian distinction between the one intentional immaterial substance and the one material inert substance with meditating entities. And 17th century natural philosopher's notion of what it is to be a subscriber to the mechanical philosophy does not always jibe with what the modern reader would expect on the basis of a reading of the programmatic writings of Descartes and Boyle. Kenelm Digby's account of growth and resurrection in his *Discourse Concerning the Vegetation of Plants* furnishes an example of the inclusiveness of mechanism. Digby is a corpuscularian. He is scornful of those who have recourse to a *vis formatrix*, or to secret instinct and sympathies, or who "fly to occult and imaginary qualities, to shroud their ignorance under inconceivable termes."[33] He is a determinist as well. One who considers, he says,

> "the whole course of nature set on foot by God Almighty for this admirable work ... not stirring ... till he have fully examined and discussed what must necessarily follow out of such or such matter, in such or such circumstances, so and so tempered, and so and so wrought upon, will evidently discern that it is throughout impossible, any thing should happen in it otherwise than just what and how it doth."

In his whole account of the growth and resurrection of plants, he says, "There is nothing more clear, nothing more evident, nothing more rigorously true."[34] But his ontology includes a "balsamick Saline juyce"[35] "aereall and suplhureous parts of ... rectified spirit" an "aethereall or wild spirit" "aereall bodies" that subsist when a plant is destroyed,[36] matter and

31 Thorndyke(1931), vol. 7, pp. 210-11.
32 Hutchison (1982). Reprinted in Dear (1997), pp. 86-106.
33 Digby (1661), p. 47f.
34 Digby (1661), p.100.
35 Digby (1661), p. 58.
36 Digby (1661), p. 81.

form, and "a tincture extracted out of the whole plant ... dryed up into a kind of Magistery, full of Fire and of Salt," namely the seed.[37]

As John Henry has exhaustively shown, many formulations of the 1660s, 1670s, and 1680s – supposedly the period of the first conversion to mechanism – span the supposed gap between older conceptions of the world soul, the divine aether, the aerial balsam, and a corporeal particles.[38] The porosity of matter and its permeation by invisible streams of particles including fire atoms, light atoms, magnetic atoms, and others producing wonderful effects is almost universally noted by Boyle's contemporaries, and the entities involved are fitted into what might be considered semi-mechanical or paramechanical explanations.[39] Digby in 1661says that the natural action of fire is to "stream out from its Center on all hands in a continued floud of extreamly rarifyed atomes ..."[40] Henry Power hopes to see magnetic effluvia as well as atoms of light with the microscope.[41] Twenty years later, Thomas Willis describes "Spirits" as "Substances highly subtil, and Aetherial particles of a more Divine Breathing, which our Parent Nature hath hid in this Sublunary World, as it were the instruments of Life and Soul, of Motion and Sense, of every thing."[42] And Newton, in the *Principia* of 1687 suggests that matter is formed by a condensation of a vaporous ether, which gives successively, water, salt, sulphur, and terrestrial substances.[43] Comets, he concluded at the end of the *Principia* may be the source of life-giving exhalations.[44] Newton is not being in the least original when he suggests that "[o]ne use of matter is to admit menstruums easily into its pores in order to [generate] new mixtures by fermentation, putrefaction, corruption, and generation."[45]

Boyle's "Lucretian" interest in the beneficial and toxic effects of corporeal effluvia persisted throughout his career as a naturalist. In the early 1670s, he began to write tracts on subterranean, aerial and celestial effluvia, beginning with *Cosmicall suspicions* (1671), followed by *Essays of the strange subtilty, great efficacy, and determinate nature of effluviums* (1673) and *Some additional suspicions especially relevant to the hidden quality of air* (1674). The species-conception of corpuscles and its affinity for quasi- and paramechanical explanation are particularly evident in these treatises.[46] Plants and animals, as well as minerals, Boyle thinks, emit streams of invisible particles. "The common air we live in, and breathe, does always abound, and for many ages has been impregnated, with the copious magnetical effluvia of the earth."[47] Gems "emit streams and expirations" and curative stones and amulets held next to the skin work by the insinuation of minute corpuscles.[48] They are wrong who

---

37 Digby (1661), p. 44.
38 John Henry suggests Bacon's latent Paracelsianism or Neoplatonism as the sources of 17th century effluvial speculation in Henry (1994), in Hunter (1994), pp. 119-138. I do not mean to exclude these possibilities, but only to point to the robust importance of aerial effluvia in traditional Lucretian atomism.
39 See also Schaffer (1987), pp. 55-85.
40 Digby (1661), p. 12.
41 Power (1664), preface.
42 Willis (1681), p.3, quoted in Clericuzio (1994), n. 44, p. 90.
43 Newton (1687), II:542. See Dobbs (1973).
44 Newton (1687), p. 506.
45 Newton, Cambridge ms. 3970, quoted in Rattansi/McGuire (1966), pp. 108-143.
46 See Henry (1994).
47 Boyle (1744g), pp. 278-9.
48 Boyle (1744f), p. 328.

say that mercury imparts the virtue of killing worms to water through a kind of irradiation rather than by means of anything corporeal.[49] When a partridge or a hare walks on the ground, "effluxions" transpire out of its feet that the hounds can detect.[50] Mixed bodies make "little atmospheres about divers of them."[51] Around and above us, the air, as this term signifies the atmosphere, not "air" strictly speaking, which consists of springy particles, "abounds with vapours and exhalations;" these are the probable causes of endemical diseases. The earth emits from its bowels both ordinary and extraordinary exhalations, and these have a profound effect on human health. People can be killed in their sleep by the fumes of a charcoal fire in a newly-plastered room,[52] and Boyle repeats Lucretius's story of the lakes that poison birds as they fly overhead.[53] The air contains saline, sulphureous, arsenical, antimonial, etc. particles that may "produce this or that determinate disease."[54] Alternatively, "fossiles" may be "enriched with medicinal and fugitive salts and spirits" that are beneficial. Tin-miners, who apparently do not suffer the ghastly diseases of other miners, must benefit from the activity of salubrious particles that prolong life and health.[55] "... [I]n some places the air is observed to be much more healthy, than the manifest qualities of it would make one expect: and in divers of these cases I see no cause, to which such a happy constitutions may more probably be ascribed, than to friendly effluvia sent up from the soil into the air: which particles by promoting transpiration ... or by hindering the production, or checking the activity of morbific ferments; or by mortifying or disabling some noxious particles, that would otherwise infest their air ... may not a little contribute to keep the bodies of those, that live in that air ... in health."[56]

> "And ... I think it very possible, that diverse subterraneal bodies, that emit effluvia, may have in them a kind of propagative, or self-multiplying power ... I will not here examine, whether this proceeds from some seminal principle, which many chymists and others ascribe to metals, and even to stones ... or ... to something analogous to a ferment, such as in vegetables enables a little sour dough to extend itself through the whole mass; or such as, when an apple or pear is bruised in one part, makes the putrified part, by degrees, to transmute the sound into its own likeness."[57]

In short: Boyle inherits an older QC manner of thinking about particles and invisible corporeal effluvia that is not firmly tied to doctrines about permissible explanations.

Effluviums, he says, "at their first parting from the bodies, whence they take wing ... may retain as much of the nature of those bodies, as we have ascribed to them."[58] This point is important: the types of effluvia correspond to the types of natural kinds, animal, vegetable,

---

49 Boyle (1744f), p. 317.
50 Ibid., p. 319.
51 Boyle (1744h), p. 328.
52 Boyle (1744h), p. 322
53 Boyle (1744g), p. 279.Cf. Lucretius (1992), I:818ff.
54 Ibid. p. 280.
55 Boyle (1744g), p. 277.
56 Ibid., p. 276.
57 Ibid., p. 282.
58 Boyle (1744i), p. 333.

and mineral that we find on and below the earth, and Boyle consistently emphasizes the determinate and specific nature of the effluvia, criticizing Aristotle for having crudely distinguished only between vapours and exhalations, as though, he says, one were to divide all animals into the horned and the two-footed. There are also four-footed beasts and fishes, and many divisions and subdivisions even within those two: sea monsters, rhinoceroses, deer, elk, unicorns, eagles, nightingales, and so on. Were our sensories "sufficiently subtil and tender," Boyle says, they might perceive differences as marked as between differing sorts of birds: hawks, partridges, sparrows and hens.[59] Elsewhere, Boyle adduces the results of experiments to show that even "fixed and solid bodies such as metals are may by art be reduced into such minute corpuscles, that, without losing their nature and all their properties they may become parts of fumes, or perhaps of invisible vapours, or even of flame itself."[60]

In the *General History of the Air* (1692), Boyle decided that air consists of a mixture of three types of particles; first, "vapours or dry exhalations" from the earth, water, animals, vegetables and minerals; second, magnetical and light particles, or particles of celestial matter, and third, "elastical particles."[61] The action of these various particles is apparently due to "the congruity and incongruity of their bulk and shape to the pores of the bodies they are to act upon;" also to "the motions of one part upon another, that they excite or occasion in the body they work on, according to its structure."[62] But they also operate by their huge number, by their "penetrating and pervading nature," by their "celerity and other modifications of motion" and, more mysteriously, in interaction with "the more catholick agents of nature."[63] These include light, magnetisms, (sic) the atmosphere, gravity, and others.[64] Boyle describes his QC particles as "friendly," as capable of "disabling" ferments, as "operative," as having "natures," "fitnesses," and "powers." "A body may either consist of, or abound with, such corpuscles as may be variously associated with those of other bodies, and exceedingly disguised with those mixtures, and yet retain their own nature."[65] Boyle explains how butter can take on a rank taste when cows feed on certain weeds,

"[i]f it be considered how many, if I may so call them, elaborate alterations the rank corpuscles of this weed must have undergone in the various digestions of the cows, stomach, heart, breasts & c ... it will scarce be denied but that vegetable corpuscles may, by association, pass through divers disguises without losing their nature, especially considering that the essential attributes of such corpuscles may remain undestroyed ..."[66]

Since the appearance of Antonio Clericuzio's paper on this subject, there is a recognition that speciated particles were actually construed by Boyle as composites of non-speciated or barely-speciated primary particles.[67] What Boyle refers to as a "sulphureous particle" or an "arseni-

---

59 Ibid., p. 331.
60 Boyle (1744g), p. 297.
61 Boyle (1744j), p. 108 .
62 Boyle (1744h), p. 321.
63 Ibid.
64 Ibid., p. 328.
65 Boyle (1744e), p. 541; SPP, p. 199.
66 Ibid, p. 541; SPP, p. 200.

cal particle" is not an indivisible unit, but a secondary "molecule" composed of some arrangement of genuinely primary particles or *minima naturalia*.[68] "[T]here are also multitudes of corpuscles, which are made up of the coalition of several of the former minima naturalia, and whose bulk is so small, and their adhesion so close and strict, that each of these little primitive concretions or clusters ... remain entire in great variety of sensible objects ... And these are, as it were, the seeds or immediate principles of many sorts of natural bodies, as earth, water salt, & c."[69] These corpuscles can be decomposed in principle into reductive atoms, so that transmutation is a theoretical possibility, but Boyle realized that it is very hard if not impossible to break or dissolve some chemically active substances.[70]

Did Boyle conceive the working of these corpuscles to be "mechanical" in the sense required by RA? If individual atoms have only primary qualities, so, too, one might suppose, do concretions of them, whether they are stuck, hinged, or hooked together. And indeed Boyle seems to realize that he has no real warrant for talking about his corpuscles as qualitatively endowed, because, a few pages later, he reaffirms the view that "body and body being but a parcel and a parcel of universal matter mechanically different, either parcel may successively put on forms in a way of circulation."[71] But this putting on of forms proceeds from the body's "mechanical affections." So the ascription of any functional powers to corpuscles is a kind of *facon de parler*, since these qualities are merely dispositional qualities and are not intrinsic to the effluvial particles. A key of a particular shape can, by motion alone, open a lock of a particular shape, that is, create some new effect ... an open lock. So we might conceive the corpuscles of determinate effluvia as like so many differently shaped keys, able to open so many differently shaped locks. But there is an obvious tension between the QC view that the role of the naturalist is to *experiment* with substances endowed with characteristic dispositions and powers that happen to be particulate, and the RA view that the role of the naturalist is to *hypothesize* about variously shaped particles and how observed effects might be explained by various key-in-lock mechanisms.[72]

The tension between RA and QC emerged in Boyle's treatment of respiration. Though he accepted the idea that the air was a mixture of qualitatively-distinct effluvia, his experimental discipline took him only so far. His propensity to see aerial effluvia dichotomously as either

---

67 Clericuzio (1990), p. 580 ff.

68 Similarly, for Lucretius, what is called an "atom" is not always an indivisible particle; it can be a composite of simpler forms:
Suppose that atoms consist of three minimum parts, or enlarge them by a few more, When by fitting on parts at the top or bottom and transposing left and right you have exhausted every shape that can be given to the whole body by all possible arrangements of the parts, you are obviously left with no means of varying its form further except by adding other parts ...
His point is that unless the atoms are allowed to grow impossibly large, their forms are limited by the limits of their "minimum parts."

69 Boyle (1744b), pp. 470-1; SPP, pp. 41-2.

70 Clericuzio (1990), p. 582.

71 Boyle (1744e), p. 542; SPP, p. 202.

72 A passage discovered by John Henry amongst Boyle's unpublished manuscripts reads as follows: "[It would] be backward to reject or despise all explications that are not immediately deduced from the shape, bigness, and motion of atoms or other insensible particles of matter ... [for those who] pretend to explicate every phenomenon by deducing it from the mechanical affections of atoms undertake a harder task than they imagine." Quoted in Henry (1994), p. 123. Henry argues that Boyle envisioned as a legitimate form of explanation "something which is rather less compatible with ... all standard accounts of the mechanical philosophy." (Henry (1994), p. 121.)

nourishing (and likely of celestial origin) or toxic (and likely of subterraneal origin) dominated his interpretations. At the same time, he wanted to understand nourishment and poisoning
"mechanically." Boyle, it appears, never accepted the combustion theory of respiration assembled by his contemporary Richard Lower and his assistants Robert Hooke and John Mayow. For the combustion theory was simply irreducible to mechanical form. The "fire particles" that Mayow considered to play the principal role in respiration were not merely agitated
particles of ordinary matter, as Boyle held, but qualitatively different from other particles and
not composed of them.

3. The Experimental Subjection of the Corpuscle: Boyle's Mixed Reaction.

In the 1660s, dispute still reigned – and would even continue to reign through the time of
Haller – over whether the function of breathing was mechanical – to force the blood around
the body by pressure (Breathing did appear to assist the passage of the blood from the right to
the left ventricle.[73]), or to cool the body through the expulsion of heat particles, or to purify it
by expelling some kind of excrement – or was nutritional – to feed the body with a *pablum
vitae*.[74] It was debated whether the death of animals and the extinction of flames in closed
vessels was caused by their suffocation by exhaust vapours, or the loss of mechanical "spring"
in the exhausted air, or starvation from a lack of something nutritive in the air.[75] Paracelsus
had referred to "an astral balsam, an invisible fire, an included air, and a tingeing spirit of
salt" as the principle of life.[76]

Boyle's *New Experiments Physico-Mechanical Touching the Spring of Air* published in
December of 1659 summarized a sequence of experiments on animals in exhausted receivers.
Why did these animals die? One possibility that occurred to him was that the fall in air
pressure caused ethereal matter to penetrate the pores of the glass to take its place leading to
an expansion in the blood that killed the animal. Another was that air was needed to carry
away waste vapours from the vital flame in the heart. Boyle was never able to settle the issue
between a "poisoning" theory and a "nourishment" theory, experimentally but his preference
for the latter was evident. "There is some use of the air," he speculated, "that we do not yet so
well understand, that makes it so continuously useful to the life of animals. *Paracelsus* indeed tells us, That as the stomach concocts meat, and makes part of it *useful* to the body,
rejecting the other part; so the lungs consume part of *the air* and proscribe the rest. ... [I]t
seems we may suppose, that there is in the air a little vital quintessence ... which serves to the
refreshment and restauration of our vital spirits."[77] Air, he decided, is not a "simple and

---

73 Frank (1980), p. 100.

74 See Douglas McKie's useful survey: McKie (1953) in Underwood (1953), vol. 1, pp. 469-87.

75 Locke, a figure strongly associated with structural corpuscularianism, took an active interest in respiration
and effluvia. He saw the function of the former as the "carrying away those vaporous excrements of ye bloud
wch are usually called fuliginous" which find a "fit receptacle in ye pores or ye aire," and which beyond that
"served for the ventilation of the blood through the admixture of aerial particles." (Quoted in Frank (1980),
p. 186.)

76 Quoted in Debus (1964), p. 45

elementary body, but a confused aggregate of effluviums from such differing bodies, that ... perhaps there is scarce a more heterogeneous Body in the world."[78]

> "It seems to me not absurd to suspect, that the subtil, but corporeal emanations [of the sun, planets, and stars] may ... reach to our air, and mingle with those of our globe in that great receptacle or rendezvous of celestial and terrestrial effluviums, the atmosphere."[79]

He thought that these emanations might have a nature "quite different from those we take notice of here about us," and might operate "after a very different and peculiar manner."[80]

Meanwhile, Boyle along with many of his contemporaries had grown deeply interested in saltpetre or nitre. In his above-cited *Disourse on the Vegetation of Plants* read to the Royal Society in 1661, Kenelm Digby described a nitrous salt that "foecundatheth the Aire ... and gave cause to the Cosmopolite to say, there is in the Aire a hidden food of life. Such Aires as are the most impregnated with this benigne fire, are healthfull to live in. Others, which about with Earthy exhalations or Marshy vapours, and have little balsamick Salt in them, are as unsound."[81] This salt was "... the food of the Lungs, and the nourishment of the Spirits." It is a "Universal Spirit" that is of the same nature as Gold and can be extracted from a "boiled decoction of crayfish, which when combined with the distillate and incubated in a moist place, would give rise to many animals."[82] Why was nitre such an interesting substance? First, there was an empirical association between nitre, growth, and preservation. Dung, sprinkled with urine and left in the fields irradiated by the sun, turns to nitre, which is a powerful fertilizer, able to promote rapid vegetation. Nitre preserves meat, turning it bright red. As a component of gunpowder, it seems to confer flammability.[83] Boyle, as he reports in his 1661 essay on saltpetre, tried dissolving and reconstituting nitre into pure crystalline form, and came to the conclusion that it was composed of two types of corpuscle, one of which was acidic and volatile, the other of which was fixed and alkalizate.[84] He found that nitre crystals formed seemingly only under the influence of air, suggesting that "whatever the Air hath to do in this experiment, I dare invite you to believe, that it is so enrich'd with variety of streams from terrestrial (not here to determine whether it receive not some also from celestial) bodies that inquiring into the further uses of it ... may very well deserve your curiosity."[85]

---

77 Boyle (1744k), p. 69.
78 Boyle (1744l), p. 463
79 Ibid.
80 Ibid.
81 Digby (1661), pp. 64-5.
82 Frank (1980), p. 127.
83 Great interest in novel methods for the manufacture of saltpetre and the awarding of commercial patents fostered investigation into this interesting and potentially lucrative substance from the mid-1640s. Speculation on the aerial nitre and its role in burning, respiration, and animal warmth and motion began with Ralph Bathurst in 1647. See Guerlac (1954), p. 248.
84 Frank (1980), p. 124. Clericuzio (1990), p.576, argues that Boyle rejected mechanical account of nitre and describes his controversy with Spinoza.
85 Boyle (1744m), p. 237.

In 1665, Robert Hooke in his chapter "Of Charcoal" in the *Micrographia,* argued that air acted as a solvent of sulphureous bodies, causing them to flame, in virtue of the air's containing something that was or was like something fixed in saltpetre. And, in 1667, Hooke struck the first experimental blow against the mechanical theory of respiration by pumping air directly into the multiply-punctured lungs of a dog; the dog remained alive even with its chest immobilized.[86] Richard Lower who had known since 1658 that arterial and venal blood differ in colour,[87] had shown that blood turns bright red when air is pumped through it. In 1670, Boyle argued on the basis of an experiment with a pressure gauge that air can become unfit for respiration without losing any of its pressure, and this was taken as evidence against Hooke's theory which, in any case, had not attracted much following in the Royal Society. Boyle became, for a time, less certain that air was necessary for the production and conservation of flame, and he tried to show that burning could take place under water. He found that a mixture of gunpowder, charcoal, sulfur and saltpetre would do so. But he thought of a rival hypothesis: namely that salt-petre might contain "little aereal particles between the very minute solid ones which those corpuscles are made up of." He was also puzzled by the observation that a bird continued to live for a time in a closed vessel in which a candle had gone out. Was this because the candle and the bird needed different nourishment, or because the "more temperate flame" of the bird needed less? Boyle – the "diffident naturalist" – seemed unwilling to pronounce definitely for the nutritional-nitre theory.

By 1668, Boyle's sometime assistant, John Mayow had synthesized reasonably conclusive evidence that the function of breathing was not macromechanical but micromechanical, that breathing extracted something from the air that was taken up by the "pores" of the lungs as revealed by the microscope, that the source of the warmth of the body was the blood, not the heart, and that the generation of this warmth was an interaction between blood particles and particles of what he called "nitro-aerial spirit."[88] Contrary to Boyle's negative results, Mayow found that something in the air is used up by respiration and burning alike: water rises in a closed vessel in which a candle or animal is allowed to extinguish itself.[89] He noted that exhausted air that has lost some of its elasticity and will not support life still has a mechanical effect on the lungs. He drew on Richard Lower's finding that artificially insufflated blood turns from dark to bright. (Mayow found significant the comparable redness of spirit-saturat-

---

86 Hooke (1667), p. 539.
87 Patterson (1931), p. 60.
88 Mayow sharply divides historians into supporters, including Thomas Beddoes, his English rediscoverer, R. T. Gunther, and Robert Frank, and detractors like Richard Westfall and T. S. Patterson who regard him as a confused obscurantist. Patterson maintained in that there was nothing in Mayow that was not previously to be found in Boyle, Hooke, and Lower. Mayow's reputation was defended by Partington (1956), pp. 217-230. On balance, it seems that although Patterson clarified the state of knowledge about respiration, combustion, and the role of the blood in England pre-Mayow, he failed to appreciate Mayow's role in synthesizing available knowledge, organizing and presenting it, and devising simple but subtle experiments. Mayow's theory of nitroaerial particles was well-known in the 18th century (Priestley and Lavoisier possessed copies of his work.), and, according to Partington, was variously admired, reviled, and treated with cautious skepticism. Frank (1980), p. 279, points out that he was widely read on the continent, though his influence faded in England. Thomas Beddoes, who issued a precis in 1790, complained that he had been unable to find a copy of Mayow's book in the public library. He was convinced that Mayow was a genius on the level of Newton. (Beddoes (1790).)
89 Mayow (1674), p. 36. There is no modern edition of Mayow's works. I have mainly relied on the German translation Mayow (1901), as the English translation of the Edinburgh Alembic Club of 1907 is not easy to come by.

ed blood and the redness of flame) and the observation that fresh arterial blood froths in a vacuum while darker blood does not. He noted that the fetus in the womb does not need to breathe, but,once having drawn breath, it can no longer survive on the supply of maternal blood. These observations indicated to Mayow that breathing served to introduce air, which he conceived as a combination of – "nitro-aerial" particles, which he described as "very subtle, movable, ethereal particles of a firey nature," with the stiff, interlaced twiglike particles with which it was normally bound, into the blood, causing a disposition to froth and brightening its colour. In the blood, the nitro-aerial particles were dislodged by friction against the blood particles, and rubbed up against particles of the body's sulfur, producing warmth and a vital fermentation. In a candle flame, nitroaerial particles are dislodged by heated sulfur and given off as heat or light.[90]

The reception of the Hooke-Mayow's theory by the Royal Society was poor. However, neither the 1668 review of the first edition of Mayow's *De respiratione*, nor the 1674 review of the *Tractatus quinque Physico-Medici* published in the *Philosophical Transactions* can be regarded as hostile. The second reviewer (perhaps Henry Oldenburg) reports that according to the author, "*Sal-Niter* is made up of a threefold Salt; whereof one the most Active, deduces its origin from the Air, and is of an Ethereal and Igneous nature; and this by its architectonical power forms to it self out of a Terrestrial matter a Salin vehicle ..."[91] He finds that air deprived of Nitro-aerial particles "looseth its Springy vertue" concluding that "something Aerial, absolutely necessary to life, passeth into the Blood of Animals by means of their Respiration, whose necessity therefore he cannot acknowledge to arise from thence only, that thereby and by the Motion of the Lungs the mass of Blood may be communicated as some have asserted."[92] Mayow's assertion that the sun is a "vast *Chaos* of nitro-aerial corpuscles, wheeled about by a perpetual and very swift rotation"[93] does not arouse comment. Nor does his notion that air is the blood of the macrocosm, that the macrocosm takes in nitro-aerial spirit, and exercises a kind of respiration.[94] Willis repeats this ancient idea seven years later. Even Mayow's assertion that the sensitive soul is "a more divine *aura*, endow'd with sense from its first creation, and co-extended to the whole world; a small portion of which being contain'd in a duely disposed subject exerts such functions, as we see and admire in the bodies of Animals"[95] is reported without the expression of skepticism.

Yet Mayow's reputation in his own time was negligible. Oldenburg wrote to Boyle "I heare some very Learn'd and knowing men speake very slightly of ye Quinque Tractatus of J. M.; and a particular friend of yrs and mine told me yesterday, yt, as farr as he had read him, he would shew to any impartiall and considering man more errors than one in every page."[96] The second review gives some clues as to why Mayow was held in such low esteem, noting the discrepancy with the universal matter theory of the "Noble R. Boyle's" as outlined in his *Excellency and Grounds of the Mechanical Hypothesis* of 1674. Mayow, according to the reviewer, "rejecting the opinion of those, that will have Fire producible by the subtile and

---

90 Hall (1969), vol. 1, pp. 329-30.
91 Anonymous (1674), p. 104.
92 Anonymous (1674), p. 105.
93 Anonymous (1674), p. 106.
94 Anonymous (1674), p. 106.
95 Anonymous (1674), p. 112.
96 Oldenburg (1965-83), XI:50, No 2514, 10 July 1674.

briskly mov'd parts of *any* matter, and declaring on this occasion his dissent from those Philosophers, that deduce all Effects of Nature from the same Uniform Matter, and the various Modifications thereof, which he thinks inconsistent with the *Phaenomena* of Fire, not at all, *in his opinion*, producible but by a certain determinat kind of particles, such as he calls Nitro-aerial. This he endeavors to prove by divers Experiments."[97]

The opening of Mayow's treatise on saltpetre, the first essay of the *Tractatus quinque* of 1674 states that "the surrounding air which escapes our eyes through its fineness, which appears to be like empty space, contains a substance with a connection to saltpetre. It is a spirit which is connected with the phenomenon of life, with burning, and with fermentation."[98] Thus far, Mayow was not arguing anything particularly controversial. But he goes on to criticize the reductive atom. Mayow sought what he called a "middle way" between that of the peripatetics who, he said, called in a new being to explain every type of phenomenon, and the Neotericks (he meant Boyle), who refer all phenomena to the "form, movement and rest of one type of matter."[99] Where Boyle maintained that diversity could arise only from complex structure, common matter "being in its own nature but one" and separated into corpuscles, and that the unitary nature of matter made general transmutation possible, Mayow argued for diversity on the simplest level: "I believe that the particles of matter are distinguished from one another in their form and thickness in such a way that no force of nature can transform one into another."[100] This conclusion was, I venture, unacceptable to Boyle and Oldenburg. For Boyle, although experimentation could never establish the truth of RA, it could not refute it either. Mayow's intimation that he *had* refuted RA experimentally threatened the foundations of Boyle's scheme. Boyle's tolerance for paramechanical explanations for the nourishing effect of celestial effluvia was considerable, but to be directly contradicted was another matter.

4. Some Historiographical Conclusions

Oliver Ellis in his *History of Fire and Flame* describes how the atmosphere appeared to our ancestors. It was "the great receptacle of things that disappeared from sight." The air retained the species of everything – it was a "sea of things invisible,"[101] replete with vapours, dews, powders, dusts, rays, tinctures, and influences, and forms.[102] Before the 17th century, these dwellers of the air belonged to a realm of discourse that was invisible-poetic; subsequently some of them would make the transition into the realm of the invisible-hypothetical by entering the space of practice with its experimental apparatuses – air pumps, bellows, closed vessels, scalpels, sutures, clamps. Other air-dwellers, like visible species and ghosts, failed to make the transition, even though attempts to bring ghosts into laboratory persisted, and at-

---

97 Anonymous (1674), p. 103.
98 Mayow (1901), p. 1.
99 Mayow (1901), p. 13.
100 Mayow (1901), p. 13.
101 Ellis (1932), p. 191.
102 Ellis (1932), p. 228.

tempts to bring parapsycholgical thoughts and influences out of the "air" and into the space of experimental practice still persist.

How far had the installation of the invisible material corpuscle as a scientific entity come by the end of the 17th century? It is fair to say that atomism's associations with hedonism and mortalism continued to foment moral-theological anxiety at the end of the 17th century. Indeed, despite the best efforts of Descartes, Gassendi, and Boyle, the Epicurean threat to theology never really diminished. Lucretius had repeatedly emphasized that believing in the atoms would take away the fear of death and the fear of Hell, and bring men off their barbarous superstitions. But moral philosophers in the 17th century were unanimous that ordinary men, at least, should not be freed of the fear of death and Hell, and if superstitions were needed to keep them in line, so be it. Spinoza tried to preserve something of the traditional theory of divinity and immortality in his otherwise Epicurean *Ethica* of 1677; Leibniz fought against RA for moral reasons on metaphysical grounds, as Berkeley fought against it on epistemological grounds. Boyle's *The Christian Virtuoso*[103] is a framing exercise, comparable to Descartes's, for making reputable the experimental-hypothetical science of manifest qualities produced by matter-in-motion, and the Boyle Lectures were established after his death to establish the compatibility of Epicurean naturalism with Scripture.

Although Boyle remained personally conflicted, he did separate systematic talk about corpuscles from systematic talk about morals and religion, and it was increasingly possible, thanks to his extended nontheological discussions of corpuscular activity, to take this science in its own terms, without worrying about the moral character of Epicureans. Does the material corpuscle enter, with Boyle, into sober, restrained, unpoetic discourse? From one perspective, the hermetic and corpuscularian conceptions of the aerial nitre allegedly represented by Digby and by Boyle are not very distinct. Boyle and Digby were both QC corpuscularians. Both were interested in palingenetical reconstitution experiments. Boyle certainly believed in quasi-vitalistic seeds, including mineral seeds and seeds of epidemic diseases.[104] Both believed that invisible bodies of terrestial and celestial origin had profound toxic and beneficial effects on human and animal physiology. Yet, for Digby, the aerial nitre is considerably more "admirable" than it is for Boyle. It furnishes the essence of gold, it induces magical Ovidian transformations; it is a gift of heaven and belongs to the realm of wonders.[105] It is poetically imagined, one wants to say, not scientifically hypothesized. Boyle invests celestial effluvia with "very different and peculiar" properties, but he is demonstrably less associationist in his thinking about nitre, or at least in the expression of his ideas about nitre.

Both Boyle and Digby performed experiments designed to reveal the properties of the aerial nitre. The difference between them is that Boyle's experiments were better designed, more pertinent to the questions being asked, performed before witnesses, and carefully reported. But Mayow – who belongs to the tradition of admirable discourse where the nitre is concerned – planned even better experiments. When marking distinctions that are relevant to understanding the development of the theory of respiration, it is misleading to set Mayow and Digby on one side, as proponents of the hermetic nitre, and Boyle on the other as a proponent

---

103 Boyle (1690).
104 His two treatises on "Seminal principles" and "Seeds of Plants, Animals, & c." were unfortunately missing when his MSS were collected.
105 Digby (1661), pp. 68-71.

of the scientific nitre. The three parameters I have singled out – detachment from morality, sober discourse, and experimental control are independent of one another.[106]

Clearly, it is tempting to regard corpuscular effluvia *either* as important constituents of the mechanical philosophy *or* as residues of an older qualitative chemistry, without recognizing that there is nothing remarkable about their being both, that we are not in the presence of two opposing mentalities or paradigms. Westfall appears to recognize the consistency between corpuscularian mechanism and Paracelsian ideas when he notes that Mayow's work on respiration "showed how easily the mechanical philosophy could function to sustain a traditional point of view in chemistry."[107] But the critical tone of the remark seems insufficiently motivated.[108] The implication is that Mayow, as an experimentalist and a corpuscularian, ought to have abandoned his belief in a world soul and fire particles. But there is nothing incongruous in Mayow's theory of the vivifying corpuscular nitre. Mayow's allegedly "traditional" beliefs did not hold him up, and there is little basis for the suggestion that Oxford physiology failed on account of incoherence in its foundations; it is more likely that, just as no one could think what to do next with the microscope after 1690, no one could think of more good experiments to do to clarify the relationship between atmospheric effluvia, breathing, the circulation of the blood, and animal heat.[109]

Conversely, Boyle's "modern" commitment to RA assuredly did not help him to understand respiration. Thomas Kuhn properly remarks that "a retrospective glance at the history of seventeenth and early eighteenth century chemistry suggests that the true progenitors of Lavoisier's chemical revolution were necessarily among Boyle's opponents."[110] Yet Boyle's acceptance of the subtlety, efficacy, and determinate nature of effluvia ought, it seems, to have brought him closer to the theory of respiration-as-combustion than they did. R. S. Westfall remarks that Boyle's commitment to the mechanical philosophy – which implied the possibility of the transmutation of anything into anything else – appeared to thwart the most promising aspects of his chemistry – his intimation that there were chemical elements, determinate species of bodies.[111] And this diagnosis seems just as accurate as Kuhn's.

---

106 Frank thus overstates – though only slightly – the case when he contrasts the "hermetic" nitre of Kenelm Digby with Boyle's "corpuscular nitre":
One can understand why Boyle's silence on such a hermetick niter was so profound. Any body found both in the air and in the earth, that fertilized plants and nourished animals, that could be transmuted into gold or reconstituted into small animals, was more in the nature of a quasi-vitalistic seed than a delimited substance definable in corpuscular terms and manipulatable by chemical experiments. Boyle was not so enthusiastic for chemistry as to see an Aristotelian system of vague concepts and jargon replaced by one equally fanciful ... Indeed the entire import of Boyle's work on saltpeter in the years after 1655 was, on the one hand, to examine its properties by a continuing programme of focused experimentation, and on the other to understand those properties by conceiving of them in corpuscular terms." (Frank (1980).)
107 Westfall (1971), p. 73.
108 Westfall complains that "[a]t least three different elements as we know them were jumbled together in Mayow's nitroaerial spirit: As the agent of combustion and life, it referred to oxygen; as the agent of vegetable life, to nitrogen; as the acidic spirit, to the hydrogen ion. Chemistry in the 17th century was not sophisticated enough to make these distinctions ..." (Westfall (1971), p. 73.)
109 Mayow's results failed to move anyone "either individually or collectively, to ingenious designs of apparatus and experiments." (Frank (1980), p. 282.)
110 Kuhn (1952), repr. in Dear (1997), pp. 212-236, p. 236.
111 Westfall (1971), p.79.

Neither observation gives as much support to the two conversions theory – one to, the other away from, RA and universal mechanism – as might appear at first glance. There is no denying the embarrassment with which some 18th century chemists, and most modern historians too, looked or look back on the figured particles and the "Cartesian" explanations of the former era. But Newton was not the first to recognize the limitations of mechanism, vivifying the "dead," "inert" RA corpuscle and making the subsequent history of science possible.[112] English experimental philosophers were little interested in the RA corpuscle, endowed with primary qualities alone. The QC corpuscle predominates in their writings, and most naturalists – Digby, Hooke, Willis, Mayow – wrote enthusiastically about this corpuscle and its curious, admirable, or frightening properties, with Boyle devoting hundreds of pages to the toxic and salutary qualities of the aerial effluvia. In short: early modern corpuscularianism did not need enrichment because its QC form, an authentic strand of ancient atomism, had always been vivid in the minds of Lucretius's many readers. And if Boyle's prejudice in favour of RA and rigid lock-and-key mechanisms induced him to reject a promising account of respiration, the wider influence of that prejudice has yet to be demonstrated.

References

Anonymous (1674): "An accompt of two Books," *Philosophical Transactions* 9 (1674), 101-113.

Aristotle: *Generation and Corruption*, 1 10 328 a 19-31.

Beddoes, Thomas (1790): *Chemical Experiments and Opinions. Extracted from a Work published in the Last Century*. Oxford: Clarendon.

Boyle, Robert (1690): *The Christian Virtuoso, shewing, that being addicted to experimental philosophy, a man is assisted, rather than indisposed to be a good Christian*. London.

Boyle, Robert (1744a): *Works*, 5 vols. London.

Boyle, Robert (1744b): "The Origin of Forms and Qualities," in Boyle (1744a), vol. 2.

Boyle, Robert (1744c): "A Requisite Digression Concerning Those that would Exclude the Deity from Intermeddling with Matter," in Boyle (1744a), vol. 1.

Boyle, Robert (1744d): "Excellency and Grounds of the Mechanical Hypothesis," in Boyle (1744a), vol. 3.

Boyle, Robert (1744e): "Some Physico-theological Considerations about the Possibility of the Resurrection," in Boyle (1744a), vol. 3.

Boyle, Robert (1744f): "Of the Strange Subtilty of Effluviums," in Boyle (1744a), vol. 3.

Boyle, Robert (1744g): "An Experimental Discourse of some unheeded Causes of the Insalubrity and Salubrity of the Air," in Boyle (1755a), vol. 4.

Boyle, Robert (1744h): "Of the great Efficacy of Effluviums," in Boyle (1744a), vol. 3.

Boyle, Robert (1744i): "Of the Determinate Nature of Effluviums," in Boyle (1744a), vol. 3.

Boyle, Robert (1744j): "The general history of the air, designed and begun," in Boyle (1744a), vol. 5.

Boyle, Robert (1744k): *New Experiments. Physico-Mechanical Touching the Spring of Air*, in Boyle (1744a), vol. 1.

Boyle, Robert (1744l): "Suspicions about some hidden qualities in the air," in Boyle (1744a), vol. 3.

Boyle, Robert (1744m): "A physico-chemical essay ... touching the differing parts and redintegration of salt-petre," in Boyle (1744a), vol. 1.

Clericuzio, Antonio (1990): "A Redefinition of Boyle's Chemistry and Corpuscular Philosophy," *Annals of Science* 47 (1990), 561-589.

Clericuzio, Antonio (1994): "Carneades and the Chemists," in Hunter (1994), 79-90.

Dear, Peter (ed) (1997): *The Scientific Enterprise in Early Modern Europe: Readings from Isis.* Chicago: University of Chicago.

Debus, Allan (1964): "The Paracelsian Aerial Nitre," *Isis* 55 (1964), 43-61.

Digby, Kenelm (1661): *A Discourse Concerning the Vegetation of Plants.* London.

Diogenes Laertius (1938): *Lives of Eminent Philosophers*, vol. 10, in Inwood/Gerson (1994).

Dobbs, B. J. T. (1973): *The Foundations of Newton's Alchemy.* Cambridge: Cambridge University Press.

Ellis, Oliver (1932): *The History of Fire and Flame.* London: Simpkin and Marshall.

Frank, Robert (1980): *Harvey and the Oxford Physiologists: Scientific Ideas and Social Interaction.* Berkeley and Los Angeles: University of California Press, 1980.

Geber: *Summa perfectionis.*

Glanville, Joseph (1665): *Scepsis Scientifica.* London.

Guerlac, Henry (1954): "The Poets' Nitre, Studies in the Chemistry of John Mayow," *Isis* 45 (1954), 243-255.

Hall, T. S. (1969): *History of General Physiology*, 2 vols. Chicago: University of Chicago.

Henry, John (1986): "Occult Qualities and the Experimental Philosophy: Active Principles in Pre-Newtonian Matter Theory," *History of Science* 24 (1986), 335-381.

Henry, John (1994): "Boyle and Cosmical Qualities," in Hunter (1994), 119-138.

Hooke, Robert (1667): "An Account of an Experiment made by M. Hook, of Preserving Animals alive by Blowing through their Lungs with a Bellows," *Philosophical Transactions* .2 no. 28, (1667), 539.

Hunter, Michael (ed) (1994): *Robert Boyle Reconsidered.* Cambridge: Cambridge University Press.

Hutchison, Keith (1982): "What Happened to Occult Qualities in the Scientific Revolution?," *Isis* 73 (1982), 233-253. Reprinted in Dear (1997), 86-106.

Inwood, Brad, and Gerson, L. P. (eds) (1994): *The Epicurus Reader.* Indianapolis/Cambridge: Hackett.

Kim, Sik Yung(1991): "Another Look at Robert Boyle's Acceptance of the Mechanical Philosophy," *Ambix* 38 (1991), 1-10.

Kuhn, Thomas (1952): "Robert Boyle and Structural Chemistry," *Isis* 43 (1952). Reprinted in Dear (1997), 212-236.

Locke, John (1975): *An Essay Concerning Human Understanding*, ed. P. H. Nidditch. Oxford: Clarendon.

Lucretius (1992): *De Rerum Natura.* Translated by W. H. D. Rouse. Cambridge, MA: Harvard University Press.

Luethy, Christoph (1997): "Thoughts and Circumstances of Sebastien Basson. Analysis, Micro-History. Questions," *Early Science and Medicine* 2 (1997), 1-73.

Mayow, John (1674): *Tractatus quinque.*

Mayow, John (1901): *Untersuchungen ueber den Salpeter und den salpetrigen Luftgeist, das Brennen und das Athmen*. Translated and edited by F. G. Donnan. Leipzig.

McKie, Douglas (1953): "Fire and the Flamma Vitalis: Boyle, Hooke, and Mayow," in Underwood (1953), 469-487.

Meinel, Christoph (1988): "Early Seventeenth-Century Atomism: Theory, Epistemology, and the Insufficiency of Experiment," *Isis* 79 (1988), 68-103. Reprinted in Dear (1997), 176-211.

Newman, W. R. (1996): "Robert Boyle's Corpuscular Philosophy," *Annals of Science* 53 (1996), 567-585.

Newton, Isaac (1687): *Principia mathematica*.

Oldenburg, Henry (1965-83): The *Correspondence of Henry Oldenburg,* 13 vols. Madison: University of Wisconsin.

Partington, J. R. (1956): "Some Early Appraisals of the Work of John Mayow," *Isis* 47 (1956), 217-230.

Patterson, T. S. (1931): "John Mayow in Contemporary Setting," *Isis* 15 (1931), 47-96.

Power, Henry (1664): *Experimental Philosophy, In Three Books*. London.

Rattansi, P. M. and McGuire, J. E. (1966): "Newton and the Pipes of Pan," *Notes and Records of the Royal Society of London* 21 (1966), 108-143.

Schaffer, Simon (1987): "Godly Men and mechanical Philosophers: Souls and Spirits in Restoration Natural Philosophy," *Science in Context* 1 (1987), 55-85.

Sennert, Daniel (1637): *Hypomnemata physica*.

Stewart, M.A. (ed) (1991): *Selected Philosophical Papers of Robert Boyle*. Indianapolis: Hackett.

Thorndyke, Lynn (1931): *A History of Magic and Experimental Science*, 8 vols. New York: Columbia University Press.

Underwood, E. A. (ed) (1953): *Science, Medicine and History: Essays on the Evolution of Scientific Thought and Medical Practice written in honour of Charles Singer*, 2 vols. Oxford.

Westfall, Richard (1971): *The Construction of Modern Science: Mechanisms and Mechanics*. Cambridge: Cambridge University Press.

Willis, Thomas (1681): *The Remaining Medical Works*. London.

Wolfgang Neuber

Systematische und kasuistische Wissensordnungen.
Mnemotechnische Prozesse im 17. Jahrhundert[1]

Die folgenden Ausführungen gehen am Beispiel der ars memorativa (Mnemonik, Gedächtniskunst) der Frage nach, wie sich systematische und kasuistische Wissensordnungen in der Frühen Neuzeit zueinander verhalten. Der Ausgang der Modellkonkurrenz ist bekannt: Die Systematik hat sich, nicht zuletzt durch die wissenschaftstheoretische, soziale und politische ordo-Fixierung des 17. Jahrhunderts, gegenüber der Kasuistik weitgehend durchgesetzt. Es geht indessen nicht darum, ohnehin bekannte Resultate historischer Ausdifferenzierung paradigmatisch und in teleologischer Absicht nochmals durchzuhecheln. Es geht vielmehr um die Frage, in welcher Weise sich der Selektionsprozeß vollzog, und um die Beschreibung jener funktionalen Residuen, die jenem Modell, das sich schließlich als geschichtlicher Sieger erweisen sollte, wenigstens zeitweilig zur Verfügung standen.

Simonides von Keos, der legendäre Begründer der ars memorativa, soll dem athenischen Feldherrn und Bezwinger des Xerxes, Themistokles, angeboten haben, ihn seine Kunst zu lehren. Darauf habe dieser geantwortet, „er wolte lieber eine Vergeßkunst lernen: Dann / sagt er / es gedencket mir mehr / als mir lieb ist / vnd kan nicht vergessen / was ich gerne vergessen wolte."[2]
    Themistokles ist damit das singuläre Gegenbild zu der langen Reihe derjenigen, die als stolze Helden des Gedächtnisses ihren Platz in der abendländischen Literatur seit der Antike gefunden haben: Pythagoras, Charmadas, Metrodoros von Skepsis, Simonides von Keos, Kyros, Mithridates VI. Eupator, Dionysos König von Pontus, Theodektes, Kineas, Hortensius, L. Licinius Lucullus, der Rhetor Lucius Annaeus Seneca, der Declamator Marcus Porcius Latro, Scipio Africanus, C. Iulius Caesar, Simplicius (der Mitschüler Augustins), Papst Gregor III., der Wüstenvater Didymus, die Hl. Gertrud, Pico della Mirandola, Girolamo Cardano, Blaise Pascal und viele andere mehr. Im Altertum dominieren unter den memoria-Athleten die Staatsmänner, Rhetoren und Philosophen, im Mittelalter Heilige, Asketen und Kirchenlehrer, in der Frühen Neuzeit sind es vor allem humanistische Gelehrte, Polyhistoren, Juristen, Astronomen und Mathematiker.[3] Die meisten, wenngleich nicht alle dieser Wundermänner des Gedächtnisses, verfügen über eine exorbitante memoria naturalis, die keiner mnemonischen Unterstützung bedarf. Vom Herrscher, dessen überragende Gedächtnisleistung als Naturgabe ihn erst herrschaftstüchtig macht bzw. ihn zur Ausübung von Herrschaft legitimiert,[4] ausgehend, sind auch die sonstigen Gedächtnisheroen weitestgehend nicht einer ars memorativa bedürftig.

---

1 Die Vortragsform der folgenden Ausführungen wurde beibehalten und um die Literaturnachweise ergänzt.
2 Garzoni (1641), zit. nach Berns/Neuber (1998), S. 41.
3 Vgl. Ernst (1997), S. 105.

Wäre dies allerdings der Normalfall der conditio humana, dann hätte es niemals eine Mnemonik gebraucht, um die Insuffizienz des natürlichen Gedächtnisses zu kompensieren. Diese Insuffizienz wird nicht erst seit christlichen Zeiten als ein durch den Sündenfall selbstverschuldetes Defizit des Menschen wahrgenommen, sondern bereits in der sogenannten heidnischen Antike.

Die Kunst des Gedächtnisses, die aus der Antike stammende ars memorativa, birgt bekanntlich zwei unterschiedliche Theoriepotentiale: erstens jenes, das die Konstruktion von Imaginationsbildern, und zweitens jenes, das die Ordnung, die Lozierung, dieser Bilder betrifft. Der erste Bereich war Gegenstand affekt-, kognitions- und zeichentheoretischer Überlegungen, was bereits in der römischen Rhetorik zu bestimmten Regulierungsvorschlägen führte. Die Erinnerungsdisposition aber blieb bemerkenswerterweise bis ins frühe 17. Jahrhundert[5] hinein dem Individuum gutteils freigestellt. Ob anhand des eigenen oder eines fremden Hauses, ob anhand eines Bildes oder anhand dessen, was sich vom Stadtbild durch tägliche Rundgänge festsetzt: Es war dem einzelnen überlassen, sich Ordnungsmuster zurechtzulegen, wodurch die Erinnerung strukturiert werden konnte.

Wie ich an anderer Stelle[6] gezeigt habe, wird die individuelle Ordnung des zu merkenden Wissens, die sich verwendungspraktisch als kasuistische Ordnung beschreiben läßt, hinfällig, sobald systematische Ansprüche den metaphorischen Prozeß der dispositio, als ,architectari' kodifiziert, begriffslogisch zu unterwandern begannen. Unter ramistischen Prämissen bezog sich seit dem 17. Jahrhundert in der Mnemonik das ,architectari' nicht mehr auf die fallbezogene Konstruktion des ordo aus empirischen Gebäudeteilen, sondern auf das in systematischer Gestalt vorliegende Wissens-,Gebäude' der Begriffe.

Man steht in der Frühen Neuzeit vor einer De-Empirisierung und gleichzeitigen Systematisierung von Erinnerungs- und Wissensordnungen.[7] Dieser Prozeß läßt sich nicht anders lesen denn als Normierung des kollektiven Gedächtnisses, an welcher die Mnemonik entscheidenden Anteil hat. (In Klammer und gegen die Thesen von Jan Assmann gesagt: Das kulturelle Gedächtnis ist wesentlich von den Kodifizierungen geprägt, die in der mnemonischen Theorie und den mnemotechnischen Praktiken verhandelt werden.)

Vorschub leistet dieser Normierung des kollektiven Gedächtnisses die Theologisierung der memoria unter christlichen Prämissen. Das christliche Gedächtnis ist tendenziell systemhaft angelegt, weil die memoria als eine facultas animae den Widerschein des göttlichen Lichtes wachhält; ohne die göttliche Herkunft der Seele ließen sich empirisch nicht erfahrbare Dinge, wie Gott selbst, wie die ewige Seligkeit, aber auch die Hölle[8], gar nicht imaginieren – und das heißt, vor allem seit Augustins Confessiones (Buch X): erinnern. Gerade auf dieser Grundlage können sich die frühneuzeitlichen Bemühungen um eine scientia universalis nur im Rahmen der mnemonischen Ordnungsmöglichkeiten artikulieren. Die Ordnung der Dinge, die Ordnung des Intellekts und die Ordnung der memoria sind unmittelbar aufeinander und ineinander abbildbar.[9] Umgekehrt formuliert: Die Universalwissenschaft setzt einen

---

4 Vgl. dazu Traninger (2000), S. 37-51.
5 Hiermit soll keineswegs Kontinuität unterstellt sein; wie Carruthers (1990), S. 153, gezeigt hat, kann nach dem Ende der Antike erst wieder ab dem 13. Jahrhundert von der Existenz einer ars memorativa gesprochen werden.
6 Vgl. Neuber: (2000).
7 Vgl. dazu allgemein Schmidt-Biggemann (1983).
8 Vgl. Berns (2000).

Ordnungsbegriff voraus, „der den Horizont sowohl der theoretischen Erkenntnisleistung und Wissenserstellung als auch der Erinnerungstätigkeit [...] darstellt: ‚memoria praesupponat ordinem rerum'."[10] (So Daniel Georg Morhof in seinem *Polyhistor*.)

Ich wage hier einen ersten Vorgriff auf den Kernpunkt meines Themas und verweise auf das, was der Ramismus mit der Rhetorik angestellt hatte: Nur die Stiltheorie und der Vortrag waren ihr zugeordnet geblieben, die inventio, die dispositio und die memoria waren auf die Dialektik übergegangen, die nun von dem Anspruch getragen ist, systematische Wissensordnungen zu erschließen – mit anderen Worten: die memoria spielt qua Topik ins Systemische,[11] der Kasuistik bleiben die Äußerlichkeiten der elocutio und der pronunciatio.

Dies hat gravierende Folgen für die in der Frühen Neuzeit alle Bildungsgänge beherrschende Schulrhetorik; die memoria wird bereits während des 16. Jahrhunderts immer stärker aus ihr ausgegliedert. Joachim Knape hat dies zu Recht auf einen veränderten Textbegriff der humanistischen Gelehrtenkultur zurückgeführt, einen Textbegriff, der nicht vom Vortrag ausgeht, sondern von der Geschlossenheit des Textes als ergon auf dem Blatt Papier. Unter der von mir eingenommenen Perspektive wird allerdings auch erkennbar, daß die rhetorisch verfaßte Rede unter der Voraussetzung auf die Würde des Gedächtnisses verzichten mußte, daß dieser die systematische heuristische Beweislast der Heilsgeschichte aufgebürdet war; der kasuistischen Wissensordnung der zu haltenden Rede kam nicht mehr die Dignität des Ganzen zu. Anders gesagt: Wenn eine frühneuzeitliche rhetorische Lehrschrift auf die memoria Verzicht leistet, dann ist ihr eine tendenzielle Geringschätzung kasuistischer Wissensordnungen zu unterstellen.

Es ist nur selbstverständlich, daß angesichts solcher universalistischer Prämissen nicht bloß der abstrakte, mit begrifflichen Stemmata operierende Ramismus Erfolg versprach, sondern daß auch vermeintlich anschaulichere Konzepte, wie sie etwa in den memoria-kosmologisch argumentierenden Entwürfen von Giulio Camillo oder Giordano Bruno vorliegen, zur Entfaltung gebracht wurden. Morhof, erneut im Polyhistor, paraphrasiert Bruno folgendermaßen: Da Bruno die clavis magna gefunden und vollendet habe, „bedurften wir weder weiterer materieller Örter [...], noch verbanden wir die Ordnung der Örter, die gemerkt werden sollten, mit einer anderen Reihenfolge, sondern uns allein an die Architektur der Phantasie anlehnend, verbanden wir die Ordnung der Örter mit der Ordnung der Dinge, die gemerkt werden sollten."[12]

Angesichts der Dominanz des Systematischen stellt sich die Frage, unter welchen Umständen denn überhaupt kasuistische Wissensordnungen denkbar sind. Bevor dem nachgegangen werden kann, will ich jedoch anhand eines konkreten Beispiels den frühneuzeitlichen Primat der Wissenssystematik vor der Wissenskasuistik erläutern. Ich beziehe mich auf Francis Bacon – und zwar deswegen, weil ich diesen Primat von einem Ort außerhalb der mnemonischen Theorien im engeren Sinne her aufweisen möchte.

---

9 Vgl. Leinkauf (1993), S. 5.
10 Ebd., – Das lateinische Zitat vgl. Morhof (1747), S. 369.
11 Analoges gilt für die Kombinatorik, besonders jene, die aus dem Lullismus heraus theoretisch begründet ist: die Kasuistik wird aus dem System erzeugt, ist allerdings im Sinne Bacons auch nur notwendig ein begrenzter Abschnitt aus der Serialität systemischer Ordnungsmöglichkeiten.
12 Morhof (1747), zit. nach Berns/Neuber (1998), S. 269.

- Oberstes Ziel aller Wissenschaft ist die Restitution des Heilsplanes (bei den Protestanten mit der Implikation, so das Ende der Zeiten herbeizuführen): Das wahre Ziel des
  Wissens „is a restitution and reinvesting (in great part) of man to the sovereignty and
  power (for whensoever he shall be able to call the creatures by their true names he
  shall again command them) which he had in first state of creation."[13]
- Das Mittel dazu ist die Suche nach der Universalität des Wissens und der Wissenschaft (als philosophia perennis und scientia universalis): „And to speak plainly and
  clearly, it is a discovery of all operations and possibilities of operations from immortality (if it were possible) to the meanest mechanical practice."[14]
- Das gewonnene ist Wissen im Sinne einer Enzyklopädie zu versammeln, um seine
  Homogenität sicherzustellen: „And it is a matter of common discourse of the chain of
  sciences how they are linked together, insomuch as the Grecians, who had terms at
  will, have fitted it of a name of *Circle Learning* [enkyklos paideia, W. N.]. [...] I mean
  it directly of that use by way of supply of light and information which the particulars
  and instances of one science do yield and present for the framing or correcting of the
  axioms of another science in their very truth and notion."[15]
- Dazu wird Imaginationskontrolle erfordert als Prozeß der bereits eingangs angespro-
  chenen kognitiven Kollektivierung, am deutlichsten faßbar in den Entwürfen utopi-
  scher Idealsozietäten, aber auch erneut bei Bacon: „And therefore without this inter-
  course the axioms of sciences will fall out to be neither full nor true; but will be such
  opinions as Aristotle in some places doth wisely censure, when he saith *These are the
  opinions of persons that have respect but to a few things*. So then we see that this note
  leadeth us to an administration of knowledge in some such order and policy as the
  king of Spain in regard of his great dominions useth in state; who [...] hath one council
  of State or last resort, that receiveth the advertisements and certificates from all the
  rest."[16]

Die Homogenisierung des Wissens ist also ein hoheitlicher Akt, ein Akt der Ausübung von
absolutistischer Macht, ein Akt der Unterwerfung. – Fragt man an dieser Stelle nach den
Möglichkeiten von Wissenskasuistiken, dann wird deutlich, daß der Begriff sich tatsächlich
in einer Weise in Opposition zum Begriff des Systems befindet, daß sein Auftreten als Regelverstoß zu verstehen wäre, als Regelverstoß, der entweder sanktioniert werden muß oder aber
zugelassen werden kann. Muß das Systematische als gegebene Norm behandelt und vorausgesetzt werden, so kann sich die Zulässigkeit des Kasuistischen – beschrieben in den Kategorien mnemotechnischer Prozesse, nicht als mögliche oder hypothetische Wissensstruktur
schlechthin – nicht anders als performativ darstellen. Die kasuistische Ordnung des Wissens
ist in einem performativen Horizont in der Tat unumgänglich und unvermeidbar, nämlich
überall dort, wo fallbezogene Beweisgänge nötig sind; das sieht auch Bacon so: „That never
any knowledge was delivered in the same order it was invented [...]."[17] Dies trifft nicht allein

---

13 Bacon (1963d), S. 222.
14 Ebd.
15 Ebd. S. 228-229. – Die Hervorhebung durch Kursivschrift entspricht dem Original.
16 Ebd. S. 230-231. – Die Hervorhebung durch Kursivschrift entspricht dem Original. Andere Hervorhebungen
von mir.
17 Ebd. S. 248.

die Predigt und die Traktatistik, sondern in einem eminenten Sinne die Mathematik ebenso wie auch die Literatur.

Ich habe mich bisher auf Bacons Frühschrift *Valerius Terminus* bezogen. *De Augmentis Scientiarum* scheint den dargelegten Thesen nicht zu folgen, wenn man sich die Rolle ansieht, die Bacon der memoria zumißt. Denn radikaler als bei Augustin, Camillo, Bruno oder Ramus werden memoria und Vernunft auf unterschiedliche menschliche Handlungsfelder bezogen: Der Historie ordnet Bacon das Gedächtnis zu, der Poesie die Phantasie und der Philosophie die Vernunft. Der Historie entspricht damit die chronologische Verlaufsordnung der geschichtlich singulären Phänomene. Der Poesie kommt der ordo artificialis zu: Sie ist diejenige, die das, „quæ in rerum natura nunquam conventura aut eventura fuissent ad libitum componat [...].“[18]

Der Philosophie schließlich ist die Ordnung „ex lege naturæ et rerum ipsarum evidentia“[19] konform. Doch sie kann ohne Operationen der memoria nicht funktionieren; ohne Erinnerungstätigkeit keine Erkenntnis. „Individuorum eorum imagines [...] figuntur in memoria, atque abeunt in eam a principio tanquam integræ, eodem quo occurrunt modo. Eas postea recolit et ruminat anima humana; quas deinceps aut simpliciter recenset; aut lusu quodam imitatur; aut componendo et dividendo digerit.“[20] Das Ordnen der Vernunft bzw. der Philosophie ist gegenüber der historischen Ordnung als Akt des Berichtigens gedacht.

Diese Abweichung von *De augmentis* läßt sich indessen mit dem *Novum organon* als scheinbar erklären. Denn erst hier ist von jenen mnemonischen Prozessen die Rede, die ich eingangs als konstitutiv für die scientia universalis eingeführt habe. Was im Sinne einer Mnemonik das Gedächtnis stärkt, das sind „ordo sive distributio [...]; item Loci in memoria artificiali, qui aut possunt esse loci secundum proprium sensum, veluti janua, angulus, fenestra, et similia, aut possunt esse personae familiares et notae, aut possunt esse quidvis ad placitum (modo in ordine certo ponantur), veluti animalia, herbae; etiam verba, literae, characteres, personae historicae, et caetera [...]. Item carmina facilius haerent et discuntur memoriter quam prosa.“[21]

Auch wenn Bacon hier tatsächlich von drei gedächtnisstützenden Prinzipien spricht, von ordo, loci artificiales und dem Vers, so ist doch deutlich, daß es sich bei allen drei Instanzen letztlich um Ordnungsmodelle handelt. Daß diese Ordnungssysteme nicht ad libitum kasuistisch konstruiert werden dürfen, macht die Fortsetzung von Bacons Argumentation deutlich. Er verlangt nämlich, die „Abscissio Infiniti“, die Abschneidung des Unendlichen: „Cum enim quis aliquid reminisci aut revocare in memoriam nititur, si nullam praenotionem ha-

---

18 Bacon (1963b), S. 494 („was in der Natur der Dinge niemals zusammentreten oder erfolgen würde, nach Belieben zusammenfügt [...].“ – Übersetzung von mir.)

19 Ebd. („nach dem Gesetz der Natur und der Evidenz der Dinge selbst“ – Übersetzung von mir.)

20 Ebd., S. 494-495. („Die Bilder der Einzeldinge [...] werden im Gedächtnis festgeheftet und gehen in selbiges von Anfang an so unversehrt hinein, wie sie auftreten. Die menschliche Seele holt sie sich danach wieder hervor und erneuert sie sich; die sie dann entweder einfach beurteilt; oder durch ein gewisses Spiel nachahmt; oder durch Zusammensetzen und Zerteilen berichtigt.“ – Übersetzung von mir.)

21 Bacon (1963c), S. 275. („Ordnung oder Aufteilung [...]; ebenso die loci in der memoria artificialis, die entweder Orte nach dem eigentlichen Wortsinn sein können, wie ein Tor, ein Winkel, ein Fenster oder etwa ähnliches, oder sie können vertraute und bekannte Personen sein oder sonst etwas nach Gefallen – sofern sie in einer gewissen Ordnung gesetzt werden –, wie etwa Tiere, Pflanzen; auch Wörter, Buchstaben, Zeichen, historische Personen usw. [...]. Ebenso werden Gedichte leichter behalten und auswendig gelernt als Prosa.“ – Übersetzung von mir.)

beat aut perceptionem ejus quod quaerit, quaerit certe et molitur et hac illac discurrit, tanquam in infinito. Quod si certam aliquam praenotionem habeat, statim abscinditur infinitum,
et fit discursus memoriae magis in vicino. In tribus autem illis instantiis quae superius dictae
sunt, praenotio perspicua est et certa. In prima videlicet, debet esse aliquid quod congruat
cum ordine; in secunda debet esse imago quae relationem aliquam habeat sive convenientiam
ad illa loca certa; in tertia, debent esse verba quae cadant in versum; atque ita abscinditur
infinitum."[22]

Die Ordnung der Dinge wird als gegeben vorausgesetzt und ist als solche infinit (sei es
tatsächlich infinit als zirkuläre Struktur der scientia universalis oder potentiell infinit im
Sinne einer ,catena rerum'). Wird das Unbegrenzte nun abgeschnitten, so kann man darin,
wie ich meine, wohl drei Aktualisierungsmöglichkeiten von kasuistischen Wissensordnungen[23] erblicken. Alle drei Arten der Begrenzung beruhen nach Bacon auf einem Vorbegriff:
erstens einem Vorbegriff hinsichtlich der gegebenen Ordnung, zweitens hinsichtlich der
Auswahl einer imago, die mit dem locus in Beziehung steht, und drittens hinsichtlich der
Konformität im Rhythmus eines gewählten Wortes mit einem Vers.

Wie man sieht, stünden kasuistische Wissensordnungen zum System hier nur in einem
Verhältnis der Partikularität, nicht aber in einem Verhältnis der abweichenden Neuordnung.
Dies mag mit einem grundsätzlichen Mißtrauen zu erklären sein, das die gesamte Frühe
Neuzeit hindurch immer wieder gegenüber der Mnemonik formuliert wurde und das letzten
Endes immer wieder auf den Verdacht hinausläuft, die Gedächtniskunst könne das natürliche
Gedächtnis überfordern und den Menschen in den Wahnsinn treiben.

Bereits Heinrich Cornelius Agrippa von Nettesheim äußert im X. Kapitel von *De incerti-
tudine et vanitate scientiarum atque artium* (1530) die Befürchtung, die Gedächtniskunst
mache verrückt. Die Mnemonik braucht, so seine Argumentation, das natürliche Gedächtnis,
„das allerdings sehr häufig von ihr mit vielen abstrusen Dingen vollgestopft wird. So kommt
es manchmal statt der erwünschten hohen Gedächtnisleistung zum geistigen Zusammen-
bruch und Wahnsinn, weil das natürliche Gedächtnis allzusehr mit Dingen und Worten belastet wird und die Leute, die sich mit den von der Natur gesetzten Grenzen nicht abfinden
wollen, durch diese Kunst in den Wahnsinn getrieben werden."[24]

Soweit Agrippa. Noch ist von einer einfachen Überlastung der memoria naturalis die
Rede. Im 17. Jahrhundert aber werden die Akzente verschoben, tritt der Konflikt zwischen
einem als ,natürlich' postulierten System-ordo der Vernunft und einem als ,artifiziell' ver-
standenen kasuistischen ordo der Mnemonik in den Vordergrund. Befürchtung und Verdacht,
die Gedächtniskunst könne verrückt machen, lassen sich somit nur vor dem Hintergrund
topischer Ordnungsmodelle verstehen.

---

22 Ebd., S. 275-276. („Denn wenn jemand etwas zu erinnern oder ins Gedächtnis zurückzurufen trachtet, dann
sucht er, wenn er keinen Vorbegriff hätte oder keine Wahrnehmung dessen, was er sucht, bestimmt und müht
sich ab und verläuft sich da und dort gleichsam ins Unbegrenzte. Denn wenn er irgendeinen Vorbegriff hätte,
dann wird das Unbegrenzte sofort abgeschnitten und der Suchlauf des Gedächtnisses geschieht mehr im Nahe-
gelegenen. In jenen drei obenerwähnten Fällen ist der Vorbegriff deutlich und gewiß. Im ersten Fall nämlich soll
es etwas sein, das mit der Ordnung übereinstimmt; im zweiten soll es ein Bild sein, das ein gewisses Verhältnis
habe oder eine Übereinstimmung mit jenen bestimmten Örtern; in dritten Fall sollen es Wörter sein, die in einen
Vers fallen; und solcherart wird das Unbegrenzte beschnitten." – Übersetzung von mir.)
23 Die noch nichts mit der Frage einer individuellen Begrenzung zu tun haben!
24 Agrippa von Nettesheim (1993), S. 46.

Jan Amos Comenius etwa verwehrt sich auf das heftigste gegen eine Interpretation seines *Orbis sensualium pictus* (EA 1654) als mnemonisch konstruiertes Lehrwerk. Er verweist darauf, daß die Abfolge der einzelnen Unterweisungsschritte gänzlich als ordo naturalis zu verstehen sei, von dem die Mnemonik doch gerade durch ihren ordo artificialis diametral abweiche.[25] Nur wenige Jahre später diskutiert Christian Thomasius in den *Monatsgesprächen*, im Zuge der Rezension von Daniel Georg Morhofs *Polyhistor*, den Nutzen bzw. Schaden der ars memorativa.[26] Als mnemotechnisch zulässig nennt Thomasius – ähnlich wie Bacon vor ihm – Metrum bzw. Reim, Hilfswörter, Symbola und imagines. Zugleich aber warnt Thomasius, die mnemonischen imagines könnten verrückt machen, weil sie den ordo naturalis bedrohen.[27] In der Folge lehnt er die geschichts- und bibelmnemonischen Konzepte von Johannes Buno und Johann Justus Winckelmann ab. Die lullistische ars combinatoria stellt Thomasius daher – entgegen Morhof, dessen Thesen er ja diskutiert – nicht zur Logik, welcher der ordo naturalis entspricht, sondern zur Mnemonik[28] mit ihrem artifiziellen ordo.

Die Invektive gegen Buno und Winckelmann trifft die Verfasser der erfolgreichsten, profiliertesten und zugleich elaboriertesten Mnemoniken der zweiten Hälfte des 17. Jahrhunderts.[29] Buno ist mit anwendungsbezogenen Mnemoniken zur Universalgeschichte, zur lateinischen Grammatik, den Digesten und der Bibel hervorgetreten, Winckelmann vor allem mit einer Kaisergeschichte. Sie folgt denselben mnemotechnischen Prinzipien wie Bunos universalgeschichtliche Mnemonik. Hier ist ein intrikates Geflecht aus unterschiedlichen Verschlüsselungsmechanismen am Werk: Zahlen werden durch Buchstaben ersetzt und ergeben Worte, die ihrerseits durch Sinn- und Klangassoziationen in verklausulierte Bilder umgesetzt werden. Das besondere an den so entstehenden Bildtafeln der genannten Werke ist eine sequentiell gereihte Folge von mnemonischen imagines, die nicht allein aus der Sicht von Thomasius einem ordo artificialis gehorcht. Auch Buno selbst hebt als Neuerung die „catena" hervor, „die den mnemonischen Bildzusammenhang einer jeden Tafel klärt. Sie enthält, wie Buno in der Einleitung erwähnt [...], ‚auch die loca und res præcipuæ mit fabulis gleich als catenis aneinander gehänget.'"[30]

Was Bunos und Winckelmanns Gegnern bedrohlich erscheint, ist genau dies: Die Universalgeschichte wird erstens durch uneigentliche, auf Zahlen bezügliche und auf Sinn- und Lautassoziationen beruhenden imagines repräsentiert, die ihrerseits und zweitens einer uneigentlichen Diskurslogik folgen, die auf einer ‚fabula' anstelle der ‚historia' beruht. Hier sind die Beschränkungsstrategien, die Bacon hatte gelten lassen, überzogen: Die sequentielle Beschränkung wird sinnlos, weil die Sequenz nicht mehr zu erkennen ist; die Ähnlichkeit von imagines und loci ist nicht gegeben.

Gefährlich ist Bunos und Winckelmanns Verfahren deshalb, weil die beiden mit ihren Werken nicht nur die historiographische Bildimagination denaturieren und die ‚historia' durch eine bildlogische ‚fabula" ersetzen, sondern vor allem weil dieser ordo artificialis als zwar kasuistische Wissensordnung dennoch allgemeine Verbindlichkeit beansprucht; schließ-

---

25 Vgl. Schaller (1976), S. 198-209.
26 Vgl. Thomasius (1972), besonders S. 606-619.
27 Vgl. ebd., S. 617.
28 Ebd., S. 610f. – Den Passus aus dem *Polyhistor* vgl. bei Morhof (1688), Bd. I, Lib. II, Cap. V.
29 Zu Buno vgl. Strasser (2000b), S. 639-660; zu Winckelmann vgl. Kocher (2001), S. 71-85; zu beiden vgl. Strasser (2000a).
30 Vgl. Strasser (2000b), S. 654.

lich handelt es sich um Lehrbücher. Was zur Disposition steht, ist das kollektive Geschichts-
gedächtnis und seine Integrität. Die Universalgeschichte ist für die Mnemonikkritiker ein zu
schützendes Gut, weil qua Imaginationsdisziplinierung das kulturelle Gedächtnis abgesi-
chert werden muß.

Und dieses kulturelle Gedächtnis ist offensichtlich das punctum saliens, wenn kasuistische Wissensordnungen verhandelt werden. Denn nicht alle Formen solcher kasuistischer
Wissensordnungen bergen eine Bedrohung für die kollektive memoria in sich, ja bisweilen
ist die Kasuistik sogar ausdrücklich aufgerufen – nämlich dann, wenn sie als individualisie-
rendes Verfahren in spezifischen Kontexten nutzbar gemacht werden kann. Die Rede ist hier
von der protopietistischen Meditation,[31] die sich selbst als mnemonischen Vorgang versteht.

Catharina Regina von Greiffenberg schreibt am 28. August 1668 an Sigmund von Birken: „Jch An Meinem Ohrt [sc. in der Abgeschiedenheit ihres adeligen Landsitzes in Nieder-
österreich, W. N.] Muß mir Die äusserlichen Wihrtschaffts-Werke, Zu Geistlichen Gedächtnus Öhrteren machen, und meine gedanken bald in diese Bald in jenes Geschöpfe und Ge-
schäffte Losieren.“[32] – Was die Greiffenberg hier anspricht, folgt verfahrenstechnisch den
Occasional meditations (1630) des Bischofs von Exeter, Joseph Hall. Sein Text wurde von
Georg Philipp Harsdörffer teilübersetzt und im Anhang seiner Schrift *Nathan und Jotham*
(1650) abgedruckt; die Greiffenberg hat diese Fassung rezipiert.[33]

In der Folge läßt sich die Greiffenberg zu zahlreichen Meditations-Gedichten[34] inspirie-
ren, indem sie ihre Gedanken in buchstäblich alle möglichen Gegenstände und Geschäfte
lociert. D. h. sie benutzt diese Gegenstände und Geschäfte als mnemonische Örter, die Erinnerung bewahren und abrufbar machen. Die Ordnung des solcherart aktivierten Wissens ist
nun in der Tat kasuistisch. Ob die Greiffenberg zu Epiphanias Eier brät, die dann in der
Pfanne Kreuzesform annehmen, ob sie die Wiederkunft des Frühlings oder des Sommers
beobachtet oder mit einem absichtlich fremden Blick eine unscheinbare Pflanze wie das Ver-
gißmeinnicht wahrnimmt – immer wird ein zufälliger Erinnerungsstimulus in eine poetische
Argumentation übergeführt, die eine kasuistische Wissensordnung demonstriert.

„Über das kleine wolbekandte Blümlein: Vergiß mein nicht

SChönes Blümlein! deine Farbe / zeigt des Höchsten Hoheit an /
als spräch sie: vergiß mein nicht / du / dem also hoch beliebet
dieser Erden Eitelkeit / die doch endlich nur betrübet.
Wisse / daß man / meiner denkend / wol vergnüget leben kan.
Von dir kleinem Sitten-Lehrer lern' Geheimnus jederman.
Deiner Blätlein fünffte Zahl / in mir die Gedächtnus übet
ihre fünff ergebne Sinn / und sie durch betrachten schiebet

---

31 Katholische Meditationsanleitungen bedürften einer eigenen Untersuchung; vgl. dazu exemplarisch: Berns
(1988).
32 Catharina Regina von Greiffenberg: [Brief an Sigmund von Birken, 28. August 1668]. Zit. nach Daly (1973),
S. 203.
33 In diesem Kontext wären auch Christian Scrivers *Zufällige Andachten* einer Untersuchung wert.
34 Vgl. dazu Thums (2000), S. 251-272; Soboth, *„HErr / mein Gedächtniß ist vom Wachs zu ...“*, S. 273-290.

in die fünff hochwehrten Wunden / welche unsre Lebens-Bahn.
Deines Krauts und Stängels grün lehret / daß wir hoffen sollen /
GOtt werd' unser nicht vergessen / ob wir wol auf Erden seyn /
unter manchem Creutz und Vnglück / werd auch bald zu sich uns holen.
Ach vergiß mein nicht / O Schöpffer! deine Hülf' auch mir erschein'.
Jst doch meiner Hoffnung Safft / her aus deinem Wort gequollen /
in dir liget grosse Weißheit / Blümlein / wärstu noch so klein."[35]

Die erste Beobachtung betrifft die Farbe der Blume; ihr Blau wird als Himmelsblau verstanden und als Verweis auf den Schöpfer gedeutet, in dessen Gedenken, also Erinnerung, man ebenso vergnügt leben könne wie in der Verfallenheit an die irdischen Eitelkeiten. Die zweite Beobachtung erschließt über die Fünfzahl der Blütenblätter analoge Fünfersysteme: die fünf Sinne und die fünf Wundmale Christi. In beiden Fällen wird die argumentative Heuristik als mnemonischer Akt vergegenwärtigt. Die dritte Beobachtung, die nun von den Quartetten in das erste Terzett überführt, wird von der grünen Farbe des Stengels geleitet. Der Erinnerungsakt wird auf Gott verschoben, der den im Unglück lebenden Menschen nicht vergessen und zu sich in die Seligkeit leiten wird. Das zweite Terzett hebt mit einem Wortspiel an: Aus dem Namen der Blume wird eine Anrufung Gottes, das memoria-Argument verschiebt sich ein weiteres Mal, hier zu der exhortatio Gottes, er möge den Menschen erinnern und ihm seine Hilfe angedeihen lassen. Mit der Lebenssubstanz der Blume, ihrem Saft, wird die Lebenssubstanz des Christen alludiert, nämlich die Hoffnung auf Erlösung, wie sie das Gotteswort verheißt. Die letzte Zeile preist die Blume für ihre Weisheit, die gleichsam nur erinnernd zu aktualisieren war.

Die dispositio der Wissensbestände – blaue Farbe und Gottesgedenken, Fünfzahl und Sinne bzw. Wunden, grüne Farbe und Hoffnung, die Lebenssubstanz der Blume und des Christen – entspricht in keiner Weise den zeitgenössischen systematischen Ordnungen, seien sie in der Theologie oder den Kräuterbüchern zu suchen. Was dieser dispositio noch am nächsten kommt, ist die allegoretische Auslegungspraxis von Pflanzen, wie sie von der Fruchtbringenden Gesellschaft betrieben wurde. In beiden Fällen geht es darum, eine durchaus kasuistische oder gar dezisionistische Wissensordnung vorzuführen, deren Sinn sich primär aus einer individuellen Heuristik erschließt. Im Falle von Greiffenbergs Gedicht tritt jedoch hinzu, daß sich diese Heuristik mnemonisch situiert; reminiszierend wird die memoria gestärkt, oder mit anderen Worten: In einer kasuistisch inszenierten Wissensordnung wird das Ich in die Lage gesetzt, sich dem kollektiven Bewußtsein von der Heilsmöglichkeit einzuschreiben.

Die kasualpoetische Meditationslyrik der Greiffenberg ist doch wohl keine Ausnahmeerscheinung. Nach den weiteren Perspektiven von kasuistischen Wissensordnungen wäre etwa anhand der loci-comunes-Sammlungen, der Emblematik, aber auch der Grammatik und damit der Kritik[36] sowie des Zeremoniells[37] zu fragen. Sie alle wären einmal in Beziehung zu setzen zu den Modellen einer applizierten Mnemonik, wie sie in gegenstandsformalen Merklehren (corpus-Modelle des menschlichen Körpers, der Hand, des Gesichts, Bäume, Tiere

---

35 Greiffenberg (1967), S. 238.
36 Vgl. dazu Jaumann (1995).
37 Vgl. dazu Berns (1984); Berns/Rahn (1995); Rahn (1997).

etc.), in mechanischen Modellen (Gebäude, Stadtpläne, Itinerare, Labyrinthe, Schiffsarchitektur, Maschinen, Theaterräume, memoria-Hallen, Chartiludien etc.) sowie inhaltsbezogenen Modellen (Logiklehren, Schreiblehren, Bibel- und Geschichtsmnemoniken etc.) vorliegen. Nach gegenwärtigem Kenntnis- und Forschungsstand scheinen sich hier in der Tat fruchtbare Felder für die Untersuchung dessen zu eröffnen, wo es selbst im ordnungsbessenenen 17. Jahrhundert funktionale Residuen kasuistischer Wissensordnungen – und das heißt wohl auch: der Selbstbehauptung von Individualität – gibt.

Literatur

Agrippa von Nettesheim, Heinrich C. (1993): *Über die Fragwürdigkeit, ja Nichtigkeit der Wissenschaften, Künste und Gewerbe*. Mit einem Nachwort herausgegeben von Siegfried Wollgast. Übersetzt und mit Anmerkungen versehen von Gerhard Güpner. Berlin.

Bacon, Francis (1963a): *The Works of Francis Bacon,* 14. Bände (1857-1874). Faksimile-Neudruck der Ausgabe von Ellis Spedding und London Heath. Stuttgart/Bad Cannstatt.

Bacon, Francis (1963b): „De Augmentis Scientiarum. Caput I", in Bacon (1963a), Bd. 1.

Bacon, Francis (1963c): „Novum Organum", in Bacon (1963a), Bd. 1.

Bacon, Francis (1963d): „Valerius Terminus of the interpretation of nature: with the annotations of Hermes Stella", Bacon (1963a), Bd. 3.

Berns, Jörg Jochen und Neuber, Wolfgang (Hg.) (1998): *Das enzyklopädische Gedächtnis der Frühen Neuzeit. Enzyklopädie- und Lexikonartikel zur Mnemonik* (= Documenta Mnemonica 2). Tübingen.

Berns, Jörg Jochen und Neuber, Wolfgang (Hg.) (2000): *Seelenmaschinen. Gattungstraditionen, Funktionen und Leistungsgrenzen der Mnemotechniken vom späten Mittelalter bis zum Beginn der Moderne* (= Frühneuzeit-Studien. N.F. 2). Wien/Köln/Weimar.

Berns, Jörg Jochen und Rahn, Thomas (Hg.) (1995): *Zeremoniell als höfische Ästhetik in Spätmittelalter und Früher Neuzeit* (= Frühe Neuzeit 25). Tübingen.

Berns, Jörg Jochen (1984): „Die Festkultur der deutschen Höfe zwischen 1580 und 1730. Eine Problemskizze in typologischer Absicht", *Germanisch-Romanische Monatsschrift* 34: 295-311.

Berns, Jörg Jochen (1988): „,Vergleichung eines Vhrwercks, vnd eines frommen andächtigen Menschens.' Zum Verhältnis von Mystik und Mechanik bei Spee", in Italo M. Baltafarano (Hg.), *Friedrich von Spee. Dichter, Theologe und Bekämpfer der Hexenprozesse.* Gardolo di Trento, 101-206.

Berns, Jörg Jochen (2000): „Höllenmeditation. Zur meditativen Funktion und mnemotechnischen Struktur barocker Höllenpoesie", in Gerhard Kurz (Hg.), *Meditation und Erinnerung in der Frühen Neuzeit* (= Formen der Erinnerung 2). Göttingen, 141-173.

Carruthers, Mary (1990): *The Book of Memory. A Study of Memory in Medieval Culture.* Cambridge.

Daly, Peter M. (1973): „Emblematische Strukturen in der Dichtung der Catharina Regina von Greiffenberg", in Gerhart Hoffmeister (Hg.), *Europäische Tradition und deutscher Literaturbarock. Internationale Beiträge zum Problem von Überlieferung und Umgestaltung.* Bern/München, 189-222.

Ernst, Ulrich (1997): „Die Bibliothek im Kopf: Gedächtniskünstler in der europäischen und amerikanischen Literatur", *Zeitschrift für Literaturwissenschaft und Linguistik* 27: 86-123.

Garzoni, Thomaso (1641): *Piazza Universale*. Frankfurt.

Greiffenberg, Catharina Regina von (1967): *Geistliche Sonette, Lieder und Gedichte*. Mit einem Nachwort zum Neudruck von Heinz-Otto Burger. Darmstadt.

Jaumann, Herbert (1995): *Critica. Untersuchungen zur Geschichte der Literaturkritik zwischen Quintilian und Thomasius* (= Brill's Studies in Intellectual History 62). Leiden/New York/Köln.

Kocher, Ursula (2001): „Emblematik, Mnemonik, Semiotik. Die ‚Gedächtnisbilder' Johann Justus Winckelmanns", in Heinz J. Drügh und Maria Moog-Grünewald (Hg.), *Behext von Bildern? Ursachen, Funktionen und Perspektiven der textuellen Faszination durch Bilder* (= Neues Forum für Allgemeine und Vergleichende Literaturwissenschaft 12). Heidelberg, 71-85.

Leinkauf, Thomas (1993): „Scientia universalis, memoria und status corruptionis. Überlegungen zu philosophischen und theologischen Implikationen der Universalwissenschaft sowie zum Verhältnis von Universalwissenschaft und Theorien des Gedächtnisses", in Jörg J. Berns und Wolfgang Neuber (Hg.), *Ars memorativa. Zur kulturgeschichtlichen Bedeutung der Gedächtniskunst 1400-1750* (= Frühe Neuzeit 15). Tübingen, 1-34.

Morhof, Daniel Georg (1688): *Polyhistor*. Lübeck.

Morhof, Daniel Georg (1747): *Polyhistor*. Lübeck, 4. Aufl.

Neuber, Wolfgang (2000): „Die vergessene Stadt. Zum Verschwinden des Urbanen in der ars memorativa der Frühen Neuzeit", in Jörg J. Berns und Wolfgang Neuber (Hg.), *Seelenmaschinen. Gattungstraditionen, Funktionen und Leistungsgrenzen der Mnemo-techniken vom späten Mittelalter bis zum Beginn der Moderne*. Wien/Köln/Weimar, 91-108.

Rahn, Thomas (1997): „Grenz-Situationen des Zeremoniells in der Frühen Neuzeit", in Markus Bauer und Thomas Rahn (Hg.), *Die Grenze. Begriff und Inszenierung*. Berlin, 177-206.

Schaller, Klaus (1976): „Johann Balthasar Schupp – Muttersprache und realistische Bildung", in Albrecht Schöne (Hg.), *Stadt – Schule – Universität – Buchwesen und die deutsche Literatur im 17. Jahrhundert*. München, 198-209.

Schmidt-Biggemann, Wilhelm (1983): *Topica Universalis. Eine Modellgeschichte humanistischer und barocker Wissenschaft* (= Paradeigmata 1). Hamburg.

Scrivers, Christian (1663): *Zufällige Andachten*. Magdeburg, EA.

Soboth, Christian: „*HErr / mein Gedächtniß ist vom Wachs zu deinen lenken" – Formen und Funktionen der memoria in den „Geistlichen Sonetten, Liedern und Gedichten" von Catharina Regina von Greiffenberg*. 273-290.

Strasser, Gerhard F. (2000a): *Emblematik und Mnemonik der Frühen Neuzeit im Zusammenspiel: Johannes Buno und Johann Justus Winckelmann* (= Wolfenbütteler Arbeiten zur Barockforschung 36). Wiesbaden.

Strasser, Gerhard F. (2000b): „Johann Bunos mnemotechnische Verfahren", in Jörg J. Berns und Wolfgang Neuber (Hg.), *Seelenmaschinen. Gattungstraditionen, Funktionen und Leistungsgrenzen der Mnemotechniken vom späten Mittelalter bis zum Beginn der Moderne*. Wien/Köln/Weimar, 639-660.

Thomasius, Christian (1972): *Freimütige, lustige und ernsthafte, jedoch vernunftmäßige Gedanken oder Monatsgespräche über allerhand, fürnehmlich aber neue Bücher*, Bd. II. Juli – Dezember 1688. Faks. Frankfurt/M.

Thums, Barbara (2000): „Zur Topographie der memoria in frühneuzeitlicher Mystik: Catharina Regina von Greiffenbergs ‚Geistliche Gedächtnisorte'", in Gerhard Kurz (Hg.), *Meditation und Erinnerung in der Frühen Neuzeit* (= Formen der Erinnerung 2). Göttingen, 251-272.

Traninger, Anita (2000): „Domänen des Gedächtnisses. Das Scheitern der Mnemotechnik an der memoria des absoluten Herrschers", in Jörg J. Berns und Wolfgang Neuber (Hg.), *Seelenmaschinen. Gattungstraditionen, Funktionen und Leistungsgrenzen der Mnemotechniken vom späten Mittelalter bis zum Beginn der Moderne.* Wien/Köln/Weimar, 37-51.

Friedrich Steinle[1]

Negotiating Experiment, Reason and Theology:
The Concept of Laws of Nature in the Early Royal Society

The concept of laws of nature was not new in 17th century natural philosophy.[2] I only mention two of the various traditions. In astronomy, "laws" counted as the central principles, like axioms, within a larger system of mutually dependent theorems. Such a use was still found in Copernicus and Kepler.[3] In a different tradition, emphasis was put instead on theological aspects: "laws" have an author, and the main use of the concept of laws of nature was to emphasize their divine authorship. The axiomatic character, however, and the position within a larger system, played no role here. One of the representative figures of that tradition is Ockham.[4]

The situation changed profoundly and rapidly in the 17th century. First, the concept came into much wider use than before, and, second, its meaning was significantly broadened. At the turn to the 18th century, the situation came much nearer to our modern understanding of the concept than formerly. Those changes took place at different rates in different parts of Europe. In this paper, I shall confine my study to London and the Royal Society, with its attempt to establish a new, empiristic, and socially organized natural philosophy. My claim is that the changes to the concept of laws of nature were essentially connected with the program of a new philosophy. What I will present is essentially a collection of material which highlights the change. To place it profoundly in a wider picture of cultural, philosophical and theological developments is an open task which requires further research.

Descartes and his reception in England

I start my England story with a French author. René Descartes was the first to give the concept of laws of nature an absolutely central and prominent place in natural philosophy. In his *Principia Philosophiae* of 1644, he started the exposition of his system with three me-

---

1 The paper results from research conducted within a working group at the Max-Planck-Institute for the History of Science in Berlin. I thank Lorraine Daston and Sophie Roux for many discussions and critiques, and the discussants of meetings in Bielefeld, Bad Homburg and Frankfurt for stimulating comments. A German version of this text appeared in Steinle (2001).
2 For existing literature on the development of the concept of laws of nature, cf. Zilsel (1942), Schimank (1949), Needham (1956), ch. 18, Oakley (1961a), Oakley (1961b), McGuire (1972), Casini (1976), Milton (1981), Schramm (1981), Klaaren (1985), Ruby (1986), reprinted in Weinert (1995), Loeck (1986) and Steinle (1995).
3 Graßhoff (1999), Graßhoff (2001), Ruby (1986) pp. 298-9 of the reprint.
4 Cf. Oakley (1961a), Oakley (1981), or Crombie (1994), I, pp. 402-3.

chanical principles of how the overall motion of the world was conserved and transmitted. He called them "Loix de la Nature" and reserved that term for those three principles: no other theorem, rule or principle was called a law in his system.[5] Such a use was new in that the points of emphasis of the two aforementioned traditions came together here. Laws had, on the one hand, a quasi-axiomatic status within his whole system, and they reflected, on the other hand, the voluntaristic tradition: Descartes emphasized that God could have created different laws.[6]

In England, Descartes' concept of laws was received in as multifarious a way as his program as a whole. One of the most important figures in England was Henry More in Cambridge, who initially welcomed Descartes' approach most emphatically, but switched, in the course of some decades, to a critical point of view and finally became a strong opponent.[7] More not only fully adopted Descartes' concept of laws of nature, but extended it significantly. In his pro-Cartesian *Epistola ad V. C.* of 1664, he spoke of "leges Naturae", of "leges mechanicae", "leges motus", very often of "Materiae Motusque Leges" and even of "Magnetismi leges".[8] There are not many texts in the 17th century in which the term was used so frequently. Even after his anti-Cartesian turn, More maintained the concept of laws. One of his main texts against Cartesian philosophy, the *Enchiridion Metaphysicum* of 1671, has a characteristic subtitle:

> "De exsistentia & natura rerum incorporearum in genere, in qua quamplurima Mundi Phaenomena ad Leges Cartesii Mechanicas obiter expenduntur, illiusque Philosophiae, & aliorum omnino omnium qui Mundana Phaenomena in Causas purè Mechanicas solvi posse supponunt, Vanitas Falsitásque detegitur"

It is significant that More mentioned mechanical laws in such a prominent place, and that he took a neutral stance: he intended to "consider" ("expendere") the natural phenomena with respect to those laws. In the text of the work, More mentioned "leges mechanicae" in many places.[9] At the same time, and in striking contrast, he harshly rejected mechanical philosophy as such.

Though More did not discuss the concept of laws in its own right, he used it in a distinct way. By referring to laws, he made sure that the new approach to natural philosophy was consistent with religion and theology. In an early defense of Descartes against the claim of exclusively concentrating on general principles while neglecting experiments, More emphasized

> "That the first and most generall principles of Nature have more of Divinity and Majesty in them then ever to suffer themselves to be Hermetically imprison'd in some narrow neck'd glasse."[10]

---

5 For details and further literature cf. Steinle (1995) or Roux (1999).

6 As to the background of Descartes' terminology, recent research suggests that Descartes drew more than hitherto realized from the Jesuit tradition which he knew from La Flèche: cf. Specht (1999) or Knebel (1992).

7 Cf. Hall (1990) or Crocker (1990), for example.

8 More (1679), vol. II, 1, pp. 109, 110, 112, 113.

9 *Ibid.*. 246, for example. The formulation of the title is taken up again in the text: "si Phaenomenon ad leges Mechanicas expendamus", cf. p. 248. See also the English translation of the *Enchiridion:* Jacob (1995).

With the second part of the phrase More referred to chemical experiments. The "first principles" he mentioned here were elsewhere called the "Laws of Mechanics".[11]

More was an important figure for the development of the concept of laws in the Royal Society. After all, he was one of its first members, and maintained a longstanding exchange with some fellows. He never participated, however, in the Society's central program: to establish an "experimental philosophy" in a Baconian sense. That program, with its empiristic and inductivistic ideal and its anti-theoretical claim "nullius in verba", differed essentially from the Cartesian, rationalistic approach. It is often overlooked that the Royal Society was extremely heterogeneous, even with respect to the way in which natural philosophy was done. That heterogeneity had consequences for the development of the concept of laws of nature.

Extending traditional meanings

In the Royal Society, the concept of laws was well known in its traditional understanding. In reviewing a book of the French astronomer Adrien Auzout, and alluding to Auzout's own terminology, Henry Oldenburg wrote in 1665:

"... all the world having hitherto persuaded that the motions of Comets were so irregular that they could not be reduced to any Laws ..."[12]

With those laws, he referred – as was done in the tradition – to general principles with axiomatic character. As in More, however, there was also a visible tendency to widen that meaning and use the term for more particular propositions. The mathematician John Wallis started a paper on tides in 1666 with general considerations about how the motion of particular bodies could be explained from "*Mechanick* Principles"; i. e. "according to the *Laws of Motion* and *Statick Principles.*"[13] In the case of the tides, Wallis mentioned three observed "periods": daily, mensual and annual, and aimed

"to give account of these three periods, according to the *Laws of Motion* and *Mechanick Principles*, ..."

One of those principles (just to give a characteristic example) postulated

"that a Body in motion is apt to continue its motion, and that in the same degree of celerity, unless hindered by some contrary Impediment".

---

10 Letter to W. Petty of 12 March 1649, transcribed in (Webster 1969), pp. 369-372, here p. 369. Whereas More sometimes spoke of "Cartesian Principles," his correspondent Petty preferred the term "Axioms" (*ibid.*, 368).
11 More (1659), 12, for example, quoted in Boylan (1980), p. 397.
12 *Philosophical Transactions* 1 (1665) 4; cf. the first (unpaginated) page of the dedication "Au Roy" of Auzout (1665/6).
13 Wallis (1666), p. 264.

One of the "Laws of *Staticks*" stated that the common center of gravity of two bodies was always on the line between them.[14] From those laws Wallis derived more particular statements. Observational regularities themselves, such as the periods of the tides, were never called laws: rather they were to be explained by laws.

In the late 1660s the collision of bodies was a prominent topic in the Royal Society. In 1668 Wallis published a paper:

"A summary Account given by Dr. John Wallis, of the General *Laws* of *Motion*, ..."

In the text, Wallis spoke of "de Motibus ... Principiis", "ex principiis Hydrostatici" and the like. An example of such a principle is the statement that something which moves in the period T along the distance L will move along the distance nL in the period nT. The last paragraph opened thus:

"Atque hae sunt ... Generales Motuum Leges; quae ad Casus particularibus Calculo sunt accomodandae."[15]

Again laws were fundamental principles from which particular cases could be explained.

Another paper on collision, by Christopher Wren, started with the announcement "Lex Naturae de Collisione Corporum".[16] The law claimed that in collisions the loss of motion (combined from the bulk and velocity) of one body was equal to the gain of motion in the other. In contrast to Wallis, however, Wren used the law concept even for more particular cases. The level of generality which his "laws" envisaged was comparable to that of Descartes' seven "rules" of collision, which Descartes had been eager *not* to call "laws".

Talking of laws in such a context was not self-evident. Christiaan Huygens published, in March 1669, a summary of his researches on the subject in the *Journal des Sçavans*. He did not use the terms "laws of motion", "laws of collision" or the like. Even general statements such as the conservation of the overall motion (m·v) in collision (§ 5), or the conservation of the sum of m·v² (§ 6), were called "règle", "proposition" or "principe." That is insofar remarkable as Descartes, to whom Huygens owed a lot, had called exactly those conservation principles "laws." Only in one case: with reference to the statement that the motion of the common center of gravity of colliding bodies was not affected by the collision, did Huygens use the term "loy de la Nature".[17] Though he gave no hint as to the background of that use, it was probably the very general scope of the statement – explicitly emphasized by Huygens – which was decisive here. In any case, it becomes clear that Huygens used the law-concept much more reluctantly than his English fellow philosophers did.[18]

Henry Oldenburg published a Latin translation of Huygens' text in the April issue of the *Philosophical Transactions*. He mainly adopted Huygens' terminology of rules, principles and propositions. At a central point, however, he changed terms. Whereas Huygens had

---

14 *Ibid.*, pp. 268 and 272.
15 Wallis (1669), pp. 864 and 866.
16 Wren (1669), p. 867.
17 Huygens (1669), cf. (Huygens Oeuvres) t. 16, pp. 179-181; in particular p. 181.
18 In later life, Huygens used the terminology of laws more freely; cf. his manuscript of centrifugal force of about 1690: (Huygens Oeuvres), t. 16.

entitled his paper "Regles de mouvement," Oldenburg presented it as "A summary account of the *Laws of Motion*".[19] And in the alphabetical index at the end of the volume, he mentioned, under the entry "Motion," Huygens' summary as "The Laws of it, by M. Huygens." It is difficult to imagine that Oldenburg did not realize those changes. In the table of contents at the beginning of the volume he had properly announced Huygens' article as "A summary account of the *Rules of Motion*." Oldenburg was obviously inclined to give the law-concept a more prominent place than Huygens was. Accordingly, in correspondence on the subject the term appeared very often.[20]

A similar tendency can be observed with Robert Boyle. In his papers on hydrostatics he often mentioned the "Principles and Laws of Hydrostaticks" or the "known laws of hydrostaticks".[21] Boyle also mentioned "optical laws," "mechanical laws" and "magnetical laws,"[22] or even the "laws of refraction."[23] Like Wallis and Wren, he used the terms "Principle" und "Law" equivalently; often he spoke of "rule".[24] A similarly broadened use can be seen in the writings of Isaac Newton. In his early (unpublished) research on collisions, he adopted Wallis' and Wren's use of the law-concept.[25] When, in 1672, still as a rather unknown figure, he submitted his first paper on light and colours to the Royal Society, he spoke of the "received laws of refraction" but called the same proposition the "Hypothesis of the proportionality of the sines of Incidence and Refraction".[26] In the controversy following his publication, the terms "Laws of Refraction" or "lex refractionis" were used by all parties without hesitation, and synonymously as "rule of refraction" or "regula refractionis".[27] Those examples indicate a gradual expansion of the traditional concept of laws of nature into other areas of research – in optics, e. g., Descartes had been eager not to call the sine-relation a "law" – and into the realm of lower level statements. The general meaning, however, according to which laws are the starting point of deductions, at the end of which particular phenomena and observations were explained, was still maintained.

New meanings

In the course of that development, however, an essential shift of meaning occurred that perhaps was barely realized. A characteristic example can be seen in Robert Hooke. In his *Micrographia* of 1665 he not only named the Snellius/Descartes sine-relation in optical refrac-

---

19 Huygens/Oldenburg (1669); all emphasis in the original.

20 Letters to Huygens, Wallis, Wren, Sluse: Hall/Hall (1965-86), V, pp. 103, 117, 358.

21 Boyle (1666), cf. Birch (1772) II, pp. 742-5; or Boyle (1663), cf. Birch (1772) II, p. 36; or *Philosophical Transactions* 1 (1666), pp. 233-4. The terminology appeared also in the review of his main publication: *Philosophical Transactions* 1 (1666), p. 174.

22 Boyle (1674), cf. Birch (1772) IV, pp. 69 and 72.

23 *Usefulness of of Mathematics to Natural Philosophy* in Boyle (1671): cf. Birch (1772) III, p. 430.

24 Boyle (1674), cf. Birch (1772) IV, pp. 68 and 71, for example.

25 Cf. his manuscript "The Lawes of Motion" of 1666-69: Herivel (1965).

26 Newton (1672), (Cohen (1958), pp. 47-59); in particular p. 3076 (reprint 48) and p. 3077 (reprint 49)).

27 *Philosophical Transactions* 7 (1672), pp. 4081, 4089 (Cohen (1958), pp. 79 and 81), or *Philosophical Transactions* 9 (1674), pp. 218 and 501(Cohen (1958), pp. 149 and 152), for example.

tion the "law of refraction" (and was probably the first to do so), but he emphasized that he found the law "experimentally".[28] The sine-relation was not, as traditionally, regarded as a statement from which observational results should be derived, but as itself being subject of empirical investigation. It could stand as a law even without being embedded in a larger system of statements. In similar manner Hooke mentioned, in discussing his newly invented baroscope, the "very odd laws" of the change of the pressure of the air.[29] Again he had in mind something like empirical regularities without reference to a larger system by which they could be explained. To call them "laws" was to invert the traditional idea. No longer were laws the central principles, but the particular phenomena and regularities which traditionally had to be derived from them: no longer the traditional *explanans*, but the *explanandum*! The position of laws within the explanatory structure of natural philosophy had been turned upside down.

Henry Oldenburg, again, adopted and pushed forward the new meaning. In reviewing Hooke's book, he mentioned both the "law of refraction" and the "odd laws" of the airpressure.[30] In a later praise of the usefulness of the new natural philosophy, he mentioned, among other instruments,

"an Instrument to measure the Refractions of Liquors of all kinds, for establishing the Laws of Refraction".[31]

That laws could – and perhaps should – be established by *instruments* is a significant indicator of the change of perspective. A similar use can be seen in Newton's texts during the controversy about his optical theory. At one point, he emphasized that the question to be investigated was "What is the Law according to which each ray is more or less refracted?"[32] His target was not the sine-relation, but a regularity indicating how the different colours were refracted – a dispersion-rule in modern terminology. And he was fully aware that such a relation could only be found empirically and that any explanation from general principles was far away. Nevertheless he called it a law. In another passage, with respect to the diverse refrangibility of rays with the same incidence, he put the question:

"Unde oriatur illa diversitas; An sit à causa aliqua incerta & irregulari, vel certâ lege, secundum quam radius quilibet aptus est determinatam aliquam refractionem pati."[33]

The contraposition of an "uncertain and irregular cause" and a "certain law" is revealing and tells a great deal about laws. Empirical statements obtained the status of laws if they represented a *distinct* and *regular* outcome. There was not much left of the traditional notion of laws as principles. Of course one could attempt to include empirical laws, once established, into a larger system – Newton's later attempts to explain the different refrangibility mechan-

---

28 Hooke (1665), preface, unpaginated page after the table (p. xxv in my counting). Ruby (1986) (reprint p. 297) ascribes the term "lex refractionis" already to Roger Bacon, but gives no reference.
29 Hooke (1665), preface, unpaginated (p. xiv in my counting).
30 *Philosophical Transactions* 1, Heft 2 (1665), pp. 31 and 32.
31 Preface to the 4th Volume of the *Philosophical Transactions*: vol. 4, Heft 45 (1669), p. 894.
32 *Philosophical Transactions* 7 (1672), p. 4004, reprint in Cohen (1958), p. 93.
33 *Philosophical Transactions* 7 (1672), p. 5015, reprint in Cohen (1958), p. 100.

ically provide a famous example. But their status as laws did not depend on such an inclusion: it was just secured by their certainty and regularity.

Neither Newton nor Hooke, however, brought forward a consistent terminology. Hooke, e. g., presented the inverse relation between pressure and volume of gases, discovered by Boyle and Townley, not as a "law", but always as "Mr. Townley's Hypothesis".[34] There was no elaborate terminology, and neither Hooke nor Newton[35] were inclined to discuss those questions in their own right. And it is not clear whether they realized how significant a step they had taken: after all, it could appear as just another step within the ongoing broadening of the terminology.

Boyle, however, was more reflective. Probably as a response to what he recognized as changing use, he presented an extended discussion of the concept of laws of nature and tried to combine the different views. He introduced a whole scale of laws: At the one end were the "fixt laws of nature, or rules of action and passion among the parcels of the universal matter",[36] being general rules with broad scope. At the other end he found empirical regularities. He compared them to "municipial laws ... that belong to this or that particular sort of bodies." Only the former ones were really "laws of nature, more properly so called," the latter ones could also be called "customs of nature".[37]

Boyle's terminology points back to the ideas of Francis Bacon. Bacon had actually used a concept of law. It was closely connected to his idea of "form" with all the latter's unclarities, and therefore not very sharply defined.[38] In Bacon's law-concept, theological considerations played no role; instead he spoke of the "law and its paragraphs" and thus brought the concept much nearer to the juridical realm than traditionally – after all, he was a lawyer and statesman. In that context he drew the distinction between "consuetudines naturae" – observed regularities – and "leges fundamentales et communes".[39] In the juridical language, an analogous terminology was well known: customs of particular regions, e.g, were sometimes called "laws of the land", and there was definitely no reference to an external commanding authority.[40] Much of the common-law tradition, so important in England, was based on "customary" and on "municipial laws", which were incompatible with the idea of an individual legislator. In the 17th century, there was a great deal of debate about the status of those laws, with immediate political consequences. Bacon was himself involved in those issues and acted as a proponent of authoritative legislation.[41] In transferring that concept to natural philosophy, Bacon was well aware that there were different types of statements involved, which played different roles within the system and had to be justified in different ways. "Consuetudines naturae" could only be recognized by empirical means and were valid even without being

---

34 He followed just Boyle's own terminology: Hooke (1665), pp. 225 and 227.

35 A famous illustration of this point is Newton's renaming of astronomical regularities from "Hypotheses" in the first edition of his *Principia* (beginning of book III) to "Phaenomena" in the second edition.

36 Boyle (1663), cf. Birch (1772), II, p. 37.

37 Hunter/Davis (1996), sect VII, § II, p. 107; cf. Birch (1772) V, p. 217.

38 He spoke, for example, of "leges fundamentales et communes, quae constituunt Formas": *Novum Organon* II, p. 5; cf. Spedding/Ellis et al. (1858) I, p. 232. For a closer discussion of Bacon's concept of law, cf. my Steinle (1995).

39 *Novum Organon* II, 2 and 5.

40 S. OED, s. v. "law", sect. II, p. 14.

41 For the common law tradition, cf. Pocock (1987), in particular p. 103. I am grateful to Laurence Brockliss for bringing that tradition to my attention.

derived from more general principles, whereas "laws" in the proper sense were always part of a larger structure of argumentation and took the status of principles. While Bacon thus restricted his concept of laws to those statements, it was Boyle who gave up that restriction. Though he took up Bacon's distinction in verbal translation, he extended the talk of laws and even included the "customs of nature" in one overarching concept.

"Laws of Nature" and metaphysics

Boyle's hierarchy of laws of nature made explicit the widening of the concept he noted in Hooke, Newton and probably others. That he could take such a step without much hesitation had to do with his general background. He analyzed the metaphysical and theological implications of the law-concept more closely and explicitly than others. Since the idea of laws referred to an author of laws, it was important to consider how that author was supposed to act in the world. Boyle, propagating his "corpuscular philosophy," stated

> "that the universe being once framed by God, and the *laws of motion* being settled ... , the phenomena of the world ... are physically produced by the *mechanical affections* of the parts of matter, and that they operate upon one another *according to mechanical laws*."[42]

Once the laws were settled, all further action was considered to be purely mechanistic. In particular, laws could not be conceived as constantly acting:

> "I must freely observe that, to speak properly, a law being but a notional rule of action according to the declared will of a superior, it is plain that nothing but an intellectual being can be properly capable of receiving and acting by a law. For if it does not understand, it cannot know what the will of the legislator is, nor can it have any intention to accomplish it, nor can it act with regard to it, or know when it does, in acting, either conform to it or deviate from it."[43]

With implicit reference to Descartes, he emphasized that it was easily intelligible that God had impressed a certain amount of motion upon the parts of matter, and also the "powers" to transfer motion between them, while the overall amount remained constant. But he hurried to add a 'caveat':

---

42 Boyle (1674), cf. Birch (1772) IV, pp. 68-69; see also another explicit passage in Boyle's treatise on the *Origin of Forms and Qualities* (1666): Birch (1772) III, p. 48.

43 This and the next quote from *On the vulgarly recieved notion of nature*, Hunter/Davis (1996), end of sect. II, pp. 24-5 (Birch (1772) V, pp. 170-1). The book was printed in 1686. While the main text written already in 1665-6, the quoted passages were added in 1680.

"I cannot conceive how a body devoid of understanding and sense, truly so called, can moderate and determine its own motions, especially so as to make them conformable to laws that it has no knowledge or apprehension of."

Given such a fundamental critique, it is surprising that Boyle himself used the terminology of laws so often and even propagated a widening of its meaning. Realizing that problem, he argued that it was the resemblance between the regular motions of inanimate bodies and the actions of lawful agents that led many (including himself) to use the notion of laws, "for brevity's sake or out of custom".[44] But that analogy had, so Boyle emphasized, to be handled carefully: even artefacts, such as a clock, or an arrow shot towards a target, showed regular behaviour. But no one would speak here of a law: what was active here was an obvious mechanic cause. "Laws," Boyle warned, had always to be understood as metaphors rather than real agents.

With those considerations, Boyle responded to a similar critique out of the circle of the "Cambridge Platonists".[45] Ralph Cudworth was, like Henry More, much opposed to Descartes' mechanistic philosophy as something that would lead to atheism. Just like More, he nevertheless adopted the concept of laws of nature and proposed a strictly verbal interpretation. God governed the world by laws, but

"Though it be true that the Works of Nature are dispensed by a *Divine Law* and *Command*, yet this is not to be understood in a *Vulgar Sence*, as if they were all effected by the mere Force of a *Verbal Law* or *Outward Command*, because Inanimate things are not *Commandable* nor *Governable* by such a *Law*."

That observation led him to far reaching conclusions:

"and therefore besides the Divine Will and Pleasure, there must needs be some other Immediate *Agent* and *Executioner* provided, for the producing of every Effect. Wherefore the *Divine Law* and *Command*, by which the things of Nature are administred, must be conceived to be the Real Appointment of some *Energetick, Effectual* and *Operative* Cause for the Production of every Effect."[46]

Cudworth called that agent a "Plastick Nature", mediating between brute matter and the spiritual realm. A similar idea of such an intermediate, immaterial substance, transferring God's intentions onto matter, had been proposed by Henry More under the label "spirit of nature", but, in contrast to Cudworth, not explicitly been connected to the concept of laws of nature.[47]

Boyle's notes of 1680 can well be understood as a reaction to Cudworth's considerations which had been published in 1678. While Boyle agreed with Cudworth that laws could not possibly be understood as real agents, he rejected the introduction of an intermediate agent

---

44 *Ibid.*, p. 24.
45 I thank Andreas Hüttemann for discussions on Cudworth's critique.
46 Cudworth (1678), p. 147; original emphasis.
47 Particularly elaborate in More (1659), (reprint More (1659)); cf. Boylan (1980).

and insisted, in contrast, on a merely metaphorical interpretation of the law talk. Their dispute strikingly shows how deeply the role assigned to laws of nature was dependent on metaphysical and theological beliefs.

Increasing use

Although those discussions pointed to fundamental problems of the concept of laws of nature, they did not lead to a reluctant use. Quite the contrary, the use developed in the other direction. Whereas in the early years of the *Philosophical Transactions,* the talk of laws had shown up here and there but had not been broadly established, it became increasingly prominent towards the end of the century. One of the main steps was the publication of Newton's *Principia* in 1686. Although not too often read in detail, the work became most influential, even for the presentation of natural philosophy to a broader public. At the very outset of the book Newton put his three "Axiomata, sive Leges Motus." His terminology directly and deliberately reflected Descartes' use of "laws" as the central principles of natural philosophy. In Newton's *Principia*, however, the concept was also present in many other places, and in various other meanings. Newton mentioned, for example, the "laws of planetary motions," "of the oscillating pendulum," "of gravitation," "of resistance" of a fluid against the motion of a projectile, or "of condensation" of an elastic fluid.[48] In his not less influential *Opticks* of 1704 he added the "Laws of refraction".[49] The whole spectrum of the use I have sketched above was present in the widely visible works of one author. At the same time, it became clear that the concept was no longer as sharp as before. "Law" could well stand for the central, axiomatic principle of a field, as well as for more or less empirically established regularities, or for anything in between. Again the terminology was not consistently used: Newton called Kepler's three discoveries, e. g., both "laws" as well as "Phaenomena".[50]

Newton was not alone in such a use. In his review of the *Principia,* Edmond Halley adopted Newton's terminology in its whole broadness, speaking of the "Principles of Natural Philosophy" as "Laws of Motion", of "Laws of reflection of Bodies in Motion after their Collision", of the "Law or Rule of the decrease or increase of the Tendency or Centripetal force", or of (different) "Laws of decrease of Density" (in different resistive media).[51] For Halley, such a use was as the more significant, as he previously, in his many articles on different subjects, never had used the law-concept, but had instead preferred a fine-grained mathematical language of Propositions, Corollaries, Lemmata etc. A similarly broad use of the law-terminology can be seen in Roger Cotes' preface to the second edition of the *Principia* in 1713.[52] The concept went far beyond a specialist audience, moreover. In his popular sermons against atheism in the early 1690s, the clergyman Richard Bentley spoke extensive-

---

48 Cf. the edition of Cohen/Koyré et al. (1972), pp. 54, 359, 428, 521, 569, 577, 759. For an extended discussion, see my Steinle (1995).
49 Book 1, part 2, Prop.3, exp.7: Newton (1952), p. 129.
50 Cohen/Koyré et al. (1972), pp. 556-63 and p. 569.
51 Halley (1687), pp. 291, 292, 295.
52 Cohen/Koyré et al. (1972), pp. 20, 24, 33.

ly of "Laws of Nature," "Laws of Motion," "Laws of Vegetation and Life," "Laws of Mechanism" and the like. He introduced universal gravitation as a

> "constant Energy or Faculty (which God hath infused into Matter) perpetually acting by certain Measures and (naturally) inviolable Laws".[53]

In those sermons the theological aspect was, of course, most prominent. Newton, from whom Bentley got all his necessary information, applauded such a use of the concept of laws. From the 1690s on, the notion was well established in England[54], and, at least in some branches of research, even outside: in all further articles on the collision of bodies, even from authors such as Malebranche, Mariotte and Leibniz, the terminology of laws was widely present.

A significant indicator of that development is the point that many of the great successes of the "new sciences" in the 17th century came to be called "laws" only late in the century.

- Kepler had introduced his three propositions in the *Astronomia Nova* (1609) and his *Harmonice Mundi* (1615) as "propositions." In 1686, Newton called them the "laws of planetary motions, detected by Kepler".[55]
- In Galilei's analysis of the free fall of bodies (1638), and in nearly his whole work, the concept of laws had not been mentioned. Boyle, in 1674, spoke of the "laws of acceleration of heavy bodies descending".[56]
- Neither Descartes nor Snellius had called the sine-relation in optical refraction a "law." Hooke, in 1665, referred to the "law of refraction."
- Boyle had introduced the inverse proportionality between pressure and volume of gases as a "Hypothesis" (1662).[57] At the end of the century the statement just counted as "Boyle's law."
- Hooke finally seems to be the first to call one of his own propositions a "law of nature": he did so in introducing the proportionality between distance and restituting force, soon to be known as "Hooke's law".[58] But that happened only in 1678.

Within a few decades, the concept of laws of nature had been given a firm and visible place. It was no longer debated, but could just be used. At the same time, however, it had lost all sharpness, and the broadness of its meanings – from a-priori insights to empirical regularities – allowed most different authors to refer to its authority.

---

53 Bentley (1693a) and Bentley (1693b). I refer to the reprint in Cohen (1958), pp. 348, 368, 369, 387, 388.
54 For further references see Zilsel (1942), pp. 270-273.
55 *Principia* III, prop.4, scholium: Cohen/Koyré et al. (1972), p. 569.
56 Boyle (1674), cf. Birch (1772), IV, p. 71.
57 Boyle (1662); cf. Hunter/Davis (1999), III, pp. 50, 58.
58 Hooke presented the law in 1678 in the last of his Cutlerian Lectures: *De Potentia Restitutiva*; s. Gunther (1923), VIII, pp. 333-334 and p. 336. Characteristically enough, he mostly used the combined phrase "Rule or Law of Nature."

Epilogue

What is the background of this development? Newton himself mentioned that new concepts were needed in order to formulate causality after the substantial forms of the tradition had been discarded.[59] But that can only be a part of the story. Not everyone associated causality with laws, and there had been obvious alternatives to the concept of laws. Historical research has not yet proceeded far enough to provide a comprehensive answer. In this last section, I want to point out some preliminary guesses as working hypotheses for further research.

It is significant that it was exactly the protagonists of the program of an organized science in the Royal Society who propagated the use and widening of the law-concept: Boyle, Hooke, Oldenburg, Wren, and Wallis. Even Henry More was one of the first members of the Society, and so was Walter Charleton, who used the terminology of laws of nature in the course of physiological considerations.[60] The program of an organized research had to establish itself and was challenged and attacked from different angles. In many of the early papers there is an apologetic mood visible.[61] In announcing his *Micrographia*, e. g., Hooke started with a discourse on the superiority of the "*Experimental* and *Mechanical* knowledge" over the traditional "Philosophy of *discourse* and *disputation*".[62] Thomas Sprat's *History of the Royal Society* of 1667 – published under the auspices of the Royal Society in the fifth year of its existence exactly for the purpose of gaining public support – was full of similar considerations, showing the usefulness of the new experimental philosophy. One of the most serious objections the Society had to face was of a theological nature: how did such a new science relate to religion and the divine authority? Many arguments were brought forward in order to show that experimental philosophy did not lead to atheism, but was, on the contrary, supportive of religion.

Oldenburg, for example, was quite active. In one of the first issues of the *Philosophical Transactions* he reviewed a book in which the clergyman and later Archdeacon of Canterbury Samuel Parker analyzed Descartes' proof of God's existence, found it insufficient, and replaced it by a physico-theological argument.[63] Oldenburg emphasized that Parker, being a Fellow of the Royal Society, discussed and plainly refuted the objection "that Philosophy is the Apprenticeship of Atheisme." His praise went so far as to compare the book with Boyle's "Usefulness of Experimental Natural Philosophy" which was written to the same purpose.[64] In an editorial note to the third volume of the *Philosophical Transactions*, Oldenburg characterized the Royal Society as

> "not taking up old fame, or flying Reports, upon too easie trust; nor straining for other kinds of wonders, than the most wise Author of Nature hath allowed, but attending closely to the strict measures of *Natural Truth*, and to the useful contrivances of *Art*".[65]

---

59 *Principia*: Auctoris Praefatio ad Lectorem: Cohen/Koyré et al. (1972), p. 15.
60 Charleton/Gassendi (1654), pp. 342-343, 348, 435.
61 Wood (1980), Hunter (1989), Hunter (1990), Stewart (1992), ch.1.
62 *Philosophical Transactions* 1 (1665), p. 27.
63 Parker (1665).
64 *Philosophical Transactions* 1 (1666) issue 18, pp. 324-325.
65 *Philosophical Transactions* 2 (1667) issue 23, unpaginated page at the beginning of the issue.

The emphasis of the "Author of Nature" was certainly deliberate.

In such an apologetic context the concept of laws of nature was very useful: it always pointed to the author of those laws – that was a good argument anyway. That the concept of laws of nature was assigned an increasingly prominent place, had to a large part to do with the apologetic situation of all programs of a "new science" in the 17th century. That function was, however, the only common denominator of the uses of the concept. Since it was connected to theology and metaphysics, it inevitably became involved into the controversies in those areas. And these were issues in which the Royal Society was all but homogeneous. In order to instrumentalize the concept, it had to be explicated in different ways. Those who envisaged a deductive system à la Descartes (More), or after the model of mathematics and astronomy (Wren, Wallis) could adopt Descartes' understanding and call the central principles of their fields "laws." The experimentalists, on the other hand, like Hooke or Boyle, could use the concept of laws of nature to give their results higher dignity: empirical regularities were nothing but an expression of God's will, even more directly given than the difficult and much debated central principles.[66] The explicit voluntaristic aspect of such a concept of law was probably the reason for Boyle to maintain the concept although he realized, more than many others, the problems to which it led in the context of his "corpuscular philosophy". The reason why the meaning of laws of nature was so much widened and why the concept became so heterogeneous in itself, was mainly a reflection of the enormous heterogeneity of those programs which had come together under the common roof of the Royal Society.

References

Auzout, Adrien (1665/6): *L'Éphémeride du comète*. Paris.

Bentley, Richard (1693a): *A Confutation of Atheism from the Origin and Frame of the World, Part II*. London: Mortlock.

Bentley, Richard (1693b): *A Confutation of Atheism from the Origin and Frame of the World: the third and last part*. London: Mortlock.

Birch, Thomas (ed.) (1772): *The Works of the Honourable Robert Boyle*. London, Johnston.

Boylan, Michael (1980): "Henry More's space and the spirit of nature," *Journal of the History of Philosophy* (18): 395-405.

Boyle, Robert (1662): *A Defence of the Doctrine Touching the Spring and Weight of the Air*. London.

Boyle, Robert (1663): *Some Considerations Touching the Usefulnesse of Experimental Naturall Philosophy*. London.

Boyle, Robert (1666): *Hydrostatical Paradoxes, made out by new Experiments*. London.

Boyle, Robert (1671): *Some Considerations Touching the Usefulnesse of Experimental Naturall Philosophy, the second tome, part 2*. London.

---

66 While it has sometimes been emphasized that the experimenters of the Royal Society claimed "to establish the glory of creation while not troubling to descend into matters of theology", Stewart (1992), p. 19, it has not yet been realized how useful the terminology of laws of nature came out to be for that purpose.

Boyle, Robert (1674): *The Excellency of Theology, compared with Natural Philosophy, to which are annexed Some Occasional Thoughts about the Excellency and Grounds of the Mechanical Hypothesis*. London.

Casini, Paolo (1976): "Loi naturelle: réflexion politique et sciences exactes," *Studies on Voltaire and the eighteenth century* 151: 417-432.

Charleton, Walter and Gassendi, Pierre (1654): *Physiologia Epicuro-Gassendo-Charltoniana: Or a fabrick of science natural, upon the hypothesis of atoms, founded by Epicurus, repaired by Petrus Gassendus, augmented by Walter Charleton*. London.

Cohen, I. Bernard (ed.) (1958): *Isaac Newton's Papers & Letters On Natural Philosophy, and related documents*. Cambridge, Mass.: Harvard University Press.

Cohen, I. B., Koyré, A. and Whitman, A. (eds.) (1972): *Isaac Newton's Philosophiae Naturalis Principia Mathematica: The third edition with variant readings*. Cambridge: Cambridge University Press.

Crocker, Robert (1990): "Henry More: A biographical essay," in S. Hutton and R. Crocker (eds.), *Henry More (1614-1687): Tercentenary Studies. With a biography and bibliography by Robert Crocker*. Dordrecht: Kluwer, 1-17.

Crombie, Alistair C. (1994): *Styles of scientific thinking in the European tradition: the history of argument and explanation especially in the mathematical and biomedical sciences and arts*. London: Duckworth.

Cudworth, Ralph (1678): *The True Intellectual System of the Universe: First Part, wherein all the reason and philosophy of atheism is confuted, and its impossibility demonstrated*. London: Richard Royston.

Graßhoff, Gerd (1999): *Natural Law, Divine Creation and Regularities in Heaven: Natural Laws in the Copernican Revolution*. Natur – Gesetz – Naturgesetz: Historische und zeitgenössische Perspektiven, Bad Homburg, (unpublished manuscript).

Graßhoff, Gerd (2001): *Keplers Naturgesetz der Planetenbewegung*. Meeting of the "Laws of Nature" study group. Berlin, (unpublished manuscript).

Gunther, R. T. (ed.) (1923): *Early Science in Oxford*. Oxford: Oxford University Press.

Hall, Alfred Rupert (1990): *Henry More: magic, religion and experiment*. Oxford: Blackwell.

Hall, Alfred Rupert and Hall, Marie Boas (eds.) (1965-86): *The Correspondence of Henry Oldenburg*. Madison: University of Wisconsin Press.

Halley, Edmond (1687): "Account of Isaac Newton's Philosophiae Naturalis Principia Mathematica," *Philosophical Transactions* 16 (186): 291-297.

Herivel, John (ed.) (1965): *The background to Newton's Principia*. Oxford: Clarendon Press.

Hooke, Robert (1665): *Micrographia*. London.

Hunter, Michael (1989): *Establishing the new science: the experience of the early Royal Society*. Woodbridge: Boydell.

Hunter, Michael (1990): "First Steps in Institutionalization: The Role of the Royal Society of London," in T. Frängsmyr (ed.), *Solomon's House Revisited: The Organization and Institutionalization of Science*. Canton: Science History Publications, 13-30.

Hunter, Michael and Davis, Edward B. (eds.) (1996): *Robert Boyle: A Free Inquiry into the Vulgarly Received Notion of Nature*. Cambridge: Cambridge University Press.

Hunter, Michael and Davis, Edward B. (eds.) (1999): *The Works of Robert Boyle*. London: Pickering & Chatto.

Huygens, Christiaan (1669): "Extrait d'une lettre de M. Hugens à l'Auteur du Journal: Regles du mouvement dans le rencontre des Corps," *Journal des Scavans* (18 Mars): 19-24.

Huygens, Christiaan (Oeuvres): *Oeuvres complètes de Christiaan Huygens, ed. par la Société Hollandaise des Sciences*. La Haye: Martinus Nijhoff.

Huygens, Christiaan and Oldenburg, Henry (1669): "A summary account fo the Laws of Motion," *Philosophical Transactions* 4 (46): 925-928.

Jacob, Alexander (ed.) (1995): *Henry More's Manual of Metaphysics. A Translation of the Enchiridion Metaphysicum (1679) with an introduction and notes*. Studien und Materialien zur Geschichte der Philosophie. Hildesheim, Olms.

Klaaren, E. M. (1985): *Religious Origins of Modern Science*. Lanham: University Press of America.

Knebel, Sven K. (1992): "Necessitas moralis ad optimum (III): Naturgesetz und Induktionsproblem während des zweiten Drittels des 17. Jahrhunderts," *Studia Leibnitiana* 24: 182-215.

Loeck, Gisela (1986): *Der cartesische Materialismus: Maschine, Gesetz und Simulation*. Frankfurt/M.: Peter Lang.

McGuire, J. E. (1972): "Boyle's Conception of Nature," *Journal of the History of Ideas* 33: 523-542.

Milton, John (1981): "The origin and development of the concept of the 'laws of nature'," *Archives Européennes de Sociologie* 22: 173-195.

More, Henry (1659): *The Immortality of the Soul, So farre forth as it is demonstrable from the Knowledge of Nature and the Light of Reason*. London: Flesher.

More, Henry (1679): *Henrici Mori Cantabrigiensis Opera Omnia*. London: Maycock.

Needham, Joseph (1956): *Science and Civilisation in China, vol.2*. Cambridge: Cambridge University Press.

Newton, Isaac (1672): "A Letter or Mr. Isaac Newton, containing his New Theory about Light and Colors," *Philosophical Transactions* 6 (80): 3075-3087.

Newton, Isaac (1952): *Opticks, based on the fourth edition 1730*. New York: Dover.

Oakley, Francis (1961a): "Christian theology and the Newtonian science: the rise of the concept of the laws of nature," *Church History* 30: 433-457.

Oakley, Francis (1961b): "Medieval theories of natural law: William of Ockham and the significance of the voluntaristic tradition," *Natural Law Forum* 6: 65-83.

Oakley, Francis (1981): "Natural Law, the Corpus Mysticum, and Consent in Conciliar Thought from John of Paris to Matthias Ugonius," *Speculum: A Journal of Mediaeval Studies* 56: 786-810.

Parker, Samuel (1665): *Tentamina Physico-Theologica de Deo: sive Theologia Scholastica ad normam novae & reformatae Philosophiae concinnata, & duobus libris comprehensa*. London.

Pocock, John Greville Agard (1987): *The ancient constitution and the feudal law: a study of English historical thought in the seventeenth century*. Cambridge: Cambridge University Press.

Roux, Sophie (1999): *Les lois de la Nature au XVIIe siècle: idée, concept, métaphore polémique*. Natur – Gesetz – Naturgesetz: Historische und zeitgenössische Perspektiven. Bad Homburg, (unpublished manuscript).

Ruby, Jane E. (1986): "The origin of scientific 'law'," *Journal of the History of Ideas* 47: 341-351.

Schimank, Hans (1949): "Der Aspekt der Naturgesetzlichkeit im Wandel der Zeiten," in J. J.-G. d. Wissenschaften (ed.), *Das Problem der Gesetzlichkeit II: Naturwissenschaften*. Hamburg: Meiner, 139-186.

Schramm, Matthias (1981): "Roger Bacons Begriff vom Naturgesetz," in P. Weimar (ed.), *Die Renaissance der Wissenschaften im 12. Jahrhundert*. Zürich: Artemis, 197-209.

Specht, Rainer (2001): "Regulae quaedam sive leges naturae," *Studia Leibnitiana-Sonderheft: Kausalität und Naturgesetz in der frühen Neuzeit*, hg. Andreas Hüttemann 31: (in print).

Spedding, J., Ellis, R. L. and Heath, D. D. (eds.) (1858): *The works of Francis Bacon*. London: Longman.

Steinle, Friedrich (1995): "The amalgamation of a concept – Laws of nature in the new sciences," in F. Weinert (ed.), *Laws of Nature: Essays on the Philosophical, Scientific and Historical Dimensions*. Berlin: de Gruyter, 316-368.

Steinle, Friedrich (2001): "Von a-priori-Einsichten zu empirischen Regularitäten: Der Gesetzesbegriff und seine Alternativen in der frühen Royal Society," *Studia Leibnitiana-Sonderheft: Kausalität und Naturgesetz in der frühen Neuzeit, hg. Andreas Hüttemann* 31: 77-98.

Stewart, Larry (1992): *The Rise of Public Science: Rhethoric, Technology, and Natural Philosophy in Newtonian Britain, 1660-1750*. Cambridge: Cambridge University Press.

Wallis, John (1666): "An Essay, of Dr. John Wallis, exhibiting his Hypothesis about the Flux and Reflux of the Sea," *Philosophical Transactions* 1 (16): 263-289.

Wallis, John (1669): "A summary account of the General Laws of Motion," *Philosophical Transactions* 3 (43): 864-866.

Webster, Charles (1969): "Henry More and Descartes: some new sources," *British Journal for the History of Science* 4: 359-377.

Weinert, Friedel (ed.) (1995): *Laws of Nature: Essays on the Philosophical, Scientific and Historical Dimensions*. Philosophie und Wissenschaft: Transdisziplinäre Studien vol. 8. Berlin/New York: Walter de Gruyter.

Wood, P. B. (1980): "Methodology and Apologetics: Thomas Spratt's *History of the Royal Society*," *British Journal for the History of Science* 13: 1-26.

Wren, Christopher (1669): "Dr. Christopher Wren's Theory concerning the same subject (i. e. the Laws of Motion)," *Philosophical Transactions* 3 (43): 867-868.

Zilsel, Ernst (1942): "The genesis of the concept of physical law," *The Philosophical Review* 51: 245-279.

Claus Zittel

„Truth is the daughter of time".
Zum Verhältnis von Theorie der Wissenskultur, Wissensideal, Methode und Wissensordnung bei Bacon

„Hic vero experimentandi modus plane irrationalis est, et quasi furiosus"

(DA, I S. 632).

1. Polyhistorie vs. Naturwissenschaft?

Wie kaum ein anderer Denker hat Bacon die vielfältigen und widersprüchlichen Strömungen seiner Zeit in seinem facettenreichen Werk aufgenommen und zugleich seine Epoche tief beeinflußt. Nicht zuletzt aufgrund dieser zahlreichen Facetten ist sich die Forschung bei der Bewertung seines Philosophierens insbesondere darüber traditionell uneins, ob Bacon nun als Herold der Moderne aufzufassen oder doch eher den Renaissancemetaphysiken zuzuschlagen ist. Diejenigen, die ihn zu den Modernen zählen, beziehen sich dabei fast ausschließlich auf seine Methodologie und sein technokratisches Ideal der totalen Naturbeherrschung[1], und dies häufig mit stark kritischem Akzent.[2] Doch mehren sich in jüngster Zeit die positiven Stimmen und Ian Hacking gab sogar „Zurück zu Bacon" als neue Losung aus.[3] Die andere Rezeptionslinie sieht Bacon hingegen mit Blick auf seine enzyklopädischen Projekte[4] stärker im Kontext magischer und metaphysischer Philosophien seiner Zeit verhaftet. Somit zeichnet die Forschung ein janusköpfiges Baconporträt, vorwärtsgewandt mit seiner Methode, konservativ und eklektizistisch[5] mit seinem Wissens-System, und mancher Interpret resümiert, Bacons Philosophie trage „die Grenze von Naturwissenschaft und Polyhistorie als Widerspruch in sich selbst."[6]

---

1 Vgl. Adorno/Horkheimer (1971), S. 8ff.

2 Kritisch zur Methode, die als diletantisch und wertlos bezeichnet wird, äußerte sich für die deutsche Bacon-Rezeption folgenreich besonders Cassirer (1971), Bd. 2, S. 11ff. Bezeichnenderweise preist sich das neue Buch von Gaukroger (2001) bereits auf dem Umschlag an als „the first truly general account of Francis Bacon as a philosopher"!

3 Hacking (1996), S. 250. Zu nennen ist aber auch Lothar Schäfer, der bilanziert: „verglichen mit Bacons Wissenschaftskonzept ist Descartes traditionell, geradezu platonisch. Die Differenz von Subjekt und Objekt wird substantiell verankert in der Unterschiedenheit von res extensa und res cogitans – bei Bacon wird sie handelnd produziert. Descartes denkt über das Wissen nach als Grundlage rationalen Handelns, bei Bacon ist Erkennen selbst ein Handeln, und Fakten werden geschaffen. Wenn denn die Moderne durch die Selbstermächtigung der Subjektivität charakterisiert werden soll, dann ist Bacon eine Leitfigur von größerer Radikalität." Schäfer (1999), S. 99.

4 Thiel (1993), hier: S. 171: „Zweifellos gehören Bacons Systementwürfe in den Zusammenhang [...] eines gewissen [...] Platonismus".

5 Z. B. Gaukroger (2001), S. 28-36.

Ich möchte im folgenden zeigen, daß so auf fatale Weise Bacons Werk zu seinen Lasten künstlich aufgespalten und die genannten Widersprüche zwischen Methodologie und Enzyklopädik nachträglich in seine Philosophie hineinokuliert wurden. Beide Rezeptionstraditionen bringen zu grobschlächtige Beschreibungskategorien für sein Philosophieren in Anschlag, unterschätzen dramatisch die Radikalität seiner Methoden- und Ordnungsentwürfe, und verabschieden zu leichtfertig die Möglichkeit, seine verschiedenen philosophischen Ansätze als ineinandergreifende Teile eines kontinuierlichen und zusammenhängenden Forschungsprojektes zu begreifen.[7]

Dem entgegen stellt sich mir somit die Aufgabe, deutlich zu machen, *daß und wie bei Bacon Geschichtsauffassung, Methodologie und Enzyklopädik untereinander in einer stringenten Verbindung stehen.* Bei diesem Vorhaben wird mir gerade jene seiner zahlreichen Naturhistorien als Fluchtpunkt dienen, die aufgrund ihrer Centurien-Ordnung am stärksten der enzyklopädischen Programmatik der traditionellen Naturhistorie verpflichtet zu scheint und daher auch am meisten als reaktionär und unwissenschaftlich in Mißkredit kam, nämlich: *Sylva Sylvarum.*[8] Denn selbst an diesem Werk läßt sich zeigen, daß die Ordnungsformen von Bacons Naturhistorien sich nicht in einen *ordo disciplinarum* integrieren und damit eben nicht am Enzyklopädieideal orientiert sind, ja dieses faktisch sogar verabschieden.

In der Baconforschung operiert man oft entweder mit einem unspezifischen Verständis von naturhistorischen Enzyklopädien und betrachtet diese als blinde Faktensammlungen oder aber mit einem zu speziellen Enzyklopädieverständis[9], das verallgemeinert und Bacon vorschnell als Leitidee unterstellt wird. Nach diesem speziellen Verständnis waren viele Enzyklopädien in der frühen Neuzeit an einem platonistischen Alleinheitsideal und christlich-heilsgeschichtlich orientiert: Durch die Rekonstruktion des nach dem Sündenfall verloren gegangenen Wissens, soll der göttliche Heilsplan offenbar gemacht und die Ordnung wiederhergestellt werden. Über die Cyklusvorstellung bei der Enzyklopädie wird als Ideal des perfekten Wissens die Einsicht in den hinter allen Dingen und Disziplinen liegenden universalen Zusammenhang supponiert. Nicht zuletzt gegen diese heilsgeschichtliche und universalwissenschaftlich ausgerichtete Bacon-Deutung, die mehr oder weniger strikt von Forschern wie Whitaker, Hattaway, Rossi, Yates, Vickers und Whitney[10] vertreten wird, werde ich in diesem Essay argumentieren und beginne folglich mit einer Diskussion von Bacons Geschichtsauffassung. Die Orientierungsmarken für den folgenden Argumentationsgang werden sein: zuerst werde ich vom Begriffspaar Zufall und Geschichte, dann von Zufall und Methode und schließlich von Zufall und Ordnung bei Bacon handeln.

---

6 Schmidt-Biggemann (1983), S. 225.

7 Leitend ist bei den bisherigen Darstellungen zumeist eine evolutionistische Geschichtsauffassung, derzufolge sich die Wissenschaft allmählich aus magischen Kontexten befreit und emanzipiert.

8 Von den 39 philosophischen Schriften Bacons sind nur wenige bekannt, in der deutschsprachigen Rezeption gerade mal eine Handvoll. Bacons Werke werden mit Siglen, Band und Seitenzahl, bzw. Paragraphennr. zitiert nach: Bacon (1857); Bacon (1990b), zitiert als N. O. mit Aphorismennummer.
Verwiesen sei ferner die (teils sehr freien) englischen Übersetzungen von *Redargutio philosopharum* (RPh), *Temporis partus masculis* (TPM) und *Cogita et Visa* (CV) in Farrington (1964) und die deutschen Übersetzungen von *De Augmentis* (= DA) in Bacon (1966) (= WF, Nachdruck der sehr schlechten und lückenhaften Übertragung von J. H. Pfingsten, 1783), sowie von Bacon (1984) (= VT); Bacon (1190a).

9 Zu Definitionsproblemen der keineswegs eindeutigen Textgattung: ‚Enzyklopädie‘ in der frühen Neuzeit insbesondere in Abgrenzung zur Naturhistorie: Schütze (2000), S. 20-28.

2. Zufall und Geschichte: Die Wahrheit ist die Tochter der Zeit

Einsetzen möchte ich mit Bacons berühmten Diktum: „Die Wahrheit ist die Tochter der Zeit", das in in dieser sententiösen Form im *Novum Organum* und in der Frühschrift *Cogitata et visa* auftaucht.[11]

Drei Deutungsvarianten dieses Satzes mit jeweils weitreichenden Folgen für die Gesamtinterpretation Bacons lassen sich unterscheiden:

1. die relativistische Deutung: Bacon erkennt in aller Schärfe die Geschichtlichkeit von Wahrheit und mithin auch der eigenen Position
2. die evolutionistische Deutung: Nach und nach, mit dem Fortschreiten der Wissenschaft wird das Entstehen von Erfindungen von seiner Zufälligkeit befreit, und sich der Wahrheit sukzessive angenähert.
3. die platonistische Deutung: Die Wahrheit wird von der Zeit geboren, d. h. aus dieser entlassen und steht dann als ewige jenseits von Raum und Zeit bzw. hinter den vergänglichen Wahrheiten steht die eine zeitlose göttliche Wahrheit.

Um mit der letzten Variante zu beginnen, da sie klar abzuhandeln ist: Nur in den erhaltenen Fragmenten der sehr verquasten Frühschrift *Temporis partus masculus* und (mit großen Einschränkungen) auch im *Valerius Terminus* finden sich Stellen, die eine solche Deutung stützen können, da Bacon dort mit einem traditionellen Einheitsideal operiert. So heißt es etwa in Hinblick auf die verfeindeten philosophischen Schulen: „Einheit ist das Gütesiegel der Wahrheit, und die Vielfalt ihrer Meinungen ist der Beweis des Irrtums."[12]

In den anderen Textpassagen, die von höheren Wahrheiten handeln und auf die sich ‚metapysische' Bacondeutungen stets berufen, erläutert jedoch Bacon seine Position gegenüber den göttlichen Wahrheiten kristallklar:

„Denn wenn jemand meinen sollte, er könnte durch Anschauen und Untersuchen der sinnlichen und körperlichen Dinge so viel Licht erlangen, das dies ihm die Natur oder den Willen Gottes enthüllen würde, *so wird er tatsächlich durch die eitle Philosophie betrogen.* Denn die Betrachtung der Geschöpfe und Werke Gottes erzeugt wohl im Hinblick auf die Geschöpfe und Werke selbst Wissen; doch in bezug auf Gott [kein vollkommenes Wissen (no perfect knowlegde), sondern][13] nur Staunen, und das ist gleichsam ein abgerissenes Wissen (quasi abrupta scientia) [...] *So entdecken die Sinne die natürlichen Dinge, die himmlischen aber verdunkeln und verschließen sie.*"[14] Über eine sukzessiv sich komplettierende Erkenntnis der natürlichen Dinge führt folglich kein Weg zur Einsicht in die höheren Zusammenhänge, wie es das enzyklopädische Ideal verheißt.

---

10 Whitaker (1968); Rossi (1957); Hattaway (1978); Vickers (1968); Yates (1975); Withney (1989). Vgl. auch den Beitrag Wolfgang Neubers in diesem Band.

11 „Omnium enim consensu veritatem Temporis filiam esse" (CV, III S. 612; N. O. I Nr. 84).

12 „Errori varietas, veritati unitas competit" (TPM, III S. 535).

13 Der Zusatz: „no perfect knowledge" des englischen Textes von *The Advancement of Learning* wurde im lateinischen Text von DA getilgt. Beim folgenden Halbsatz wird in der lateinischen Version durch das zusätzliche ‚quasi' der Gleichnischarakter betont (s. folgende Fn.).

14 „Si quis enim ex rerum sensiblium et materiatarum intuitu tantum luminis assequi speret quantum ad patefaciendam diviam naturam aut voluntatem sufficiet, *nae iste decipitur per inaniam philosophiam.* Etenim com-

Im *Valerius Terminus* erklärt dies Bacon noch genauer: „There is no proceeding in invention of knowlegde but by similitude; and God is only self-like" (VT, III S. 218). Schaden droht in Bacons Augen gerade durch die Vermischung von menschlichem und göttlichen Wissensbereich (ebd.). Bacon geht es hierbei jedoch weniger darum, falsche Vorstellungen von Gott zu verhindern, als diesen aus dem Bereich des menschlichen Wissens zu exkludieren. Seine Strategie ist hier primär darauf ausgerichtet, durch die Trennung der Bereiche dem „natürlichen und gesetzmäßigen Wissen" keinerlei religiös-moralisch motivierten Beschränkungen auferlegen zu lassen, damit Forschungen auch ohne den Segen der Bibel unbegrenzt vorangetrieben werden können. In dieser frühen Schrift spricht Bacon zwar noch davon, daß hinter all den wechselnden Einzelfällen „the highest generality of motion or summary law of nature God should still reserve within his own curtain" (ebd., S. 220). Doch ist dieser supponierte Allzusammenhang für den Menschen unerkennbar und damit, das ist ganz entscheidend, für den Menschen auch bedeutungslos. Ohne erkennbares Bedauern geht Bacon sofort dazu über, den dem Menschen zugänglichen Bereich zu preisen; für diesen allein sei der menschliche Verstand *wie ein Spiegel* angemessen angepaßt (VT III, S. 221). Es gebe keine Konkruenz zwischen göttlicher Ordnung und menschlichem Verstand, der als ein Zauberspiegel („enchantet class", VT III S. 241) keine wahren Abbilder der Natur liefert. Entgegen vielen Interpretationen[15] stellt die zweite ‚Spiegel'-Passage des *Valerius Terminus* klar, daß diese durch den Spiegel bedingten Irrtümer angeboren und irreversibel sind: „the native and inherent errors in the mind of man which have coloured and corrupted all his notions and impressions" (ebd.). Sie gehören daher zu den Idola Tribus und bedingen die grundsätzliche Spaltung zwischen menschlicher Erkenntnis und universaler Ordnung.[16]

Das wahre Ziel des Wissens besteht daher auch schon im *Valerius Terminus* nicht in einer reinen Erkenntnis der Natur oder im Erreichen eines in sich harmonischen und kohärenten Wissenssystems, was sowieso unmöglich sei, sondern in der Beherrschung der Natur. Wissen heißt nun: Beherrschen, und man hat nur insofern etwas erkannt, als man im aktiven Vollzug

---

templatio creaturum, quantum ad creaturas ipsas, producit scientiam; quantum Deum, admirationem tantum, quae est quasi abrupta scientia. [...] *Sensus humanos solem referre, qui quidem revelat terrestrem globum, coelestem vero et stellas obsignat.*" (DA, I S. 436; Adv. III S. 267 vgl. a. N. O. I, Vorrede). Die lateinische Variante klingt weniger eschatologisch: abrupta scientia ist abgerissenes Wissen, also ein abgebrochenes *Geschehen*, ‚broken knowledge' betont wie die deutsche Übersetzung ‚unvollständiges Wissen' (WF, S. 45) stärker den Fragment*zustand* und der heischt eher nach Wiedervervollständigung. Die Parallelstelle im *Valerius Terminus* ersetzt ‚knowledge' durch ‚contemplation' („contemplation broken off", also ‚abgebrochenes Nachdenken'; III S. 218) und betont so auch eher den unterbrochenen Verlauf.

15 Vgl z. B.: Park (1984) und den Beitrag Reicherts in diesem Band.

16 So Bacon später unter Verwendung der Spiegelmetapher in DA, I S. 643ff. und im *Neuen Organon*, wo es von den Idola Tribus heißt, sie seien in der menschlichen Natur selbst, dem Stamme oder der Gattung des Menschen begründet. Daher sei es ein „Irrtum zu behaupten, der menschliche Sinn sei das Maß der Dinge; ja das Gegenteil ist der Fall; alle Wahrnehmungen der Sinne wie des Geistes geschehen nach dem Maß der Natur des Menschen, nicht nach dem des Universums." Immer wieder vergleicht Bacon um diese Inkonkruenz zu veranschaulichen den menschlichen Verstand mit einem verfälschenden Spiegel, der „die strahlenden Dinge nicht aus ebener Fläche zurückwirft, sondern seine Natur mit der der Dinge vermischt, sie entstellt und schändet." („Idola Tribus sunt fundata in ispsa natura humana, atque in ipsa tribu seu gente hominum. Falsum enim asseritur, sensum humanum esse mensuram rerum, quin contra, omnes perceptiones tam sensus quam mentis sunt ex analogia hominis, non ex analogia universi. Estque intellectus humanus instar speculi inaequalis ad radius rerum, qui suam naturam naturae rerum immiscet, eamque distorquet et inficit." N. O. I, Nr. 41)

die Natur manipuliert, um „die Hoheit und Macht des Menschen" über alle Geschöpfe zu erreichen sowie die Entdeckung „aller Tätigkeiten und Möglichkeiten von Tätigkeiten."[17]

Bereits im *Valerius Terminus* kommt Bacon daher zum Schluß, daß die Menschen sich bislang verfehlte Aufgaben gestellt und falsche Ziele gesteckt haben, indem sie Bestimmungen und Kriterien des Wissens formulierten, die ihren Kompetenzbereich übersteigen. Die Liste der falschen Wissensideale ist lang und umfaßt nicht wenige, die Bacon von seinen Interpreten gelegentlich zugeschrieben werden: „That in deciding and determining of the truth of knowledge, men have put themselves upon trials not competent. That antiquity and authority; common and confessed notions; the natural and yielding consent of the mind; the harmony and coherence of a knowledge in itself; the establishing of principles with the touch and reduction of other propositions unto them; inductions without instances contradictory; and the report of senses; are none of them absolute and infallible evidence of truth, and bring no security sufficient for effects and operations" (VT, III S. 242).

Der Grund für die menschliche Erkenntnis-Inkompetenz sei, daß die „Subtiltität" der Dinge viel feiner sei, als die der Worte, Argumente, Begriffe und des Sinnesapparates, daher von diesen auch nicht eingefangen werden können (ebd., vgl. a. N. O. I Nr. 10 u. Nr. 50). Schwerwiegende Wahrnehmungs- und Erkenntnisprobleme verbindet Bacon mit dem Begriff der Subtilität, weil er diesen im Unterschied etwa zu Bestimmungen in Cardanos *De subtilitate*, mit dem Okkulten, nicht unmittelbar Erfaßbaren identifiziert. Das ist keineswegs unbedingt zwingend, denn wie bei Cardano könnte Subtiles ja auch nicht Verborgenes sein, und Verborgenes müßte nicht per se subtil sein.[18]

---

17 „but it is a restitution and reinvesting (in great part) of man to the sovereignty and power [...] which he had in in his first state of creation. And to speak plainly and clearly, it is a discovery of all operations and possibilities of operations from immortality (if it were possible) to the meanest mechanical practise." (III S. 223, vgl. S. 233) Daß Bacon hier auch noch davon spricht, es würde so dem Menschen die Macht, „die er im Urzstande der Schöpfung hatte" wiedergegeben, scheint heilsgeschichliche Deutungen wie die Hattaways zu stützen. Doch nach dem vorherigen Ausschluß des göttlichen Bereichs erscheint dies hier als nur rhetorisch inszeniertes Bekenntnis. Andernorts wird anhand des gleichen Bildes klargestellt, daß der Sündenfall ein moralischer war, der die Erkenntnis von Gut und Böse betraf und nicht die Erkenntnis, d. h. die Beherrschung der Natur. Diese war und ist moralisch neutral und das ist hier Bacons Punkt. (Vgl. DA, I S. 465 und N. O. I Vorrede)). Auch ist die wichtige Passage in Kap. 8 des VT hinzuzunehmen, in welcher Bacon sich mit antiken Vorstellungen universalen Wissens und insbesondere dem traditionellen concatenatio-Gedanken auseinandersetzt, und die auch Neuber in seinem Beitrag in diesem Band als Beleg anführt (VT, III S. 228f.). Bacon zitiert diese antike Vorstellung zwar, doch grenzt er sich gleich darauf explizit von ihr ab: „And it is a matter of common discourse of the chain of sciences how they are linked together, insomuch as the Grecians, who had terms at will, have fitted it of a name of *Circle Learning*. Nevertheless I that hold it for a great impediment towards the advancement and further invention of knowledge, that particular arts and sciences have been disincorporated from general knowledge, do not understand one and the same thing which Cicero's discourse and the note and conceit of the Grecians in their word *Circle Learning* do intend." (ebd.) Er wendet sich zwar seinerseits scharf gegen jedes Spezialistentum, doch ist sein Argument nicht metaphysisch bzw. ontologisch, sondern strikt praktisch und funktionalistisch. Denn Bacon zielt mit dem Begriff des ,universalen Wissens' nicht auf ein zu errichtendes Gesamtgebäude allen Wissens als Zweck an sich, sondern auf die Meta-Ebene der Konstruktions- und Korrekturprinzipien (ebd., S. 229). Daran anschließend reflektiert er die Interdependenzen und möglichen Interaktionen zwischen Einzelwissenschaften: „for sciences distinguished have a dependance upon universal knowledge to be augmented and rectified by the superior light thereof, as well as the parts and members of a science have upon the *Maxims* of the same science, and the mutual light and consent which one part receiveth of annother."

18 Vgl. dazu: Schütze (2000), S. 28ff. Zum Begriff der Subtilität bei Bacon: Wilson (1997), Kap. 2.

Gleich wird noch davon die Rede sein müssen, wie diese Probleme von Bacon methodisch angegangen werden.[19] Es sei jetzt nur darauf hingewiesen, daß schon beim frühen Bacon der entscheidende Unterschied zwischen seinen und traditionellen Vorstellungen okkulter Ursachen ist, daß er die okkulte Form nicht positiviert. Anders als bei Agrippa oder Cardano wird die latente Form nicht essentialistisch gedacht, sondern funktionalistisch. Sie bleibt nur ex negativo über ihre Wirkungen faßbar, und kontrolliert man die Wirkungen, hat man die Form. Bacon vollzieht einen erkenntnistheoretischen Perspektivenwechsel, weg vom ,warum' hin zum ,wie'. Gesucht werden nicht mehr die zugrundeliegenden Ursachen, sondern das Interesse verlagert sich auf die Beschreibung und Beherrschung der Wirkungen.

In der Früh-Schrift *Redargutio philosophiarum* findet Bacon bereits eine schöne Metapher für die Vorzüge der Mannigfaltigkeit gegenüber der Einheit. Die griechische Weisheit „war vielfältig. Und obgleich Vielfalt nicht mit der Wahrheit in Einklang stehen mag, zuletzt bewahrt sie doch den Irrtum davor, fixiert zu werden. Vielfalt verhält sich wie der Regenbogen zur Sonne. Von allen Bildern ist der Regenbogen das vergänglichste und flüchtigste, doch es ist ein Bild. Aber die Vielfalt der Griechen wurde ausgemerzt durch Aristoteles, der selbst ein Grieche war."[20]

Dieser Hinweis Bacons auf den Beitrag des Aristoteles für den philosophischen Fortschritt leitet zur Diskussion der anderen beiden Lesarten des Diktums von der *Wahrheit als der Tochter der Zeit* über, der skeptisch-relativistischen und der evolutionistischen, die ich nun gemeinsam abhandle, wobei ich mich zuerst dem Problem des sozialen Kontextes und dann dem der Überlieferung von Wissen zuwende.

Bacon entfaltet, um die Abhängigkeit der Wahrheit von historischen Umständen aufzuweisen, bereits in seinen frühen Schriften sowie in seinen Hauptwerken ein wissenssoziologisches Untersuchungsprogramm[21]. Die verschiedenen Wissensformen werden aus ihren gesellschaftlichen Bedingungen abgeleitet. Dazu entwickelt er regelrecht eine dynamische Theorie der Wissenskultur, mit welcher aufgezeigt werden kann, wie etwa bestimmte Umstände Entdeckungen begünstigen oder nicht, aber auch, daß neue Entdeckungen auf die Wissensform selber zurückwirken[22], also nicht nur einen quantitativen Fortschritt von Wissen dar-

---

19 Schon dem frühen Bacon zufolge braucht man eine Methode, die angesichts der Subtilität der Dinge nicht vorschnell abstrahiert. Doch sei es auch kein Weg von einer Einzelheit Licht auf die nächste fallen zu lassen: „in case where one particular giveth light to annother; but there were particulars induce an axiom or observation, which axiom found out discovereth and designeth new particulars" (VT, III S. 242). Bacons Terminologie ist hier verwirrend, insbesondere wenn man einen fundamentalistisch-essentialistischen Axiombegriff im Hinterkopf hat. Unter Axiom versteht Bacon offenbar eine nach mehreren Einzelbeobachtungen aufgestellte Hypothese, die sich jedoch nicht an den Fällen bewähren soll, anhand derer sie gebildet wurde, sondern die neue, andere Einzelheiten oder Instanzen entdecken muss. Falls ein Axiom in dieser Hinsicht nicht ,fruchtbar' ist, ist es „vain and untrue". (ebd.) Auch noch im letzten Werk *Sylva Sylvarum* hält Bacon an diesem Axiomverständnis fest, wenn er schreibt: „The eye of the understanding is like the eye of sense: for as you may see great objects through small crannies, or levels: so you may see great axioms of nature through small and contemptible instances" (Sylv. Nr. 91).

20 „Ea enim varia fuit; varietas autem ut veritati non acquiescit, ita nec errorem figit; sed ad veritatem est instar iridis ad solem, quae omnium imaginum est maxime infirma et quasi deperdita, sed tamen imago. Verum et hanc quoque varietatem nobis extinxit (Graecus et ipse) Aristoteles" (RPh, III S. 561).

21 Zum ,wissenssoziologischen' Projekt Bacons insbesondere in seinen Frühschriften vgl. das schöne Buch von Krohn (1987), S. 35-42 und 107ff und Rossi (1968), S. 44-51.

22 „Auch ist es nicht gering einzuschätzen, daß durch die weltweiten Fahrten zu Wasser und zu Lande, die in unserer Zeit so zugenommen haben, sehr vieles in der Natur entdeckt und aufgefunden worden ist, was über die

stellen, sondern neue Konzeptualisierungen verlangen. Darüber hinaus führt das Bewußtsein des Wandels und der Vergleich von Kulturen zur Einsicht in die Geschichtlichkeit philosophischer Wahrheit und diese Einsicht muß laut Bacon die Art und Weise zu philosophieren radikal verändern.

Als erste Konsequenz bricht Bacon radikal mit der Antike als Vorbild und bringt dies im *Advancement* auf die paradoxe Formel: „Antiquitas saeculi, juventus mundi". Denn, so führt er im *Neuen Organum* aus, „das Altertum ist doch in Wahrheit für das Greisen- und Großväteralter der Welt zu halten; und dieses muß von unserer Zeit ausgesagt werden und nicht von jenem jüngeren Zeitalter der Welt, in dem die Alten lebten. Denn dieses ist zwar mit Rücksicht auf unsere Zeit älter und entfernter, in bezug auf die Welt selbst aber neuer und jünger. Wie wir eine größere Kenntnis der menschlichen Verhältnisse und ein reifes Urteil mit Recht von einem Greis als von einem Jüngling erwarten [...], so kann man auch von unserer Zeit [...] weit mehr als von den alten Zeiten erwarten." (N. O. I, Nr. 84). Die Bedeutung dieses Bruchs ist kaum zu überschätzen. Viele Gelehrte im 16. und 17. Jahrhundert waren selbstverständlich davon überzeugt, daß die Antike ein vollkommeneres Wissen hatte, und alle späteren nur „Zwerge auf den Schultern von Riesen" seien[23].

Bacon macht sich indes daran, die griechische Kultur in ihrer kindlichen Beschränktheit vorzuführen. Seine Strategie zielt darauf ab, die Philosophen der Tradition nicht direkt zu widerlegen, sondern sie vor dem Hintergrund ihrer sozialen und kulturellen Bedingungen zu beurteilen.

„Alle griechischen Philosophen saßen im gleichen Boot. Ihre Irrtümer waren vielfältig, die Ursachen ihrer Irrtümer aber die gleichen."[24] Die Griechen „lebten in einem Zeitalter, das von Fabeln umgrenzt war, hatten nur dürftig Kenntnisse in der Geschichte, waren kaum durch Reisen und Kenntnisse der Natur informiert oder aufgeklärt." Ohne Unterschied hätten sie alle im Norden lebenden Menschen Skythen genannt, alle im Westen Kelten und die Forschungsreisen eines Demokrit, Platons oder Pythagoras mögen denselben zwar wundervoll vorgekommen sein, seien jedoch schwerlich mehr gewesen als Stadtrandexkursionen (RPh, III S. 564). Auch wenn einzelne Griechen überragende Fähigkeiten besessen hätten, könnte dies doch keinesfalls den Vorteil heute Lebender aufwiegen, 2000 Jahre Geschichte und 2 Drittel der Erdoberfläche vergleichen zu können. (RPh, III S. 563f.) Ihr mentaler und professoraler Habitus sei der Weisheit und Wahrheit feindlich gewesen. Sie seien immer Kinder geblieben, denn ihre Philosophie war schnell dabei mit Geschwätz und Argumenten, aber unfähig etwas zu bewirken (ebd.). Im Einzelnen klassifiziert Bacon dann die griechischen Philosophen strikt soziologisch nach der Art ihrer Lehre und unterscheidet herumreisende Sophisten, Schulgründer und zurückgezogene Naturforscher (ebd., S. 564).

Philosophie ein neues Licht ausbreiten kann. Es wäre ja auch eine Schande, wenn die Verhältnisse der materiellen Welt, nämlich die der Länder, Meere, Gestirne, zu unserer Zeit bis ins Äußerste eröffnet und beschrieben worden sind, die Grenzen der geistigen Welt indes auf die Enge der alten Entdeckungen beschränkt bleiben sollten." (N. O. I, Nr. 84)

23 Zu den merkwürdigen Verquickungen des Bacon-Paradoxes mit dem Motiv der Zwerge auf den Schultern von Riesen in der Baconnachfolge siehe: Merton (1980), Kap. 24-28. Im Speddings Kommentar (I S. 458) zu dieser Stelle in DA wird darauf verwiesen, daß Bacons Paradox bereits vorher, z. B. in Brunos *Cena di Cenere* auftaucht. Vgl. a. Simone (1949). Bei Bruno jedoch ist dieses Motiv eingebunden in eine zyklische Geschichtsauffassung und entbehrt völlig der baconschen Schärfe.

24 „Una enim quasi navis philosphiae Grecorum videtur, atque errores diversi, causae errandi communes" (RPh, III S. 570).

Dann kommt es aber zu einem Kulturvergleich, die an der Antike erprobte Kulturkritik wird grundsätzlicher formuliert und auf die eigene Zeit angewendet. Zugleich wird die Kritik dadurch selbstreflexiv und das vom Antikenparadoxon noch suggerierte lineare Entwicklungsmodell fraglich. Von dem Moment an, wo man sprechen lernt, so Bacon, müsse jeder Mensch einen Mischmasch von Irrtümern aufnehmen und assimilieren, doch gewinnen diese Irrtümer ihre Kraft nicht allein durch die Alltagsgebräuche. „Sie sind vielmehr sanktioniert durch akademische Institutionen, Kollegien, Orden, und sogar durch Staaten."[25]

Bacon diagnostiziert: „In der Tradition und Organisation von Akademien, Kollegien und anderen Institutionen, welche als Orte des Lernens und des Austauschs von Ideen gedacht sind, kann alles was dem Fortschritt der Wissenschaften feindlich ist, angetroffen werden."[26] Er macht keinen Hehl daraus, daß in seinen Augen die traditionellen Institutionen ein Bündnis zur Unterdrückung all jener Wissenschaftler darstellen, die unabhängig forschen und urteilen, die mit Begeisterung und unkonventionell vorgehen. Solche Forscher werden ausgeschlossen, durch Isolation wirkungslos gemacht und um ihre Karriere gebracht (ebd.). Außerdem lebe man „in einer Periode, in welcher religiöse Fragen den Verstand monopolisiert haben." (RPh, III S. 572)

Konkrete praktische Forderungen hingegen, die die Mängel beheben sollen, sind die Einrichtung eines freien Kollegiats, in welchem auch neue Disziplinen wie Geschichtswissenschaft, moderne Sprachen, Gesellschaftstheorie und Politikwissenschaft vertreten sein sollten. Für die Naturwissenschaften müssten vorallem praktische Forschungseinrichtungen wie Experimentierlabore, Sammlungen von Instrumenten und Karten sowie botanische Gärten eingerichtet werden. Außerdem solle man dafür sorgen, daß man „keinen Mangel an toten Körpern zu anatomischen Beobachtungen hat"(DA, I S. 489).[27] Die Lehrpläne müßten reformiert werden, die Studenten sollten aus didaktischen Gründen mit dem Studium der Künste beginnen und nicht mit Logik und Rhetorik, denn das entspräche mehr der natürlichen Entwicklung der Kinder. Insbesondere sollte wie im alltäglichen Leben die Trennung von Gedächtnis und Erfindung aufgehoben werden. Außerdem sei es nötig, Rahmenbedingungen für einen freien Erfahrungsaustausch aller Gelehrten Europas zu schaffen. (Adv., III S. 325ff.). Politisch wünschenswert sei ein Kräftegleichgewicht zwischen den Staaten im Friedenszustand (CV, BF 94f.).

Bacon macht aber eine bezeichnende Einschränkung in bezug auf die Beseitigung der Hemmnisse des Fortschritts.

„Es gibt einen Unterschied zwischen Politik und den Künsten. Ein neues Licht und Innovationen sind in dem einen Fall weniger gefährlich als im anderen. In der Politik sogar Verbesserungen aufgrund ihres Verwirrungspotentials eher verdächtig."[28] Damit fordert Ba-

---

25 „Neque haec tantum consensu singulorum firmata, sed et institutis academiarum, collegiorum, ordinum, fere rerumpublicarum, veluti sancita est" (RPh, III S. 562).

26 „in moribus et institutis Academiarum, Collegiorum, et similum conventuum, quae ad doctorum hominum sedes et operas mutuas destinata sunt, omnia progressui Scientarum in ulterius adversa inveniri." (CV, III 597; vgl. N. O. I Nr. 90)

27 Der Hintergrund für Bacons Forderungen war, daß es zu dieser Zeit in England im Unterschied zum Kontinent an den Universitäten keinerlei Studienangebote für Naturgeschichte gab und bis 1620 keinen universitären botanischen Garten.Vgl. dazu: Findlen (1997), hier S. 242ff.

28 „non enim idem periculum a nova luce ac a novo motu instare; verum in rebus civilibus, motum etiam in melius suspectum esse ob perturbationem; cum civilia authoritate, consensu, fama, opinione, non demonstratione et veritate constent" (CV, III S. 597, vgl. N. O. I Nr. 90).

con ein asymmetrisches Verhältnis zwischen Politik und Wissenschaft, doch wird der zu bewahrende Zustand des Staates bezeichnenderweise nicht als der wahrere verabsolutiert und dadurch dem Wandel der Zeit enthoben. Bacon geht hier deskriptiv vor und argumentiert strikt konventionalistisch, wie seine Erläuterung verdeutlicht: „Die politische Kontrolle beruht auf Autorität, Zustimmung, Reputation, Meinung, und nicht auf Demonstration und Wahrheit." (ebd.)

Tritt man nun einen Schritt zurück und überschaut die Vorschläge, so ergibt sich als philosophische Schwierigkeit für die evolutionistische Lesart, daß die Überlegenheit der von Bacon vorgeschlagenen Veränderungen sich erst später zeigen kann und, schärfer noch, daß die Wahrheit der eigenen Position ja relativ zu den bisherigen schlechten Bedingungen gedacht werden muß. Diese muß daher in ihrer Kritik und ihren Forderungen ihrerseits als korrumpiert angesehen werden. Die evolutionistische Position hat darüber hinaus zu verfechten, daß man die kontingenten Faktoren beim Wissen durch die sukzessiven Verbesserungen nach und nach ausschalten kann, die relativistische wird behaupten, daß man zwar neue, und den gerade gemachten Entdeckungen adäquatere Rahmenbedingungen schaffen kann, damit die prinzipielle Kontextualität jedes Wissensanspruchs aber nicht tangiert ist.

In welches Lager Bacon gehört, läßt sich klar entscheiden, wenn man seine Auffassung über die Überlieferung von Wissen mit in Betracht zieht. Der *Veritas filia temporis* – Gedanke tritt stets im Kontext des Sinnbildes der ‚Zeit als Strom' auf und gerade diese Flußmetapher ist hochinteressant, denn Bacon stellt mit ihr eindrucksvoll klar, daß es für ihn keine echte Kontinuität gehaltvollen Wissens gibt: „time is like a river which carrieth down things which are light and blown up, and sinketh and drowneth that which is sad and weighty." (VT, III S. 227; CV, III S. 599, u. N. O. Vorr.)

Andernorts benennt Bacon, was seiner Ansicht nach auf dem Flußbett verloren liegt. Untergegangen seien insbesondere die „veritatis inquisatores Heraclitus, Democritus, Pythagoras, Anaxagoras, Empedocles" (TPM, III, S. 535; vgl. a. *De principis* III S. 84). Sehr hübsch ist auf der anderen Seite seine Erklärung dafür, warum und wie die Überlebenden sich im Fluß der Zeit behaupten konnten. Bacon kommt hierbei zu erstaunlich nüchternen und modernen Einsichten: In der Geschichte der philosophischen Überlieferung setzte sich nicht das bessere Argument durch (VT, III S. 227); es gab daher defacto auch keinen kontinuierlichen Wissensfortschritt. Wenn sich eine Position durchsetzte, dann aus theorie-externen Gründen. Polemisch, aber wissenssoziologisch konsequent, erläutert Bacon den nachhaltigen Erfolg von Aristoteles gerade nicht mit der überlegenen Erklärungskraft von dessen Theorie.[29] Zu Zeiten des hochzivilisierten römischen Reichs seien noch viele Schriften der alten Griechen vollständig präsent gewesen; Aristoteles „wäre niemals erfolgreich bei seiner Vernichtung der anderen griechischen Philosophen gewesen, wären ihm dabei nicht Attila, Geiserich und die Goten zu Hilfe gekommen. Erst dann, als die menschliche Bildung (doctrina) Schiffbruch erlitten hatte trieben die Planken der Philosophie des Aristoteles, die leichter waren und weniger solide in der Substanz, an die Oberfläche der Wellen und der Tod seiner Rivalen wurde zur allgemein akzeptierten Sicht."[30] Diesen Vorgang führt Bacon als ein Bei-

---

29 „er schuf eine Kunst oder ein Handbuch der Verrücktheit und machte uns zu Sklaven von Worten" (TPM, III S. 529f.).

30 „Atque satis constat, sub tempora excultiora Romani plurimos antiquorum Graecorum libros incolumes mansisse. Neque enim tantum potuisset Aristoteles (licet voluntas ei non defuerit) ut ea deleret, nisi Attila et Gensericus et Gothi ei in hac re adjuctores fuissent. Tum enim postquam doctrina humana naufragim perpessa

spiel für die zufälligen „Geburten der Zeit" (ebd.) an. Hier scheint er am weitestgehensten sich der Position genähert zu haben, daß jeweils das wahr ist, was zu einer bestimmten Zeit für wahr gehalten wird.

Es gibt viele Passagen, vor allem im *Novum Organum*, in welchem er die eigene Zeitbedingtheit selbstreflexiv einbekennt. Eine radikales Beispiel dafür ist, wenn er erklärt, sein eigenes Schaffen sei „mehr das Geschenk des Glücks als das der Geschicklichkeit und mehr die Geburt der Zeit als die eines Genies."[31] In Hinblick auf diese Selbstcharaktierisierung verdient ein weiterer Aspekt erwähnt zu werden.

An zwei Stellen identifiziert Bacon die Zeit ebenso traditionell wie fälschlich mit Kronos (DA, I S. 458; Adv. III S. 291[32], dort überraschenderweise jedoch nicht um die Emanzipation von der Herrschaft des Vaters durch die Revolte des Technokraten Zeus zu preisen. Stattdessen wird die frühere Sequenz des Mythos als Vergleichspunkt gewählt:

Bei den Wissenschaften bestehe „die erste Krankheit in einem übermäßigen Begehren zweier äußerst gegensätzlicher Dinge, nämlich der Antike und der Neuheit; wodurch die Töchter der Zeit nach der Natur und Bösartigkeit ihres Vaters geraten. Denn wie die Zeit die Nachkommen verschlingt, so versuchen die Töchter [also die Wahrheiten, C. Z.[33]] sich gegenseitig aufzufressen. Während die Antike neue Errungenschaft mißgönnt, gibt sich die neue Zeit nicht damit zufrieden neues hinzuzufügen, wenn sie nicht zugleich das Alte gänzlich ausrottet und verwirft."[34]

Auch wenn hier der Gedankengang eine andere Richtung nimmt, – durch die Konnotation von Zeit und Kronos bekommt die für sich genommen schon bemerkenswerte Selbstrelativierung Bacons aus dem *Novum Organum* eine bedrohliche Färbung. Die eigene Wahrheit ist nicht nur abhängig von der Zeit, sondern unter der Drohung geboren, sogleich von ihr wieder verschlungen oder aber durch die konkurrierenden Wahrheiten vernichtet zu werden.

Im Lichte der unaufhebaren Kontingenz und Zufälligkeit der historischer Überlieferung verändert alles, was bisher als gelehrtes Wissen galt, radikal seinen Status: Es wird nicht einfach verworfen, sondern es ist jetzt nur Bestandteil eines allgemeinen Erfahrungsfundus, dessen Elemente erst noch auf ihre Wissenstauglichkeit geprüft werden müssen. Bacon befand sich gleichsam bereits in der postmodernen Situation: alles was der Fluß der Geschichte als Treibgut anschwemmt, wird nun gesammelt und ahistorisch nebeneinandergestellt, und

---

esset, tabula ista Aristotelicae philosophiae, tanquam materiae alicujus levioris et minus solidiae, servata est, et extinctis aemulis recepta. At quod de consensu homines sibi fingunt, in et infidum et infirmum est." (RPh III S. 567f.) Überliefert wurde nur daher nur ein „Fragment der griechischen Philosophie". Noch dazu wurde „diese Kreatur nicht genährt auf den Lichtungen und im Dickicht der Natur, sondern in Schulen und Zellen wie ein zu mästendes Haustier." (eamquem certe minime in saltu aut sylvis naturae nutritam; sed in scolis et cellis, tanquam animal domesticum saginatum; RPh III S. 561)

31 „Itaque haec nostra [...] foelicitatis cujusdam sunt potius quam facultatis, et potius temporis partus quam ingenii" (N. O. I Nr. 122).

32 Im lateinischen Text ist von Töchtern der Teit (Temporis filiae) im englischen von Kindern, die deutsche Übersetzung tilgt die Anspielung, indem sie nur noch von Töchtern der Zeit, die ihrer Mutter (!) übel nacharten spricht, und gänzlich den Sinn, wenn sie fortfährt: „denn wie eine Zeit die andere frißt ..." (WF, S. 92).

33 Wenngleich an diesen beiden Stellen, im Unterschied zu den anderen zitierten, nicht die Wahrheit als Tochter der Zeit von Bacon direkt genannt wird, so geschah dies jedoch zwei Abschnitte zuvor (DA, I S. 458).

34 „Horum primus est immodicum studium duorum extremorum, Antiquitatis et Novitatis; Ut enim Tempus prolem devorat, sic haec se invicem; dum Antiquitas novis invideat augmentis, et Novitas non sit contenta recentia adjicere, nisi vetera prorsus eliminet et rejiciat." (DA, I S. 458)

gleichermaßen auf seine Nützlichkeit getestet. Trotz des polemischen Neuerungs-Pathos gilt dann aber auch, daß es nicht klar ist, welche Einzelerklärung sich als beste herausstellt, es könnte auch eine alttradierte sein. Das ist der Ausgangspunkt für Bacons Methodologie- und Enzyklopädiekonzept, von hier aus sind die zu lösenden Aufgaben definiert.

Bacons Enzyklopädiekonzept beruht also auf der Annahme, daß die Verfügbarkeit von Wahrheit prinzipiell problematisch ist, daß es in der Geschichte keine Tendenzen gibt, die garantieren, daß die Wahrheit ans Tageslicht befördert wird, daß die traditionellen Annahmen – Wahrheit bei den Alten, Wahrheit als gemeinsamer Nenner verschiedener Positionen etc. – falsch sind und nichts Vergleichbares an ihre Stelle treten kann. Diese Einsicht relativiert jede Erkenntnis, die wir haben können. Sie relativiert zwar nicht den Wahrheitsbegriff selbst, sofern man darunter die göttliche Wahrheit versteht, doch da wir vom göttlichen Wahrheitsideal strikt getrennt sind, wie sich bei der Prüfung der ersten Deutung von „veritas filia temporum" ergeben hat, ist das Wahrheitsideal, da unerreichbar, auch irrelevant. Wir müssen damit leben, daß das, was wir für wahr halten, immer relativ zu unseren Bedingungen ist.

3. Zufall und Methode: Die Jagd Pans

Eine erste Konsequenz aus der Einsicht in die Geschichtlichkeit des Wissens, ist die eigene Art der philosophischen Darstellung zu verändern und nicht mehr durch die Form zu suggerieren, man könnte universales Wissen erlangen. Deshalb beginnt Bacon aphoristisch zu schreiben, wofür er drei Gründe anführt: erstens, sei es „offener und ehrlicher" wie die Vorsokratiker Beobachtungen „in knappen, scharf umgrenzten Sätzen zusammenzufassen, ohne sie methodisch miteinander zu verketten; sie täuschten nicht vor, noch behaupteten sie, die ganze Kunst zu erfassen." (N. O. I Nr. 86). Zweitens sei das aphoristische Schreiben progressiver: „so knowledge, while it is in aphorisms and observations, is in growth; but when it once is comprehended in exact methods, it may perchance be further polished and illustrate, and accomodated for use and practise, but it increaseth no more in bulk and substance." (Adv., III S. 292) Drittens hätten die kurzen unverbundenen unmethodischen Sentenzen der Vorsokratiker das kritische Denken angeregt und veranlassten die Leser über sich selbst nachzudenken und zu urteilen.[35]

Angesichts der Zeitverhältnisse bedarf es zum zweiten einer besonderen Strategie bei der eigenen Wissensvermittlung: „Bei Wahnsinnigen wird es schlimmer, wenn man ihnen heftig entgegentritt, doch können sie durch die Kunst verführt werden. Dies gibt uns einen Hinweis, wie wir in dieser universellen Verrücktheit vorgehen sollten."[36]

Drittens muß man sich vergegenwärtigen, vor welcher neuen Aufgabe Bacon nun bei seinem eigenen Ordnungsvorhaben stand. Die Grundvorausetzung war völlig verschieden von der der traditionellen Systementwürfe. Denn Bacon ging nicht mehr von einem tatsächlich

---

35 „atque hominum ingenia et meditationes ad judicandum et ad inveniendum simul excitabant" (CV, III S. 593f.).
36 „Ut enim phreneticorum deliramenta arte et ingenio subvertuntur, vi et contentione efferantur, omnino ita in hac universali insania mos gerendus est" (TPM, III S. 529).

oder im Prinzip geordneten Kosmos aus, sondern von einer chaotischen Welt, der der Mensch je nach historischem Ort seine jeweils kontingenten und fiktiven Ordnungen überstülpt. Natur ist für ihn zerstreut und vielgestaltig, sie ist „physica sparsa" (DA, I, S. 548).

Das Problem ist daher zum einen, zu vermeiden, vorschnell Ordnungen zu supponieren und zum andern bessere Alternativen finden. Folglich geht es jetzt darum, die physica sparsa als Wirklichkeit zu akzeptieren, dann eine adäquate Darstellung für das Chaos zu finden, um es schließlich handhabbar zu machen. Wie radikal Bacon dies denkt, wird meist unterschätzt. Er spricht von einer Labyrinthstruktur der Natur oder von einem Dickicht der Erfahrungen[37]. Hier muß man sich ganz bewußt den präzisen Sinn der von Bacon gewählten Metaphorik vor Augen führen. Er will einen Faden für das Labyrinth geben oder einen Pfad ins Dickicht schlagen, doch dadurch ändert sich überhaupt nichts an der grundsätzlichen Labyrinth- oder Dickichtstruktur der Natur! „Der Bau des Weltalls aber erscheint seiner Struktur nach dem Menschengeist, der es betrachtet, wie ein Labyrinth, wo überall unsichere Wege, täuschende Ähnlichkeiten zwischen Dingen und Merkmalen, krumme und verwickelte Windungen und Verschlingungen der Eigenschaften sich zeigen. Dabei muß der Weg bei dem unzuverlässigen, bald aufleuchtenden und bald verschwindenden Lichte der Sinne fortwährend durch das Dickicht der Erfahrungen und einzelnen Dingen gebahnt werden." (N. O. I Vorr.)

Schwierig ist die Sache auch deshalb, weil nach Bacon der menschliche Geist grundsätzlich dazu tendiert, Wirklichkeit zu fälschen und „leichthin in den Dingen eine größere Ordnung und Gleichförmigkeit" vorauszusetzen, als man in ihnen findet. Denn „obgleich vieles in der Natur einzeln und voller Ungleichheit ist, so fügt der Verstand dennoch Gleichlaufendes, Übereinstimmendes, und Bezügliches hinzu, was es in Wirklichkeit nicht gibt" (N. O. I Nr. 45; vgl. DA, I S. 644).

Interessanterweise führt dieser Gedanke Bacon dahin, die Wissenschaften und die Magie (N. O. I Nr. 85) analog zu kritisieren: Beide unterstellen vorschnell Ordnungen, seien es nun Naturgesetze, Analogien und Sympathie-Antipathie-Relationen, Mikrokosmos-Makrokosmos-Verhältnisse und setzen fälschlicherweise dadurch Ungleiches gleich.[38]

Es ist daher gar nicht leicht, Vielfalt vorzuführen, ohne nicht doch unter Hand sofort vorzuselektieren. Allein daß Bacon dies bedenkt, zeugt von einem eminenten, und zu seiner Zeit singulären, Problembewußtsein. Ihm war klargeworden, daß Fakten nicht einfach nur vorliegen, sondern mit der Entscheidung was als Fakt zugelassen oder als Untersuchungsresultat gewünscht wird, bereits der weitere Weg der Forschung vorstrukturiert ist. Dies gilt etwa für aristotelische Wissenschaftsauffassungen, denen zufolge alles nicht Regelmäßige als Abweichung vorab herausfällt oder für interessegeleitete Experimente, bei denen man „bei aller Mühe darauf bedacht gewesen" sei, „in unangebrachter Geschäftigkeit auf ganz bestimmte Ergebnisse sich festzulegen." (N. O. I Vorr.). Doch sind es laut Bacon gerade die Abweichungen, die eine Theorie einfangen und erklären muß. Wunder und Ausnahmeerscheinungen der Natur, z. B. Mißgeburten, sind daher die Testfälle, an denen sich die Erklä-

---

37 Vgl. N. O. I Nr. 82; N. O. I Nr. 124 und die Werktitel *Filum labyrinthi* und *Sylva Sylvarum*. Zu Bacons Metaphorik: Krohn (1994).

38 „Die antike Meinung, daß der Mensch ein *mikrokosmos* ist, ein Auszug oder Modell des Universums sei, ist von Paracelsus und den Alchimisten in phantastischer Weise überspannt worden, so als würde man im menschlichen Körper gewisse Korrespondenzen und Parallelen finden, die sich auf die ganze Vielfalt der in der großen Welt vorhandenen Dinge, seien es nun Sterne, Pflanzen oder Minerale, beziehen sollen." (DA, III S. 370)

rungskraft der Methode zu bewähren hat.[39] Diese Fälle wirken auf die Methode in produktiven Sinne zurück, denn sie verlangen die bisherigen Verfahrensweisen kritisch zu reflektieren und zu verändern, bis sie die Abweichung in Blick bekommen. Die Abweichungen „machen den Verstand frei vom Gewohnten und helfen, allgemeine Formen zu finden. Denn auch hier soll man immer gründlicher untersuchen, bis man die Ursache solcher Abweichungen gefunden hat. Diese Ursache trifft freilich nicht die Formen selbst, sondern nur den noch verborgenen Prozess zur Form". (N. O. II, Nr. 29)

Bacon legt daher seine Materialsammlung zunächst strikt deskriptiv an, alle Sorten von Beobachtungen und Berichten werden gesammelt und gemäß vager Verwandtschaften in einer ersten Ordnung zusammengestellt: „Diese Sammlung ist rein geschichtlich ohne voreilige Beobachtung oder überspitze Unerscheidung anzulegen" (N. O. II, Nr. 11).[40] Das ist der erste Schritt. Dann wird selektiert, indem fehlende Übereinstimmungen bei verwandten Fällen festgestellt werden, also per Falsifikation. Im Prinzip ist hier denkbar, daß je nach Experimenttyp auch je andere Selektionen sich ergeben und somit nicht aus der vorliegenden Menge die eine gültige Bestimmung gefunden wird. Bacon macht diesen Schritt zu einem kontextualistischen Wissensbegriff hier nicht explizit, der Sache nach ist diese Möglichkeit jedoch da und liegt als Konsequenz nahe. Dies wird später u. a. die Untersuchung von Bacons Formbegriff ergeben. Für diese Deutung spricht auch Bacons liberaler Einsatz von *instantiae crucis,* also Fällen des Scheidewegs. Diese versteht er nämlich nicht als Entweder-Oder-Entscheidung, sondern man kann auch wieder den zuerst gewählten Weg zurückgehen und den anderen einschlagen.[41]

Konkret vermeidet Bacon das vorschnelle Ordnen, indem er den Begriff der verwandten Fälle sehr weit faßt, und aus Prinzip eher nominalistisch als analogisch verfährt. ‚Nominalismus als Methode', das heißt, auch unähnliche Fälle, die nur unter dem gleichen Begriff stehen, wie im Englischen Hitze bei Feuer und bei Gewürzen, nebeneinanderzustellen. Dazu kommt, daß kein Status oder Kriterium für die Beobachtung der zu vergleichenden Fälle festgeschrieben wird. Bei den 41 Fällen der Tafel der Grade der Wärme (N. O. II Nr. 13) etwa finden sich Berichte von unmittelbaren Sinneswahrnehmungen (13) gleichberechtigt neben oder auch in Kombination mit geplanten und experimentell überprüften Beobachtungen, Erzählungen, Reiseberichten und Legenden, Hypothesen und unüberprüfbaren Spekulationen (18) und Hinweisen auf noch zu untersuchende Fälle (21).

Diese Mischung hält Bacon bei. Sein Verfahren ist daher nicht einfach unsystematisch, sondern systematisch unsystematisch.

Es ist daher nicht sinnvoll, Bacon vorzuhalten er hätte keinen rechten Begriff von Wärme gehabt. Ein solcher soll ja gerade nicht vorab das Ergebnis präjudizierend in Anschlag gebracht werden.

---

39 Vgl. dazu: Daston (1994), (1991) und (1998); Daston/Park (1997).

40 Bacon war sich sehr wohl der Unzuverlässigkeit vieler seiner Beispiele bewußt und erklärt: „wie ich statt sicherer Wahrheit und klarer Fälle, oft gezwungen bin, das, was die Überlieferung und Erzählung bieten, aufzunehmen, wiewohl berechtigte Zweifel an Berichten oder Mitteilungen von mir immer betont worden sind. Oft war ich genötigt, Zusätze von der Art zu machen: man versuche es, oder man erforsche es weiterhin'." (N. O. II, Nr. 14). Dieses Bekenntnis sollte man aber nicht so verstehn, daß Bacon dann alles Sonderbare herausfiltern will. Entscheidend ist, daß die Beobachtungen glaubwürdig bestätigt werden. Man vergleiche etwa den 11. Hinweis zum 6. Fall in N. O. II, 12!

41 Das hat sehr schön Hacking (1996), S. 411ff. herausgearbeitet.

Daß auch *auf der Ebene der Selektion* des Materials, *der Zufall nicht ausgeschlossen, sondern produktiv genutzt werden soll*, zeigt auch die nähere Betrachtung von Bacons Methodologie. Auf den ersten Blick scheint dem zu widersprechen, daß Bacon gelegentlich erklärte, seine Methode würde wie durch eine „Maschine vorangetrieben" (N. O. Vorrede; I 152), wobei die *Machina intellectus inferior* die äußeren Merkmale des untersuchten Objektes sammle, die *Machina intellectus superior* dagegen das Verborgene entdecken und bis zu den Formen selbst vordringen soll.[42] Immer wieder hat man diese Maschinen-Metapher zur Charakterisierung seines Verfahrens verwendet[43] und nicht selten spöttisch gegen ihn gekehrt.

Doch Bacons Methode funktioniert keineswegs mechanisch wie eine Maschine. Zieht man insbesondere hierzu seine Konzeption der *literata experientia*[44], also der gelehrten oder besser wissenschaftlichen Erfahrung heran, die viel präziser in *De Augmentis* als im *Novum Organum* (N. O. I Nr. 103) dargestellt wird, ergibt sich ein komplexeres Bild. Dort heißt es „die wissenschaftliche Erfahrung, oder die Jagd Pans, behandelt die Art Versuche anzustellen".[45] Pan selbst ist wird in *De sapientia veterum* als Inbegriff des Jägers charakterisiert. Dabei wird er bezeichnenderweise nicht nur trotz, sondern wegen seines planlosen Aufstöberns zum positiven Vorbild für Bacons Wissenschaftler: „Wie er jagen die Wissenschaften und Künste ihren Werken nach." Während aber der Sage nach alle anderen Göttern Ceres vergeblich suchten, „so eifrig sie auch bemüht waren", fand Pan sie zufällig bei der Jagd. Bacon interpretiert den Mythos so: „Die Entdeckung der für das Leben nützlichen Dinge, wie etwa des Getreides, darf nämlich nicht von den abstrakten Philosophien, gleichsam höheren Göttern erwartet werden, auch wenn sie ihre ganze Kraft darauf verwenden, sondern nur von Pan, d. h. von scharfsinnigen Versuchen und der allgemeinen Kenntnis der Natur, die häufig zufällig, als sei sie auf der Jagd nach anderen Dingen über solche Entdeckungen stolpert."[46]

Dennoch kann man die Jagdweise Pans anleiten. „Begegnet man der Erfahrung so obenhin, nennt man sie Zufall, sucht man sie, nennt man sie Experiment." (N. O. I Nr. 82) Wie man dem Zufall auf die Sprünge hilft, erklärt Bacon en detail nur in *De Augmentis*. Er unterscheidet 8 Suchweisen. Im Einzelnen erfolgt die Jagd durch Variation des Experiments (per variationem), durch seine Wiederholung und Verlängerung (per productionem), durch seine Verlagerung (per translationem), Umkehrung (per inversionem) oder durch äußerstes Zwingen (per compulsionem); durch seine Anwendung (per applicationem), seine Verknüpfung mit anderen Experimenten (per conjunctionem) und durch glücklichen Zufall (sortes experimenti). (DA, I S. 623f.) Bacon beschreibt diese Methoden an diesem Ort alle sehr eingehend. Für meine Zwecke ist besonders sein letztes Beispiel des glücklichen Findens von Interesse. Diese Art Versuche nennt er „ganz unvernünftig und gleichsam verrückt: da man etwas versuchen will, nicht weil die Vernunft oder ein anderes Experiment dazu hinführt,

---

42 Vgl. *Filum labyrinthi sive Inquisitio legitima de motu*, III S. 634ff.

43 So stellvertretend für viele Steven Shapin in seinem ausgezeichneten Buch Shapin (1998), S. 108.

44 Vgl. frühe Texte, in welchen ironisch als Mysterium oder gleichsam als ‚Gott aus der Maschine' die *literata experientia* als „der wahre Pfad von den Sinnen zum Verstand" hervorgezaubert werden: „Neque tamen nos peregrinum quiddam, aut mysticum, aut Deum Tragicum ad vos adducimus. Nil enim aliud nostra via, nisi literata experientia. Atque ars sive ratio naturam sincere interpretandi, et via vera a sensu ad intellectum." (RPh, III S. 573; vgl. CV, III S. 618).

45 „Litera experientia, sive Venatio Panis, modos experimentandi tractat" (DA, I S. 624). Vgl. zur *experientia literata* aus anderer Perspektive: Jardine (1990) und speziell zur Jagd Pans: Eamon (1994), S. 281-291.

46 Bacon (1990b), S. 25 f.

sondern einfach deshalb, weil diese Sache bislang noch nie versucht wurde." Doch könnte dahinter etwas anderes großes verborgen liegen. „Denn die großen Dinge der Natur liegen allgemein abseits der gewohnten Wege und ausgetretenen Pfade, sodaß auch die gänzliche Absurdität einer Sache manchmal sich als nützlich zeigt."[47]

Wie man sieht, handelt es sich hierbei um alles andere als mechanische Prozeduren. Diese Vorgehensweisen erfordern Phantasie[48] und Spürsinn, den produktiven Einsatz von Ahnungen, und das Herstellen von Analogien. Gelehrte Erfahrung sei daher „mehr Scharfsinn und eine Fährten-witternde Jagdweise als Wissenschaft".[49]

Die Analogie wird jedoch von Bacon nicht als Gesetz oder Struktur der Natur supponiert, sondern als heuristische Erkenntniskategorie eingesetzt (N. O. II Nr. 42), als *scientiae analogia*[50]. Zu ihr wird ergänzend Zuflucht genommen, wenn man mit der Sinneserkenntnis nicht mehr weiter kommt.[51] Unter anderem soll mit ihrer Hilfe das verborgene Gesetz der Natur, die okkulte[52] Form der Dinge erkannt werden. Die Analogie ist aber, anders als im alchemistischen oder magischen Weltkonzeptionen, nicht selbst dieses Gesetz. Dies ist entschieden gegenüber Deutungen wie der Hattaways herauszustellen, der aus dem bloßen Verwenden von Metaphern aus dem alchemistischen Kontext und Bacons Operieren mit Analogien folgert, dieser teile die magischen Auffassungen und sei deshalb ein typisch konservativer Renaissancemetaphysiker.[53]

---

47 „Hic vero experimentandi modus plane irrationalis est, et quasi furiosus; cum aliquid experiri velle animum subeat, non quia aut ratio aut aliquod aliud experimentum te ad illud deducat, sed prorsus quia similis res adhuc nunquam tentata fuit [...] Magnalia enim naturae fere extra vias tritas et orbitas notas jacent, ut etiam absurditas rei aliquando juvet" (DA, I S. 632). Vgl. a.: „Etiam ex casus vi et natura hujusmodi divinatioem sumpsit. Casum nimirum proculdubio multis Inventis principium dedisse, sumpta ex natura rerum occasione." (CV, S. 614).

48 Vgl. DA, I S. 615f.: „Die Phantasie tut eine Art Botendienst vom einen zum anderen der beiden großen Bezirke; wie ein Unterhändler geht sie unaufhörlich vom einen zum anderen hin und her. Die Sinne nämlich liefern der Phantasie Bilder aller Art und auf Grund der Bilder urteilt die Vernunft. Dabei übersetzt man diese Bilder zunächst in die Sprache der Phantasie. [...] Auch bei der Willensregung geht allemal die Phantasie voraus, so daß also die Phantasie beiden, sowohl der Vernunft als auch dem Willen, zum gemeinsamen Werkzeug dient. [...] Doch ist die Phantasie kein bloßer nackter Bote, sondern eine nicht geringe Autorität fällt ihr teils zu, teils nimmt sie dieselbe in Anspruch; denn diese in Anspruch genommene Autorität liegt außerhalb der einfachen Übermittlung eines Auftrags."

49 „*Sagacitas* potius est et odoratio quaedam venatica, quam *Scientia*" (DA, I S. 633).

50 Mit der *scientiae analogia* umreißt Bacon ein Verfahren, daß er gegen Paracelsus' bloße Koinzidenz zwischen Erfahrung und einigen unfundierten Hypothesen ins Feld führt. Genuine Wahrheit aber, so heißt es noch in TPM, sei einheitlich und selbstreproduzierend, Glückstreffer seien widersprüchlich und stünden alleine. Wäre das Schießpulver nicht durch Glück, sondern durch richtige Anleitung entdeckt worden, würde es nicht alleine stehen, sondern wäre von einer Menge hervorragender und wünschenswerter Erfindungen begleitet worden. (TPM, III S. 538)

51 Vgl. N. O. II, Nr. 42. Die Analogie, so heisst es an dieser Stelle weiter, stellt Nicht-Sinnliches sinnlich dar, indem verwandte Körper zum Vergleich herangezogen werden: „Ea fit cum deducitur non-sensibile ad sensum, non per operationes sensibiles ipsius corporis insensibilis, sed per contemplationem corporis alicujus cognati sensibilis."

52 Zum Begriff des Okkulten zur Bezeichnung von sinnlich nicht wahrnehmbaren Ursache von allerdings beobachtbaren Effekten und seinem Schicksal vgl. Hutchinson (1982); Eamon (1994); Wilson (1997), Kap. 2.

53 Hattaway (1978), S. 184: „Bacon worked largely by correspondences and analogies". Eine ausgezeichnete Kritik an Hattaways Deutung, die einer vollständigen Demontage gleichkommt, gibt: Horton (1982). Es ist bemrkenswert, daß dennoch Hattaways Lesart bis heute zu den einflußreichsten zählt.

Allerdings bleibt zu klären, was Bacon dann mit okkulter *Form* meint.[54] Dies ist besonders wichtig, da er das Wissen von den Formen als das höchste menschliche Wissen bestimmt und dieForm mit dem Gesetzesbegriff gleichsetzt.

Zu Bacons Formbegriff gibt es in der Forschung äußerst kontroverse Ansichten. Ich kann diese hier nicht ausführlich diskutieren, sondern möchte nur einen Aspekt an ihm beleuchten, der aus der Perspektive meiner bisherigen Ausführungen sich womöglich in deutlicherem Licht zeigt.[55]

Wäre Form essentialistisch als Substanz aufgefasst oder als hinter allen Dingen verborgen liegendes Strukturgesetz, so würde mit ihrer Entdeckung erstens ein positives universales Gesetz der Natur aufgewiesen, daß zweitens von der Philosophie in einem entsprechend strukturierten Ordnungssystem rekonstruiert werden könnte. Enzyklopädisch-heilsgeschichtliche Interpretationen könnten sich hier andocken. Doch Bacon verwendet den Formbegriff nicht als metaphysischen Substanz-, sondern als pragmatischen Funktionsbegriff.

Im 17. Aphorismus des zweiten Buches des *Neuen Organon* identifiziert Bacon Form mit Gesetz und stellt dabei klar, daß er damit nicht die abstrakten Formen meint, sondern „nur jene Gesetze und Bestimmungen des reinen Vorgangs zu verstehen, wodurch eine einfache Eigenschaft hervorgebracht und bewirkt wird, z. B. die Wärme, das Licht, das Schwere, wie sie in jeder geeigneten Materie bestehen. Deshalb ist die Form des Warmen und die Form des Lichtes genau dasselbe, wie das Gesetz des Warmen und das Gesetz des Lichts". (N. O. II 17) Bacons Formbegriff wird nicht über die Identität gleicher Fälle gebildet, sondern über die ähnliche Wirkung verschiedener Fälle. Die Aufindung des Gesetzes setzt somit nicht auf der Ebene der Ursachen an, sondern bei den beobachteten Wirkungen. Hat man deren Form erkannt, sozusagen die Form vergleichbarer *Wirkprozesse*, beherrscht man die Wirkungen, und beherrscht man die Wirkung, hat man das Gesetz. Vielleicht wäre daher am besten, zu sagen, daß Gesetze oder Formen *Wirkäquivalente* sind. Für Bacon gilt damit auch, daß alles was wirkt, wirklich ist, und Wirklichkeit nicht auf Realität, als traditionell-szientifisch fassbares Dasein unter Gesetzen der Naturwissenschaft reduziert werden darf.

Man muß sich Bacons Form-Idee viel weiter vorstellen, als häufig von Interpreten, die es gut mit Bacon meinten, angenommen. Wenn etwa jemand nach dem Konsum von Opium einschläft und ein anderer durch einen Schlag auf den Kopf, so hat man das Gemeinsame der betäubenden Wirkung zu erkennen, und als handhabares Gesetz zu formulieren. Wenn dies gelingt, dann ist ihre gleiche Form erkannt. Bacon war sich dieser Zumutung bewußt: „Meine Formen werden manchem ziemlich abstrakt vorkommen, nicht zuletzt, weil sie sehr buntfaltige Dinge miteinander mischen und zusammenstellen." Er nennt als Beispiel hier u. a. diverse gewaltsame und natürliche Todesarten. Doch sei es „ganz sicher, daß diese Dinge bei all ihrer Verschiedenheit und Fremdartigkeit in jener Form oder jenem Gesetz zusammenkommen, daß dem [...] Tode zugrundeliegt" (N. O. II 17). Nur dadurch gewänne der Mensch die Macht die Natur vollständig zu beherrschen.

---

54 Vgl. Schon *Cogita et visa* behauptet Bacon, die Endeckungen des Schießpulver, der Seidenspinnerei, des Seekompasses, Zucker, Glas beruhten auf den verborgenen Eigenschaften der Dinge („occultis rerum probrietatibus") (CV, III S. 615).

55 Meine Überlegungen stützen am ehesten die Interpretation von Perez-Ramos, welcher Bacons Formbegriff zwischen dem aristotelischen Begriff der substantiellen Form und dem internen Strukturgesetz der Korpuskulartheorie ansiedelt. Vgl. Antonio Pérez-Ramos (1996).

Bei Aristoteles war die Form das, was einem Ding in allen verschiedenen akzidentiellen Relationen in gleicher Weise eigen war und was deshalb seine stabile Substanz ausmachte. Bacon dreht dies um: bei ihm sind es gerade die (in traditioneller Sicht) akzidentiellen Relationen, die die Form ausmachen. (Denn jenseits dieser Wirkungsverhältnisse gibt es nichts, was einen Gegenstand an sich selbst ausmachen könnte.) Bacons Formbegriff schließt folglich auch ein, daß ein und dasselbe Ding verschiedene Formen hat, wenn es verschiedene Wirkungen hat. Der Schlag auf den Kopf, der betäubend wirkt, hat eine andere Form als der gleiche Schlag, wenn er belebend wirkt.

Bacon löst so das Problem der okkulten, also nicht beobachtbaren Ursachen, die weder metaphysisch supponiert noch sonstwie als Abstraktionen positiviert werden. Es gibt für ihn hier keinen Dualismus zwischen verborgenen wahren Ursachen und sichtbaren Wirkungen, sondern die Differenz zwischen beiden wird handelnd konstituiert.

„Daher wird man von einem wahren und perfekten Grundsatz des Wissens folgende Aussage machen und zu ihm folgende Vorschrift machen müssen: man entdecke eine andere Eigenschaft, welche mit einer gegebenen Eigenschaft vertauschbar ist und dennoch ein Sonderfall der bekannteren Eigenschaft ist, also gleichwohl ein treues Abbild der wahren Gattung darstellt. Beide Aussagen, die für das Handeln wie die für das Betrachten, sind ein und dieselbe Sache und was im Tätigsein am Nützlichsten ist, ist im Wissen reine Wahrheit." (N. O. II Nr. 4) Und verzichtet man auf die willkürlichen Abstraktionen, dann sind hier „Wahrheit und Nutzen diesselben Dinge" (N. O. I Nr. 124).[56]

Wenn Bacon in diesem Kontext schreibt: „Wer aber die Formen kennt, der begreift die Einheit der Natur in den unähnlichsten Dinge" (N. O. II Nr. 3), so ist nun klar, daß „Einheit" hier nicht als All-Einheitsideal einer *unitas naturae* vorgeben wird, sondern lediglich als bestimmter praktisch konstituierter Wirkzusammenhang, was bedeutet, daß es einen Pluralismus vieler, je nach Kontext ganz verschiedener konkreter Einheiten gibt und nicht *die* Einheit als Metainstanz.

Um diese dem Menschen so nützliche Entdeckung der Formen in den verschiedensten Fällen auffinden zu können, entwickelte Bacon seine Experimentalmethode und nur an dieser Zielvorgabe wird richtig klar, warum er die Vielfalt der *experientia literata* benötig. Auffinden ist hier nicht im Sinne von Entdecken eines Vorliegenden zu verstehen, sondern als konstruktivistisches Invenieren. Bacon ist auch hier hochreflektiert und mißtraut der unmittelbaren Erfahrung. Invenieren heißt bei ihm Intervenieren. Von alleine gibt die Natur ihre Geheimnisse nicht preis, sie muß „durch die Tat unterworfen" (N. O. I, Distr.), zerschnitten (N. O. I Nr. 51) und durch Kunst gezwungen (per vexationes artes) werden, um sie dadurch für die Erfahrung erst zuzurüsten (N. O. I Nr. 98). „Aus ihrem eigenen Zustand herausgetrieben, gepresst und geformt" können der so gefesselten und mißhandelten Natur ihre Geständnisse abgepresst werden[57]: „Wie Proteus nur dann verschiedene Gestalten annahm, wenn man ihn in Fesseln schlug, so zeigt sich die durch die künstliche Mittel angeregte und gefangene Natur offenbarer als wenn sie sich frei und überlassen bleibt."[58]

---

56 Vgl. dazu auch Krohn (1987), S. 124-134.

57 Vgl.: „conficimus historiam non solum naturae liberae ac solutae (cum scilicet illa sponte fluit et opus summ peragit) [...] sed multo magis naturae constrictae et vexatae; nempe, cum per artem et ninisterium premitur et fingitur" (N. O. I, Distr.).

58 „neque Proteus se in varias rerum facies vertere solitus est, nisi manicis arcte comprehensus; similiter etiam natura arte irritata et vexata se clarius prodit, quam cum sibi libera permittur." (DA, I S. 500). Vgl. zur Moder-

IV. Zufall und Ordnung: Höhere Magie im Dickicht der Erfahrung

Resümierend fordert Bacon im *Neuen Organon*, man müsse zuerst „eine Zusammenstellung oder eine besondere Naturgeschichte aller wunderlichen Naturerzeugnisse schaffen, ebenso brauchen wir eine Sammlung von allem Neuen, Seltenen und Ungewöhnlichen." (N. O. II Nr. 29). Eine solche Sammlung hat er dann insbesondere mit seiner letzten Schrift *Sylva Sylvarum* vorgelegt. Dieses Werk steht keineswegs im Widerspruch zu Bacons Methodologie, sondern erhält von dieser aus den wohldefinierten Auftrag, Material zu sammeln, das Anlaß für ein produktives Voranschreiten des Wissens geben kann. Daher geht auch der Vorwurf fehl, diese Schrift sei „ein Sammelsurium völlig unwissenschaftlichen Materials"[59], denn die Wissenschaftlichkeit steckt nach Bacon nicht im Material selber, sondern in der neuen Untersuchungsweise, die durch es provoziert ist. Auch die Auffassung, *Sylva Sylvarum* sei ein bloßes Kuriositätenkabinett oder sei gar auf den hinter derartigen Wunderkammern stehenden enzyklopädischen Ordo-Gedanken verpflichtet, erweist sich als inadäquat[60]. Ebenso ist der dritte häufig erhobene Einwand, Bacon verfahre hier eklektizistisch[61], verfehlt, denn er verdankt sich erstens einem uninformierten Verständnis, was zu Bacons Zeit Quellen waren und mißachtet daher zweitens den Sinn des Zusammenstellens von Material.

Es gibt in Bacons Schriften wie bei den meisten Texten der frühen Neuzeit eine dichte Intertextualität und damit Kontextualität mit anderen Diskursen. Die Art und Weise wie die vermeintlichen Quellen im neuen Kontext verarbeitet und für die eigene Argumentation eingesetzt sind, ist daher entscheidend, und nicht die Herkunft.[62] Keineswegs ergibt sich aus dem bloßen Befund, daß Bacon zahlreiche Beispiele von anderen Autoren aufgreift, ein simples Einflußverhältnis, wie es die Quellenmetaphorik suggeriert, noch ein willkürlicher oder gar humanistischer Eklektizismus[63]. Die Quellenfunde sind für ihn buchstäblich auf Tauglichkeit zu prüfendes Material. Dies macht ein erhellender Vergleich Bacons aus dem *Parasceve* klar, der mit handfest merkantiler Metaphorik anstelle hehrem Kunstkammerfeinsinns operiert, und der fulminant mit folgender Aufforderung vorbereitet wird:

„Zuerst fort mit den Altertümern und all den Zitaten und Zeugnissen von Autoren, und genauso mit den Disputen und Kontroversen und unterschiedlichen Meinungen, kurz: mit allem, was philologisch ist." Und dann heißt es: „Kein Mensch, der Material sammelt und lagert um ein Schiff zu bauen, wird daran denken dieses elegant anzuordnen und für das Auge wohlgefällig auszustellen wie in einem Shop. All seine Sorge ist, daß das Material

nität von Bacons Verfahren vorallem Hacking (1996), S. 250ff., 280ff. und bes. 406ff. sowie Schäfer (1999), S. 95 ff. Bacon favorisiert dabei insbesondere auch den unterstützenden Einsatz von Instrumenten, z. B. des Mikroskops. Im ersten Teil des *Neuen Organon* noch eher skeptisch, im zweiten dann allerdings entschieden (N. O. II, Nr. 39). Vgl. zur Bedeutung des Mikroskops für die wissenschaftliche Diskussion der frühen Neuzeit das überaus instruktive Buch von Wilson (1997) sowie ihre Aufsätze Wilson (1998) und (1988).

59 Vickers (1988), S. 88.
60 Z. B. Krohn (1987), S. 56f.; Findlen (1997); Bredekamp (1993), S. 63-76; Whitaker (1968), S. 39.
61 Vgl. Whitaker (1968), S. 29; Hattaway (1978), S. 197.
62 Vgl. dazu: Neuber (1994), hier S. 254.
63 Gaukroger (2001), S. 28-36 wendet z. B. diese Eklektizismus-Charakteristik nicht weniger verfehlt ins Positive, indem er Bacon, besonders mit *Sylva Sylvarum* in die Tradition eines humanistischen Ekletizismusprogramms und der commonplace books seit Plinius stellt, deren vermeintliche Linie er mit rhapsodisch herbeigesuchten Zitaten zu zeichnen versucht.

einwandfrei und gut ist und so gelagert wird, daß es möglichst wenig Raum im Warenhaus beansprucht. Und das genau ist es, was hier getan werden soll."[64]

Bereits bei der Deutung des eigentümlichen und rätselhaften Titels *Sylva Sylvarum* findet man sich sofort in eine analoge unnötige Kontroverse verstrickt: Ellis vermutet, daß *Sylva* für *hylae* steht, und somit etwa ‚Urstoff der Urstoffe' bedeutet, und diese Lesart schlüge Bacons Werk der essentialistischen Tradition zu. Es gibt jedoch auch die antiessentialistische Lesart ‚Wald der Wälder', was der Bedeutung ‚Sammlung der Sammlungen' entspräche. Beide Varianten stehen in den seltenen Erwähnungen der *Sylva* bei den Interpreten oft kommentarlos nebeneinander.[65] Bacon selbst verwendet den Ausdruck ‚Sylva' in beiden Bedeutungen, als *hylae* z. B. im *Advancement* (III S. 326) und als ‚Wald' oder ‚Dickicht' im *Neuen Organon*. Im letztgenannten Fall verwendet er die Metapher im Kontext einer Diskussion um die richtige Ordnung der Erfahrung im Sinne von „Dickicht" oder „Wäldern von Erfahrungen" („per experientiae sylvas" N. O. I, Nr. 82; vgl. a. N. O. Vorr.).[66] Hier setzt er diese Metapher also terminologisch ein und verleiht ihr bezeichnenderweise einen positiven Sinn, wenn er erklärt, daß sein Verfahren durch die Wälder zu den Lichtungen der Lehrsätze führe. Dies ist, wie gleich noch klarwerden soll, eine adäquate Beschreibung der in *Sylva Sylvarum* de facto vorgestellten Vorgehensweise.

Darüberhinaus jedoch ist Ellis und den späteren an ihm orientierten Bacon-Kommentaren komplett entgangen, daß der Begriff ‚Sylva' zunächst in der Antike eine traditionelle Bezeichnung für eine heterogene Gedichtsammlung, später auch für eine Sammlung von Prosatexten gemischten Inhalts ist. ‚Sylva Sylvarum' bedeutet somit die Steigerung dieser Art von Kollektion zum Prinzip. Einzig Wolfgang Adam hat, ohne jede Resonanz in der Baconforschung, Bacons Schrift in diesen Gattungskontext eingeordnet. Auch er diskutiert die verschiedenenen Bedeutungen des Wortfeldes Sylva/Wald/hylae. Silva kann mit mit Mischwald, Forst, - hylae als Stoff, Rohstoff, Material, Materie oder Vorrat übersetzt werden und ist in dieser Hinsicht vom philosophischen Sprachgebrauch, welcher hylae als Urstoff bezeichnet, zu unterscheiden.[67] Auch können schnell hingeschriebene Gelegenheitsentwürfe als Sylvae bezeichnet werden.[68] Adam nennt als (dann historisch und systematisch zu differenzierende) Hauptkennzeichen für Sylvendichtungen, daß sie eine bunte Themenvielfalt unter Unterhaltungs- oder Nützlichkeitsaspekten in Gestalt eines ordo neclectus gegliedert arrangieren. Zu Bacons Lebzeiten waren Sylven als Ordnungsformen durchaus bekannt. Beispiele sind Ben Jonsons *The Forrest* und *The Under-woods*[69]. Zusammenfassend kann man sagen, ihr Prinzip sei ‚order in variety'. Wie jedoch sieht es bei Bacons Sylva, die im Titel bereits die alte Gattungsbezeichnung programmatisch überbietet, aus?

---

64 „Primo igitur facessant antiquitatis et citationes aut suffragia authorum; etiam lites et controversiae et opiniones discrepantes, omnia denique philologica [...] Nemo enim qui materialia ad aedificia vel naves vel huiusmodi aliquas structuras colligit et reponit, ea (officinarum more) belle collocat et ostenat ut placeant, sed in hoc tantum sedulus est ut proba et bona sint, et ut in repositorio spatium mimimum occupent. Atque ita prorsus faciendum est." (I, S. 396, engl.: IV 254f.) Es gilt im übrigen für den ersten Teil von *De Augmentis*, das auch dieser schon den Stoff der Tradition nur als Material behandelt.

65 Vgl. Krohn (1987), S. 56f.

66 Rawley spricht in seinem Vorwort zu *Sylva Sylvarum* von der Gefahr, sich in einem „vaste wood of experience" zu verlieren (II S. 337).

67 Adam (1988), S. 64-68.

68 Ebd., S. 78f.

Auf den ersten Blick zeigt dieser Wald ein frappierendes Übermaß an Ordnung: *Sylva Sylvarum* besteht aus 10 Teilen, Centurien genannt, da sie jeweils 100 Paragraphen vereinigen. Jeder dieser 1000 Paragraphen nennt eine oder mehrere Tatsachen, und gibt dazu unterschiedlich ausführliche Kommentare, die die beobachteten Fälle erläutern. Die jeweiligen thematischen Gruppen werden stets eingeleitet mit der Überschrift *„Experiments in …"*. Wie im *Novum Organum* entstammen die aufgeführten Beispiele äußerst unterschiedlicher Provenienz und changieren in ihrem Status. Bacon fand sie unter anderem bei Aristoteles, Plinius, Porta, Cardano, oder in Reiseberichten. Daneben werden eigene Beobachtungen und mündliche Berichte angeführt.

Wie man gleich sieht, ist der quantitative Umfang der einzelnen Paragraphen und folglich auch der Centurien sehr unterschiedlich. So ist z. B. die 8. Centurie mehr als doppelt so lang als die 5. Hier gibt es keine Symmetrien. Ohnehin hatte sich Bacon im *Novum Organum* lustig gemacht über jede Art von Symmetriezwang, denn, so fragt er dort spöttisch, warum soll es ausgerechnet 4 Elemente geben und nicht drei? Weshalb soll man zu 10ner-Ordnungen und nicht zu irgendwelchen anderen greifen, die gerade der tatsächlichen Anzahl der Fälle entsprechen? (vgl. N. O. I, Nr. 45). Die 10ner-Einheiten in *Sylva Sylvarum* können daher gar nicht als in irgendeiner Weise höhere oder tatsächliche Ordnungen repräsentierende oder rekonstruierende Ordnungskategorien angelegt sein. Es gibt hier bezeichnenderweise keinerlei Entsprechungen zwischen der äußeren Form und dem Inhalt. Weder ist eine schlüssige chronologische oder zyklische Abfolge zu konstatieren, noch ein Übereinstimmen von thematischen Gruppen und den Zehnereinheiten. Zwar gibt es immer wieder solche inhaltlichen Verbindungen von benachbarten Paragraphen, doch entsteht daraus keine Sachordnung. Außerdem werden selbst diese Gruppen von völlig anderen Beispielen durchsetzt, und es kommt zu abrupten Übergängen zu anderen Themen. Nach den „Einflüßen des Mondes" wird der Essig untersucht, danach Tiere, die Winterschlaf halten usw. (Sylv. Nr. 897f.). Hinzu kommt, daß jeder einzelne Paragraph in sich äußerst uneinheitlich ist. Gleich im ersten Paragraphen etwa geht Bacon von praktischen Anweisungen über zu historischen Exkursen über die Fehler, die Caesar bei vergleichbaren Experimenten machte usw.

Insbesondere fehlen auch interne ontologische Hierarchien, wie sie etwa für Cardanos Naturgeschichte *De subtilitate* trotz ihrer scheinbar chaotischen Materialpräsentation charakteristisch sind.[70] Die *Sylvae* beginnen unvermittelt mit einer direkten Leseranweisung zu Experimenten, die sich auf die Durchlässigkeit von Körpern beziehen: „Dig a pit upon the sea-shore, somewhat above the high water mark …" etc. Dieser Anfang ist sowohl von der Sache her als auch von der Wertigkeit völlig beliebig. Wichtig ist stattdessen der Aufforderungscharaker der direkten Anrede an den Leser, und die ihm offerierten Anleitungen zum selbständigen Überprüfen der Experimente. Der Leser wird rhetorisch als Mit-und Weiterarbeiter am offenen Forschungsprojekt kooptiert.

---

69 Ben Jonson: „The Forrest", in C. H. Herford Percy und Evelyn Simpson (Hg.), *Ben Jonson*, 8 Bde., S. 91-122; „The Under-wood", ebd. S. 123- 295; „Timber, or Discoveries", ebd. S. 561-649; Pero Mexia: *Silva de varia lecions …*, (1556). Wenig spätere Beispiele: Phinaes Flechters *Sylva* (1633), Miltons *Sylvarum Liber* (1645). Adam (1988) nennt eine Fülle weiterer Texte.
70 Zu Cardanos ontologischen Hierarchien: Schütze (2000), S. 30.
71 Adam (1988), S. 238f. Adam beruft sich auch auf Rawleys Vorwort, welches von einer „secret order" (II S. 337) spricht, und bezieht dies auf die thematischen Zusammenhänge der Stoffkomplexe. Rawley bleibt hier

Daher ist auch die *Sylva*, die durch ihre äußere Centurien-Form gerade die enzyklopädische Tradition fortzusetzen und ein positives Ordnungsideal zu affirmieren scheint, tatsächlich eine zutiefst ordnungsskeptische Schrift, die Bacons Mißtrauen gegen jede Form von abstrakter Ordnung exemplifiziert. Es wird in ihr in traditionellem Sinn nichts mehr geordnet.

Dennoch ist es vorschnell, wie Adam zu urteilen, daß *Sylva Sylvarum* „in erster Linie eine Stoffsammlung" sei, die überdies in auffallender Weise der von Bacon selbst propagierten Vorstellung von einer empirisch verfahrenden Wissenschaft[71] widerspreche.

Diese Stoffsammlung folgt nämlich erkennbar einem Prinzip, welches lautet: ‚variety as order'. Paradoxerweise[72] ist es gerade die starre Äußerlichkeit der Centurienordnung, die es Bacon erlaubt, seine offene Methode vorzuführen und die größte Mannigfaltigkeit und Disparatheit des Materials einzufangen; je starrer die Ordnung, desto chaotischere Inhalte können präsentiert werden. Von diesem Ordnungsprinzip der Centurien ist es nicht mehr weit zur geistlosesten aller möglichen Ordnungen, die sich bald als Prinzip der Lexika durchsetzen wird, nämlich der alphabetischen. Bei Bacon hat dieses Anordnen jedoch, wie anhand der Darstellung von Bacons Methodologie gezeigt, einen präzisen forschungsstrategischen Sinn und dient gerade in seiner Willkürlichkeit der Investigation neuer Phänomene, es beabsichtigt und ermöglicht die überraschende Kombination auch unähnlicher Fälle. Diese Kombination wird von Bacon selbst nur ansatzweise durchgeführt, es geht ihm vielmehr darum, das Material zu liefern und so zu disponieren, daß die nachfolgenden Generationen damit weiter offen experimentieren können, d. h. ohne daß ihre Vorgehensweisen durch vorgebene starre Ordnungszusammenhänge und einem metaphysisch befrachteten Systemgedanken vorstrukturiert, festgelegt, eingeengt, und ausgerichtet werden. Bacons Konzeption ist daher keineswegs enzyklopädisch in dem Sinne, daß der Kreis des überlieferten Wissens durchschritten und geordnet würde, sondern zukunftsoffen. Sie entspricht mit dem Prinzip variety as order exakt dem systematisch unsystematischen Vorgehen, wie es im *Novum Organum* und *De augmentis* gefordert und geplant wurde.

Selbst wenn man die Centurienordnung in der soeben skizzierten Weise versteht, muß man sich wundern, daß Bacon ausgerechnet Centurien, die erst wenige Jahrzehnte zuvor in der frühen Neuzeit als Ordnungsform aufgekommen waren, als seine Einteilungsstruktur wählt. Denn es liegt nahe zu vermuten, daß Centurienordnungen in dieser Zeit entwickelt wurden, weil sie durch ihre zyklische Anlage für platonisierende Enzyklopädieprojekte attraktiv waren. So z. B. bei Camerarius[73]. Ein kurzer Vergleich ist daher sehr erhellend.

Das ab 1590 sukzessiv in Teilen erscheinende und dem Späthumanismus zuzuordnende Emblembuch *Symbola und Emblemata*[74] des Arztes und Botanikers Camerarius ist in 4 Centurien aufgeteilt. Er legt damit konsequenter als alle anderen Emblembuchautoren das ordo-Denken der Anlage seines Werkes zugrunde. Den Centurien entsprechen in absteigender Hierarchie die Reiche der Pflanzen, der Tiere der Erde, der Tiere der Luft und schließlich der niederen Tiere des Wassers und Reptilien, die jeweils einen vollkommenen Zyklus repräsen-

---

indes unbestimmt, betont aber entgegen Adams Einschätzung die empirische Grundausrichtung der Schrift. Die secret order besteht in meinen Augen eher im Varianzprinzip (s. u.).

72 Anzunehmen, daß diese paradoxale Anordnung gewollt ist, legt eben die Verbindung von den weit auseinanderliegenden Sylva- und Centurienordnungen nahe, die wie ein kalkuliertes manieristisches Concetto erscheint.

73 Ich danke Wolfgang Neuber für diesen Hinweis.

74 Camerarius (1986).

tieren. Die genauen und sorgfältigen, an den empirischen Fakten[75] orientierten Naturbe-
schreibungen stehen bei Camerarius im Zeichen der sich in der Natur offenbarenden göttlichen Schöpfungsordnung.

Man sieht, daß Bacon diese Centurienform noch herbeizitiert, aber sie schon nicht mehr
erfüllt und insbesondere ihren zyklischen Charakter marginalisiert. Er verleiht ihr stattdessen die gegenteilige Funktion, nämlich möglichst heterogenes Material möglichst unstruktu-
riert zu sammeln.[76]

Damit liefert Bacon gerade *keine Naturgeschichte* traditionellen Zuschnitts mehr. Er kor-
rigiert innerhalb der *Sylva* daher selbst ausdrücklich den eigenen Untertitel seiner Schrift
(„A Natural History in Ten Centuries"), von dem sich Forschung seither hat blenden lassen.
In einer Schlüsselstelle, die wie eine frühe Definition des Dekonstruktivismus als höherer
Magie klingt, präzisiert er:

> „For this writing of our ‚*Sylva Sylvarum*‘ is, to speak properly, not natural history, but
> a high kind of natural magic. For it is not a description only of nature, but a breaking
> of nature into great and strange works." (Sylv. Nr. 93)

In seiner Begründung nimmt er wieder die Begriffe des Okkulten und der Subtilität auf, mit
deren Hilfe er jetzt eine genauere Abgrenzung gegenüber der konventionellen Magie vornehmen kann:

> „The knowledge of man hitherto hath been determined by the view of sight; so that
> whatsoever is invisible, either in respect of the fineness of the body itself, or the smal-
> lness of the parts, or of the subtility of the motion, is little inquired. And yet these be
> the things that govern nature principally; and without which you cannot make any
> true analysis and indications of proceedings of nature. The spirits or pneumaticals,
> that are in all tangible bodies, are scarce known." (Sylv. 98)

---

[75] Im einzelnen sind Camerarius' Beschreibungen, trotz der schöpfungstheologischen Ordnung, sogar zuver-
lässiger als die von Bacon in *Sylva Sylvarum* gegebenen. Man vergleiche z. B. Bacons Bericht über den torpedo
marina (Sylv. 998) mit Camerarius Darstellung des Rochens (L., IV 41 u. 42 und I, 41).

[76] Auch ein Vergleich mit den späteren Ordnungsauffassungen der frühen *Royal Society* ist aufschlußreich und
verdeutlicht die Sonderstellung Bacons: Im Programm der *Royal Society* führte Thomas Sprat eine Verquickung
baconischer Methodik und platonistischer Enzyklopädik durch. Ich bringe ein längeres Zitat, weil dadurch sehr
klar wird, daß das jedenfalls nicht Bacons Programm war, sondern vielmehr sogar einem methodologischen
Rückfall hinter die methodische Radikalität Bacons gleichkam:
„Such is the dependence amongst all the orders of creatures; the inanimate, the sensitive, the rational, the natural, the artificial: that the apprehension of one of them, is a good step towards the understanding of the rest: And
this is the highest pitch of *humane reason* ; to follow all the links of this chain, till all their secrets are open to
your minds; and their works advanc'd, or imitated by our hands. This is truly to command the World; to rank all
the *varieties* and *degrees* of things, so orderly one upon another; that standing on the top of them, we may
perfectly behold all that are below, and make them all serviceable to the quiet, and peace, and plenty of Man's
life. And to this happiness, there can be nothing else added: but that we make a second advantage to this *rising
ground*, thereby to look the nearer into heaven." Sprat (1667), S. 110. Zur Metapher Sprats von der ‚Great Chain
of Being‘ vgl. Lovejoy (1985), S. 280. Eine solche Kette alles Seins wird im Dickicht der *Silvae* weder gesucht
noch gefunden, die einzelnen Entdeckungen werden eben nicht sofort in einen klar gegliederten Kosmos eingeordnet, wie es Sprat sich wünscht, sondern verweigern gerade eine solche Einbindung.

Die üblichen wissenschaftlichen Verfahren versagen hier: „and for the more subtile differences of the minute parts, and the posture of them in the body, which also hath great effects,
they are not at all touched: as for the motions of the minute parts of bodies, which do so great
effects, they habe not been observed at all; because they are invisible, and incur not to the eye;
but they are to deprehended by experience" (Sylv. 98).[77]

Bacon nennt als Beispiele für Vorgänge, die man bislang wahrnahm, weil man sie mit
irreführenden Namen belegte und von „Kräften, Naturen, Handlungen und Leidenschaften"
und „anderer solcher abstrakter Worte (logical words)" sprach: „Trocknen, Verflüssigen,
Kreiieren, Reifen" („arefaction, colliquation, concoction, maturation"). (Sylv. 98)

Bei der Beschreibung von solchen Vorgängen greift Bacon immer wieder auf magische
Berichte zurück,[78] doch macht er hier hohe Auflagen:

„The relations touching the force of imagination, and the secret instincts of nature, are so
uncertain, as they require a great deal of examination ere we conclude upon them" (Sylv. Nr.
986). Im Falle der Waffensalbung bleibt er skeptisch, will aber nichts für die Zukunft aus-
schließen: „It is constantly received and avouched, that the anointing of the weapon that
maketh the wound, will heal the wound itself." Bacon führt für dieses Experiment „men of
credit" als Zeugen an, gesteht aber, daß er selbst bis jetzt nicht vollständig geneigt sei, es zu
glauben. (Sylv. 998)[79]

Oder um ein anderes einprägsames Beispiel zu zitieren:

„The authors of natural magic report, that the heart of an ape, worn near by the heart,
comforteth the heart, and increaseth audacity. It is true that the ape is a merry and bold
beast. And that the same heart likewise of an ape, applied to the neck or head, helpeth
the wit; and is good for the falling sickness: the ape also is a witty beast, and hath a dry
brain: which may be some cause of attenuation of vapours in the head. Yet it is said to
move dreams also. It may be the heart of man would do more, but that it is more
against men' s minds to use it; except it be in such as wear relicts of saints." (Sylv.
978)

So abstrus dieses Beispiel klingen mag, es zeugt von einem neuen Umgang mit der Magie.
Magische Phänomene wird weder fraglos affirmiert noch ausgeschlossen, sondern es werden
Tests gefordert, die experimentell ihren Nutzen erweisen sollen.[80] Die Frage nach der ‚Wahr-

---

77 Vgl. N. O. II, Nr. 40.
78 Z. B. geht er auf verborgene Beziehungen der Sympathie oder Antipathie ein. Bacon nennt hier als Beispiel
die „vielbezeugte" Beziehung zwischen dem Tod eines nahen Blutsverwandten, und dem inneres Gefühl mit dem
der Angehörige diesen Tod auch von fern spürt. Bacon berichtet hier, daß er selbst als er in Paris weilte den Tod
seines Vaters zwei oder drei Tage zuvor in einem Traum angekündigt bekam. (Sylv. 986) Vgl. auch als Beispiele
für *actio in distans* den Paragraphen zum „torpedo marina" (Sylv. 993).
79 Er ist verwirft daher keineswegs grundsätzlich die Waffensalbung. Vgl. dagegen Shapin (1998), S 57. Im
Übrigen wurden solche Überlegungen im 17. Jh. wissenschaftlich ernst genommen. Es reicht sich zu vergegenwärtigen, daß Kenelm Digbys Traktat über das durch Sympathie heilende Wundpulver (Digby (1658)) noch
Jahrzehnte später als theoretisch schlüssige Konzeption galt und in der Royal Society auch für abenteuerliche
Vorschläge zur Bestimmung des Längengrades aufgegriffen werden konnte. Vgl. dazu Sobel (1999).
80 Vgl. a.: Sylv. Nr. 950: „for the experiments of witchcraft are no clear proofs; for that they may be by a tacit
operation of malign spirits: we shall therefore be forced, in this inquiry, to resort to new experiments." Diesem

heit' der Magie stellt sich hier gar nicht. Wenn z. B. magische Heilerfolge zuverlässig be-
zeugt sind, dann braucht man nach Bacon keine Theorie, sondern ein Set von Übertragungs-
regeln, die gewährleisten, daß eine erfolgreiche Praxis in anderen Fällen eingesetzt werden
kann. Bringt man die Heilung zustande, beherrscht man die Methode. Mehr Wissen braucht
man nicht und mehr Wissen gibt es nicht. Scheitert das Übertragen, taugt das Verfahren
nicht, und man kann das Affenherz wegwerfen, bzw. versucht es dann doch mit menschlichen Herzen.

Interessanterweise wird der traditionellen Magie vorgeworfen, daß obgleich ihre Verfahren experimentell seien, ihre Theorie sich jedoch nicht durch Experimente falsifizieren ließe.
Und dies ist sein entscheidender Kritikpunkt, mit dem er Pierre Duhems These[81] über die
Funktion von Hilfshypothesen in der Wissenschaft vorwegzunehmen scheint: „Denn der Alchemist nährt eine immerwährende Hoffnung; und wo die Sache nicht glückt, so beschuldigt
er nur sein eigenes Irren [...] er schilt sich, daß er in seinem Verfahren schwierige Punkte und
Momente übersehen habe; daher wiederholt er die Versuche immer und immer wieder."[82]

Überaus erfolgreich, erlebte *Sylva Sylvarum* im Zeitraum von 1626-1685 20 Auflagen, 16
englische, 3 lateinische und 1 französische und war damit nach den Essays das zweiterfolg-
reichste Buch Bacons. Es ist aber offensichtlich, daß dieser überwältigende Erfolg im „Fluß
der Rezeptionsgeschichte" nicht Bacons Wissenschaftsprogramm zuschreiben ist, sondern
der Eigen-Attraktivität der als wissenschaftliches Testmaterial in der *Sylva* gesammelten
Fälle. Obleich als Kritik von Kuriositätensammlungen angelegt, las man dieses Buch als
Wunderkammer abenteuerlicher Berichte, früher mit Begeisterung, heute um es abzutun:
Truth is the daughter of time.

Literatur

Adam, Wolfgang (1988): *Poetische und kritische Wälder. Untersuchungen zu Geschichte
und Form des Schreibens ‚bei Gelegenheit'*. Heidelberg.

Adorno, Theodor W. und Horkheimer, Max (1971): *Die Dialektik der Aufklärung*. Frankfurt.

Bacon, Francis (1857): *The Works of Francis Bacon*. Hrsg. von James Spedding und Robert
L. Ellis. London.

Bacon, Francis (1966): *De Augmentis: Über die Würde und den Fortgang der Wissenschaften*.
Darmstadt.

Mangel wurde erst 55 Jahre später abgeholfen, als der Propagandist der Royal Society John Glanvill ein neues,
auf jeden Einzelfall bezogenes Prüfverfahren für Hexerei vorstellte: *Saducismus triumphatus: or full and plain
evidence concerning witches and apparitions*, London 1681. Aufschlußreich in dieser Hinsicht ist auch Bacons
Auseinandersetzung mit der Naturalmagie in *Aditus Ad Titulos in Proximus: Historia sympathiae et antipa-
thiae rerum* (II S. 81), in welcher er kritisiert, daß diese Wunder als Erklärungen akzeptiert, statt die Wunder zu
erklären.

81 Zu Duhem vgl. Hacking (1996), S. 414ff.

82 „Alchymista enim spem alit aeternam, atque ubi res non succedit errores proprios reos substituit [...] aut in
practicae suae crupulis et momentis aliquid titubatum esse, unde experimenta in infinitum repetit;" (N. O. I Nr. 85).

Bacon, Francis (1984): *Valerius Terminus. Von der Interpretation der Natur.* Anmerkungen von Hermes Stella. Engl./dt. eingeleitet und hrsg. von Franz Träger. Aus dem Englischen von Hildegard Träger und Franz Träger. Würzburg.

Bacon, Francis (1990a): *Die Weisheit der Alten.* Hrsg. und übersetzt von Philipp Rippel. Frankfurt.

Bacon, Francis (1990b): *Neues Organon* (Novum Organum). Lat./Dt. 2 Tle. Vorwort und hrsg. von Wolfgang Krohn. Aus dem Lateinischen von M. Buhr. Hamburg.

Bredekamp, Horst (1993): *Antikensehnsucht und Maschinenglaube. Die Geschichte der Kunstkammer und die Zukunft der Kunstgeschichte.* Berlin.

Camerarius, Joachim (1986): *Symbola und Emblemata* (1590ff.). Nachdruck in 2 Bdn. Hrsg. von Wolfgang Harms und Ulla-Britta Kuechen. Graz.

Cassirer, Ernst (1971): *Das Erkenntnisproblem in der Philosophie und Wissenschaft der Neueren Zeit* (1922), Bd. 2. Nachdruck Darmstadt.

Daston, Lorraine J. (1991): „Wunder, Naturgesetze und die wissenschaftliche Revolution des 17. Jahrhunderts", *Jahrbuch der Akademie der Wissenschaften* (1991), Göttingen, 99-122.

Daston, Lorraine J. (1994): „Baconian facts. Academic Civility and the Prehistory of Objectivity", in A. Megill (Hg.), *Rethinking Objectivity.* Durham/London, 37-63.

Daston, Lorraine J. (1998): „Wunder und Beweis im frühneuzeitlichen Europa", in G. Smith und M. Kroß (Hg.), *Die ungewisse Evidenz. Für eine Kulturgeschichte des Beweises.* Berlin, 13-68.

Daston, Lorraine J. und Park, Katharine (1997): *Wonders and the Order of Nature, 1150-1750.* Cambridge.

Digby, Kenelm (1658): *Late discourse ... touching the cure of wounds by the powder of sympathy.* London.

Eamon, William (1994): *Science and the Secrets of Nature.* Princeton.

Farrington, Benjamin (1964): *The Philosophy of Francis Bacon.* Liverpool.

Findlen, Paula (1997): „Disciplining the Discipline: Francis Bacon and the Reform of Natural History in the Seventeenth Century", in D. Kelley (Hg.), *History and the Disciplines in Early Modern Europe.* Rochester, 239-260.

Gaukroger, Stephen (2001): *Francis Bacon and the Transformation of Early-Modern Philosophy.* Cambridge UP.

Glanvill, John (1681): *Saducismus triumphatus: or full and plain evidence concerning witches and apparitions.* London.

Hacking, Ian (1996): *Einführung in die Philosophie der Naturwissenschaften.* Stuttgart.

Hattaway Michael (1978): „Bacon and ‚Knowledge Broken'. Limits for Scientific Method", *Journal of the History of Ideas* 39 (1978), 183-197.

Horton, Mary (1982): „Bacon and ‚Knowledge Broken'. An Anser to Michael Hattaway", *Journal of the history of ideas* 43/3 (1982), 487-504.

Hutchinson, Keith (1982): „What happend to Occult Qualities in the Scientific Revolution?", *Isis* 73 (1982), 233-253.

Jardine, Lisa (1990): „*Experientia literata* oder *Novum Organum?* The Dilemma of Bacon's Scientific Method", in William A. Sessions (Hg.), *Francis Bacon's Legacy of Texts.* New York, 47-68.

Krohn, Wolfgang (1987): *Francis Bacon.* München.

Krohn, Wolfgang (1994): „Die Natur als Labyrinth, die Erkenntnis als Inquisition, das Handeln als Macht: Bacons Philosophie der Naturerkenntnis betrachtet in ihren Metaphern", in Lothar Schäfer und Elisabeth Ströker (Hg.): *Naturauffassungen in Philosophie, Wissenschaft, Technik,* Bd. 2: Renaissance und frühe Neuzeit. München, 59-100.

Lovejoy, Arthur O. (1985): *Die große Kette der Wesen.* Frankfurt.

Merton, Robert (1980): *Auf den Schultern von Riesen.* Frankfurt.

Neuber, Wolfgang (1994): „Topik und Intertextualität. Begriffshierarchie und ramistische Wissenschaft in Theodor Zwingers *Methodus apodemica*", in Wilhelm Kühlmann und Wolfgang Neuber (Hg.), *Intertextualität in der frühen Neuzeit.* Frankfurt, 253-278.

Park, Katherine (1984): „Bacons's ‚Enchanted Glass'", *Isis* (1984), 290-302.

Pérez-Ramos, Antonio (1996): „Bacon's forms and maker's knowledge", in Markku Peltonen (Hg.), *The Cambridge Companion to Bacon.* 99-120.

Rossi, Paolo (1957): Francesco Bacone: Dalla magia alla scienza. Bari (engl.: Chicago 1968).

Rossi, Paolo (1968): *Francis Bacon. From Magic to Science.* London.

Schäfer, Lothar (1999): *Das Bacon-Projekt.* Frankfurt.

Schmidt-Biggemann, Wilhelm (1983): *Topica Universalis.* Hamburg 1983.

Schütze, Ingo (2000): *Die Naturphilosophie in Girolamo Cardanos De subtilitate.* München.

Shapin, Steven (1998): *Die wissenschaftliche Revolution.* Frankfurt.

Simone, Franco (1949): „Veritas filia temporis", *Revue de litterature comparée* XXII (1949), 508-511.

Sobel, Dava (1999): *Längengrad.* Berlin.

Sprat, Thomas (1667): *History of Royal Society.* London.

Thiel, Detlef (1993): „Schrift, Gedächtnis, Gedächtniskunst. Zur Instrumentalisierung des Graphischen bei Francis Bacon", in J. J. Berns und W. Neuber (Hg.), *Ars memorativa.* Tübingen, 170-205.

Vickers, Brian (1968): *Francis Bacon and Renaissance Prose.* Cambrigde.

Vickers, Brian (1988): *Francis Bacon.* Berlin.

Whitaker, Virgil K. (1968): „Francis Bacon's intellectual Milieu", in B. Vickers (Hg.), *Essential Articles for the Study of Francis Bacon.* Hamden/Conn, 28-50.

Wilson, Catherine (1988): „Visual Surface and Visual Symbol: the Microscope and the Occult in Early Modern Science", *Journal of the History of Ideas* 49 (1988), 85-108.

Wilson, Catherine (1997): *The Invisible World: Early Modern Philosophy and the Invention of the Microscope 1620-1720.* Princeton University Press, 2nd ed.

Wilson, Catherine (1998): „The Eye and the Microscope in Early Modern Science", in T. Borsche, J. Kreuzer und C. Straub (Hg.), *Bild und Reflexion.* München, 13-30.

Withney, Charles (1989): *Francis Bacon. Die Begründung der Moderne.* Frankfurt.

Yates, Frances A. (1975): *Aufklärung im Zeichen des Rosenkreuzes.* Stuttgart.

Klaus Reichert

In diesem Herbst der Welt.
Francis Bacons Begründung der Wissenschaft aus dem Geist
der apokalyptischen Verheißung[1]

Läßt sich eine Verbindung von Apokalyptik und Wissenschaft herstellen? Läßt sich zeigen, daß zumindest bei der Entstehung der neuzeitlichen Wissenschaften im 17. Jahrhundert die Apokalyptik die Stichworte geliefert und sogar Pate gestanden hat? Aber wozu forschen – und zwar mit einer in der Geschichte bisher nie dagewesenen Beschleunigung –, wenn die Welt ohnehin auf ihr Ende zusteuert? Stehen in Krisenzeiten wirklich alle Optionen zur Verfügung oder gibt es Deutungsmuster, die in dem Maß an alten Orientierungshilfen hängen wie die Überforderung durch das Neue zu groß wird? Oder gibt es gar eine ursächliche Verknüpfung zwischen Wissenschaft und Apokalyptik[2], wobei die theologische Altlast jetzt zu verstehen ist als Eintritt in ein Weltalter, das sich von den bisher geltenden Gesetzen von Natur und Zeit gelöst hat? Diesen Fragen möchte ich mich zuwenden. Ich beginne mit Isaac Newton, dem ersten Giganten des neuzeitlichen Forschens, zu dem Pope der Zweizeiler einfiel: „Nature and Nature's Laws lay hid in night / God said: *let Newton be!* and all was light." Ich gehe dann zurück durch das 17. Jahrhundert mit seinen millenarischen Erwartungen und komme zu Francis Bacon, dessen Wissenschaftsverständnis der Flucht- und Angelpunkt für das Forschen der nachfolgenden Generationen gewesen ist.

Newton hatte seine großen Entzifferungswerke des Buchs der Natur – die Beschreibung der Himmelsmechanik nach dem von ihm entdeckten Gravitationsgesetz und die Optik – mit dem Ziel geschrieben, die Wahrheit des Buchs der Bücher nachzuweisen. Mit analog gedachter, aber strikt getrennt gehaltener methodischer Präzision ging er gleichzeitig daran, den historischen – oder genauer: den chronologischen – Sachgehalt der biblischen Bücher selber zu entschlüsseln.[3] Durch Rückrechnung der Äquinoktien samt ihrer jährlichen Differenz

---

1 Eine erste Fassung des Vortrags wurde auf einer Frankfurter Konferenz über „Wissensideale und Wissenskulturen in der frühen Neuzeit" (1.-3. Dezember 2000) vorgestellt. In erweiterter Form wurde der Vortrag am Zentrum für Literaturforschung am 13. Dezember 2000 in Berlin gehalten. Noch einmal erheblich erweitert (besonders in den Schlußthesen) wurde die Arbeit in der hier vorliegenden Form auf Einladung der Carl Friedrich von Siemens-Stiftung am 29. Mai 2001 in München vorgetragen.
2 Nach Abschluß meiner Arbeit erschien das grundlegende Buch von Fried (2001). Fried zeigt mit einer geradezu erdrückenden Anzahl von Belegen, wie die Dauerpräsenz der Vorstellungen vom Ende der Welt (als Bedrohung *und* Verheißung) immer genauere astronomische, astrologische und mathematische Berechnungen verlangten, woraus schließlich die modernen Naturwissenschaften hervorgingen. Das heißt auch, daß die Annahme eines ‚Bruchs', eines Neubeginns der Naturwissenschaften in der Frühen Neuzeit zu modifizieren, wenn nicht zu korrigieren ist und wir eher die Kontinuität des apokalyptischen Denkens anzusetzen haben. Das war auch mein Grundgedanke bei der Verortung Bacons als Apokalyptiker, allerdings ohne genauere Kenntnisse der Vorgeschichte.
3 Zu Newton allgemein, aber auch zu seinen apokalyptischen Spekulationen nach den Manuskripten, siehe Westfall (1980) – Zu den Spekulationen mit Abdruck einiger Manuskripte in extenso: Manuel (1974).

von 50 Sekunden versuchte er nicht nur das Alter der Schöpfung genauer zu bestimmen als andere vor ihm, sondern auch – flankiert durch die zusätzliche Annahme der Durchschnitts-herrschaftszeit von Königen – zu zeigen, daß die Hebräer das älteste Volk waren und die anderen Hochkulturen nichts weiter als Abkömmlinge, Abspaltungen, Transformationen des Schöpfungsvolkes. Newton hat seine Berechnungen nicht veröffentlicht. Moderater Skepti-ker, der er war – auch in seinem mathematisch-physikalischen Werk spricht er eher von „Rules" und „Propositions" als von zweifelsfrei beweisbaren Wahrheiten (gerade die Gravita-tion konnte er nicht ursächlich begründen[4]) –, war ihm die Sache doch vielleicht noch nicht schlüssig genug, zumal erheblicher theologischer Sprengstoff darin steckte – so hatte er zum Beispiel Noah als Gründervater der Religion ermittelt und die Patriarchen und Moses nur als dessen Abkömmlinge, die nicht mehr im Besitz der ‚ganzen Wahrheit' waren, gelten lassen.[5] Der Text erschien posthum 1728 unter dem Titel *The Chronology of Ancient Kingdoms Amen-ded* und erregte keine große Aufmerksamkeit – außer einer gewissen Irritation bei Voltaire – , weil er nicht ins Bild des großen Mannes paßte. Ebenfalls posthum 1733 erschien ein fast noch erstaunlicherer Text, *Observations upon the Prophecies of Daniel and the Apocalypse of St. John.* Es ist der Versuch, die jahrhundertelangen Berechnereien der Endzeit – wenn die Welt einen Anfang hatte, mußte sie auch ein Ende haben – zunächst einmal dadurch zu durchkreuzen, daß er Kohärenz im Sinnwust der Bildangebote herstellte (apokalyptische Tiere zum Beispiel bedeuten immer Reiche), zweitens aber rechnerisch gesicherte Daten zu extrapolieren, also Prophetie in Prognostik zu verwandeln.

Newton wollte sicherstellen, daß „the language of the prophets will become certain and the liberty of wresting it to private imaginations be cut off." In einem Nachlaßfragment heißt es, „... the longer they (= die Prophezeiungen) have continued in obscurity, the more hopes there is that the time is at hand in which they are to be made manifest. If they are never to be understood, to what end did God reveale them?"[6] Newton vergleicht die Unfähigkeit der Juden, in Jesus den Messias zu erkennen, mit der Abwehr einiger seiner Zeitgenossen gegen-über der Apokalyptik: „why should we think he will excuse us for not searching into the Prophesies which he hath given us to know Antichrist by? For certainly it must be as dange-rous and as easy an error for Christians to adhere to Antichrist as it was for the Jews to reject Christ."[7] Der Antichrist sitzt natürlich immer noch in Rom, und der junge Newton bezieht die vom Propheten Daniel verschlüsselt angegebene, von Newton errechnete Zeit von 1260 Jahren bis zur Wiederkunft des Messias auf die Geschichte der römischen Kirche: 1200 Jahre sind bisher verstrichen, die Wiederkunft wird sich also in absehbarer Zeit ereignen, und es heißt immer wieder, „the time is at hand", „die Zeit ist herbeigekommen", in deutlicher

---

4 Der zweiten Auflage seiner *Principia Mathematica* (1713) fügte er ein ‚Scholium Generale' an, in dem es heißt: „But hitherto I have not been able to discover the cause of those properties of gravity from phenomena, and I frame no hypotheses (hypotheses non fingo); for whatever is not deduced from the phenomena is to be called an hypothesis; and hypotheses, whether metaphysical or physical, whether of occult qualities or mechanical, have no place in experimental philosophy." Newton sah sich zu der Erklärung veranlaßt, weil Leibniz (im Briefwech-sel Leibniz/Hartsoeker, der auszugsweise im Frühjahr 1712 in England veröffentlicht worden war) ihm vorge-worfen hatte, daß die *Principia* „deserts Mechanical causes, is built upon Miracles, and recurs to Occult quali-tys." (Cohen (1975), S. 243, Anm. 14)

5 Westfall (1986), S. 230. Bezieht sich vor allem auf die Yahuda-Manuskripte in der Jewish National and University Library, Jerusalem.

6 Manuel (1974), S. 107 (Yahuda Ms. 1).

7 Ibid., S. 109.

Anspielung also auf die Parusieerwartung. Was genaue Zahlenangaben betrifft, wird Newton im Gegensatz zu vielen seiner Zeitgenossen zunehmend vorsichtig, hält aber daran fest, „... it was Gods designe in these prophesies to typefy and describe not trifles but the most considerable things in the world during the time, time of the Prophesies."[8] Newton sucht in den Apokalyptiken nach einem Prinzip – mit dem Wort ‚Gesetz' geht er sehr vorsichtig um –, das so ‚einfach' ist wie das in seiner Physik entwickelte: „To ‹prefer› choose those ‹interpretations› constructions which without straining reduce things to the greatest simplicity ... Truth is ever to be found in simplicity, and not in the multiplicity and confusion of things. As the world, which to the naked eye exhibits the greatest variety of objects, appears very simple in its internall constitution when surveyed by a philosophic understanding, and so much the simpler by how much the better it is understood, so it is in these visions."[9] Newtons Vorbehalte gegen einen geschlossenen Uhrwerkmechanismus, die Möglichkeit einer jederzeitigen Intervention Gottes in einem absolut gedachten Raum, den Newton das Sensorium Gottes nennt, die Unerklärbarkeit der berechneten Gravitationsbalance, die unbestreitbar gedachte aber nicht begründbare Annahme einer Fernwirkung, all das wird durch seine Überlegungen zur Apokalyptik zumindest gestützt, wenn nicht veranlaßt. Noch zwei Generationen nach Newtons Tod wird der schwäbische Pietist Friedrich Christoph Oetinger, der apokalyptischem Denken nahestand, in Newton denjenigen sehen, der die beiden Grundprinzipien der Welt – das Reich der Finsternis und das der Schöpfung oder des Lichts – als antinomische Struktur wissenschaftsfähig gemacht hat: „... die Neutonianer heisen den Neuton einen schöpferischen Geist. Warum? weil er der erste ist, der die zwei widerwärtige Central=Kräften Contripetam und Centrifugam in die Grundweißheit [das heißt Philosophie bzw. Wissenschaft] eingeführt."[10] Attraktion und Fliehkraft also als Manifestationen Gottes und des Teufels in der Himmelsmechanik, auf ewig – oder für die Dauer der Menschenzeit – aneinander gebannt. So strikt Newton Theologie und Physik methodisch auch auseinanderhielt, ist es für diesen Giganten neuzeitlicher Wissenschaft doch bemerkenswert, daß in die Erklärungslücken immer wieder Gott einspringen mußte: Gott ist die Ursache der unerklärlichen Gravitation. So ausdrücklich sagt er das zwar nicht, nimmt aber doch zumindest an, daß der Raum dasjenige ist, worin der Wille Gottes die physische Welt lenkt und kontrolliert.[11] Leibniz bemerkt dazu anerkennend in der Theodizee, Newton habe die direkte Fernwirkung eines Körpers auf den anderen, die von modernen Philosophen verworfen worden sei, rehabilitiert und in Verbindung mit dem in den Evangelien berichteten Geschehen gebracht.[12] So erscheint das vom 18. und 19. Jahrhundert aus gesehen epochal Neue immer wieder und immer noch zurückgebunden an die alte Form der christlichen Wahrheit. Und wenn das Buch der Bücher und das Buch der Natur zwei Seiten einer Medaille sind, müßte sich der neue Anspruch der Wissenschaft, Gesetzmäßigkeit z. B. über Prognostizierbarkeit von Naturtatsachen zu begründen, auch in den heiligen Schriften einlösen lassen, eben in der Prophetie. Im noch immer wenig bearbeiteten Nachlaß Newtons in Jerusalem und Cambridge liegen zahllose Manuskripte, die seine lebenslange Beschäftigung mit der Provokation durch Prophetie und Apokalypse dokumentieren. Die Zeit ist also noch für Newton keineswegs zur Zukunft

---

8 Ibid., S. 119
9 Ibid., S. 120. Die Stelle gehört zu den „Rules for methodising (construing) the Apocalyps".
10 Oetinger (1987), S. 534
11 Thayer/Randall (1960), S. 185.
12 Nach Freudenthal (1982), S. 50.

hin offen. Nur schwanken naturgemäß die Berechnungen des Endes. In einer ganz späten Prognose nennt er das Jahr 2060 für das Endgericht, fügt aber hinzu: „It is not for us to know the times and seasons which God has put in his own breast."[13]

Mit Newton kommt eine Denkbewegung zu ihrem vorläufigen Höhepunkt, die im 17. Jahrhundert die Debatten zumindest in England beherrscht hatte: die Bindung der Neuen Wissenschaft an die Theologie im allgemeinen, die Apokalyptik im besonderen. Die Gründe dafür sind komplex, und sie werden kontrovers diskutiert, seit in den 30er Jahren des 20. Jahrhunderts Abschied genommen wurde von der internen Erfolgsgeschichte einzelner Disziplinen – der Physik, der Mathematik, der Chemie – und stattdessen ihre multikausalen Entstehungsbedingungen in den Blick kamen, also die Ökonomie, der Gegensatz zwischen halboffiziellen Wissenschaftszirkeln und den Universitäten, die Legitimationskrise des Staates, und eben die Theologie. Bis heute diskutiert wird die sogenannte Merton-These, derzufolge es die Puritaner waren, denen die Wissenschaft ihren Aufschwung und ihre Professionalisierung verdankte.[14] Ihre Ablehnung jeglicher Dogmatik, Autorität oder der Ölgötzen der Überlieferung à la Aristoteles schien ihnen den Blick geöffnet zu haben für Neues. Ihr Pragmatismus erkannte die Bedeutung der mechanischen Künste für Navigation, Kartographie, Handel, Landwirtschaft, die alle zum Wohl der Menschheit entwickelt werden mußten und die doch gleichzeitig in der Artistenfakultät der Universitäten keine Rolle spielten. Ihre Überzeugung von der Offenbarung Gottes in seinen Werken machte es nötig, ihn eben dort zu erforschen und nicht mehr nur in seinem offenbarten Wort, denn Religion (faith) legitimierte sich durch Rationalität (reason), nicht durch schlichten Glauben (belief). Charles Webster hat an eine zeitgenössische Kennzeichnung der Puritaner als der „‚impatient' or ‚hotter sort' of protestants" erinnert.[15] Womit hing diese Ungeduld zusammen? Webster konstatiert, daß im Umkreis von Bürgerkrieg und Revolution (also grob gerechnet in den Jahren von 1640 bis 1660), aus denen die Independentengruppe der Puritaner ja kurzfristig siegreich hervorging, sich die Entwicklung der Wissenschaften beschleunigte. Warum das so war, hing mit der Erwartung des Milleniums zusammen, das als unmittelbar bevorstehend gedacht war. Chaos und Krieg, wie im übrigen auch der Blick auf Europa zeigte, sowie die beispiellose Enthauptung eines Königs, der sich die Oberhoheit über die Religion angemaßt hatte, deuteten ja klar darauf hin. Es kam daher alles darauf an, die Ermöglichungsbedingungen für die Wiederkunft Christi, des Messias, so schnell es ging zu schaffen. Warum das so war, hing sicher mit der Aufwertung der vita activa zusammen und dem Prosperieren weltlicher Dinge, also auch der Wissenschaften, als einem Zeichen der Wohlgefälligkeit in den Augen Gottes. Es hing aber auch mit einer apokalyptischen Verheißung zusammen, daß unmittelbar vor der Wiederkunft der prälapsarische Zustand wieder erreicht wäre, was sich in einer nie geahnten Entfaltung der Wissenschaften und Künste – Künste im weitesten Sinne verstanden – zeige. Solche Erwartungsbeschleunigungen waren nicht auf den Fortschritt der Wissenschaften beschränkt. Der Amsterdamer Rabbi Menasse Ben Israel schickte zwei Petitionen an das Cromwellsche Parlament und bat um Wiederzulassung der Juden in England, mit der Begründung, erst wenn in jedem Land der Erde Juden seien, könne der Messias kommen, und England war das einzige Land, in dem es seit 1290 keine Juden gab. Menasses Gesuch wurde stattgegeben,

---

13 Westfall (1980), S. 816 (nach Yahuda Ms 7.3g, f. 13).
14 Merton (1978).
15 Webster (1986), S. 193. Webster resümiert (und modifiziert) darin einige Thesen seines grundlegenden Werks Webster (1975).

allerdings mit der im Parlament geäußerten Maßgabe, die Zugewanderten noch schnell konvertieren zu müssen.[16]

Doch zurück zur Wissenschaft. Man muß sich klarmachen, daß von organisierter Forschung noch keine Rede sein kann. Gewiß, es gab Gresham College, von dem Gründer der Royal Exchange gestiftet, und dort wurden technologische Projekte in pragmatischer oder ökonomischer Absicht diskutiert. Aber das eigentliche Forschen, das heißt vor allem ein Experimentieren in Bacons Sinne, fand in kleinen und kleinsten Zirkeln statt und wurde in der Regel per Korrespondenz diskutiert. Robert Boyle, der Begründer der Chemie, sprach in diesem Zusammenhang von einem „Invisible College", in dem man die Keimzelle der Royal Society gesehen hat. Aber es gab andere Gruppen und Grüppchen, die sich zum Teil überschnitten, zum Teil nichts miteinander zu tun hatten, außer dem gemeinsamen Nenner eines wünschbaren kooperativen Forschens und, vorsichtig gesagt, einer Veränderung der Lebensumstände. Nicht immer wurde die Absicht so prononciert geäußert wie von einigen radikalen Puritanern, die dabei keine Sozialrevolutionäre im Sinne der Ranters und Diggers waren, die ein Jahr vor dem Sturz der Republik aktiv ein Paradies auf Erden planten, eine Erneuerung der Kirche und eine Organisation der Wissenschaften im Hinblick auf ein unmittelbar bevorstehendes Zeitalter von „Universall love, and Universall Commerce, in mutuall peace and Nobel Communications", wie John Beale am 28. November 1659 an Samuel Hartlieb schrieb, die beide alles andere als Schwarmgeister waren und bald Mitglieder der Royal Society wurden.[17] Webster interpretierte die millenarischen Tendenzen unter der Republik so: „There was no expectation of the restoration of any monarchy other than the monarchy of Christ's personal reign on earth, and for these reformers the restoration of the Church of England and formation of a Royal Society must have seemed contrary to the course of history."[18]

Die These, daß die Puritaner die treibende Kraft hinter dem Fortschritt der Wissenschaften waren, ist natürlich längst modifiziert worden. Zunächst einmal ist das Bild, das die Puritaner bieten, viel zu buntscheckig, als daß sich eine klare Linie abzeichnete, und reicht von moderaten, nüchternen Pragmatikern, die sich für nützliche Wissenschaften interessierten, bis zu den sogenannten Enthusiasten, den Schwarmgeistern, die sich auf die unterschiedlichsten Eingebungen verließen. Zweitens gab es zahlreiche Anglikaner verschiedener Zuordnung – High Church, Low Church, Broad Church –, die eine bedeutende Rolle spielten. Thomas Sprat, der erste Historiograph der Royal Society, wurde später anglikanischer Bischof, Joseph Glanvill, der *seine* Geschichte der Royal Society unter dem Baconschen Titel *Plus Ultra*[19] erscheinen ließ, war anglikanischer Geistlicher mit latitudinarischen, d. h. liberalen Tendenzen. Zwei Katholiken, Sir Kenelm Digby und Thomas White legten die ersten systematischen Ausarbeitungen einer mechanischen Philosophie vor. Schließlich ist nicht immer klar, in welchem Lager jemand zu suchen war. Boyle wurde „a puritan at heart if ever there was one" genannt, andere nannten ihn einen moderaten Royalisten und Episkopalisten, wieder andere einen frommen Anglikaner, noch andere einen Anhänger der Cambridger Platonisten mit ihren kabbalistischen Neigungen. Diese Liste soll nichts anderes sein als ein Hinweis, daß Wissenschaft ohne Theologie denkunmöglich war. Allerdings – und dieser Zusatz ist wichtig, denn sonst hätte es keine Wissenschaft im neuzeitlichen Sinn gegeben –

---

16 Katz (1982).
17 Webster (1986), S. 212 f.
18 Ibid., S. 213.
19 Vgl. hierzu meinen Aufsatz Reichert (1994).

hatten beide methodisch strikt getrennte Wege zu gehen, auch wenn, wie im Fall Newtons, Parallelen denkbar, Begründungsnachhilfen möglich waren. Bacons Forderung einer Trennung von Theologie und Wissenschaft und der Begründung der Einzeldisziplinen aufgrund ihrer je eigenen Methode, die durch das, was er Experiment und Induktion nannte, zu erarbeiten war, ist vielleicht sein wichtigstes Vermächtnis. Erst dann, nach der methodisch gelenkten Entzifferung des Buchs der Natur, ließ sich sehen, daß sie auch den Schlüssel zum Wort Gottes lieferte, der auf dieser höheren Erkenntnisstufe um so mehr glorifiziert werden konnte. Oder anders gesagt: das Buch der Natur und das Buch der Bücher boten andere Wahrheitsbegründungen und andere Gewißheiten, die in ein Verhältnis zueinander zu setzen waren.

Boyle etwa war einer der genauesten Beobachter und Experimentatoren zum Beispiel zum Problem des Vakuums, der in der Wissenschaft jede theologische Spekulation vermied.[20] Zugleich war er überzeugt, daß es in der Wissenschaft keine absolute, sondern nur eine eingeschränkte Gewißheit geben könne, das was er Hypothesen nannte. ‚Indubitable certainty' gab es für ihn nur in der Religion (faith). Boyles bedingte oder moderate Skepsis in Dingen der Naturgesetzlichkeit hing mit dem Willen Gottes zusammen: auch wenn wir dieses oder jenes als Naturgesetz erkannt und durch Experiment bewiesen haben, wäre es eine Einschränkung Gottes, von ihm zu erwarten, daß er sich auch daran hält: „... if we consider God as the author of the universe, and the free establisher of all the laws of motion ... we cannot but acknowledge, that, by with-holding his concourse, or changing these laws of motion, which depend perfectly upon his will, he may invalidate most, if not all the axioms and theorems of natural philosophy."[21] So wendet sich auch Boyle zur Offenbarung als – im Gegensatz zur Naturforschung – „an impeachable source of truth", allerdings mit der Einschränkung des skeptischen Chemikers, „in our present human situation we may not be able to understand such truths." Und so wendet er sich lieber den eingeschränkten Gewißheiten seiner chemischen Experimente zu, die er methodisch strikt von theologischen Spekulationen getrennt hält. Auch Newton hielt, wie gesagt, beide Bereiche strikt getrennt, sah sich aber genötigt, der zweiten Ausgabe der *Principia* ein „Scholium Generale" anzufügen, in dem er die Bedeutung der Theorie für die Religion („God's dominion and power"), aber eben auch das Unzureichende dieser Theorie, herausstellte.

Überblickt man also das 17. Jahrhundert, so kann an der wechselseitigen Bedingtheit von Religion und Wissenschaft, bei getrenntem Procedere, kein Zweifel sein. Auch der gelegentlich geäußerte Atheismusverdacht gegen diesen und jenen hatte mit dem, was spätere Jahrhunderte darunter verstanden, nichts zu tun. Atheist war zum Beispiel einer, der die leibliche Auferstehung der Toten oder die Existenz von Hexen leugnete. Auch dies letzte, die Existenz von Hexen, war ein Problem der Naturforschung: an der *Wirkung* von Hexen konnte kein Zweifel sein – der Anglikaner Glanvill führte zahlreiche Beispiele an –, ihre Verursachung aber mußte erforscht werden – wie die Ursache einer Krankheit, um sie entsprechend rational zu behandeln –, wollte man sie nicht nur der Veranlassung durch eine Einwirkung des Antichrist überlassen. Aber noch Newton kann , wie gesagt, das Problem der Fernwirkung nicht

---

20 Shapin/Schaffer (1985).
21 Boyle (1675). Zitiert nach van Leeuwen (1963), S. 93. Ähnlich äußert sich Clarke, der Briefpartner von Leibniz (und dessen Verbindungsmann zu Newton): „there is no such thing as what men commonly call the course of nature, or the power of nature. [It] is nothing else but the will of God producing certain effects in a continual, regular, constant, and uniform manner." Zitiert in Jacob (1986), S. 246.

lösen. Die Balance zwischen Zentripetal- und Zentrifugalkräften war im jetzigen Zustand des Universums für ihn nachweisbar, nicht aber deren Verursachung. Die lag dann, hilfsweise und vorläufig, bei Gott, wie die der Hexen beim Teufel. Der Vorwurf des Atheismus deckt also ein weites Feld unter den Spielformen der Neuen Wissenschaft. Der, der den Vorwurf am häufigsten zu hören bekam, war Thomas Hobbes. Gewiß eignete er sich als Kandidat, weil er im 1. Buch des *Leviathan* vier Gründe für die Entstehung der Religionen genannt hatte: Angst, Unkenntnis von Kausalursachen für Naturerscheinungen, ein Für-wahr-Halten von Fantasmen, etwa in Träumen, und die Interpretation zufälliger Naturereignisse – Kometen, Sonnenfinsternisse, Erdbeben – als Prophetien. Doch im 3. und 4. Buch des gleichen *Leviathan* geht es um nichts anderes als um die Vindizierung der wahren christlichen, freilich antirömischen Religion und ihre Legitimationsgrundlage für eine christliche Herrschaft. Auch Hobbes ist vermutlich Apokalyptiker, allerdings nicht im Sinne seiner Landsleute, die mit der Revolution das Millenium gekommen sahen. Der *Leviathan* erschien 1651 im Exil in Paris, also mitten im Interregium. Aus seiner Position waren Chaos und Unordnung der Schwarmgeister denkbar ungeeignet, im Heilsplan eine Rolle zu spielen. Und so heißt es nach langem close reading der prophetischen Schriften im Bezug zum Geschichtsverlauf: „The Antichrist is not yet come"[22] oder „wee are yet in the Dark"[23], mit der Gewißheit, daß sich das Dunkel einmal – er sagt nicht wann – lichten werde. So muß also vermutlich auch Thomas Hobbes, der Begründer der radikal säkularen, die Religion der Herrschaft untergeordneten, auf Zukunft hin entworfenen Staatstheorie der Frühen Neuzeit vor dem Hintergrund der Auseinandersetzung mit der Apokalyptik gesehen werden.[24]

Hobbes ist ein typisches Paradox in einer Zeit, die das Paradox – das Zusammenspannen von Unvereinbarem – als Denkform konstatierte wie keine Zeit zuvor. In einer Definition von 1616 heißt es noch, das Paradox sei eine Meinung, die einer allgemein akzeptierten Meinung zuwiderliefe, wie wenn einer behaupte, daß die Erde sich in einem Kreis bewege und die Himmel stillstünden. Hier wird noch das einander Ausschließende nebeneinander gestellt wie bei Zenons Paradox von Achilles und der Schildkröte. Eine Generation später spricht der Cambridger Platoniker und zeitweilige Cartesianer Henry More (1633), der zugleich erheblichen Einfluß auf die „Theologie" Newtons (und Leibniz') hatte, von „That pleasant and true Paradox of the Annual Motion of the Earth."[25] Das heißt: das offensichtlich Wider-

---

22 Hobbes (1959), Teil 3, Kap. 42, S. 302. Der Satz bezieht sich darauf, daß viele antirömisch Gesinnte im Papst den Antichrist sahen. Wäre er es, wie in Matthäus 24 angekündigt, müßte „such tribulation" folgen, „as was not since the beginnings of the world ... But that tribulation is not yet come; for it is to be followed immediately by a darkening of the Sun and Moon, a falling of the Stars, a concussion of the Heavens, and the glorious coming again of our Saviour in the cloudes. And therefore *The Antichrist* is not yet come; whereas many Popes are both come and gone."
23 Ibid., Teil 4, Kap. 44, S. 332.
24 So säkular ist aber sein Staat auch wieder nicht. Im Kapitel über ‚The World to come' (Teil 3, Kap. 38) heißt es „that the kingdom of God is a civil commonwealth, where God himself is Sovereign, by vertue first of the *Old*, and since of the *New* Covenant, wherein he reigneth by his vicar, or lieutenant; the same places do therefore also prove, that after the coming again of our Saviour in his Majesty, and glory, to reign actually, and Eternally; the Kingdom of God is to be on Earth." Ibid., S. 244. Wir sehen hier das Motiv eines postapokalyptischen, keineswegs absolutistischen, irdischen Paradieses, das sich analog zu Bacons kommender Welt im Zeichen der durch die Apokalyptik in Gang gesetzten neuen Wissenschaft lesen läßt. Der Übergang ist bei Hobbes vermutlich markiert durch den Bürgerkrieg, obwohl er das (Anm. 22 und 23) zu leugnen scheint.
25 Beide Zitate zum Paradox stammen aus dem *Oxford English Dictionary*.

sprüchliche – Augenschein gegen Naturgesetz – wird in seiner Gleichwertigkeit im Sinne unterschiedlicher, inkompatibler Verstehensweisen der Natur stehengelassen, allerdings noch nicht in seiner zerreißenden Spannung für die von Pascal sogenannte condition humaine diskutiert. Wir haben angedeutet, wie Newton das Paradox, das für ihn vermutlich keines war – also die Vereinbarkeit von Physik und Evangelium bzw. Prophetie –, durchdachte, daß er aber zu vorsichtig war, die Teile seines spekulativen Forschens, die ihn sein Leben lang beschäftigten, selber zu veröffentlichen. Wir werden sehen, wie Newtons Zeitgenossen und die Naturforscher vor ihm in der Vereinbarkeit des Unvereinbaren offenbar keinen Widerspruch sahen und das eine das andere bedingte und beflügelte. Es handelt sich in dieser Geniezeit der Neuen Wissenschaft um das Paradox der allenthalben konstatierbaren Beschleunigung der Erkenntnis und des Wissens mit dem Ziel eines Wirkens *in der Zeit*, etwa zur Verbesserung der Lebensbedingungen auf der Erde, und *zugleich* der Erwartung des *Endes der Zeit*, also ein Heraustreten *aus* der Zeit, angesichts der millenarischen Verheißung. Wir können davon ausgehen, daß die Debatten in England, sofern sie sich auf Religion und Wissenschaft beziehen, nicht zu lösen sind von einer affirmativen oder skeptischen, offenen oder verdeckten Auseinandersetzung mit den Spielarten des Milleniums im besonderen und der Apokalyptik im allgemeinen. Das ist, wie angedeutet, auch keineswegs auf die Puritanerkreise beschränkt. Als die Royal Society wegen irreligiöser Tendenzen unter Beschuß durch die High Church gerät, bedienen sich ihre anglikanischen Apologeten durchweg millenarischer Rechtfertigungen und beschreiben ihre Forschungsgemeinschaft als die Verwirklichung von Bacons House of Salomon auf *Nova Atlantis*. Es ist also Francis Bacon, der Vater der Neuen Wissenschaft, der Patron der Royal Society, der mit seinen Forschungsprogrammen als erster in der Neuzeit, wie mir scheint, die Verbindung von Apokalyptik und Wissenschaft in Gang gesetzt hat. Ihm wenden wir uns jetzt zu.

Bacon, 1561 geboren, gelernter Jurist, wurde unter Jakob I einer der mächtigsten Männer Englands, zuletzt als Lordkanzler. Bacon stellte das Wissen, Learning, auf eine neue Grundlage, um überhaupt erst so etwas zu ermöglichen, was Wissenschaft, scientia, science, genannt werden konnte. Dazu setzte er eine strikte Trennung von Theologie und Naturforschung voraus, die methodisch verschiedene Wege zu gehen hätten, bis sie am Ende in der Erkenntnis der Werke und des Wortes Gottes wieder konvergieren oder sie gegenseitig stützen könnten. Zweitens demontierte er das, was er Vorurteilsgötzen nannte, Idola, die jahrhundertelang die Entwicklung des Wissens blockiert hätten. Dazu gehörte die Macht der Autoritäten, allen voran die des Aristoteles[26], gehörte das auf Bücher fixierte Wissen, gehörten Denktraditionen, Meinungen und Gewohnheiten, gehörte das, was Wittgenstein später die Verhexung durch Sprache nannte, also die Beschränkung der Erkenntnismöglichkeiten durch die Struktur der Grammatik, in der sie zu formulieren waren, und durch die Unschärfe der dabei verwendeten Wörter ohnehin. Schließlich, und vielleicht am wichtigsten, gehören hierzu die Vorurteile der Gattung: der Mensch sieht die Natur, nach dem alten homo-mensura-Satz, nur in Bezug auf sich selbst, oder im Bezug zu Gott, wohingegen es darauf ankäme,

---

26 Die Perspektiven wechseln, die Namen bleiben. Der Bischof von Paris, ein Anti-Aufklärer, verurteilte 1277 219 (z. T. angebliche) Thesen des Aristoteles (siehe Flasch (1989)); der anti-orthodoxe, aber eher wissenschaftsfeindliche Martin Luther sprach von Aristoteles als einem heidnischen ‚narrentreyber' gegen die heilige Schrift (Fried (2001), S. 177 und 246, Anm. 485); für den Frühaufklärer Bacon ist Aristoteles der Antichrist (s. u.); und Hobbes, sozusagen die Anti-Inkarnation des Bischofs von Paris, spricht von „the vain and erroneous Philosophy of the Greeks, especially of Aristotle" (Hobbes (1959), Teil 4, Kap. 44, S. 332).

sie für sich, in ihrer Eigengesetzlichkeit zu sehen, um sie dann allerdings, womöglich, dem Zugriff des Menschen zu öffnen, das heißt, durch den Menschen beherrschbar zu machen. Man sieht, daß hier viele Voraussetzungen formuliert sind, die für die Herausbildung der modernen Wissenschaften konstitutiv werden sollten. Für Bacons Pragmatik bedeutete dies, das Feld der Forschung zu erweitern durch Bereiche, die von den Universitäten ausgeschlossen waren. Daher sein Interesse für die mechanischen Künste, für die Handwerker, für die Schmiede und Hydrauliker, die Instrumentenbauer und Landvermesser, die Kartographen und Navigatoren. Sie alle hatten durch praktische Erfahrung eine Kenntnis von der Natur, die den Spekulationen der Gelehrten überlegen war, und die es zu systematisieren galt, um den Gesetzen der Natur auf die Spur zu kommen. Niemand vor Bacon hat mit solcher Unbeirrbarkeit die Bedeutung des Experiments für die Wissenschaft propagiert, diese Schnittstelle zwischen Empirie und Theorie. Bacon selbst hat nicht experimentiert, er hat auch kein einziges Gesetz formuliert, weil seine Kenntnisse in keiner der Disziplinen dazu ausreichten. Sein Rang liegt darin, die methodischen Rahmenbedingungen aufgestellt zu haben, hinter die zukünftige Generationen nicht mehr zurückgehen konnten. Diese kurze Skizze mag ausreichen, um an das zu erinnern, wofür Bacon und der sogenannte Baconianismus für das 17. Jahrhundert stand: das Neue, Entwicklung, Fortschritt, der Aufbruch in eine offene Zukunft. Sehen wir zu, ob es dazu eine Gegengeschichte gibt, bei einem Mann, der gern im Frühlingsregen spazieren fuhr, um sich dadurch den vitalistischen Samen des Nitrats in der Luft und damit dem „universal spirit of the world" auszusetzen.[27]

Auf dem Titelblatt von Bacons Fragment gebliebenem Hauptwerk, der *Instauratio Magna*, sind zwei riesige Säulen zu sehen und dazwischen ein Schiff mit geblähten Segeln auf offener See. Darunter steht der Spruch: „multi pertransibunt et augebitur scientia". Man hat das immer gelesen, und so legen es Bild und Unterschrift ja nah, als den Aufbruch zum Neuen, als das Durchbrechen der Grenzen der alten Welt, die durch die Säulen des Herkules markiert waren, als das bildgewordene „Plus Ultra", wie das Motto Bacons und Karls V. lautete. Schauen wir uns aber das Schiff genauer an, so fällt auf, daß es so eindeutig nicht ist: die Wimpel, die Legitimationszeichen, scheinen zwar auf Ausfahrt zu stehen, aber die Segel zeigen die Rückkehr an, eine paradoxe Doppelbewegung also. Es wird sich zeigen, daß genau diese Doppelbewegung Bacons Denken bestimmt, also Aufbruch und Rückkehr in einem. Der Spruch selbst ist ein halbiertes Danielzitat. Der Zusammehang der Stelle ist folgender: Der Prophet berichtet von den Offenbarungen, den Apokalypsen, die ihm zuteil geworden sind: Es wird eine Zeit des Kampfs der Reiche gegeneinander sein, der Bürgerkriege, der Gotteslästerungen und Verwüstungen, eine Zeit großer Trübsal. Dann heißt es: „Aber zu jener Zeit wird dein Volk errettet werden ... Aufwachen zum ewigen Leben oder zu ewiger Schmach und Schande." Dann kommt der von Bacon halbierte Vers: „tu autem Danihel clude sermones et signa librum usque ad tempus statutum / pertransibunt plurimi et multiplex erit scientia." (12, 4) Hier ist also vom Verschließen der offenbarten Worte und vom Versiegeln des Buches die Rede, bis hin zum tempus statutum, was immer das heißt, bis zur festgesetzten, bestimmten, gegebenen Zeit (die Authorized Version übersetzt „even to the time of the end", Luther übersetzt „bis auf die letzte Zeit", wodurch also das offenlassende ‚statutum' vereindeutigt wird). Man kann das auch übersetzen: „bis an die angehaltene Zeit", und so kommt es dem Hebräischen mit den unverbunden nebeneinander gestellten Wörtern „Zeit"

---

27 Die Geschichte erzählt in Aubrey (1978), S. 179.

„Ende" ('et qez) vielleicht am nächsten, und so übersetzten es die englischen Protestanten 1560 in Genf: „til the end of time". Man kann sich nur wundern: hier „Ende der Zeit", dort „Zeit des Endes"; dazwischen liegen fünfzig Jahre Theologie. Für die Genfer ist das Ende der Zeit, gut augustinisch, unbestimmt, zur Zukunft hin offen; in der von König Jacob 1611 autorisierten Bibel ist die Endzeit gemeint, und daß Endzeit und Jetztzeit zusammenfielen, werden die folgenden Jahrzehnte zu erweisen suchen. An diesem Datum jedenfalls, Endzeit oder Zeitende, das ist nicht ausdrücklich gesagt, aber durch die Nebeneinanderstellung impliziert, wird das Buch entschlüsselt werden und dann „pertranisbunt plurimi et multiplex erit scientia". Was Bacon also zur Markierung des Neuanfangs kupiert, ist der apokalyptische Zusammenhang. Aber heißt das, daß die Taue wirklich – programmatisch! – gekappt sind oder heißt das nur, daß die Verkürzung des Zitats ausreicht, weil seine Kenntnis ohnehin unterstellt werden kann? Für den zeitgenössischen Leser und das ganze Jahrhundert hindurch kann dieser apokalyptische Zusammenhang nur mitgedacht gewesen sein: wir befinden uns jetzt in der dem Propheten verheißenen letzten Zeit mit der Implikation ihres Endes. So eindeutig ist die Stelle übrigens nicht, wie Bacon sie macht. Man kann den von ihm anzitierten Satz nämlich auch so übersetzen: „Viele werden *es* dann durchgehen, durchforschen, und große Erkenntnis finden." Das bezöge sich dann auf das geschriebene und versiegelte Buch der Offenbarung Daniels, das zu entschlüsseln in der Tat die Theologen des 16. und 17. Jahrhunderts – und nicht nur sie, wie das Beispiel Newtons zeigt – nicht müde wurden, meist unter der Annahme, daß das tempus statutum gekommen sei. Bacon hätte darauf geantwortet, daß Gott zwei libri geschaffen habe und hier das Buch der Natur gemeint sei. Er verschiebt also durch Bild und Kupierung die Buchmetapher zur Schiffahrtsmetapher. Eine zweite Umdeutung ist vielleicht nicht unerheblich: in der Vulgata steht: „multiplex erit scientia", also vielfältig wird das Wissen, die Erkenntnis sein, die ja immer schon als Bestand da ist, nur eben bis jetzt verborgen, versiegelt war. Bacon, der immer aus dem Gedächtnis zitiert, macht daraus: „et *augebitur* scientia", das heißt das Wissen wird vermehrt werden, es kommt also vieles hinzu, und wodurch soll es hinzukommen, wenn nicht durch die Selbstermächtigung des Menschen als des Kapitäns auf dem Schiff. Die drei für Bacon größten Entdeckungen – Schießpulver, Druckerpresse, Kompaßnadel – sind ja so in der Natur nicht vorgesehen, sind ihr höchstens abgelistet und fügen ihr etwas hinzu.

Das Schiffsbild mit der Danielunterschrift nimmt spät im Leben Bacons sichtbare Gestalt an; der Gedanke aber hat ihn schon früh beschäftigt. Im ältesten Systementwurf, dem von ihm nicht veröffentlichten *Valerius Terminus of the Interpretation of Nature* von 1603 benutzt er das Bild einer organischen Entfaltung des Wissens: alles Wissen sei von Gott gepflanzt, jetzt aber komme es endlich zur Reife und trage Früchte, und dieser Zeitpunkt „... not only by a general providence but by a special prophecy was appointed to this autumn of the world: for to my understanding it is not violent to the letter, and safe now after the event, so to interpret that place in the prophecy of Daniel where speaking of the latter times it is said, *many shall pass to and fro and science shall be increased*; as if the opening of the world by navigation and commerce and the further discovery of knowledge should meet in one time or age."[28] Hier legt Bacon also den Akzent zunächst – nach der verbreiteten Vorstellung von der senectus mundi – auf die Endzeit, den Herbst der Welt, als ein Ereignis, das bereits eingetreten ist: wir leben schon im Millenium, und es kommt jetzt gewissermaßen nur noch

---

28 Bacon (1989), Bd. III, S. 221.

darauf an, das verheißene Wissen des Menschen auf Erden aktiv zu befördern. In dem zwei Jahre später, 1605, veröffentlichten und dem neuen König James I. gewidmeten *Advancement of Learning* ist der Hinweis auf den Herbst der Welt und das eingetretene Millenium dann vorsichtshalber gestrichen, vor allem wohl, um bei dem ängstlichen und melancholischen König keine allzu deutlichen Assoziationen an die mit Apokalypsen verbundenen Turbulenzen entstehen zu lassen; die hatte er im Land schon genug , konnte sie im gleichen Jahr auf der Bühne an *King Lear* entfalten finden, und brauchte nicht noch deren heilsgeschichtliche Rechtfertigung. Nicht gestrichen hat Bacon die Doppelbewegung, daß nämlich die Öffnung der Welt durch die Schiffahrt – hier fehlt allerdings jetzt der Handel – und das Anwachsen des Wissens für den gleichen Zeitraum verheißen waren.[29] So hat Bacon dann auch die Stelle in sein Buch mit dem antiaristotelischen Titel, das *Novum Organum* von 1620 übernommen.[30] Zum Fortschrittsbild tritt im *Advancement* noch ein Gedanke hinzu, der die fast blasphemische Schärfe einer Selbstermächtigung des Menschen erst verliert, wenn wir mitdenken, daß Bacon an anderer Stelle auch die Möglichkeit einer Annäherung des Menschen an die Gottebenbildlichkeit durch Wissen, durch das Offenbarwerden seiner Geheimnisse, anvisiert. Es habe ja geheißen, schreibt Bacon, der Blitz sei unnachahmbar („non imitabile fulmen"); jetzt, seit der Erfindung des Schießpulvers, sei er es nicht mehr. Ebenso sei der Himmel jetzt imitabile geworden: nämlich „in respect of the many memorable voyages after the manner of heaven about the globe of the earth."[31] Nebenbei wird aus dieser Stelle klar, daß Bacon alles andere als Kopernikaner war, obwohl englische Astronomen seiner Zeit es waren und sogar schon, vor Galilei, mit dem Fernrohr experimentierten.

Bacons Fortschrittsvorstellung ist sowohl vorwärts wie rückwärts orientiert; daher die Doppeldeutigkeit des Schiffsbildes. Rückwärtsgehen als Ziel des Wissens heißt, wie es schon im *Valerius Terminus* steht: „it is a restitution and reinvesting (in great part) of man to the sovereignty and power ... which he had in the first state of creation"[32] und „my intention ... is to increase and multiply the revenues and possessions of man"[33]. Beabsichtigt ist also die Rückkehr zum prälapsarischen Zustand. Den Einwand, daß der erste Mensch ja auch habe wissen wollen, und darum gefallen sei, entkräftet er damit, Adam habe vom Baum der Erkenntnis des Guten und Bösen gegessen, habe sich also „moral knowledge" angemaßt, um Gottes Willen zu ergründen, das heißt, Gottes Gebote möglicherweise in Zweifel zu ziehen („to dispute God's commandments"). Das allein sei sündig – anders gesagt: es gehört in den Zuständigkeitsbereich der Theologie, die methodisch von der Wissenschaft ja ausdrücklich abgetrennt wurde –, während die Erforschung der Natur und in Folge davon ihre Beherrschung dem Menschen ja gerade aufgegeben worden sei. Es kommt also alles darauf an,

---

29 „And this proficience in navigation and discoveries may plant also an expectation of the further proficience and augmentation of all sciences; because it may seem they are ordained by God to be coevals, that is, to meet in one age. For so the prophet Daniel speaking of the latter times fortelleth, *Plurinii pertransibunt, et multiplex erit scientia*: as if the openness and through-passage of the world and the increase of knowledge were appointed to be in the same ages; as we see it is already performed in great part ..." (Bacon (1986), S. 78).

30 I, 93: „Neque omittenda est prophetia Danielis de ultimis mundi temporibus: *Multi pertransibunt et multiplex erit scientia*: manifeste innuens et significans esse in fatis, id est in providentia, ut pertransitus mundi (qui per tot longinquas navigationes impletur plane aut jam in opere esse videtur) et augmenta scientiarum in eandem ætatem incident." (Bacon (1989), Bd. I, S. 200)

31 Bacon (1986), S. 77 f.

32 Bacon (1989), Bd. III, S. 222

33 Ibid., S. 233.

sowohl diesem Schöpfungsbefehl endlich nachzukommen, wie auch, wenn möglich, das alte Wissen, das den Schöpfungsgeheimnissen noch nahe stand, wieder zu gewinnen. Wiederholt gebraucht Bacon für die Geschichte des Wissens das Bild eines Flusses: obenauf schwimmt sichtbar und leicht das Unnütze und dreht sich hin und her, schimmert und gluckst, in der Tiefe aber ruht seit alters verborgen das Schwere, Gewichtige, auf das einzig es ankommt.[34] Teil des Baconschen Programms ist es somit, den ameisenhaft zusammengesammelten Plunder angeblicher Fakten einerseits, die spinnenhaft aus sich selbst ausgeschiedenen Gespinste angeblicher Systeme andererseits zu beseitigen und sowohl von vorn anzufangen wie auf Ältestes zurückzugreifen. (Insofern kann er das Bild der Reifung im Herbst der Welt jetzt nicht mehr gebrauchen.) Den Rückgriff rechtfertigt er an einer Stelle in *De Augmentis* mit einer erstaunlichen Parallele, wohin sein Reformkonzept sich hätte ausweiten können: „‚Things are preserved from destruction by bringing them back to their first principles', is a rule in Physics; the same holds good in Politics (as Machiavelli observed), for there is scarcely anything which preserves states from destruction more than the reformation and reduction of them to their ancient manners."[35] Zum ältesten, was ihm einfällt, gehören die Vorsokratiker. Das war nicht weiter neu, gehörten sie doch schon in der über hundertjährigen Geschichte der Suche nach einer prisca theologia zu den Favoriten. Aber was ihn daran interessierte, war ihre auf Beobachtung gegründete Naturphilosophie, die durch die auf sie folgenden Meisterdenker in Vergessenheit geraten oder durch eine nicht in Anschauung gegründete Kritik zum Verschwinden gebracht worden waren. Schuld an der Zerstörung des wahren Wissens und mithin daran, den Menschen einen Irrglauben gegeben zu haben, ist vor allem einer: Aristoteles. Ja, Aristoteles ist groß, heißt es in der unveröffentlichten *Redargutio Philosophiarum* (und in etwas abgemilderter Form anderswo[36]): „Oh, great without a doubt, but no greater than the greatest of imposters. For this is the prerogative of imposture, and in a special of the Prince of Imposture, the Anti-Christ."[37] Bacon bezieht sich auf eine Johannes-Stelle, in der Jesus sagt: „Ich kam im Namen meines Vaters und ihr empfingt mich nicht; wenn aber einer in seinem eigenen Namen kommen wird, den werdet ihr empfangen." Bacon fährt fort: „Now if any man in philosophy comes in his own name, Aristotle is that man. He is his own authority throughout."[38] Von allen Identifizierungen des Anti-Christ – Aristoteles ist neu. Ihn mit dem schlimmsten aller Namen zu bezeichnen, paßt natürlich genau in die apokalyptische Rahmung des Baconschen Projekts. Zugleich ist es bemerkenswert für die Übertragung der theologischen Anti-Christ-Debatten in die Wissenschaft: der Kampf gegen den Widersacher als Aufgabe der Wissenschaft. Auch im *Advancement* und anderswo polemisiert er gegen die Erzhäretiker, falschen Propheten und Betrüger (gemeint sind die katholischen Aristoteliker, aber auch die aristotelische Kanondominanz an den englischen Universitäten), die mit „detestable and extreme pleasure" sich die Herrschaft über das Denken der Menschen angemaßt haben: „But as this is that which the author of the revelation calleth the depth or Profound-

---

34 Zum Beispiel Bacon (1986), S. 33
35 Übersetzung von *De Augmentis* (hier III, 1) in Bacon (1989), Bd. IV, S. 338. *„Interitur rei arcetur per reductionem ejus ad principia*, regula est in Physicis; eadem valet in Politicis (ut recte notavit Machiavellus), cum illa quae interitum rerum publicarum maxime prohibent nihil aliud fere sint quam reformatio earum et reductio ad antiquos mores." (Bacon (1989), Bd. I, S. 541)
36 De augmentis III, 4 (in Bacon (1989)).
37 Die Stelle war mir nur in der englischen Übersetzung zugänglich, in: Farrington (1966), S. 113.
38 Ibid.

ness of Satan, so by argument of contraries, the just and lawfull sovereignty over man's understanding, by force of truth rightly interpreted, is that which approacheth nearest to the similitude of the divine rule."[39] Hier ist der recht geleitete – man kann auch sagen: der Bacons Vorstellungen folgende – Forscher also derjenige, der der Gottebenbildlichkeit am nächsten kommt.

Seit der Intervention der Schlange im Paradies ist der Mensch auf die falsche Fährte gesetzt und tappt in einem Labyrinth: „the universe to the eye of the human understanding is framed like a labyrinth", heißt es im Vorwort zur *Great Instauration*.[40] Dafür braucht es einen Ariadne-Faden: „our steps must be guided by a clue. (Vestigia filo regenda sunt.)" Im gleichen Zusammenhang richtet Bacon ein Gebet an die Trinität, daß „they will vouchsafe through my hands to endow the human family with new mercies ... and that from the opening of the ways of sense and the increase of natural light there may arise in our minds no incredulity or darkness with regard to the divine mysteries; but rather that the understanding being thereby purified and purged of fancies and vanity ..."[41] Am Ende der Skizzierung des Gesamtwerkes heißt es: „God forbid that we should give out a dream of our own imagination for a pattern of the world; rather may he graciously grant to us to write an apocalypse or true vision of the footsteps of the creator imprinted on his creatures."[42] Hier kommt eine Bedeutung des Wortes Apokalypse ins Spiel, die man über der Fixierung auf das Ende der Welt vergessen hat und für die es auch im Kanon der Heiligen Schriften keinen Anhaltspunkt gibt: Apokalypse als Offenbarung der Schöpfungsgeheimnisse. Daß es dazu eine reiche apokryphe Literatur gibt, war Bacon bekannt; wieviel er davon kannte, wissen wir allerdings nicht. An einer Stelle erwähnt er das Buch Henoch.[43] Henoch war ein Sohn Adams und Evas, der nicht starb, sondern direkt in den Himmel entrückt wurde und dem auf dieser Jenseitsreise die Geheimnisse der Schöpfung offenbar wurden. Das Buch ist eine der vielen Apokalypsen, die Anfang und Ende der Welt gleichermaßen zum Thema haben. Es mag für Bacon auch darum wichtig gewesen sein, weil Henoch als Adamsohn vielleicht noch im Besitz uranfänglichen Wissens gewesen war, vor allem aber, weil er der siebente und letzte Sohn Adams war, in dem die Zeit sich erfüllt. Später wird es bei dem schon zitierten Pietisten Oetinger heißen: „In der güldenen Zeit in den 1000 Jahren wird der Geist Henochs wieder offenbar werden, es wird sich anfahen mit dem Gericht über die Gottlose im Thal Josaphat sichbarer Weise."[44] Auch Henoch stünde also für Anfang und Ende zugleich – ältestes Wissen, das jetzt erst sichtbar wird. Sicher läßt sich sagen, daß Bacon sein Projekt auch in diesem Sinne als Offenbarungswerk verstand. Daß derlei Schriften in den Kanon nicht aufgenommen wurden, und nur Spuren davon in die Salomo zugeschriebenen, interpretationsbedürftigen Schriften ein-

---

39 Bacon (1986), S. 57.
40 Bacon (1989), Bd. IV, S. 18. „Ædificium autem hujus universi structura sua intellectui humano contemplanti, instar labyrinthi est ..." (*Instauratio Magna*, 1620, Bd. I, S. 129).
41 Ibid., S. 20, lat. S. 131; wo in der Übersetzung ‚new mercies' steht, hat das Original ‚novis suis eleemosynis'.
42 Ibid., S. 32 f., lat. S. 145: „Neque enim hoc siverit Deus, ut phantasiae nostrae somnium pro exemplari mundi edamus: sed potius benigne faveat, ut apocalypsim ac veram visionem vestigiorum et sigillorum creatoris super creaturas scribamus." Der Herausgeber verweist auf die häufige Verwendung der ‚vestigia' bei den Scholastikern und zitiert eine Stelle bei Thomas: „In rationalibus creaturis est imago Trinitatis, in caeteris vero creaturis est *vestigium* Trinitatis ..." Ibid. Ob mit ‚sigillum' auch auf das ‚signa' der Daniel-Stelle angespielt ist?
43 Bacon (1989), Bd. V, S. 8.
44 Oetinger (1987), S. 315.

gegangen sind, hat damit zu tun, daß dem Menschen in dieser Welt die Geheimnisse verborgen bleiben sollen, wie es nicht nur bei Daniel geschrieben steht.

Im ersten Korinther-Brief des Paulus (13, 12) heißt es: „videmus nunc per speculum in enigmitate / tunc autem facie ad faciem": „Now we see through a glass darkly – Luther sagt: ‚in einem dunklen Wort' – but then face to face". Bacon wollte aus dem tunc ein nunc machen: *Jetzt schon* sollen wir sehen, und das heißt nichts anderes als Paradise Now. Bevor wir ‚facie ad faciem' sehen, ist alles Wissen und ‚unser prophetisches Reden' nur Stückwerk; „wenn aber kommen wird das Vollkommene (quod perfectum est), so wird das Stückwerk aufhören." Sollen wir darauf warten müssen? Vielleicht läßt sich das mit der Neuen Wissenschaft beschleunigen. Bacon verwendet immer wieder das Bild des Spiegels (speculum) für das menschliche Bewußtsein. Aber dieser Spiegel ist so uneben, daß er verzerrt, und muß geradegerichtet werden, er ist verschmutzt und muß gereinigt werden. Im *Advancement* heißt es von diesem Spiegel: „wherein nevertheless there seemeth to be a liberty granted, as far forth as the polishing of this glass, or some moderate explication of this aenigma."[45] Vom paulinischen Kontext her gesehen ist das mehr und anderes als nur eine Metapher: „Darum, wenn wir an deinen Werken im Schweiße unseres Angesichts arbeiten, wirst du uns teilhaben lassen an deiner Vision und deinem Sabbath." Also schon hier: Erlösung durch Arbeit. Die Korintherstelle steht in einem Zusammenhang, in dem von Zeitlichkeit und Unzeitigkeit die Rede ist. Was zeitlich ist, wird es einmal nicht mehr geben: die Prophezeiungen werden leer werden oder vernichtet werden (evacuabuntur), die Sprachen werden aufhören, sogar Wissen und Erkenntnis wird zerstört werden (scientia destructur). Was aber niemals aufhört, weil es den Gesetzen des Irdischen nicht unterliegt, sind Glaube, Hoffnung und Liebe: „maior autem his est caritas", wobei der Komparativ (maior, meizón) im Englischen und Deutschen immer mit dem Superlativ übersetzt wurde: caritas ist die größte unter ihnen. Diese Liebe scheint es zu sein, die erst das Sehen und Erkennen in der Unzeitigkeit – facie ad faciem – verbürgt. Läßt sich daraus etwas für Bacons Programm ableiten? Bacon hat nie ein Forschen um des Forschens willen propagiert, wenn man einmal davon absieht, daß ein ‚reiner' Erkenntnisgewinn der Erkenntnis Gottes, d. h. dem von ihm Geschaffenen, zugute käme. Forschung sollte zum Nutzen des Menschen betrieben werden, „for the benefit of mankind". Man hat darunter die Verbesserung der Lebensbedingungen des Menschen auf dieser Erde verstanden. Vom Korintherbrief her gesehen bekommt dieses Ziel jedoch eine andere Tönung: es ist zum Wohl der Menschheit als einer erlösungsbedürftigen Gattung, daß die Forschungstätigkeit etwas beschleunigt herbeiführt, was erst für ein Danach verheißen ist. Eng mit dem Ausdruck ‚benefit of mankind' taucht immer wieder das Wort ‚charity' auf, und auch dies hat man im landläufigen Sinn als eine Art segensreiche Wohltätigkeit oder gar Nächstenliebe des Forschens im Auftrag der Gesellschaft verstanden. Aber man sieht leicht, daß dahinter ‚caritas' steckt, das in der James-Bibel mit ‚charity' wiedergegeben wurde. Es scheint, denke ich, diese ‚caritas' zu sein, die Bacon mit ‚charity' im Auge hatte: sie ist das Gelenkstück, das die Zeitlichkeit mit der Unzeitigkeit, der Ewigkeit verbindet, nur durch sie sehen wir ‚facie ad faciem'. In einer Marginalie der Genfer Bibel wird begründet, warum die Liebe die größte unter den Dreien ist: Glaube und Hoffnung sind auf dieses Leben beschränkt, aber die Liebe „Serueth bothe here & in the life to come." Pascal wird eine Generation nach Bacon schreiben: „L'unique objet de l'Écriture est la charité.", und eben dieses Ziel im beschriebenen Sinn hat Bacons Wissenschaft.

Dieser hohe und höchste Anspruch von Bacons Wissenschaftsverständnis wird an einer fast verborgenen Stelle in *De augmentis* explizit gemacht: Bei den Erscheinungen Gottes in seiner Herrlichkeit auf seinem Thron singen nur zweimal die Seraphim das dreifache ‚sanctus‘. Das erstemal bei der Berufung Jesaias zum Propheten: „Sanctus, sanctus, sanctus Dominus, deus exercituum, plena est omnis terra gloria eius." Aber dem Propheten, spricht Gott, wird man seine Botschaft nicht glauben. Auf die Frage Jesaias: „Wie lange?", erhält er zur Antwort: erst wenn die Städte und Felder verwüstet und die Menschen vernichtet sind, kann wie aus einem Reis aus einem abgehauenen Baum etwas Neues entstehen und die Glorie gesehen werden. Das ist eine sogenannte kleine Apokalypse. Das zweitemal taucht die dreifache Anrufung vor dem göttlichen Stuhl in der Johannes-Apokalypse auf, worauf die Entsiegelung des Buches durch das Lamm folgt. „Sancuts, sanctus, sanctus Dominus Deus omnipotens, qui erat, et qui est, et qui venturus est." Es ist eine strategische Pointe, daß Bacon die Dreiheit – sollen wir sagen die Dreieinigkeit – der Wissenschaft mit dem dreifachen Attribut der Heiligkeit verbindet: „sancte, sancte, sancte": die Naturgeschichte, d. h. die Experimentalwissenschaft, ist heilig, „For God is holy in the multitude of his works"; die Physik ist „holy in the order or connexion of them"; und die Metaphysik ist „holy in the union of them".[46] Diese Bestimmungen bekommen erst ihr ganzes Gewicht aus den Kontexten, denen sie entstammen: Erkenntnis als Gotteserkenntnis, aber eben im Zeichen der Apokalypse. Man stelle sich vor: der spiritus rector der neuen Wissenschaft als Gerichtsvollzieher der alten Prophetie!

Eine der zentralen Absichten der Reformatoren, dann der Anglikaner und Puritaner war es, hinter die Kirchengeschichte zurückzugehen zu den Anfängen des Christentums. Abgesehen von der hier zu entdeckenden Dehierarchisierung, die für die Ablehnung der Bischofskirche durch die Puritaner wichtig werden sollte, aber auch für die Bacon zugeschriebene Demokratisierung des Wissens nicht unerheblich gewesen sein dürfte, scheint es die Naherwartung des Endes und der Wiederkunft Christi gewesen zu sein, wovon die ersten Christen überzeugt waren. Luther arbeitete in der Gewißheit der Naherwartung, dann wieder die Puritaner, bei deren Berechnungen fast jedes Jahr zwischen 1640 und 1666 genannt ist. Das Problem, mit dem die Christen (und auch die Juden) der ersten Jahrhunderte fertig zu werden hatten, war, daß die Erwartung nicht in Erfüllung ging oder mindestens auf sich warten ließ. Aus der Rechtfertigung dieser Heilsverzögerung entwickelte sich bekanntlich die Theologie. Augustinus bot die Kompromißformel an, daß die Wiederkunft jederzeit geschehen könne, daß es aber gottlos sei, sie genau vorhersagen, also berechnen zu wollen, wie es die Leser der Johannes-Apokalypse glaubten tun zu können. Nun, Bacon und die Anglikaner, vielleicht ernüchtert durch die nicht erfüllten Hoffnungen Luthers, dachten ähnlich vorsichtig. Bacon entwarf eine ‚History of Prophecy‘ „for better instruction and skill in the interpretation of those parts of prophecies which are yet unfulfilled; allowing nevertheless that latitude which is agreeable and familiar to divine prophecies, that the fulfilment of them are taking place continually, and not at the particular time only. ... and though the height or fulness of them is commonly referred to some age or particular period, yet they have at the same time certain gradations and processes of accomplishment through divers ages of the world."[47] Das kann

---

45 Bacon (1986), S. 207
46 Übersetzung von *De Augmentis* (III, iv) in Bacon (1989), Bd. IV, S. 362. Das lateinische Original: „Sancte, sancte, sancte. Sanctus enim Deus in multitudine operum suorum, sanctus in ordine eorum, sanctus in unione." (Bd. I, S. 567)

also heißen, daß wir im Prozeß der Erfüllung der Prophetie längst – sagen wir seit dem
Endzeitabenteurer Kolumbus, der mehrfach genannt wird – drinstecken, nur daß es eben
unabsehbar ist, wann der Prozeß durch das Jüngste Gericht an sein Ende kommt. Im *Advancement* hatte Bacon vier Dinge benannt, die nur Gott bekannt seien. Die ersten beiden lauteten: „the mysteries of the kingdom of glory" und „the perfection of the laws of nature".[48]
Diese beiden hat er in *De Augmentis* weggelassen, weil sie ihm durch sein großes Projekt
lösbar erscheinen. Übrig blieb nur zweierlei, das „unknown to man" bleiben müsse: „the
secrets of the heart and the successions of time".[49] Das ist gut augustinisch gesagt, vielleicht
aber auch diplomatisch, um den König, um dessen Gunst es ihm ging, nicht durch den Verdacht, er vertrete eine Puritanerposition, zu irritieren. Dabei ist seine Einstellung gegenüber
dem, wie mit der doch finit gedachten Zeit umzugehen sei, durchaus zweideutig: einerseits
konstatiert er eine historisch beispiellose Zunahme und Beschleunigung des Wissens – „knowledge daily increases", oder wie der wichtigste Baconianer, Joseph Glanvill, in seinem *Plus
Ultra* schreiben wird: jeden Tag werden neue Geometrien erfunden – und er schreibt diese
Form des Fortschritts auch dem notwendigen Druck zu, unter dem geforscht wird, denn der
Mensch kommt nie zu besseren Entdeckungen, als wenn er „in trouble" ist[50], so als müßte
der Teufel, der einem bekannten Spruch zufolge keine Zeit hat, darin noch unterboten werden. Andererseits muß gleichzeitig, als sei die Zeit zur Zukunft hin eben doch offen, das
geduldige, methodisch gelenkte Forschen, das viele Induktionsreihen erforderlich sein lassen
mag, gewährleistet sein. Im *Novum Organum* heißt es dazu: „Man soll also den menschlichen Geist nicht mit Schwingen beflügeln, sondern mit bleiernem Gewicht ihn zurückhalten
von allem Schwunge. Geschieht dieses (bisher geschah es nicht), so darf die Wissenschaft
hoffen."[51] Freud wird später sagen, Heine zitierend, „Was man nicht erfliegen kann, das muß
man sich erhinken."

Es ergibt sich also ein Paradox gegenüber der dem Menschen verfügbaren Zeit: millenarisch im Sinne der Beschleunigung der Wiederkunft und zukunftsorientiert im Sinne neuzeitlichen Forschens. Dahinter verbirgt sich die alte Theologenfrage, was besser sei: Heilsbeschleunigung oder Heilsverzögerung. Ich möchte dieses Paradox abschließend noch etwas
pointieren. Die Erfahrung der Beschleunigung dürfte etwas historisch Neues gewesen sein.
Was gestern noch galt, gilt morgen vielleicht nicht mehr. Wie soll man sich zurechtfinden in
Zeiten des Umbruchs, in denen Geltungsansprüche miteinander konkurrieren oder einander
überbieten?[52] Man mag resignieren angesichts der Kontingenzerfahrung, die der Dichter
John Donne in die berühmte Zeile gefaßt hat: „'tis all in pieces, all coherence gone"[53], oder
man mag ahnen, daß damit eine Höherentwicklung des Wissens verbunden ist und daß dadurch die Menschheit insgesamt, wie Pascal schreibt, sich in einem ‚continuel progrès' be-

47 *De Augmentis*, II, Kap. XI, in Bacon (1989), Bd. IV, S. 312 f.

48 Bacon (1986), S. 207.

49 *De Augmentis*, IX, Kap. I, in Bacon (1989), Bd. V, S. 117. Lat.: „Secreta nimirum Cordis, et Successiones
Temporis." (Bd. I, S. 835)

50 Dieser Gedanke wird schon für das Mittelalter von Johannes Fried mit zahllosen Beispielen belegt in Fried
(2001).

51 Bacon (1981), S. 79. (Es handelt sich um *Novum Organum* I, 104).

52 Daß sich die alten Fragen heute verschärft wieder stellen, fördert vielleicht unser Verständnis der früheren
Umbruchzeit. Zur Gegenwartsanalyse siehe Meier (2001), besonders die Einleitung.

53 Siehe seine „Anatomy of the World", Vers 213, in: Donne (1986), S. 276.

fände, ‚à mésure que l'univers vieillit', ‚in dem Maße, in dem das Weltall älter wird.'[54] Nur: zu welchem Ende? Wie lassen sich Kontingenz und Ordnung, Perfektibilisierung und Weltende, Entschlüsselung und ineins damit Entdeckung immer neuer Rätsel, Zeitverkürzung und -verzögerung zusammendenken? Man muß das nicht tun, sondern kann das Unvereinbare nebeneinander stehenlassen, das Paradox auszuhalten versuchen, wie Pascal es tat. Man kann aber auch auf erprobte Sinnraster rekurrieren, und dazu bot sich die apokalyptische Vorstellung vom Ende der Zeit geradezu an. Beim Evangelisten Matthäus heißt es (24, 21 f): „Denn es wird alsdann eine große Trübsal sein, wie nicht gewesen ist von Anfang der Welt bisher und wie auch nicht werden wird. Und wo diese Tage nicht werden verkürzt, so würde kein Mensch selig; aber um der Auserwählten willen werden die Tage verkürzt." Im 1. Korintherbrief (7, 29) steht: „tempus breve est", was vom griechischen Original her genauer: „die Zeit drängt" heißen müßte. „Die Verkürzung der Zeit" schreibt Reinhard Koselleck, „ist ein Vorzeichen für die Erlösung aus dieser Welt."[55] Koselleck hat auch die Umbesetzung bezeichnet, wodurch der alte Glaube im neuen Kontext virulent werden konnte: „Die von der Apokalypse vorformulierte Zeitverkürzung wird zu einer Metapher der Beschleunigung, die mit dem 16. Jahrhundert andere und neue Inhalte zur Sprache bringt, als im Horizont christlicher Eschatologie gemeint waren."[56] Und: „aus der Hoffnung wird ein Erfahrungssatz"[57], wobei hinzuzufügen wäre, daß die Inhalte, als andere, als neue, in ihrer Tragweite für eine nach vorn hin offene Geschichte noch nicht gesehen werden konnten. Was als das Paradox des Nichtzusammengehörigen erschien – wissenschaftliche und technologische Beschleunigung einerseits, apokalyptische Zeitverkürzung andererseits –, läßt sich jetzt auf einen Nenner bringen, denn beiden gemeinsam ist eine Denaturalisierung der Zeiterfahrung. Sie ist durch das apokalyptische Ende der irdischen, der natürlichen Zeit ebenso verheißen, wie sie jetzt durch die wissenschaftlichen Entdeckungen und Erfindungen technisch möglich wird oder werden soll. (Und wenn der Wille Gottes die Naturgesetze jederzeit ändern kann, wie Boyle glaubte, dann kann er natürlich auch das Ganz-Andere verfügen, das allem bisher Erkannten zuwiderläuft.) Der neue Mensch, wie Bacon ihn anvisiert, steht außerhalb der natürlichen Gegebenheiten, das heißt außerhalb der Zeit. Am Ende der *Nova Atlantis*, die den Abschluß der *Instauratio magna* hätte bilden sollen, steht eine Liste mit 33 Forschungsdesiderata. Die ersten drei lauten: „The prolongation of life. The restitution of youth in some degree. The retardation of age."[58] Man kann auch das natürlich rückwärts gewandt lesen als eine Wiederherstellung der im Buch Genesis aufgelisteten Langlebigkeit der Menschen vor der Sintflut. Aber Bacon hat in einem ganzen Buch – *Historia vitae et mortis* – so detaillierte medizinische, diätetische, hygienische Vorschläge zur Verlängerung des Lebens ausgearbeitet – er rechnete mit der Möglichkeit von mindestens 100 Jahren in Ausnahmefällen von 200 –, daß es müßig erscheint, hier noch Regression und Progression unterscheiden zu wollen:

---

54 In Pascal (1954), S. 533 f.
55 Koselleck (2000), S. 185.
56 Ibid., S. 153.
57 Ibid., S. 188.
58 Abgedruckt im Anhang von Bacon (1986), S. 249. Der Herausgeber bemerkt dazu, daß viele der ‚Projekte' sich bereits bei Agrippa und della Porta finden, also in magischen Kontexten. Man sollte hinzufügen, daß sie auch bei Naturforschern des Mittelalters zu finden waren, etwa bei dem von Bacon gelesenen und hochgeschätzten Namensvetter Roger Bacon, oder daß speziell über die Verlängerung des Lebens auch so unterschiedliche Zeitgenossen spekulierten wie Campanella oder Comenius.

sie sind ein und dasselbe. Und wenn wir unter den Desiderata dann noch von der Veränderung der Menschengestalt hinsichtlich der Länge und Kürze, der Gesichtszüge oder des Leibesumfangs lesen, vom „Making of new species" oder „Transplanting of one species into another" oder „Drawing of new foods out of substances not now in use", dann ist die Zeit und mit ihr das, was einmal Natur hieß, vollends denaturalisiert, und wir sind in der Gegenwart angekommen. Die ‚ungeahnten' Möglichkeiten sind im doppelten Sinn die apokalyptischen: das ‚incipit vita nova', das die Zerstörung des alten zur Voraussetzung hat.

Apokalyptik und Neue Wissenschaft - daß die eine die andere, zumindest in England, aber auch in Holland, in Deutschland, beflügelt, wenn nicht sogar ihre Ermöglichungsbedingungen geschaffen hat, ist kaum zu bezweifeln, auch bei solchen Forschern nicht, die, ohne Apokalyptiker zu sein, auf ihren Strudel reagierten. Es bleibt die Frage, ob sich daraus Schlüsse ziehen lassen, das heißt ob erst unter dem Druck von Turbulenzen, die man als Heilsverheißungen deuten kann wie die Millenaristen, oder als Katastrophen, deren destruktive Energie sich hochrechnen läßt, die Wissenschaft provoziert wird, nach etwas Neuem zu suchen, das dann allerdings Folgelasten zeitigen mag, die nicht voraussehbar und nicht abschätzbar sind und die es glaubhaft erscheinen lassen, apokalyptische Erwartungen immer wieder und immer neu im Horizont des Möglichen zu sehen.

Literatur

Aubrey, John (1978): *Brief Lives*. Hrsg. von Oliver Lawson Dick. Harmondsworth: Penguin.

Bacon, Francis (1981): *Neues Organ der Wissenschaften*. Übersetzt und herausgegeben von Anton Theobald Bruck (1830). Darmstadt: Wissenschaftiche Buchgesellschaft, 1981, S. 79.

Bacon, Francis (1986): *The Advancement of Learning and New Atlantis*. Hrsg. von Arthur Johnson (1974). Oxford: Clarendon Press.

Bacon, Francis (1989): *The Works of Francis Bacon*. Faksimile-Neudruck der Ausgabe von Spedding, Ellis und Heath, London 1857-1874. Stuttgart: Frommann-holzboog.

Boyle, Robert (1675): *Some Considerations About the Reconcileableness of Reason and Religion*. London.

Cohen, I. Bernard (1975): Introduction to *Isaac Newton's Principia*. Cambridge: Cambridge UP.

Donne, John (1986): *The Complete English Poems*. London: Penguin Classics.

Farrington, Benjamin (1966): *The Philosophy of Francis Bacon*. Mit Übersetzung unveröffentlichter Texte Bacons. Chicago: University of Chicago Press.

Flasch, Kurt (1989): *Aufklärung im Mittelalter? Die Verurteilung von 1277*. Mainz: Dieterich.

Freudenthal, Gideon (1982): *Atome und Individuen*. Frankfurt.

Fried, Johannes (2001): *Aufstieg aus dem Untergang. Apokalyptisches Denken und die Entstehung der modernen Naturwissenschaft im Mittelalter*. München: C. H. Beck.

Henry G. David L. Lindberg und Ronald L. Numbers (Hg.): *God and Nature. Historical Essays on the Encounter between Christianity and Science*. Berkeley: University of California Press

Hobbes, Thomas (1959): *Leviathan* (1651). Everyman-Ausgabe.

Jacob, Margaret C. (1986): „Christianity and the Newtonian Worldview", in Lindberg/Numbers (1986).

Katz, David S. (1982): *Philo-Semitism and the Readmission of the Jews to England 1603-1655*. Oxford: Clarendon Press.

Koselleck, Reinhard (2000): *Zeitschichten. Studien zur Historik*. Frankfurt: Suhrkamp.

Lindberg, David L. und Numbers, Ronald L. (Hg.) (1986): *God and Nature. Historical Essays on the Encounter between Christianity and Science*. Berkeley: University of California Press.

Manuel, Frank E. (1974): *The Religion of Isaac Newton*. Oxford: Oxford UP.

Meier, Christian (2001): *Das Verschwinden der Gegenwart*. München: Hanser.

Merton, Robert K. (1978): *Science, Technology and Society in Seventeenth-Century England* (1937). Neuausgabe. New Jersey: Humanities Press.

Newton, Isaac (1713): *Principia Mathematica*.

Oetinger, Friedrich Chr. (1987): *Biblisches und Emblematisches Wörterbuch*. 2. Nachdruck der Ausgabe 1776. Hildesheim: Olms.

Pascal, Blaise (1954): „Préface pour le traité du vide", in *Œuvres complètes*. Hrsg. von Jacques Chevalier. Paris: Pléiade.

Reichert, Klaus (1994): „Joseph Glanvill's *Plus Ultra* and Beyond: Or How to Delay the Rise of Modern Science", in Y. Ezraki, E. Mendelsohn und H. Segal (Hg.), *Technology, Pessimism, and Postmodernism*. Dordrecht: Kluwer, 39-52.

Shapin, Steven und Schaffer, Simon (1985): *Leviathan and the Air-Pump. Hobbes, Boyle, and the Experimental Life*. New Jersey: Princeton UP.

Thayer, H. S. und Randall, J. H. (1960): *Newton's Philosophy of Nature*. New York: Hafner.

Van Leeuwen, Henry G. (1963): *The Problem of Certainty in English Thought 1630-1690*. The Hague: Nijhoff.

Webster, Charles (1975): *The Great Instauration. Science, Medicine and Reform 1626-1660*. London: Duckworth.

Webster, Charles (1986): „Puritanism, Separatism, and Science", in Lindberg/Numbers (1986).

Westfall, Richard S. (1980): *Never at Rest. A Biography of Newton*. Cambridge: Cambridge UP.

Westfall, Richard S. (1986): „The Rise of Science and the Decline of Orthodox Christianity: A Study of Kepler, Descartes, and Newton", in Lindberg/Numbers (1986).

Wolfgang Detel

Scepticism and Scientific Method – the Case of Gassendi

1.

Concerning the way Gassendi felt about the structure and methods of science, and in particular of physics, it is common wisdom among scholars that Gassendi, proceeding from sceptical premises, developed a picture of a scientia probabilis based on observation and empirical experiments. In this respect, he is seen, correctly, I think, to have been in sharp opposition to ideas on method Descartes was entertaining, although our picture of these Cartesian ideas on method has recently become a lot more complex.[1] Tullio Gregory, for instance, identified the "fundamental structure" (as he put it) of Gassendi's entire philosophy as the "human way of philosophising", i. e. as reliably describing the appearances and recognizing the provisional nature of assumptions about the causes and nature of things.[2] In an influential study Oliver Bloch has rejected any attempt of systematizing Gassendi's philosophy, emphasizing instead many different orientations being present in Gassendi's thinking – orientations that were, according to Bloch, often incompatible with each other. And yet even Bloch gives an account of what he sees as constants in Gassendi's works in terms of his epistemology and methodology of science – his model of knowledge, his scepticism and agnosticism concerning the res occultae, and his nominalism and empiricism.[3] This general view can also be discovered in more recent studies on Gassendi. Margaret Osler, for example, puts it this way: "The ultimate ground for our knowledge, sensory knowledge, is (according to Gassendi) true and valid and needs no justification besides ist own immediacy ... The structure based on sensory experience ... involves reasoning and various assumptions ... It is here that error can creep into the picture; consequently this science of appearances is neither demonstrative nor certain, but at best probable."[4] Brundell seems to outline the same interpretation by saying: "One can have knowledge of appearances, Gassendi maintained, and so a science of appearances was possible ... But one could not attain to the necessary cause that determined the appearances to be so, and one consequently could not provide a necessary demonstration that proved why they could not but be so."[5] There is an important and telling difference, though, between these two views in that Osler thinks that what Gassendi called the science of appearances is itself only probable, whereas Brundell takes this limited science to provide full knowledge. Later in this paper, I will come back to this problem.

---

1 See, for instance, Gaukroger (1995), Garber (1992), Detel (2000).
2 Gregory (1961).
3 Bloch (1971).
4 Osler (1983), p. 558.
5 Brundell (1987), p. 99.

While there is a remarkable agreement in the literature at least about the general lines of this picture, there is, at the same time, considerable disagreement among scholars about the intellectual and historical background of Gassendi's epistemology. This is a question that is important for a number of reasons; one is that it points to the way Gassendi felt about scepticism and Epicurean philosophy, which in turn goes right to the heart of the problem why and how Gassendi defended an empiricist and probabilistic account of knowledge and science. Another reason why this question is important is that the way we answer it has impacts for determining Gassendi's philosophical development and the crucial task he set himself to accomplish, for instance whether we agree with the assumptions, mandated by many scholars, that Gassendi started as a radical pyrrhonic sceptic in the *Exercitationes* and moved to moderate scepticism in the *Syntagma* and, furthermore, that he saw his central historical role in making Epicurean philosophy acceptable to Christianity.

As is well known, Gassendi himself makes it clear how important is was for him, at the beginning of his philosophical career, to read Vives, Charron, Ramus and Pico della Mirandola. In one of the crucial passages mentioning this point Gassendi puts special emphasis on Charron and Ramus by saying "my own Charron and Ramus especially."[6] This is telling since Charron was much more outspoken as a sceptic than Vives by explicitly advocating Pyrrhonism, and because Ramus linked scepticism to anti-Aristotelianism.[7] Indeed, as Brundell has shown convincingly, the young Gassendi felt upset about the dogmatism of the Aristotelians and, while emphasizing the need to study nature on the basis of observation and experiment, used scepticism to attack dogmatic Aristotelians.[8] It is on the basis of this anti-dogmatism that Gassendi, by approving and defending the way Kepler and Galilei, among others, studied nature, became involved in the intellectual development that produced a new physics.

Richard Popkin has argued powerfully that Gassendi's scepticism had a deeper significance in that it was a response to what Popkin calls the sceptical crisis of the seventeenth century. According to Popkin, this crisis originated in a theological movement that tried to defend Christian faith against rational attacks by casting sceptical doubts on rationality and rational belief-formation processes by launching sceptical counterattacks. But, as Popkin sees it, what began as a strictly theological strategy soon developed into a general assault against the foundations of every kind of knowledge, thus eventually provoking a comprehensive sceptical crisis that proved to be crucial for the origin of the new physics.[9] Popkin looks at Gassendi as one of the key figures in this crisis who not only supported the sceptical attacks but who also responded to the crisis by elaborating a moderate and constructive scepticism and became thereby one of the pioneers of modern science.

Recently some scholars have put even more emphasis than Popkin did on the role of theological considerations for explaining why Gassendi felt attracted by scepticism and why he developed a combination of empiricism and probabilistic methodology. One of the most prominent of these scholars is Margeret Osler. She claims that "theological issues played a formative role in Gassendi's ideas about natural philosophy and scientific method."[10] Her

---

6 P. Gassendi Op. om. III, p. 99; see also the list in ibid. VI, p. 2.
7 See C. B. Schmitt (1972).
8 Compare Brundell (1987), chapter 1 for evidence.
9 Popkin (1964), especially pp. 82-88 and 132 ff.
10 Osler (1983), especially p. 549.

main reason is that Gassendi advocated, not an intellectual, but a voluntarist theology. While intellectualists in theology certainly maintain different positions, they agree in assuming that there exists a rational order in the universe that is either independent of God or, even if God created it freely and arbitrarily, after having created it is bound by this order (for instance by natural laws or natures of things) because it will last without change until the end of all days. Voluntarists, however, emphasize God's absolute power and freedom. They think that God can change the course of things at any time by changing their nature or the natural laws that so far governed them. Contrary to the intellectualists, the voluntarists maintain that the universe is in the course of all its parts at any time dependent on God's activity and power. This is why Osler thinks that intellectual theology favors a realist ontology and a rationalist methodology whereas voluntarist theology is committed to nominalism, empiricism and probabilism concerning the nature of things and thus, in general, must be linked to moderate scepticism. And since it can be shown, according to Osler, that Gassendi was clearly an adherent of voluntarist theology, this explains why he defended moderate scepticism, i. e. empiricism and probabilism.[11]

I cannot discuss here Osler's general claim that voluntarism in theology has always been closely connected with empiricism, nominalism and scepticism; it seems to me, though, that in some cases this might be a plausible assumption. Concerning the case of Gassendi, however, the trouble with her interpretation is simply that the textual evidence is completely missing. To be sure, the claim that Gassendi was attracted by voluntarist theology can barely be doubted. But the further claim that he favored moderate scepticism, nominalism and probabilism *because* he adopted voluntarism, i.e. *as a consequence* of voluntarism, cannot be supported by textual evidence. At least I know of no such passages, and certainly Osler is, in her published papers, not able to quote one single passage that could support her reading.

To some extent, this is true for Popkin's grand thesis too, at least as Gassendi is concerned. Popkin is certainly right in thinking that Gassendi's mitigated scepticism helped to promote the new scientific outlook, but there is little textual evidence that Gassendi did this *because* he intended to respond to a sceptical crisis even if we grant that there *was* a sceptical crisis at the beginning of the seventeenth century. There *is* sufficient evidence, however, to support the view that attacking the Aristotelians and adopting scepticism was, for Gassendi, closely linked. For instance, in a passage from the preface to the *Exercitationes* I already pointed to he writes:

E1 "When I read Vives and my own Charron my spirits rose and all my fears left me. I saw that there was nothing wrong in suspecting that the Aristotelians were not always correct just because most people approved of them. But my boldness rose as I read Ramus especially, and also Mirandola."[12]

---

11 In another paper, Osler relies on the same account to explain the way Gassendi resisted Descartes' view about eternal and mathematical truths. As she sees it, Gassendi argued form a voluntarist point of view that there cannot exist some immutable and eternal nature (mathematical or non-mathematical) other than that of omnipotent god (Op. om. III 374), see Osler (1995).

12 Op. om. III p. 99.

A little later in the same preface he says that he believes that the Aristotelians would have made much progress in the discovery of truth if they had not been so sure that they already possessed it. And then he tells us:

> **E2** "After I had come to see how great a gulf there is between the human mind and Nature's genius, what else could I conclude than that the intimate causes of natural phenomena totally escape the power of human discernment? Therefore I became upset and ashamed at the foolishness and arrogance of the dogmatic philosophers who boast that they have attained a knowledge of nature and propound it with such rigour."[13]

It is in this spirit that Gassendi intended to launch a full scale sceptical attack against the Aristotelians in the *Exercitationes paradoxicae*. As the passages I just quoted show, this enterprise was inspired by three goals: attacking successfully the Aristotelians, advancing sceptical ideas, and promoting the study of nature as far as that is possible for the human mind. This is broadly the way Brundell sees things though he tends to favor an instrumentalist understanding of Gassendi's adopting scepticism by saying that Gassendi settled on scepticism as a weapon to attack the Aristotelians.[14] This sounds as if Gassendi would have adopted another position had he thought that it is an even better instrument for fighting the Aristotelians. It seems to me, however, that Gassendi's own remarks indicate that his basic methodological belief was that an anti-dogmatic methodology is right for at least two reasons: first, it is adequate in respect to the human epistemic condition (compared, for instance, to God's epistemic condition), and second, it promotes, and is even necessary for, progress in the study of nature as far as this is available for humans. And when he read the ancient and modern sceptics he realized that scepticism was, not only a good instrument for attacking dogmatism, but primarily just the right epistemology to defend and sharpen his own anti-dogmatism. In short, for Gassendi it is not that anti-dogmatism as anti-Aristotelianism was the goal and scepticism the means to reach this goal, but rather that scepticism was the best realization and justification of anti-dogmatism.

2.

My little survey of some of the recent literature, while certainly being incomplete, nevertheless indicates that scholars tend to discuss the big questions about Gassendi's epistemology and even his entire philosophy. For example: What are the motives and the background of his scepticism? How did he feel about the history of philosophy, and in particular, what is his Epicurean project all about? Is there a move from radical Pyrrhonic scepticism to moderate scepticism, and if so, does this move mean a substantial change in his epistemological position? Did he really contribute to the new physics, and if so, in which way? To be sure, these are important questions, but it seems to me that scholars so far did not bother about the details

---

13 Ibid.
14 See Brundell (1987), chapter 1.

of his epistemology and philosophy of science and so did not determine what exactly he contributed to the history of methodology and epistemic practices. To see this more clearly it will help to develop adequate answers to the big questions. In the remainder of this paper I will first try to bring into sharper focus his philosophy of science by looking not only at what he *says* about this topic but also at what he actually *does* in his works about physics; doing this will then help us to see a problem concerning his empiricism and sensualism that is usually not acknowledged in the literature, although the way Gassendi tried to solve the problem has an important impact for our understanding of his epistemology. It goes without saying that I can examine, within the limits of this paper, only a few examples.

My first example is Gassendi's discussion of the horror-vacui-theory.[15] His overall aim is, of course, to refute this theory. But like many of this contemporaries working on this topic he proceeds from raising the question how extended vacua can be technologically produced. The first part of his argument can be reconstructed in the following way. The basic premise P, established in the preceding chapter, is that matter is discrete, i. e. contains at least little vacua. From P it follows trivially that if material particles can be compressed within a certain space ceteris paribus, i. e. if no further particles can enter this space, a larger vacuum will be produced within this space (H1). It is, moreover, empirically known that for each body its force to resist compression is finite (H2). Hypotheses H1 and H2 imply, in turn, that if the ceteris-paribus-condition in H1 is satisfied, finitely strong forces will be sufficient to produce larger amounts of empty space (H3). This is true especially for water and air particles (H4), i. e. for those kinds of stuff for which at that time it could be made sure that the ceteris-paribus-condition of H1 is satisfied. And H4 is incompatible with the horror-vacui-theory at least in the strong version that claims that any artificial production of vacua is impossible.[16]

So far this argumentation is primarily theoretical, i. e. based on atomistic premises and pretty trivial conclusions from these premises. In the second part of his argument Gassendi discusses some simple empirical observations that threaten to falsify P and H1–H4. One observation is that a gas like air or a liquid like water seems to tend to occupy every empty space they can get to, which seems to confirm the Aristotelian claim that there is no vacuum at all. Another observation is that if we move, for instance, our arms through the air we do not feel any resistance that is predicted by H2–H4. Gassendi responds to these observations by explaining them theoretically. In gases and liquids the surfaces at which their atoms touch each other are very small (H5). So the vacua interspersa are in these kinds of stuff very large and their resistance against compression is so small that we cannot feel it (H6). Gassendi shows, therefore, that there is a theoretical explanation that is compatible with the observations just indicated as well as with atomism and the thesis that vacua can be produced artificially. He goes on to discuss a third objection that is supposed to attack (H6). This objection is itself based on a theory, namely Aristotelian physics, and says that this theory can also explain the second observation, for according to Aristotle particles of air tend to move upwards so that if we move through the air there is no resistance at all since the air gives way by

---

15 Op. om. I, p. 196 b2 ff., i.e. chapter IV of book II (*De loco et Duratione Rerum*) in the firsr section on physics in the *Syntagma Philosophicum*.

16 This is a reconstruction of the passage Op. om. I 196b2-197al. Note that Gassendi calls, in this passage, the finitely large resistance against compression *naturae propensio ad tuendam inane interspersum* and that he connects the ceteris-paribus-condition that I relate to the general hypothesis H1 to the special cases of air and water, i. e. to H4.

moving upwards. Facing this objection Gassendi finally introduces a further theoretical assumption that will play a crucial role in the remainder of his argument, namely the assumption that air has weight and tends to exert pressure on the things sitting under the air particles (H7). At this point, he justifies this assumption only theoretically by pointing out that it directly follows from atomistic physics. There is a last objection Gassendi mentions in this part of his argument, this time again based on observation and directed against (H7). If air had weight, we would be able to feel its weight at the surface of the earth, but that is not the case. Gassendi rejects this objection by remarking that the observation can be explained by (H2) and by the fact that we are adapted to this sort of environmental condition (H8).

In the third part of his argument Gassendi examines observations and experiments that the Aristotelians took to be confirmations of the horror-vacui-theory. For instance, the parts of bellows that are empty of space and the opening of which is closed cannot be pulled apart, and it was predicted that if one tries to do so one will have to destroy them. Likewise, in a sort of sprinkler that was used at that time by physicians and was constructed like a suction pump the cylinder if pressed down, the opening on the lower side being closed again, cannot be moved upwards again without destroying the device. It is pretty telling how Gassendi deals with this line of argument. He does not deny that such observations confirm the claim that there is something like a sort of force that seems to work against producing artificial vacua which in turn seems to support the horror-vacui-theory. But he denies that this theory is the only way and, what is more, the best way of explaining them. The alternative is to postulate that it is the pressure of the air sitting above of the devices that is responsible for the phenomena. And Gassendi hastens to add that this alternative theory implies the prediction that one will be able to pull the parts of the bellows apart and the cylinder of the sprinkler upwards if only the stuff they consist of has sufficient tensile strength. Typically, in Gassendi's view, the Aristotelians never conducted experiments to test their theory; otherwise they would have realized that their own predictions are false and the prediction of the alternative theory comes out true. Thus, the theory of air pressure is not only an alternative theory that can explain the quoted observations as adequately as the horror-vacui-theory, rather the theory of air pressure proves to be superior since it explains more than its rival.[17]

At this stage of the argument the reader is well prepared to understand, and to approve of, the fourth and last step of Gassendi's refutation of the horror-vacui-theory. He points, of course, to the well known and at his time much discussed phenomenon that in big suction pumps the water raises only up to a certain height (about 10 meters). Gassendi does not bother to discuss the ad hoc assumption, defended by Galilei and many other physicists of the time, that the horror vacui has only a limited force; rather he stresses that the horror-vacui-theory cannot explain this phenomenon whereas the atomistic theory of air pressure can. Gassendi treats the observation about suction pumps as an experimentum crucis.[18]

My second example is Gassendi's discussion of a famous experiment that concerns the issue of inertia and the movement of the earth and that he himself conducted several times. I am talking about the experiment of dropping a stone from the mast of a swiftly moving ship and Gassendi's remarks about this experiment in two letters by the title *De Motu Impresso a Motore Translato*, published in 1642. It is important to see that one of the objectives of these

---

17 See Op. Om. I 199 a2–199 b1.
18 See Op. Om. I 199 b2–203 a2.

letters is to convince some physicists and friends that the reported outcome of the experiment, i. e. that the stone falls to the foot of the mast, can be trusted and is true.[19] And the way Gassendi tries to achieve this is to offer a good theoretical explanation of the experiment. Of course he first points out that the experiment *falsifies* the alleged justification the Aristotelians used to propose for the claim that the earth does not move, namely that if a stone is thrown upwards in a straight line it comes down at the same point form which it was thrown. This argument implies, of course, that if this throwing were conducted on a moving system the stone would fall down behind the place from which it had been thrown, and it is this conclusion that is falsified by the experiment on the ship.[20] But then Gassendi goes on to explain the experiment theoretically, i. e. to show which theory is *confirmed* by the experiment. He points out, for example, that if a system S moves uniformly then all movements of things in S look for an observer who is part of S the same way as if S were in rest (H1). This explains some phenomena on moving systems but must, at the same time, be itself explained by the assumption that every uniformly moving system translates its movement to all its parts (H2). H1 follows from H2 if movements and forces can supervene on each other and can be combined to one resulting movement or force (H3). But H3 is true only if the inertia is preserved (H4). It is the set H2–H4 of hypotheses that explains every experiment and observation that has been discussed in this connection, including those ones that the Aristotelians used to use as evidence in favour of Aristotelian physics. According to Gassendi, it is this explanatory success that confirms the hypotheses H2–H4.[21]

It seems to me that these examples show at least four things:

- First, in talking about observations, experiments, and appearances that are important in physics Gassendi means to talk about things having perceptible properties *in the external world.*
- Second, in offering explanations for appearances Gassendi does not use the method of signs, i. e. does not look at appearances reported by observations as signs from which we may infer assumptions about their causes; rather, he follows pretty closely what is nowadays called the hypothetical-deductive method and stresses, in particular, the important role of experimenta crucis.
- Third, Gassendi thinks that the trustworthiness of experiments and observations can sometimes be improved by explaining them theoretically.
- Fourth, while offering theories about causes of things in a spirit that is in principle probabilistic and fallibilistic, Gassendi seems nevertheless quite confident that the theories he settles on are the right ones or are at least definitely superior to their Aristotelian rivals; Gassendi's probabilism is fully compatible with assuming that even under human epistemic conditions we have often very good reasons to think that a certain theory is, if not definitely true, still at least definitely better than any of its rivals.

---

19 Compare Op. Om. III 478 b and Gassendi's letter to Valois in 1641, Op. Om. VI 108 b f.
20 Op. om. III 500 b1.
21 Compare the passage Op. om. III 478–484 and the formulation of the principle of inertia ibid. III 495 b.

I am not claiming that this specific version of probabilism, i. e. hypothetical-deductive falli-bilism, is prominent throughout Gassendi's physics, but there are several examples, among them crucial ones from the point of view of the new physics, that can and must be read as instantiations of this version of probabilism.

3.

As is well known, an important part of what Gassendi offers as physics consists of attempts to apply atomism in a more direct way to explanations of perceptible phenomena, most promi-nently the explanation of the so called qualities.[22] Many of the explanations Gassendi offers are, from our modern point of view, much less convincing and much more traditional than his thoughts about issues of the new physics. Some of them are, for instance, clearly circular: Gassendi gives, for a quality q and generalizations about q, a redescription in terms of atom-istic physics that is derived from these generalizations and then explains the same generali-zations by pointing to the atomistic redescription. Thus, bodies are empirically dense if their parts are close to each other; from this and atomistic physics it follows that body A is more dense than body B if the volume ratio of the sum of atoms and the occupied space is for A greater than for B. This is supposed to explain, according to Gassendi, why water is more dense than air.[23] Such a reasoning looks certainly circular. But often Gassendi tries at least to derive, in addition, empirical generalizations from atomistic redescriptions that are not among the empirical premises that these redescriptions had been derived from. For instance, after having defined, in atomistic terms, transparency in terms of density and atomistic theory of light he derives, and explains thereby, the fact that bodies with specifically regular atomic structure such as salt (a fact that he took to be independently established) do not affect trans-parency.[24] So even in this context the criterion of explanatory power, this time applied to atomistic physics as a whole, seems still to be in play. This helps to understand why Gassendi, contrary to what Epicurus is doing, does not even try to deduce the basic theorems of atomis-tic physics from empirical observations and methodological claims. And it also helps to un-derstand why Gassendi and others took atomic physics to be a promising scientific founda-tion for incorporating the occult qualities into physics.[25] Furthermore, in talking about qual-ities of things and trying to explain them, Gassendi certainly assumes that these qualities can be perceived and observed, but the explananda are statements that describe, not inner epi-sodes as mental states, but rather, again, external things as possessing perceivable qualities.

---

22 See Physica Sectio Prima, book VI *De Qualitatibus Rerum* in Op. Om. I.
23 Op. om. I 375 b–376 a.
24 Op. om. I 379 a–380 a.
25 Keith Hutchinson has argued, convincingly, that this promise explains in turn the success of corpuscular physics in the 17th century, see Hutchinson (1982). It is in this connection that the microscope came into play, see Catherine Wilson (1988).

In short, to a considerable extent the following picture of scientific method seems applicable to Gassendi's physics:

(a) Theories should be approved insofar as they entail, explain, and are confirmed by, empirical observations and experiments.
(b) To form an empirical basis for the test of theories, sentences about observations and experiments must attribute perceivable qualities to things in the external world.
(c) Theoretical assumptions about causes of perceivable states of the external world and about their intrinsic nature remain at most probable and in principle fallible.
(d) Claims (a)–(c) are a specification of, and thus a result of, moderate scepticism for the case of natural sciences.

The set of these four claims raises a problem for our understanding of Gassendi's epistemology, however. One way of describing this problem is to say that claim (d) seems to be inconsistent with claims (a)–(c) if we look more closely at the way Gassendi outlines his fundamental sceptical position. Many passages support the reading that according to Gassendi

(e) the infallible empirical basis of our knowledge consists of judgements of the form
(*) It appears to S that P(a) (such that (*) *is uttered by S*)

In the *Philosophiae Epicuri Syntagma* he says, for instance:

E3 "Perception does not fail because it does not claim that the tower is such and such, but is only in a passive state in that it only receives a kind (of impression) and refers only to that what appears to it."[26]

Likewise in the *Exercitationes Paradoxicae* we read:

E4 "I know that the honey appears to me sweet rather than sour and that the fire appears to me hot rather than cold."[27]

Gassendi stresses, for instance in his quarrel with Descartes on epistemological issues, that statements of form (*) are the only ones that are clear and distinct and are therefore not affected by Cartesian doubts:

E5 "What is it that we can take to be true if not that what appears to each of us?"[28]

Gassendi thinks that this is the sceptical position[29], but he also makes it clear that he fully endorses this point of view:

---

26 Op. om. III 6 a2 f. Compare the letter to Valois where Gassendi remarks: "... since perception reports nothing else than that the tower appears to be such and such because of a kind of round (impression)," Op. om. VI 149 b1.
27 Op. om. III 192 a2.
28 Op. om. III 314 b2.
29 See, for instance, Op. om. I 70a1, 73a1.

E6 "Therefore I too, imitating the sceptics, grant such a sort of truth that for my taste the honey seems to be sweet, that is, that what I profess to know is that I experience this sweetness."[30]

In this respect, Gassendi does not find any difference between dogmatics and sceptics.[31]

So Gassendi certainly thinks that the empirical basis of all our knowledge claims consists of mental states that have propositional content and can therefore be described by using concepts and sentences of natural languages. But the problem that obviously follows from the claim that the empirical basis of all claims to knowledge consists of sentences of form (*) is that these sentences cannot serve as an empirical basis for testing theories in physics. To test theories we must rather be able to attribute perceivable qualities to things in the external world, for instance to say that the tower *is* round. What is more, in evaluating and defending theories in physics, Gassendi himself talks, as we have seen, in such a way that he relies on observations and experiments that can be and are actually reported in this objective manner, i. e. as reliable reports about external things *having* certain perceptable properties. One might think, therefore, that the way Gassendi's sceptical position is related to his probabilistic methodology in physics looks fatally inconsistent. I cannot see that this problem is acknowledged, not to say solved, in the recent literature on Gassendi's epistemology. The reason for this is, I think, that scholars do not look sufficiently closely to what exactly, according to Gassendi, the empirical basis and the methodology in physics is supposed to be.

The first move in addressing this problem must be to examine a couple of passages in which Gassendi sets out a more complex picture of the empirical foundation of claims to knowledge. In one of these passages, for instance, he says (in Canon XI of the Institutio Logica):

E7 "Certainly, it often happens that those things which are perceived by the senses appear different or in different manner than they actually are ... For this reason, in order that we may have an undoubted, true and accurate idea, we must carefully examine whether the appearance corresponds to the reality ... Indeed experience based on the senses should be the highest criterion to which we must, if there are doubts about something, resort, nevertheless not any experience whatsoever is to be so regarded but only that which has been purified from every importunity and doubt and which is so evident that all things considered it cannot reasonably be denied."[32]

In another similar passage, Gassendi explains more clearly what the purification of experience amounts to:

E8 "Whenever one is uncertain about a thing that can be judged by the sense, ... one must indeed get back to the sense and must proceed from the evidence that comes from it. I say evidence which we have whenever there is no obstacle, or if there is one, after it has been removed, for instance the distance."[33]

---

30 In *De Veritate*, Op. om. III 413 a1.
31 Op. om. I 73 a1, VI 145 a2, I 13 b2.
32 Op. om. I 96 b1.
33 Op. om. I 122 a1; compare also I 85 a2 f.

Similarly we read in the *Philosophiae Epicuri Syntagma*:

> **E9** "Evidence of the senses I call that kind of sensation and phantasy which, after all obstacles for a judgement have been removed, like distance, movement, medium and other similar things, cannot be denied."[34]

These passages clearly respond to the sceptical challenge that even our senses may deceive us. Gassendi's response is obviously compatible with sticking to the claim that the senses are infallible, i. e. produce infallible observation reports of the form (*). But according to the more complex picture there is an additional claim, roughly of the form

> (**) If (*) is the case and if there are no obstacles for judging about perceivable external things, then this is sensual evidence for the claim that P(a) (i. e. then we may conclude that P(a) is the case).

Remarkably, this is not all Gassendi has to say about this issue. For he makes it clear, in addition, that

> (***) If (**) is the case, then claiming that (*) is sensual evidence for P(a) is itself a matter of reason and is, therefore, defeasible.

Strictly speaking, therefore, if something goes wrong it is reason that fails here.[35] So reason can correct the senses, albeit in a defeasible way. According to Gassendi, trying to determine whether there are obstacles for getting evidence from the senses can be sometimes itself be a matter of knowing the causes of appearances:

> **E10** "For the sense simply reports the appearance which is truly such and such but which also has truly a physical and necessary cause why it is such and such ... Therefore, since the appearance because of manifold causes can be manifold in the sense and the sense itself is affected always in a true way, it is the reason that may fail in such a way that, although it does not know the causes of the appearance, nevertheless does not dispel it but forms a hasty belief."[36]

But Gassendi does not conclude from this that since we can never know the causes of the appearances we should never trust alleged evidence, i.e. that we should never endorse judgements of the form (**), but only of the form (*). On the contrary, he says that

> **E11** "It is possible to know why this or that appears in a certain way[37] ... since for instance position, interval, quantity, frequency produce a necessary cause why a thing appears as such and such and it can therefore be known."[38]

---

34 Op. om. III 7 a2. The same formulation (almost) applied to Epicurus in I 54 a1).
35 See, for instance, Op. om. I 81 b1, 85 a.
36 Op. om. III 472 a2 f. There are several other passages that make the same point, see Op. om. III 282 b1, I 53 a2 f., I 54 a1, I 84 b2, VI 148 a2.

I conclude from these and similar passages

- that Gassendi was well aware of the distinction between observation reports of form (*) and form (**),
- that he realized that it is only evidence of form (**), not introspection of form (*) that can serve as a basis for testing empirically theories in physics,
- that he took it that appearances of form (*) generate evidence of form (**) if normal conditions obtain,
- and that he suggested that it is reason that must determine whether normal conditions of perception obtain and what the causes of appearances are, i. e. whether they are fallacious or not.

Knowing in a probabilistic and fallible way the causes of appearances is what Gassendi calls the "science of appearances"[39] (not just knowing that we have some appearances). We have to understand, though, that according to Gassendi this sort of science can only provide probable results, at most. It follows, then, that for him the empirical foundation of testing theories in physics is itself fallible (as summarized already in (***)). Probabilism must be extended to observation reports of form (**) that claim to make judgements about external things as having certain perceivable properties.[40]

4.

It seems obvious that Gassendi's ideas about moderate scepticism and scientific method proved, historically speaking, to be on just the right track since what they envisage is something like a hypothetical-deductive method that entails a reflective empricism that tries to do justice to the theory-ladenness of observation and the defeasibility of observation reports as test basis for scientific theories. To be sure, there are passages and remarks in Gassendi's works that do not seem to fit perfectly with this picture, but what we can firmly say is, I think, that this kind of picture is gradually emerging from some of the things Gassendi articulates explicitly about methodology as well as from some of his procedures in doing physics himself.

But this picture, although solving, in a sense, the problem of characterizing adequately what the empirical basis for testing theories in physics is supposed to be against the overall background of scepticism, generates a new serious problem that is, by the way, still with us. This is the problem of epistemic circularity. For obviously, determining whether normal conditions for perception and observation obtain and explaining the causes of appearances so

---

37 Op. om. I 84 b2.

38 Op. om. VI 148 a2; see also the examples concerning the appearance of a round tower in I 53 a2 f. and the effects of light in I 84 b2. Particularly telling, however, are passages like Op. om. III 505 a1 and III 505 b1-2 where Gassendi talkes about experiences we were for a long time adjusted to and that were even, in some sense, taught (like the experience that the earth is in rest) but that need to be corrected by reason.

39 See for instance Op. om. III 192 a2.

40 In Op. om. VI 150 a Gassendi calls evidence more probable than anything else.

that we can trust the appearances consists of statements which must themselves be confirmed by perception and observation. I do not see Gassendi addressing this problem in any direct way; but I want to call attention to a sort of strategy that seems at least to mitigate this problem a little and is interesting in itself.

This strategy consists basically of three elements. One is to maintain that evidence relies, first of all, on a *causal* contact between *individual* external things and the *surface* of our perceptual apparatus (such as the retina). This contact brings about ideas of individuals in the mind.[41] General ideas about common properties of these individuals are produced by the mind through comparing and seperating the individuals. So if it seems to us that P(a) is the case, this appearance is a causal effect of a thing a such that the mind classifies it as a P-thing. To test the entire appearance including the property P, however, we have to move to the second part of the strategy, namely to vary and change those parameters that we theoretically know to influence our appearances, like distance, position, frequency and the like. If in most of these varied conditions P is perceived *consistently* as a property of a, we have good reasons to think that the appearance is reliable. A tower might appear square from one angle and round from another angle, and a stick that is placed half in and half out of water appears crooked.

E12 "To test, as Gassendi goes on to remark, whether the tower is round we move close to the tower itself, and to determine whether the stick is crooked we take it out of the water, and so on for the rest."[42]

Finally, the third part of the strategy is to give a scientific account of the physical and psychological processes underlying acts of perception and observation. It seems that Gassendi was aware of, and participated consciously in, a new trend in seventeenth-century considerations of cognition.[43] Aristotelians conceived of cognition as one single process by which the species of objects were transmitted to the cognitive faculty. In the seventeenth century, however, this process was increasingly being looked at as consisting of several parts that could, and were supposed to be, studied seperately, i. e. by optics, physiology, and psychology. Maybe Brundell is right in claiming that in his own work on these issues of a naturalized theory of cognition Gassendi showed little interest in thinking about the validity or reliability of the processes studied by cognition theory.[44] But some of the topics Gassendi discussed extensively do fit with the other two parts of the strategy mentioned above, for instance his work *On the apparent magnitude of the sun when it is low and when it is high.*[45] Obviously, at this point the science of appearances comes in, and this *is* a move that is, as we have seen, for Gassendi part of our struggle to determine which appearances are to be trusted. In short, it seems to me that it is not unreasonable to look at this threefold strategy as addressing at least in an indirect way the problem of epistemic circularity.

41 See Institutio Logica Canon IV
42 Institutio Logica Canon XI. We may read Gassendi's remark that we should trust only those appearances that all things considered cannot be contradicted as pointing to this criterium of consistency, compare Op. om. I 96 b1, VI 150 a2.
43 See Op. om. II 371 a-b and Brundell (1987), p.87.
44 Brundell (1987), p.87.
45 Published in 1642 in a series of four letters, see Op. om. III 420-477.

By way of conclusion, let me try first to outline a general diagnosis concerning the way Gassendi related scepticism to scientific method and then to say a few things about his way of looking at ancient ideas about scientific method.

The way Gassendi related scepticism to scientific method can be brought into sharper focus if we compare it with the way Descartes reacted to scepticism. Broadly speaking, and simplifying things to some extent, we should, I think, emphasize three points.

The first is that Descartes tries to face the hardest possible sceptical challenge, that is to say the evil demon hypothesis that assumes a *systematic* unreliability of our senses and thoughts, something like that we are all brains in a vat. Gassendi, however, proceeds from a less radical sceptical premiss that can be called an *unsystematic* unreliability hypothesis: we do have the knowledge about what appears to us, and appearances can be reliable for forming observational judgements about external things if the circumstances are right, i. e. if normal conditions obtain, but they can also be unreliable if the world does not do us a favour.

The second point is that Descartes in his attempt to overcome radical scepticism goes for absolute epistemological certainty, in my view mainly due to the fact that in working out analytic geometry he discovered algebraic algorithms producing new truths and hoped to be able to generalize this method to every branch of science and knowledge.[46] Gassendi, however, advances probabilism on two different levels; on the level of appearances, he thinks that unsystematic unreliability can be relieved by looking and varying the conditions of perception and examining consistency between them; on the level of scientific theories about the nature of things he maintains, at least in his best moments, a fallibilism and recommends to proceed hypothetical-deductively.

Finally, the third point is that, epistemologically, Descartes proceeds exclusively in an internalist way by entertaining the introspective criteria of clarity and distinctness for determining whether a person is a knower. Gassendi, on the contrary, takes into account an externalist strategy that points to, and analyzes, reliable causal (physiological and psychological) belief forming processes.[47]

In short, for Descartes radical scepticism provoked an internalist search for absolute epistemological certainty, whereas Gassendi was convinced that only mitigated scepticism makes sense and requires a pretty moderate externalist probabilism concerning observational reports about external things as well as a fallibilistic hypothetical-deductive methodology for developing theories in physics.

In studying what the 17th century contributed to the history of scientific methods and cultures most scholars look at Descartes or Galilei or other leading pioneers of the new physics. Looking at Gassendi and maybe at other figures that were, like Gassendi himself, highly respected at the time but are almost forgotten today shows, at least, that the picture is more complex. More importantly, if I am right, concerning the three points just mentioned Gassendi's epistemological position probably strikes most of us (including myself), from our own perspective at least, as just right and Descartes' position as profoundly misguided.

---

46 See Detel (2000).

47 As far as Descartes is concerned, I discuss this topic at length in Detel (2000). Thomas M. Lennon seems to see these things in a similar way. He argues that Gassendi, Locke and others anticipated Kripke's insight that we can refer directly to things without knowing what essentially identifies them (which is certainly a sort of externalism), see Lennon (1995).

Many scholars see the key of Gassendi's project in the attempt to simply rehabilitate Epicurean philosophy and doing his best to make Epicurean philosophy compatible with Christian faith and theology.[48] But if my discussion of some epistemological issues in Gassendi's philosophy are on the right track then it seems clear that Gassendi changed Epicurean physics and methodology in a rather significant way. For it can be argued, although I don't have the space here to go into this issue, that the sort of probabilistic empiricism and hypothetical-deductive method Gassendi seems to be entertainig and following in his own work in physics cannot be found in Epicurus, as it is certainly extravagant, from a historical point of view, to look at Epicurus as a moderate sceptic. What is more, we have good reasons to think that the way Gassendi transformed parts of Epicurean methodology responds to insuperable difficulties connected with Epicurean epistemology.[49] In my view, then, Gassendi's Epicurean project was to choose a promising philosophical system that could most easily be transformed into a basic physics and scientific methodology that served best the purpose of avoiding dogmatism and of promoting progress in natural science.

There is another final point that needs to be addressed briefly if we look at the history of scientific method. In proposing a probabilistic methodology Gassendi is responding, as everybody agrees, to what he sees as stubborn dogmatism of the Aristotelians who pretend to know for sure that every bit of Aristotelian physics is true. In pretty much the same spirit Descartes complains that Aristotelian science and method suffers from total infertility. Both Gassendi and Descartes may be right, but they also hasten to add, in some passages, that there may be a difference between Aristotelians and Aristotle, without however clarifying what exactly they have in mind. Historically, the interesting point is that we now are beginning to see that Aristotle indeed did not think that human scientists can grasp the foundations and the essence of things with absbolute certainty; rather, Aristotle felt that human scientists can go wrong in a number of different ways and can in some cases not make sure once and for all that they have grasped the truth or nature of the phenomena under consideration. At the same time Aristotle seems to have thought that to make this very point we need a conceptual grasp of what an ideal or perfect knowledge or scientific theory would look like.[50] And pretty much the same might go for Plato too.[51] From this point of view it is a lot more understandable that middle Platonists like Carneades read Plato and Aristotle in a sceptic mood and that Plato's Academy turned sceptical. Even so called dogmatists like the Stoics found it hard if not impossible to get at the true causes of things. If I am right about Gassendi then Gassendi, and not Descartes[52], can be seen to continue this important branch of ancient ideas about scientific method except that he, like later ancient sceptics, did not find it theoretically necessary to rely on an explicit conception of perfect science that, while not being within human reach, must be nevertheless part of setting out, and defending, versions of a probabilistic methodology guiding natural sciences.

---

48 See, for instance, Osler (1983) and Osler (1994).
49 Compare Detel (1978).
50 See Detel (1993) and Detel (1998).
51 See Becker, Detel, Scholz (forthcoming 2001), in particular the introduction and the contributions by Rowe, Becker, and Detel.
52 See Detel (2000).

References

Becker, A., Detel, W., Scholz, P. (eds) (forthcoming): *Ideal and Culture of Knowledge in Plato*. Stuttgart: Steiner.

Bloch, O. R. (1971): *La philosophie de Gassendi*. La Haye.

Brundell, B. (1987): *Pierre Gassendi. From Aristotelianism to a New Natural Philosophy*. Dordrecht.

Detel, W. (1978): *Scientia Rerum Natura Occultarum. Methodologische Sudien zur Physik Pierre Gassendis*. Berlin/New York.

Detel, W. (1993): *Aristoteles, Analytica Posteriora*. Übersetzt und erläutert von W. D., 2 Bd. Berlin (bes. 263-334).

Detel, W. (1998): "Aristotle's *Posterior Analytics* and the Path to the Principles," in N. Avgelis and F. Peonidis (eds.), *Aristotle on Logic, Language and Science*. Thessaloniki, 155-182.

Detel, W. (2000): "Descartes und der wissenschaftstheoretische Fundamentalismus," in Horn and H. Schnädelbach (eds.), *Descartes im Diskurs der Neuzeit*. Frankfurt, 230-258.

Garber, D. (1992): *Descartes' Metaphysical Physics*. Chicago.

Gaukroger, St. (1995): *Descartes. An Intellectual Biography*. Oxford.

Gregory, T. (1961): *Scetticismo ed empirismo – studio su Gassendi*. Bari.

Hutchinson, K. (1982): "What Happened to Occult Qualities in the Scientific Revolution?," *Isis* 73: 233-253.

Lennon, Th. M. (1995): "Pandora; or, Essence and Reference: Gassendi's Nominalist Objection and Descartes' Realist Reply," in R. Ariew and M.Grene (eds.), *Descartes and His Contemporaries*. Chicago/London: Chicago UP, 159-181.

Osler, M. (1983): "Providence and Divine Will in Gassendi's View on Scientific Knowledge," *Journal of the History of Ideas* 44: 549-560.

Osler, M. (1994): *Divine Will and the Mechanical Philosophy: Gassendi and Descartes on Contingency and Necessitiy in the Created World*. Cambridge.

Osler, M. (1995): "Divine Will and Mathematical Truth: Gassendi and Descartes on the Status of the Eternal Truths", in R. Ariew and M. Grene (eds.), *Descartes and his Contemporaries*. Chicago/London, 145-158.

Popkin, R. (1964): *The History of Scepticism from Erasmus to Descartes*. New York.

Schmitt, Ch. B. (1972): *Cicero Scepticus*. The Hague.

Wilson, C. (1988): "Visual Surface and Visual Symbol: The Microscope and the Occult in Early Modern Science," *Journal of the History of Ideas* 49: 85-108.

Alexander Becker

Das Wissen von der Musik im 16. Jahrhundert

1. Wissen von der Musik

Fände sich ein Musiktheoretiker des 16. Jahrhunderts an einer Universität des ausgehenden 20. Jahrhunderts wieder, würde er, bei allem Fremden und Neuen, das ihn umgibt, wohl erfreut feststellen, daß wenigstens seine Disziplin immer noch existiert, daß auch heute noch „Musikwissenschaft" (oder *musica scientia*, wie er sie nennen würde) gelehrt und Musik offensichtlich immer noch als ein Gegenstand gilt, über den man wissenschaftlich forschen kann. Seine Freude würde allerdings bald großer Irritation weichen, wenn er bemerkt, daß seine moderne Kollegin in ihrem Seminar nicht etwa die Eigenschaften von Zahlenproportionen untersucht, sondern sich um die Deutung der Madrigale von Gesualdo bemüht. Kompositionen, so wird er denken, mögen ja gut für ein paar Beispiele in der Kontrapunktlehre sein (wenn auch nicht gerade die exzentrischen Schöpfungen Gesualdos), aber wenn hier überhaupt etwas Gegenstand der Wissenschaft ist, dann doch bestenfalls die Regeln des Kontrapunkts und nie und nimmer einzelne Werke.

Es ist also keineswegs klar, wovon eigentlich die Rede ist, wenn man vom „Wissen von der Musik" spricht. Ich möchte meine historische Untersuchung daher mit einem knappen Blick auf eine grundsätzlichere Frage eröffnen: Was können wir überhaupt von der Musik wissen? Die Antworten lassen sich, so glaube ich, in drei große Gruppen zusammenfassen:

- Es kann sich um handwerkliches Wissen handelt, also um Wissen, das für die Komposition und Produktion von Musik relevant ist. Mit „Musik" sind in diesem Fall hörbare Artefakte gemeint. Das Wissen wird zum Teil die Form von Regeln haben; seinen Status als Wissen wird es aus verschiedenen Quellen beziehen, die praktische Bewährung handwerklicher Kenntnisse dürfte aber notwendig sein
- Gemeint sein kann auch ästhetisches Wissen. Wieder ist Musik wesentlich etwas, was wir hören können. Diesmal geht es aber um alles, was mit der Bewertung und der Wirkung von Musik auf uns zu tun hat: beispielsweise Standards, nach denen Kompositionen beurteilt werden können, Überzeugungen, die als Gründe für Geschmacksurteile dienen können, Kenntnisse über die emotionale Wirkung von Musik, oder auch Interpretationen einzelner Werke und ihre Voraussetzungen, die darauf zielen, den „Gehalt" eines Musikwerks herauszuarbeiten.
- Schließlich kann Wissen von der Musik wissenschaftliches Wissen sein. In diesem Fall ist unter „Musik" ein Phänomen in der Welt zu verstehen, das sich von gewöhnlichen akustischen Erscheinungen unterscheidet. Das, was wir in einem Konzert oder in der Kirche hören können, zählt sicherlich dazu, die Wahrnehmbarkeit durch uns ist aber keine notwendige Bedingung, um etwas als Musik zu betrachten – die Harmonie

der Sphären hat noch kein Mensch gehört, doch hat dieser Umstand unzählige Musiktheoretiker nicht daran gehindert, sie als das musikalische Phänomen par excellence
zu betrachten. Als Unterscheidungsmerkmal zwischen Musik und gewöhnlichen akustischen Phänomenen kann die Strukturierung durch Zahlenverhältnisse ebenso dienen wie das physikalische Eigenschaft von Schwingungsvorgängen, Teiltöne zu bilden. Ob eine Aussage als Wissen gilt, wird hier davon abhängen, ob sie in einen
größeren Kontext wissenschaftlicher Welterklärung eingebettet werden kann.

Man wird in allen Phasen der europäischen Musikgeschichte alle drei Arten von Wissen
(selbstverständlich in unterschiedlichen Ausprägungen) antreffen. Auch heute widmen Har-
monielehren wenigstens eine Seite der Obertonreihe als (vermeintlich) natürlicher Grundlage der Harmonik; auch im Mittelalter hat man sich über die Wirkung der Musik auf die Seele
Gedanken gemacht. Aber es scheint evident, daß sich zwischen dem 14. und dem 18. Jahrhundert ein tiefgreifender Wandel vollzogen hat: Debatten über Musik verlagerten sich vom
Bereich des wissenschaftlichen Wissens mehr und mehr in den des ästhetischen Wissens; das
wissenschaftliche Wissen von der Musik wurde allmählich von der physikalischen Disziplin
der Akustik abgelöst; das handwerkliche Wissen wurde zunehmend zu einem Spezialwissen,
dessen Status als Wissen obendrein fragwürdig wurde (bis hin zu Helmholtz' berühmtem
Vergleich der Musiktheorie mit der Kochkunst). Kurz gesagt, das musikwissenschaftliche
Paradigma von „Musik" scheint sich von der Harmonie der Sphären zum Werk, und zwar
zum Werk aus der Perspektive des Rezipienten, gewandelt zu haben.
    Es ist klar, daß ein derartiger Prozeß stets sehr viel komplexer und viel weniger eindeutig
abläuft als es solche Formeln suggerieren, die nicht nur verkürzen, sondern auch verzerren,
selbst wenn sie für eine Beschreibung stehen, die im großen und ganzen nicht unangemessen
ist. Im folgenden möchte ich auf einen kleinen Ausschnitt dieses Prozesses einen genaueren
Blick werfen, nämlich auf die Situation des Wissens von der Musik im 16. Jahrhundert.
    Diese Phase ist sowohl für den erwähnten historischen Umbruch als auch für die Frage
nach dem Verhältnis der verschiedenen Arten des Wissens von der Musik von besonderem
Interesse. Weder davor noch danach wurden wissenschaftliche, handwerkliche und ästhetische Konzeptionen je so eng zusammengebracht, waren so eng miteinander verflochten und
bargen zugleich so viel Konfliktpotential. Die Entwicklungen der musikalischen Praxis – der
frei durchimitierte mehrstimmige Satz, die Entstehung temperierter Stimmungen, die „Wie-
derbelebung" der antiken chromatischen Musik, die neue Wort–Ton–Konzeption im Madrigal, schließlich die Monodie und die Oper – finden nicht abseits oder unter Mißachtung der
Theorie statt, die durch das gesamte 16. Jahrhundert hindurch von der mittelalterlichen Auffassung vom Primat des wissenschaftlichen Wissens geprägt blieb. Vielmehr werden kompositorische Neuerungen mit Reformen der Theorie verbunden, oder sie werden gezielt als
Versuche aufgefaßt, ihre Freiräume auszunutzen, oder sie geschehen auf der Grundlage und
als Bewährungsprobe neuer theoretischer Entwürfe. Man kann das 16. Jahrhundert als eine
Art von „Engführung" im Übergang vom mittelalterlichen zum neuzeitlichen Verständnis
von „Wissen von der Musik" betrachten. Im 17. Jahrhundert entflechten sich die Fäden, die
im 16. Jahrhundert noch in einem schwer durchschaubaren Knäuel miteinander verbunden
sind , aber sie laufen auch auseinander: Generalbaß und Akustik entwickeln sich zu eigen-
ständigen Disziplinen, die erst durch Rameau im 18. Jahrhundert wieder aufeinander bezogen werden.[1] Daher ist gerade das 16. Jahrhundert nicht nur für die Geschichte der Musik-

theorie, sondern auch für eine Untersuchung, wie sich verschiedene Arten des Wissens von der Musik zueinander verhalten, ein besonders geeigneter Gegenstand.

2. Der Ausgangspunkt: Das späte Mittelalter

Will man sich mit der Musiktheorie im 16. Jahrhundert befassen, ist es nahezu unvermeidlich, sich zunächst mit der Lage im späten Mittelalter, also etwa seit dem Ende des 13. Jahrhunderts, vertraut zu machen. Zum einen kristallisierte sich in dieser Zeit ein Theoriebestand heraus, der im wesentlichen bis zum Ende des 16. Jahrhunderts bestimmend blieb. Zum anderen ist es nötig, auch vom Verhältnis von wissenschaftlichem Wissen („musica theorica" oder „speculativa") und handwerklichem Wissen („musica practica") im Mittelalter ein differenzierteres Bild zu gewinnen, bevor man sich daran macht, die Neuerungen des 16. Jahrhunderts in den Blick zu nehmen.

Ausgangspunkt eines typischen musiktheoretischen Traktats dieser Zeit (wie die „Notitia artis musica" (1321) des Johannes de Muris, dessen „Musica speculativa" (1323) zu den verbreitetsten Traktaten des späten Mittelalters gehörte[2]) ist die Feststellung, die Musik (also die musica scientia) handle „vom auf Zahlen bezogenen Ton" („de sono relato ad numeros", de Muris 1321, S. 49). Damit sind die zwei wichtigsten Stichworte genannt: Musik ist einerseits ein physikalisches Phänomen, andererseits ein mathematisches. Töne kommen durch Erschütterungen der Luft zustande; dies erklärt ihre Wahrnehmbarkeit. Diese Erschütterungen sind zählbar; ihre Zahl bestimmt die Höhe des Tons. Folglich sind es die Zahlen und ihre Verhältnisse, die eine präzise Beschreibung des Gegenstands „Musik" ermöglichen. Diesen Vorbemerkungen folgt eine ausführliche Darstellung des mathematischen Kerns der Musiktheorie, der Lehre von den Zahlenproportionen, in der die Gattungen und Arten der Proportionen unterschieden wurden.[3] Anschließend werden diejenigen Proportionen vorgestellt, aus denen konsonante Klänge hervorgehen, nämlich alle Proportionen, die sich mit Hilfe der Zahlen 1, 2, 3 und 4 bilden lassen (also die Oktave (1 : 2), Quinte (2 : 3), Quarte (3 : 4), Doppeloktave (1 : 4) und Duodezime (1 : 3)). Alle anderen Zahlenproportionen ergeben

---

1 Diese Behauptung muß selbstverständlich unter Vorbehalte gestellt werden, denn auch im 17. Jahrhundert entstehen große theoretische Entwürfe (beispielsweise von Mersenne und Kircher), die versuchen, die drei genannten Arten des Wissens von der Musik zusammenzufassen, und die nicht bloß alte Bestände zusammenfassen, sondern die Entwicklung der Musiktheorie wesentlich vorantreiben. Solche Unternehmungen stehen sicherlich in der Tradition des 16. Jahrhunderts (ohnehin ist der Wechsel des Jahrhunderts eine völlig ungeeignete Epochengrenze). Aber es ist auffällig, daß etwa in der Mitte des 17. Jahrhunderts das Interesse an einer Weiterentwicklung der Theorie nachläßt (so in Italien, vgl. dazu Groth (1989)) oder das Bedürfnis nach einer wissenschaftlichen Fundierung der Satzlehre schwindet (so in Frankreich, vgl. Seidel (1986), bes. S. 88ff.), während die herausragenden Komponisten nicht mehr den Austausch mit Theoretikern suchen oder sich gar selbst in das Getümmel der Debatten stürzen (so wie Vicentino, de Rore oder Monteverdi im 16. Jahrhundert).
2 Vgl. die Übersicht in Bernhard (1990), S. 73f.
3 Gemäß Boethius' Einteilung der mathematischen Wissenschaften in der „Institutio arithmetica" (I, 1) waren feste kontinuierliche Größen Gegenstand der Geometrie, kreisförmig bewegte kontinuierliche Größen Gegenstand der Astronomie, zählbare Größen per se Gegenstand der Arithmetik und die Verhältnisse zählbarer Größen Gegenstand der Musik.

dissonante Intervalle, zu denen somit auch die Terzen (4 : 5 und 5 : 6) und Sexten (3 : 5 bzw. 5 : 8) zählen. Warum nur die genannten Intervalle als konsonant galten, dafür wurde entweder gar keine Begründung gegeben oder auf ihren göttlichen Ursprung verwiesen (vgl. de Muris 1323, S. 108). In weiteren Schritten werden wichtige Eigenschaften dieser Proportionen erläutert, etwa ihre Verhältnisse untereinander oder die Frage, wie sie geteilt werden können. Schließlich wird gezeigt, wie aus diesen Intervallen ein Tonsystem, nämlich die diatonische Skala mit sieben Stufen innerhalb der Oktave, aufgebaut werden kann. de Muris (1321, S. 59) gibt folgendes Beispiel:

$$G \quad : \quad A \quad : \quad B \quad : \quad C \quad : \quad D \quad : \quad E \quad : \quad F \quad : \quad G$$
$$5184 \quad : \quad 4608 \quad : \quad 4096 \quad : \quad 3888 \quad : \quad 3456 \quad : \quad 3072 \quad : \quad 2916 \quad : \quad 2592$$

In dieser Skala weisen die Oktave, die Quinten und Quarten reine Proportionen (also 2 : 1, 3 : 2 und 4 : 3) auf und klingen konsonant, die Terzen und Sexten dagegen werden durch unreine Proportionen gebildet (die Terz C – E beispielsweise die Proportion 192 : 243, die um ein pythagoreisches Komma (80 : 81) größer als die reine Terz ist, so daß sie dissonant klingt). Die Konstruktion eines Tonsystems war der einzige Schritt, der von der mathematischen Lehre hin zur musikalischen Praxis unternommen wurde, denn die siebenstufige diatonische Leiter bildete die fundamentale Materialleiter der mittelalterlichen Musik. Mit der Konstruktion eines Tonsystems wurde gezeigt, daß die mathematische „musica speculativa" zu Recht die musikalische Disziplin schlechthin war, die alle Erscheinungsweisen der Musik – von der Sphärenharmonie bis hin zu dem, was man täglich in der Kirche hören konnte – umfassen und unter einheitliche Prinzipien bringen konnte.

Kaum ein Musiktheoretiker des Mittelalters sah sich genötigt zu begründen, daß die Musik eine mathematische Disziplin war – eine solche Begründung wäre ohnehin tautologisch gewesen, da der primäre Gegenstand der *musica scientia* eben Zahlenverhältnisse waren. Daran änderte sich im Grunde nichts, als mit dem Siegeszug der aristotelischen Philosophie auch neue Kriterien der Wissenschaftlichkeit aufkamen, die reflektiertere Musiktheoretiker zu berücksichtigen versuchten. Johannes de Grocheo, der Ende des 13. Jahrhunderts schrieb, allerdings nur spärlich rezipiert wurde, erkennt zwar, daß man nicht einfach die Proportionen als Prinzipien der Musik postulieren kann. Sein Vorschlag lautet, die Proportionen der Konsonanzen als *Formen* dessen, was physikalisch realisiert und wahrgenommen wird, zu betrachten (de Grocheo 1290, S. 42f.). Damit verändert sich zwar das Objekt musiktheoretischer Erkenntnis von den Zahlenverhältnissen auf wahrnehmbare Phänomene, dies blieb aber ohne Auswirkungen auf den Inhalt der Musiktheorie.[4] Daß die Aristotelische Philosophie faktisch nicht zu einer Neuorientierung führte, zeigt sich u. a. daran, wie seine These, daß jede Erkenntnis mit sinnlicher Wahrnehmung anfängt, in den Theoriebestand integriert wurde. Diese These zwang die Autoren musiktheoretischer Traktate immerhin, auf das Verhältnis zwischen sinnlicher Wahrnehmung und mathematischer Beschreibung der Intervalle einzugehen. Den generellen Rahmen hierfür hatte jedoch bereits Boethius, die überragende

---

4 Diese veränderte Orientierung schlägt sich nicht nur in Grocheos Definition der Musik nieder („musica est ars vel scientia de sono numerato, harmonice sumpto ad cantantum facilius deputata. Dico autem scientiam, in quantum principiorum tradit cognitionem, artem vero, in quantum intellectum practicum regulat operando. De sono vero harmonico quia est materia propria, circa quam operatur. Per numerum etiam eius forma designatur", S. 46), sondern auch darin, daß er Bemerkungen zum Musikleben seiner Zeit in seinen Traktat aufnimmt (vgl. S. 47).

Autorität in der mittelalterlichen Musiktheorie, vorgegeben: Die Sinne sind unzuverlässig und daher zur Erkenntnis der Wahrheit weniger geeignet als der Verstand (Boethius Kapitel I, 1; vgl. de Muris 1323, S. 90). Um nun den Weg von der Wahrnehmung zur rationalen Erfassung anschaulich zu machen, bot sich die „Schmiedehammerlegende" an, die ebenfalls von Boethius überliefert wird und daher zum Kernbestand mittelalterlicher Musiktraktate gehörte. Sie berichtet, wie Pythagoras die Musik entdeckte: Eines Tages sei er an einer Schmiede vorbeigekommen und habe in den Hammerschlägen der Schmiede verschiedene Intervalle wahrgenommen. Daraufhin sei er in die Schmiede gegangen, habe zuerst einen der fünf Hämmer aussortiert, der nur dissonante Intervalle produzierte, dann die verbliebenen vier Hämmer gewogen und festgestellt, daß ihre Gewichte sich wie 1 : 2 : 3 : 4 verhalten. Er habe weiter mit anderen Instrumenten und Materialien experimentiert, sei stets auf die gleichen Proportionen gestoßen, so daß aus seinen Erfahrungen allmählich eine allgemeine Vorstellung – nämlich die Proportionen der Konsonanzen – entstanden sei; erst sie sei für ihn zum Prinzip der Kunst und der Wissenschaft geworden.[5] Für einen Autor wie de Muris bedurfte es keiner weiteren Erläuterung, warum die Prinzipien einer Wissenschaft allgemein sein mußten, warum die Wahrnehmung solche Prinzipien niemals liefern und daher nur eine heuristische Funktion übernehmen konnte; zu gut fügte sich das Resultat in die vertrauten Schemata ein. Faktisch trat die Anektode über Pythagoras bei ihm an die Stelle einer Begründung für den mathematischen Charakter der Musiktheorie.

Die musikalische Praxis, die vom elementaren Gesangsunterricht für Chorknaben bis zur Komposition komplexer mehrstimmiger Motetten reichte, hatte ein von der „musica speculativa" in weiten Teilen unabhängiges System der Beschreibung der Musik ausgebildet. Grundlage war, wie bereits erwähnt, die diatonische Skala, die zunächst aus sieben, später aus acht Stufen innerhalb der Oktave bestand. Der verwendete Tonraum umfaßte etwa 2 1/2 Oktaven. In diesem Tonraum wurden die sogenannten Modi (die „Kirchentonarten") unterschieden, bei denen es sich um Oktavausschnitte aus der gesamten Skala handelte. Diese Modi legten den Beginn (z. B. die Anfangsimitation) und die Schlußklausel eines Werks, den Ambitus der Stimmen sowie die Ordnung der Klauseln fest, also „äußerliche" Merkmale von Tonsystem und Kompositionen, die die Beziehungen der Tonstufen untereinander nicht berührten.[6] Diese Beziehungen, die Qualität der einzelnen Tonstufen, wurden mit Hilfe der von Guido von Arezzo im 11. Jahrhundert eingeführten Hexachordlehre bestimmt. In der Form, in der sie seit dem 12. Jahrhundert feststand, sah sie drei Arten von Hexachorden vor: das *hexachordum naturale* von c – a, das *hexachordum durum* von g – e und das *hexachordum molle* von f – d. Die Bedeutung der Tonstufen ergab sich aus den Solmisationssilben ut – re – mi – fa – sol – la, die jeden Ton durch seine Lage bzw. Umgebung im Hexachord bestimmten; mi und fa kam dabei eine zentrale Rolle zu, da zwischen ihnen der charakteristische Halbtonschritt lag. In der von de Muris bis Zarlino gültigen Form sah das Hexachordsystem folgendermaßen aus[7]:

---

5 De Muris (1321), S. 57: „Ex quibus experimentis ad memoriam concurrentibus accepit unum universale, quod sibi fuit principium artis et scientiae."
6 Insofern unterscheiden sich mit Modi grundsätzlich von den modernen Tonarten, die genau diese Aufgabe erfüllen, indem sie dynamische Beziehungen zwischen den Stufen der Skala herstellen (vgl. Dahlhaus (1968a), S. 201). Daß die Modi dies nicht leisteten, zeigt sich nicht zuletzt darin, daß in den Klauseln, also dort, wo dynamische Relationen zwischen Tonstufen zum Tragen kamen, Akzidentien, also Abweichungen vom Tonbestand der modalen Skala, üblich waren.

Hexachorde

	durum	naturale	molle
e"	la		
d"	sol		la
c"	fa		sol
h'/b'	mi (h)		fa (b)
a'	re	la	mi
g'	ut	sol	re
f'		fa	ut
e'	la	mi	
d'	sol	re	la
c'	fa	ut	sol
h/b	mi (h)		fa (b)
a	re	la	mi
G	ut	sol	re
F		fa	ut
E	la	mi	
D	sol	re	
C	fa	ut	
H	mi		
A	re		
G	ut		

Es ist leicht zu erkennen, daß dieses System der drei Hexachorde, die „Anschauungs- und Darstellungsform des Tonsystems im späteren Mittelalter"[8], an zwei Stellen die diatonische Heptatonik durchbrach: Die mi-Stufe des *hexachordum durum* war das *b quadratum* bzw. *durum*, die fa-Stufe des *hexachordum molle* das *b rotundum* bzw. *molle*. Das Nebeneinander der beiden Varianten stellte jedoch keine Überschreitung des diatonischen Systems dar: Die drei Hexachorde wurden nicht als Transpositionen (um die es sich real ja handelt), das b nicht als abgeleitet aufgefaßt; die Diatonik umfaßte seit dem späten Mittelalter acht, nicht sieben Stufen. Da b und h streng auf den Kontext ihrer Hexachorde verwiesen waren, war es selbstverständlich verboten, b und h unmittelbar aufeinanderfolgend, als Umfärbung der gleichen Tonstufe, zu verwenden. Hexachordwechsel, sog. *mutationes*, die im Laufe einer Komposition aufgrund des begrenzten Tonvorrats der einzelnen Hexachorde unumgänglich waren, blieben (innerhalb einer Stimme) auf solche Stufen beschränkt, denen mindestens zwei Solmisationssilben zugeordnet waren[9] (insofern war es durchaus möglich, b und h im glei-

---

7 Vgl. de Moris (1321), S. 64a, Zarlino (1573), S. 121. Trotz der allgemeinen Akzeptanz war natürlich auch das System der drei Hexachorde nicht gegen Kritik und Reformversuche gefeit. Gerade um die Wende vom 15. zum 16. Jahrhundert tobte ein heftiger Streit: Während die einen (beispielsweise John Hothby) die Stellung des hexachordum molle anfochten, indem sie es als Transposition auffaßten und sich damit zur ursprünglichen Version Guidos zurückbewegten, bemühten sich andere wie Ramos von Pareja um seine Vereinfachung, indem sie neue Solmisationssilben vorschlugen, so daß jede Stufe in der Oktave eine eigene Silbe erhielt. Siehe dazu Rempp (1989), S. 56ff.

8 Dahlhaus (1968a), S. 156.

9 Zur mutatio vgl. de Vitry (1320), S. 17.

chen Werk zu verwenden). In mehrstimmiger Musik konnten die Stimmen sich auch gleichzeitig in verschiedenen Hexachorden bewegen, etwa wenn die eine Stimme die andere im Quintabstand imitierte, so daß die Stimmen von vornherein auf verschiedene Hexachorde festgelegt waren.[10]

Das Nebeneinander von b und h, das einzig einem Regelwerk geschuldet war, das den Unterricht und die Aufführung von Musik anleiten sollte, überschritt zwar die siebenstufige Leiter, in der sich, wie erwähnt, Theorie und Praxis trafen, aber sie brach nicht mit der Vorstellung eines diatonischen Tonsystems, in dem die Intervallabstände mit Hilfe ausgezeichneter Zahlenproportionen festgelegt werden konnten. Dies gilt im Grunde auch für ein anderes Phänomen der musikalischen Praxis vom 14. bis zum 16. Jahrhundert, die sogenannte „musica ficta", obwohl sie faktisch die durch die pythagoreische Stimmung geregelte Diatonik sprengte und daher – wie ihr Name schon andeutet – von der „wahren Musik" ausgeschlossen werden mußte.

Unter „musica ficta" verstand man vorübergehende Änderungen einzelner Tonstufen, die auch als „Akzidentien", d. h. als etwas, das zufällig hinzutrat, aber nicht zum Wesen der Musik gehörte, bezeichnet wurden. Die Einführung weiterer Tonstufen außerhalb der Oktatonik ergab sich zunächst aus satztechnischen Zwängen, zur Vermeidung dissonanter Zusammenklänge (vor allem des Tritonus). Auch einige weitere Regeln, etwa die, die Stufe über la als fa zu singen oder die „regola delle terze e seste", d. h. die Regel, daß Terzen und Sexten durch einen Halbtonschritt in der einen und einen Ganztonschritt in der anderen Stimme in eine Konsonanz überführt werden sollen, machte die akzidentielle Erhöhung oder Erniedrigung einzelner Töne notwendig, so daß bereits Philippe de Vitry im 14. Jahrhundert die *musica ficta* als „musica vera et necessaria" bezeichnete.[11] Die bestehende Praxis forderte damit eine theoretische Fundierung, um einen eklatanten Widerspruch zweier für die Praxis gleichermaßen gültiger Bezugssysteme – des Tonsystems, durch das sie als falsch, und der Kontrapunktlehre, durch die sie als notwendig eingestuft wurde – zu vermeiden. Die Deutung, die Philippe de Vitry anbot, ist exemplarisch auch für die spätere Zeit[12]:

„Est ficta musica quando de tono facimus semitonium, et e converso de semitonio tonum. Omnis enim tonus est divisibilis in duo semitonia, et per consequens signa semitonia designantia in omnibus tonis possunt applicari. ... Ubi igitur invenimus b rotundum, dicimus istam vocem fa, et ubi invenimus h quadratum, dicimus illam vocem mi."[13]

---

10 Demnach steht eine Komposition wie Josquins „Fortuna d'un gran tempo", in der der einen Stimme nichts, der zweiten ein b, der dritten zwei b vorgezeichnet sind, keineswegs außerhalb des Hexachordsystems (zum Doppel-b-System und satztechnischen Fragen s. u.).
Eine andere wichtige und für den heutigen Leser oftmals verwirrende Konsequenz aus der Einbindung von b und h in das Hexachordsystem war die Verwendung der aus den beiden abgeleiteten Vorzeichen als Solmisationszeichen: b bezeichnete eine fa-Stufe, h (d. h. das heutige Auflösungszeichen bzw. Kreuz, zwischen beiden wurde bis ins 15. Jahrhundert nicht streng unterschieden) eine mi-Stufe, so daß das im Notenbild Erscheinende nicht notwendig mit der Tonhöhe übereinstimmt.
11 Vgl. de Vitry (1320), S. 18.
12 Vgl. dazu Rempp (1989), S. 78ff.
13 De Vitry (1320), S. 26. („Es handelt sich um musica ficta, wenn wir aus einem Ganzton einen Halbton machen, und umgekehrt aus einem Halbton einen Ganzton. Denn jeder Ton ist teilbar in zwei Halbtöne, und

Tatsächlich bietet de Vitry zwei Erklärungen an, die Vertauschung der Halbton-Ganzton-Folge und die Änderung der Solmisation. Beide zielen auf eine Hexachordtransposition ab, wobei die erste eher einen lokal begrenzten Charakter einer solchen Transposition nahelegt, denn da die Alterierung eines Tones nur die Vertauschung der ihn eingrenzenden Intervalle bedeutet, bleibt der Einfluß der Alterierung auf die unmittelbar benachbarten Töne beschränkt. Außerdem schlägt sie eine Brücke zur Musiktheorie: Da jeder Ganztonschritt teilbar war, blieb die *musica ficta* durch Zahlenproportionen faßbar.[14] Die zweite Erklärung steht dagegen im Zusammenhang mit Bemühungen, die akzidentiellen Töne durch die Einrichtung von Hexachorden über anderen als den drei bekannten Tönen C, F und G zu erklären. In einem anonymen Traktat des 15. Jahrhunderts etwa werden von jenen drei Hexachorden vier weitere über A, B, D und E[s] abgegrenzt, die als *musica ficta* bezeichnet werden[15]: Hexachorde an irregulären, letztlich irrealen Orten. Der Status der Transposition war ein abgeleiteter und sekundärer; im Gegensatz zum Dur-Moll-System, in dem die Transposition einen essentiellen Bestandteil bildet und in der Gleichberechtigung aller Tonarten auch die Gleichberechtigung aller 12 Stufen angelegt ist, war im Hexachordsystem der Vorrang der Diatonik ungebrochen.

Im Prinzip ergibt sich aus der von de Vitry genannten Möglichkeit[16], jede diatonische Stufe als fa oder mi zu solmisieren, eine doppelte Teilung aller diatonischen Ganztöne: Aus c, d, f, g und a als mi-Stufen (e und h sind bereits natürliche mi-Stufen) entstehen cis, dis, fis, gis und ais; aus d, e, g und a als fa-Stufen (b,f und c sind natürliche fa–Stufen) entstehen des, es, ges und as. Zu jeder diatonischen Stufe wird ein oberer und ein unterer Halbton gebildet (die, da es sich jeweils um große Halbtöne handelt, mit denen der benachbarten Stufen nicht zusammenfallen), so daß sich ein (fiktives) 17-stufiges Tonsystem, nämlich die Skala c–cis–des–d–dis–es–e–f–fis–ges–g–gis–as–a–ais–b–h, ergibt. In dieser Skala besaßen selbstverständlich nur die Halbtonschritte c–des bzw. cis–d (usw.) Realität; weder c–cis noch gar cis–des waren existente Intervalle.

Die Hexachordlehre entstand als ein Regelwerk, das die musikalische Praxis erfassen, regulieren, anleiten und leichter lehrbar machen sollte. Sie bezog sich anfangs auf die siebenstufige diatonische Leiter und blieb damit in dem Rahmen, den die Musiktheorie als mathematisch bestimmbar vorgab. Der kurze Blick auf die spätere Form der Hexachordlehre und vor allem auf die musica ficta, die, wie gesehen, aufs engste mit der Hexachordlehre verknüpft war, hat gezeigt, daß dieses praktische Regelwerk eine autonome Entwicklung durchlief, die es in ein Spannungsverhältnis zu den Vorgaben der Theorie brachte, da sie zu einer Skala führten, die im Rahmen der pythagoreischen Stimmung als ein in sich stimmiges Tonsystem nicht mehr darstellbar war. Diese Spannungen führten jedoch bis zum 16. Jahrhun-

---

folglich können die Zeichen, die Halbtöne bezeichnen, bei allen Ganztönen angewendet werden ... Wo wir daher ein b finden, nennen wir diese Stufe fa, und wo wir ein h [Auflösungszeichen] finden, nennen wir diese Stufe mi.")

14 Da es sich nur um ein lokales, vorübergehendes Phänomen handelt, war es nicht nötig, die gesamte Skala auf die neuen Töne abzustimmen derart, daß alle neu entstehenden Quinten und Quarten rein waren – die meisten dieser Intervallbeziehungen bestanden für die Komponisten faktisch nicht.

15 Anonymus I de Lafage, zitiert in Dahlhaus (1968a), S. 208 und (1968b), S. 165; dort deutet Dahlhaus das im Text angegebene E als Es (die beiden Hexachorde ergeben bezeichnenderweise die Stufe gis/as). – Noch einen Schritt weiter ging John Hothby, der darüberhinaus Hexachorde auf Fis, As, H, Des und E gestattete (vgl. Rempp (1989), S. 81).

16 Vgl. Dahlhaus (1968a), S. 163ff.

dert weder zu einem Bruch noch zu einer Revision der Theorie. Dies dürfte verschiedene Gründe gehabt haben: Das erweiterte Tonsystem stand den Komponisten nur jeweils in kleinen Ausschnitten, nie insgesamt zur Verfügung; Stimmungsprobleme konnten in vokaler Musik flexibel gehandhabt werden; das theoretische Fundament war ohne Alternative und daher nicht anfechtbar; die Kritik der Theorie und polemische Debatten, wie sie im 16. Jahrhundert gepflegt wurden, waren über lange Zeit noch nicht üblich. Diese Faktoren änderten sich im 16. Jahrhundert.[17]

---

17 Ein anderes, sehr interessantes Beispiel für das Wechselverhältnis von Theorie und Praxis im Mittelalter ist die Lehre von der musikalischen Zeitgestaltung. Der Zeitverlauf im einstimmigen Choral folgte im wesentlichen der Deklamation des Textes; es gab daher keine spezifisch musikalische Zeitgestaltung. Das Bedürfnis nach einer solchen kam erst mit der Mehrstimmigkeit, also der Gleichzeitigkeit mehrerer, voneinander unabhängiger Stimmen auf. Anfangs zog man hierzu die aus der Sprachmetrik stammende Unterscheidung einer Länge (longa) und einer Kürze (brevis) heran, die auf eine fundamentale Zeiteinheit, das tempus, bezogen wurden und deren Verhältnis so auf 2 : 1 festgelegt wurde. Aus der Länge und der Kürze konnten additiv verschiedene rhythmische Modi gebildet werden. (Diese Darstellung gibt die Systematisierung durch Johannes de Garlandia aus der Mitte des 13. Jahrhunderts wieder, die am Ende der Entwicklung der Modalnotation steht und insofern vermutlich nicht die historische Genese nachzeichnet. Vgl. dazu Walter (1994), S. 155 sowie seine ausführliche Darstellung der Lehre Garlandias S. 167-196, der ich hier folge.) Kürzere Tondauern als die brevis kamen in der Musik zwar vor, wurden aber nicht mit Hilfe von Zahlverhältnissen reglementiert, sondern lediglich jeweils zu einer brevis, d. h. einem tempus, zusammengefaßt. Denn die Vorstellung einer durchgängigen hierarchischen Gliederung der musikalischen Zeit durch mathematische Proportionen existierte zu diesem Zeitpunkt noch nicht. Die Grundlage hierfür wurde von Franco von Köln in seinem Traktat „De musica mensurabilis" (1280) gelegt. Anstelle der Grammatik wählte Franco als theoretischen Bezugsrahmen die Proportionenlehre, die aus der Theorie des Tonsystems bekannt war. In diesem Kontext war die Drei als perfekte Zahl ausgezeichnet; folglich wurde jetzt die Proportion 3 : 1 zum tragenden Baustein (vgl. Franco (1280), S. 36). Franco blieb noch unsicher bezüglich der Organisation von Notenwerten, die kürzer als eine brevis waren. Eine vollständige Systematisierung findet sich etwa fünfzig Jahre später in de Muris' „Notita artis musicae" (1321): Eine maxima longa dauerte drei longae perfectae, eine longa perfecta drei breves perfectae, eine brevis perfecta drei semibreves perfectae, eine semibrevis drei minima – was die Proportionenreihe 81 : 27 : 9 : 3 : 1 ergibt (vgl. de Muris (1321), S. 79). Binäre Proportionen wurden in der Musik natürlich weiterhin verwendet, wurden aber als „imperfekt" abqualifiziert. Die strenge Reglementierung mit Hilfe von Zahlenverhältnissen erlaubte aber auch ihre präzise Beschreibung (eine „zweizeitige" longa war eben 18 minima lang; vgl. l. c.), so daß die gleichberechtigte Behandlung von pefekten und imperfekten Verhältnissen von Tondauern in dem etwa dreißig Jahre nach der „Notitia" ebenfalls von Johannes de Muris verfaßten „Libellus cantus mensurabilis" nurmehr eine Glättung verbliebener Unebenheiten im System der Mensuralmusik darestellte. Der „Libellus cantus mensurabilis" erlangte bald kanonischen Rang; das dort beschriebene System blieb in wesentlichen Punkten bis ins 16. Jahrhundert gültig. Denn die Eingliederung der Zeitorganisation in das allgemeine wissenschaftliche Beschreibungssystem, die Proportionenlehre, schaffte bis dahin unbekannte Variationsmöglichkeiten in der Kombination mehrerer Stimmen und war daher eine Voraussetzung für den großen Aufschwung der mehrstimmigen Musik seit dem 14. Jahrhundert. Es ist also auch hier kaum hilfreich, die Entwicklung der musikalischen Zeitorganisation im Mittelalter mit Hilfe einer Entgegensetzung von unreglementierter, aber fortschrittlicher Praxis und hemmender Theorie, die sich an musikfernen Prinzipien orientiert, zu beschreiben. Tatsächlich waren die Verhältnisse noch weitaus komplizierter als hier umrissen. Erstens war die Lehre von der Zeitorganisation eng mit der Entwicklung der Notenschrift verbunden, so daß Traktate über die musica mensurabilis auch Anleitungen zum richtigen Notenlesen waren bzw. Anleitungen zum Lesen der tatsächlich notierten Werke geben mußten. Zweitens blieb die Lehre trotz ihres einfachen Gerüsts mit einer erheblichen Unklarheit belastet: Für zwei- und dreizeitige Notenwerte stand jeweils nur ein Zeichen zur Verfügung, und es mußte durch Hilfszeichen angegeben werden, welche Proportion gemeint war; das schränkte die Kombinierbarkeit von Notenwerten natürlich erheblich ein. Diese Unklarheit in der Notation wurde möglicherweise deshalb nicht so sehr als ein theoretisches Problem angesehen, weil die Notenwerte, gleich ob drei- oder zweizeitig, als *eine* Spezies betrachtet wurden und somit einheitlich theoretisch behandelt

3. Zur Musiktheorie im 16. Jahrhundert

Überblickt man aus der Distanz die musiktheoretische Literatur des 16. Jahrhunderts, ergibt sich ein irritierendes Bild. Auf der einen Seite hat sich das äußerliche Erscheinungsbild deutlich verändert: Die Traktate werden umfangreicher und zu Fundgruben der Gelehrsamkeit; neu verfügbar gewordene antike Texte werden ausführlich kommentiert, interpretiert und diskutiert, um die antike Musik zu rekonstruieren; immer wieder brechen polemische Debatten auf, deren Heftigkeit in keinem nachvollziehbaren Verhältnis zur Bedeutung der strittigen Inhalte steht. Es scheint, kurz gesagt, daß die Musiktheorie sich aus einem Zustand der Trägheit, sogar Starre im Kern, zu einer sehr beweglichen Disziplin gewandelt hat, in der nichts davor sicher ist, in Frage gestellt zu werden. Auf der anderen Seite dominieren das Gerüst der mittelalterlichen Theorie und ihre Kerndoktrinen weiterhin, bis hin zu Zarlinos monumentalen und äußerst einflußreichen „Istitutioni harmoniche", die erstmals 1558 veröffentlicht wurden und bereits bis 1573 vier Neuauflagen erfuhren. Einen eindeutigen Trend hin zu einer neuen Theorie, gar zu einem neuen Konzept von Theorie, gibt es nicht. Diese Situation stellt den Historiker vor eine schwierige Aufgabe. Veränderungen machen sich in Details, häufig nur in Akzentverschiebungen bemerkbar, sie verbinden sich mit alten Vorstellungen und sind von ihnen nicht abzulösen; ihre Bedeutung ist selbst dann, wenn sie weitreichende Implikationen haben, nicht an langfristigen Entwicklungen ablesbar, weil sie solche Entwicklungen nicht initiiert haben.

Ich will versuchen, diese Lage der Musiktheorie anhand einiger weniger Schlaglichter etwas deutlicher zu machen. Gut geeignet hierfür ist die Einstufung von Terzen und Sexten als Konsonanzen. In der mittelalterlichen Theorie galten sie als Dissonanzen, weil ihre Proportionen nicht mit den Zahlen von 1 bis 4 darstellbar waren; in der (wenigstens in der Theorie) allein gültigen pythagoreischen Stimmung wurden sie durch Proportionen gebildet, die in der Tat dissonante Intervalle erzeugten. In der Kompositionspraxis zählten sie dagegen längst zu den Konsonanzen, da in der vierstimmigen Musik des 15. Jahrhunderts Kadenzen oft auf Dreiklängen endeten, die auch Terzen und Sexten einschlossen. Ein Vergleich der Lösungen, die dieses Problem bei Zarlino und bei Lodovico Fogliani, einem anderen prominenten Musiktheoretiker des 16. Jahrhunderts, erfuhr, ist aufschlußreich.

Zarlino eröffnet seine „Istitutioni harmoniche" mit einer Darstellung des wissenschaftstheoretischen Hintergrund seiner Version der Musiktheorie, die dem verblüffend ähnlich ist, was bereits bei de Muris in rudimentärer Form anzutreffen war. Jede Erkenntnis habe ihren Ursprung in den Sinnen, die Musiktheorie folglich den ihren im Gehörsinn; eine andere Quelle der Erkenntnis (cognitione) existiere nicht. Als Beleg für den ausgezeichneten Rang des Gehörsinnes führt Zarlino wieder die Schmiedehammerlegende an, und wieder dient sie dazu, den Übergang von der Sinneswahrnehmung zur (grundsätzlich verschiedenen) Vernunfteinsicht zu beschreiben: Motivation für die Entdeckung der Zahlenproportionen war

---

werden konnten (vgl. de Muris (1321), S. 66 und dazu Gallo (1984), S. 276). Drittens spielen in die Entwicklung der Theorie und der kompositorischen Praxis weitaus mehr Faktoren hinein als hier erwähnt. Viertens gab es auch abweichende nationale Sonderentwicklungen, vor allem in Italien, auch wenn die hier genannten Texte von Franco von Köln und Johannes de Muris in ganz Europa sehr verbreitet waren (vgl. Gallo (1984), S. 267 zu Franco und 300ff. zum „Libellus cantus mensurabilis"). Zur Entwicklung der musikalischen Zeitorganisation und Notationslehre generell siehe Gallo (1984) und Dahlhaus (1986).

der Wohlklang der Hämmer („gli movea l'udito con dilettatione") – Zarlino betont das sinnliche Wohlgefallen, um zu zeigen, daß Wahrnehmung und Verstandeserkenntnis in einem harmonischen Verhältnis zueinander stehen. Aus der Entdeckung der Zahlenverhältnisse – sie seien „evidentissime" – sei eine „perfetta e certa scienza" entstanden, die aufgrund ihrer Gewißheit Teil der mathematischen Wissenschaften geworden sei: eine „capacità di verità delle cose che sono e di loro natura non sono mutabili".[18] Das Problem der Terzen und Sexten löst Zarlino nun, indem er an Stelle Zahlen 1 bis 4 den „senario", also die Zahlen 1 bis 6, zum Prinzip der Musiktheorie erklärt (durch die Zahlen 5 und 6 lassen sich auch Terzen und Sexten darstellen). Ausführlich legt Zarlino dar, warum der *senario* diese ausgezeichnete Rolle spielt: Die Zahl 6 ist die kleinste perfekte Zahl; von den zwölf Zeichen des Tierkreises sehen wir immer nur sechs gleichzeitig in unserer Hemisphäre; Platon unterscheidet sechs Positionen im Raum, usw. – die Zahl sechs ist also überall in der Natur anzutreffen, folglich ein Prinzip der Schöpfung und somit auch Grundlage der Musik.[19] Zarlino reagiert also auf das Problem, das durch die veränderte Kompositionspraxis gestellt wurde, indem er das Begründungsmuster der mittelalterlichen Musiktheorie beibehält. Er tauscht lediglich das Prinzip aus und reichert die Argumentation durch eine beträchtlich gesteigertes Maß an Gelehrsamkeit an.

30 Jahre vor den „Istitutioni harmoniche" veröffentlichte Fogliani seine „Musica theorica", deren Aufbau gleichfalls an mittelalterliche Traktate erinnert. Fogliani beginnt mit der Bestimmung des „subjectum musicae" als „numerus sonorus" und ordnet der Musik eine Mittelstellung zwischen Mathematik und Naturphilosophie zu, denn sie hat nicht allein mit Zahlen, sondern auch mit dem „sonus", der durch die Bewegung definiert wird, zu tun.[20] Daraufhin stellt er ausführlich die Proportionenlehre sowie die sich aus den Proportionen ergebenden Intervalle und ihre Eigenschaften dar; schließlich zeigt er, wie ein Tonsystem aufgebaut werden kann. Im Unterschied zu Zarlino und den meisten seiner mittelalterlichen Vorläufer kommt es Fogliani jedoch auf die Differenz zwischen numerus und sonus bzw. forma und materia an: Das *corpus sonorum*, die natürliche Grundlage der Musik, stelle ein Kontinuum dar und bleibe dies auch, so daß der (diskrete) *numerus* nicht die „causa intrinseca" der Musik sei[21]: „Quia omnis proportio quae invenitur in discretis invenitur in continuis licet non e converso: erit igitur numerus causa tantum cognoscendi et artificialiter inveniendi proportiones consonantiarum ... quas (= proportiones) tamen sine usu numeri in continuis invenire omnino foret impossibile: unde in musicis adeo valent numeri".[22] Diese deutliche Einschränkung der Bedeutung der Zahlenproportionen – ihnen wird der ontologische Rang abgesprochen – gestattet Fogliani ungewöhnliche Folgerungen: Die ehrwürdige Beschrän-

---

18 „Eine Fähigkeit der Dinge, die sind und die aufgrund ihrer Natur unveränderlich sind." Dieses und alle vorangehenden Zitate Zarlino (1573), S. 6.
19 Vgl. Zarlino (1573), S. 28ff.
20 Fogliani (1529), fol. 1r.
21 Zarlino übernimmt zwar die Behauptung, daß die Zahl nicht „causa intrinseca" sei, setzt an ihre Stelle jedoch die Proportion: „La proportione adunque è la causa formale, intrinseca e principesca delle consonanze, e il numero è la causa universale, estrinseca e remota" (Zarlino (1573), S. 64).
22 Fogliani (1529), fol. 14v - 15r. („Da jede Proportion, die zwischen diskreten Größen anzutreffen ist, auch in einem Kontinuum gefunden wird, das umgekehrte jedoch nicht gilt, wird die Zahl nur die Ursache des Erkennens und der künstlichen Erfindung der Proportionen der Konsonanzen sein ... Da es unmöglich wäre, die Konsonanzen ohne den Gebrauch der Zahlen zu finden, deshalb gelten die Zahlen in den Dingen der Musik so viel.")

kung der Zahl der Konsonanzen auf fünf weist er kurzerhand zurück, da sie der sinnlichen Erfahrung widerspreche („Sed heac positio licet maxima innitatur auctoritate nihilominus mihi videtur falsa: quum sensui contradicat"[23]) und es keinen Grund für die Bevorzugung bestimmter Zahlen oder Proportionen gebe; Kriterium der Konsonanz wird die Klangverschmelzung.[24] Letzten Endes kommt Fogliani nicht zu einem anderen Resultat als Zarlino; doch schlägt er einen Weg ein, der ihn, für kurze Zeit jedenfalls, weit von den Grundlagen der mathematischen Musiktheorie wegführt.

Wenn Terzen und Sexten als Konsonanzen gelten, kann dies nicht ohne Folgen für die adäquate Stimmung der diatonischen Skala bleiben. Die pythagoreische Stimmung (s. o.) kam für Fogliani nicht mehr in Betracht, da sie den Terzen und Sexten nicht ihre reinen Proportionen zuwies. An ihrer Stelle favorisierte Fogliani die „reine Stimmung", die die reinen Proportionen für alle Konsonanzen beachtete und „in nudis purisque numeris ratione tantum dirigente sensu derelicto" war.[25] Für eine Stimmung, in der nicht nur Oktave, Quinten und Quarten, sondern auch Terzen und Sexten tatsächlich als Konsonanzen erklangen, war jedoch ein hoher Preis zu zahlen, denn Fogliani war gezwungen, zwei Tonstufen zu verdoppeln, wie in folgender Darstellung leicht zu sehen ist:

C	Des	D	D	Es	E	F	Ges	G	As	A	B	B	H	C[26]
3600	3456	3240	3200	3000	2880	2700	2592	2400	2304	2160	2025	2000	1920	1800

$$25{:}24 \quad 16{:}15 \quad 81{:}80 \quad 16{:}15 \quad 25{:}24 \quad 16{:}15 \quad 25{:}24 \quad 27{:}25 \quad 25{:}24 \quad 16{:}15 \quad 16{:}15 \quad 81{:}80 \quad 25{:}24 \quad 16{:}15$$

Die Verdopplung der Stufen D und B war jedoch wenigstens für diejenigen Musiker, die Instrumente mit festgelegten Tonhöhen wie Clavichord oder Cembalo verwenden, inakzeptabel. Daher schlägt Fogliani ein geometrisches Verfahren zur Halbierung des Kommas (d. h. des Intervalls 81 : 80) zwischen den beiden D bzw. B vor, ausgehend von der Erfahrung, daß das Ohr eine Abweichung von der reinen Proportion um ein halbes Komma noch toleriert. Er gelangt zum Entwurf einer temperierten Stimmung.[27]

Foglianis Diskussion der reinen Stimmung und ihrer Temperierung gibt einen wichtigen Hinweis auf einen Veränderung im Status der Theorie. Einerseits ist für Fogliani nach wie vor nur dasjenige Gegenstand der Theorie, was sich mit Hilfe von Zahlen erfassen läßt, weil nur die Zahlen eine wissenschaftliche Einteilung und Erklärung gestatten. Deshalb merkt er an, daß die Temperierung eigentlich nicht Gegenstand des Theoretikers ist.[28] Andererseits ist sich Fogliani darüber im klaren, daß sein Verfahren der Temperierung selbst ein wissenschaftliches Verfahren ist, das dem Temperieren durch Ausprobieren der „Praktiker" überlegen ist:

---

23 l. c., fol. 11v. („Aber diese Position scheint falsch, obwohl sie im höchsten Maße durch die Autorität gestützt wird: denn sie widerspricht dem Sinn").

24 l. c., fol. 15r („consonantia est … auribus amica commixtio"), 17v.

25 l. c., fol. 33r („… in nackten und reinen Zahlen unter der Leitung des Verstandes allein, nachdem die Wahrnehmung aufgegeben worden ist").

26 Vgl. l. c., fol. 35v. Ein generelles arithmetisches Verfahrung zur Halbierung von Intervallen gibt es nicht, solange man im Bereich der rationalen Zahlen bleibt, da die Hälfte eines Intervalls der Proportion x : y durch den Quotienten der Wurzeln von x und y gebildet wird und Wurzeln meistens irrationale Zahlen sind. Fogliani beruft sich für sein geometrisches Teilungsverfahren auf Euklid (l. c., fol. 36). Die Halbierung ist möglich, weil das Material der Musik ein Kontinuum ist (l. c., fol. 35v.).

27 Fogliani war übrigens nicht der erste, der eine Temperierung erörterte, sondern Ramos de Pareja in seiner

„At musici practici, quum nihil omnino sciunt de commate, nec quod talis augmentatio vel diminuitio fiat per dimidietatem commatis, *quum talia scire & speculari sit proprium Theorici musici*, quomodo per sensum venantur illud medium punctum participationis."[29]

In diesem Zitat kündigt sich ein neues, „technisches" Verständnis von Theorie an. Die Theorie erhält ihre Aufgabe, das Material, das sie zu erfassen hat, von der Natur bzw. der musikalischen Erfahrung; ihr Ziel besteht darin, dieses Material handhabbar zu machen – nämlich, ein optimales Verfahren zur Herstellung einer Stimmung der Skala zu entwickeln, das in der musikalischen Praxis verwendet werden kann. Dieses technische Verständnis steht quer zur tradierten Unterscheidung von wissenschaftlichem und handwerklichem Wissen von der Musik: Es nutzt Verfahren – nämlich mathematische –, die in ersterem beheimatet sind, entspricht in seiner Orientierung aber dem zweiten.

Fogliani legt sich auf keines der beiden Theoriekonzepte fest. Er hätte vermutlich das erste favorisiert; der große Raum, den er Verfahren der Temperierung widmet, spricht jedoch dafür, daß sein Interesse auch beim zweiten Fuß gefaßt hat. Der Status der Theorie bleibt so in der Schwebe zwischen der Auffassung, daß Musik ein natürliches Phänomen ist, das wissenschaftlich erklärt werden kann und dessen wissenschaftliche Erklärung zu Normen für die artifizielle Herstellung von Tönen führt, und der Auffassung, daß Musik etwas ist, das hergestellt wird und dessen Herstellung auf eine möglichst „wissenschaftliche" Weise erfolgen sollte. Dieser Zwiespalt findet sich bei vielen Musiktheoretikern des 16. Jahrhunderts, und er fällt je nach persönlichem Temperament des Autors mehr zugunsten des Versuchs aus, zwischen beiden Zweigen zu moderieren, oder zugunsten der Betonung des Konfliktpotentials.

Zur ersten Gruppe zählt ohne Zweifel Zarlino.[30] Seinen Versuch, der neuen Rolle von Terzen und Sexten gerecht zu werden, habe ich bereits erwähnt. Auch Zarlino plädiert entsprechend für die reine Stimmung; auch er erkennt, daß reine Stimmungen auf Instrumenten mit fester Tonhöhe nicht zu realisieren sind; auch er diskutiert Verfahren der Temperierung und kommt dabei sogar zu Ergebnissen, die einen Fortschritt gegenüber Fogliani darstellen. Für ihn ficht all dies aber nicht die Überzeugung an, daß die Musik ein von Natur aus durch einfache Zahlenverhältnisse geordnetes System sei. Temperierungen sind eine Unvollkommenheit, die nur der Verwendung von künstlichen Instrumenten geschuldet ist. Dort, wo die Natur mehr zu ihrem Recht kommt, in der Vokalmusik nämlich, bleibt die reine Stimmung

---

„Musica practica" von 1482. Für einen Überblick über diese Diskussion im 16. Jahrhundert vgl. Lindley (1986), S. 130ff.

28 l. c., fol. 35v: „Hanc vero participationem (= Temperierung), si forte musicus speculativus demonstrative cognoscere contendat, id nisi transgrediatur limites nequaquam consequi poterit; & ratio est, quia haec participatio non es de eius consideratione…"

29 l. c., fol. 35v. (Hervorhebung A. B.) („Aber die praktischen Musiker dürften irgendwie durch den Sinn zu jedem mittleren Punkt der Temperierung kommen, weil sie überhaupt nichts vom Komma wissen, noch, daß eine solche Vergrößerung oder Verringerung durch die Teilung des Kommas geschieht, denn solche Dinge zu wissen und darüber nachzudenken, ist Sache des theoretischen Musikers.").

30 Zu Zarlinos Bemühungen, sich gleichermaßen als Praktiker (Komponist) wie als gelehrter Theoretiker zu etablieren und beide Bereiche äußerlich zusammenzuführen, vgl. auch Judd (2000), S. 184ff. Ein solches Image war offenbar wichtig, um die attraktive Position des Kappelmeisters von San Marco in Vendig zu erlangen, die Zarlino sei 1565 als Nachfolger seines Lehrer Willart auch einnahm.

die einzig richtige.[31] So kann die Theorie weiterhin zugleich beschreiben, was in der Natur vorliegt, und die Grundlage für die musikalische Praxis liefern.

Der prominenteste Vertreter der zweiten Gruppe ist Vincenzo Galilei, der sich mit Zarlino gegen Ende des 16. Jahrhunderts heftige publizistische Schlachten lieferte. Wieder ging es vor allem um die korrekte Stimmung. Zarlino hatte in den 1588 erschienenen „Sopplimenti musicali" erneut betont, daß die reine Stimmung die natürliche Stimmung sei, da sie allein die wahren Formen der Intervalle verwirkliche, und jede artifizielle Stimmung demgegenüber mangelhaft sei. Darauf antwortet Galilei in seinem „Discorso intorno alle opere di Gioseffo Zarlino" von 1589: Natur und Kunst seien gleichwertig; beide seien „cause effizienti" und der Perfektion in ihrem „genere" fähig[32]. Vor allem aber gebe es überhaupt keine natürlichen Stimmung: Da „il subietto della Musica che è la voce & il suono" eine „quantità continua & non discreta" sei[33], ist jede Stimmung ein ganz und gar künstliches System („la natura poi per non havere ne mani ne bocca non gli è conceduto ch' ella suoni ne ch' ella canti, et il sonare & il cantar' nostro è tutt' arte"[34]). Die Natur stellt ein qualitätsloses, folglich beliebig quantifizierbares und neutrales Betätigungsfeld dar: „la Natura nelle sue operationi non ha rispetto a questo o a quell' altro nostro comodo & fine, perche opera senza cognitione ... che quella che noi cantiamo non è compresa dalla Sesquialtera questo non importa alla Natura piu che gli importa che una cornacchia o un corvo viva trecento e quattro cento anni, et un huomo viva solo cinquanta & sessanto; ne di ciò merita esser la Natura ripresa...".[35] Die Natur liefert nur die „materia", die „forma" ist dagegen Produkt der Kunst.[36]

Galileis Polemik gegen die Grundlagen von Zarlinos Theorieverständnis rührt unter anderem aus einigen Experimenten her, die Galilei zu der Einsicht gebracht hatten, daß die

---

31 Vgl. Zarlino (1573), S. 148: „Ma se tali inconvenienti ... si trovano negli Istrumenti arteficiali, nondimeno tra le Voci ... non si trovano tali rispetti: essendo che riducono ogni cosa nella sua perfettione, come è il dovere: poiche la Natura, nel fare le cose, è molto superiore all`Arte, & questa nello imitare fa ogni cosa imperfetta." Zarlinos Optimismus bezüglich der reinen Stimmung wurde allerdings bereits in den sechziger Jahren des 16. Jahrhunderts (also kurz nach der Veröffentlichung der ersten Auflage der „Istitutioni harmoniche") durch den Mathematiker Giovanni Battista Benedetti getrübt, der in zwei Briefen an den Komponisten Cipriano de Rore zeigte, daß die getreue Einhaltung der harmonischen Stimmung zu dramatischen Veränderungen in der absoluten Höhe der einzelnen Tonstufen führen kann.

32 Galilei (1589), S. 93.

33 l. c., S. 113. („Der Gegenstand der Musik, der Stimme und Ton ist, ist eine kontinuierliche und nicht eine diskrete Größe.")

34 l. c., S. 20f. („Da die Natur weder Hände noch Mund hat, ist es ihr nicht gewährt, zu spielen oder zu singen, und unser Spielen und Singen ist gänzlich Kunst".) Die daraus folgende Möglichkeit, jedes Intervall in beliebig viele gleich große Abschnitte zu teilen, erlaubt Galilei auch, sich nachdrücklich für das Aristoxenische „sistema incitato" (ein System, in dem der Ganzton in zwei gleiche Hälften geteilt wird, so daß sich eine temperierte zwölfstufige Skala ergibt) einzusetzen; in ihm sei nichts Irrationales zu finden (!) (S. 109ff.).

35 l. c., S. 116f. („Die Natur nimmt in ihrem Wirken keine Rücksicht auf diesen oder jenen Vorteil oder Ziel von uns, denn sie operiert ohne Verstand... ob das, was wir singen, in der Proportion der Sesquialtera ist oder nicht, das schert die Natur nicht mehr als es sie schert, daß eine Krähe oder ein Rabe drei– oder vierhundert Jahre lebt, ein Mensch aber nur fünfzig oder sechzig"). Ebenso S. 93: „È tanto naturale il Ditono contenuto dalla super 17 partiente 64, si come è naturale l'accordare dell' ottava drento la dupla quanto è naturale il dissonare della settima drento la super 4 partiente quinta." Konsequent weist Galilei auch die Auffassung zurück, beim Gesang würden natürlicherweise die reinen Proportionen der Intervalle gesungen: „Le consonanze non son prodotte dalle voci nelle lor vere forme naturalmente; ma artifizialmente per la lunga prattica appresa dall' arte del ben cantare" (S. 84).

36 l.c., S. 79f.

berühmte Schmiedehammerlegende (deren Scharnierfunktion in der Musiktheorie bereits erwähnt wurde) falsch ist. Galilei hatte nämlich festgestellt, daß Gewichte, die zwei Saiten spannen, im Verhältnis 4 : 1 stehen müssen, um eine Oktave zu erzeugen, während Pythagoras doch angeblich herausgefunden habe, daß die Gewichte der Hämmer, die eine Oktave erzeugen, im Verhältnis 2 : 1 stehen.[37] Das Volumen von Körpern wie Orgelpfeifen muß sogar im Verhältnis 8 : 1 stehen. In seinem „Discorso particolare intorno alla diversità delle forme del diapason" zieht Galilei daraus den Schluß, daß Formen weder ontologisch noch epistemisch primär sein können, wenn der gleiche hörbare Effekt, nämlich die Oktave, durch verschiedene Formen (d. h. Proportionen) zu erzielen ist. Welche Form jeweils die Oktave realisiert, hängt vom zur Klangerzeugung gewählten Gegenstand oder Material ab – es könnte sich sogar um irrationale Zahlen handeln, denn schließlich produziert auch die Proportion 2 : 1 in einigen Gegenständen bzw. Materialien dissonante Intervalle.[38] Allerdings gelangt auch Galilei nicht zur Formulierung einer kohärenten Gegenposition zum traditionellen Theoriekonzept. Am Ende seines „Discorso" spricht er wieder von einer „vera e perfetta distributione", bei der es sich um eine Stimmung mit beweglichen Tonstufen handeln soll, in der die Intervalle in ihrer „suprema perfettione" wahrgenommen werden können.[39] So sehr Galilei Kunst und Natur, und damit das technische und das spekulative Theoriekonzept, voneinander trennt, am Ende möchte auch er nicht darauf verzichten, der Kunst natürliche Prinzipien zugrundezulegen.

Ungeachtet solcher Schwankungen kann die deutliche Trennung von Kunst und Natur bei Galilei als Vorbereitung und insofern als weiteres Indiz für ein „technisches" Theorieverständnis gewertet werden. Immerhin war Galilei ein bekannter Lautenist und so unmittelbar mit den Problemen „natürlicher" Stimmungen (wie der pythagoreischen oder der reinen Stimmung) konfroniert.

Ein ganz anderer Aspekt dieses Verständnisses geht aus den Schriften Pietro Pontios hervor, dem „Ragionamento di musica" von 1588 und dem „Dialogo" von 1595, die – in der Gestalt von Dialogen zwischen einem Experten und einem bzw. zwei Laien – an den adligen Dilettanten gerichtet sind, der sein Urteilsvermögen in Sachen Musik ausbilden möchte.[40] Offiziell nimmt Pontio eher eine moderierende Position ein.[41] Die Musik, um die es eigentlich geht und über die der adlige Laie informiert werden möchte, ist aber allein die Kunstmusik, die er täglich hören kann: Es ist nicht mehr der numerus sonorus, sondern „una modulatione della voce *fatta* con ragione".[42] Informiert wird er zunächst durch eine knappe Darstellung des Lehrbestands der alten Musiktheorie, an die sich eine sehr viel ausführlichere Dar-

---

37 Vgl. Galilei (o. J. a.). Galilei erwähnt diese Experimente und die daraus folgende Fehlerhaftigkeit der Schmiedelegende bereits im „Discorso intorno alle opere di Gioseffo Zarlino" (l. c., S. 103f.), dort allerdings eher beiläufig und ohne Folgerungen daraus zu ziehen.

38 Vgl. Galilei, Discorso intorno alla diversità delle forme del diapason, fol. 47r und 53v, Anmerkung: „... sicome ho detto che la dupla contiene tra alcuni corpi intervalli dissonantissimi, non sarebbe meraviglia alcuna che tra altri si ritrovasse non solo in proportione superpartiente ma tra numeri et parti inrationali si come io spero di dimostrare."

39 l. c., S. 123. S. 129 spricht Galilei sogar davon, daß die „vera forma della Quinta sia la Sesquialtera"; er setzt dies bezeichnenderweise jedoch nicht dogmatisch voraus, sondern beschreibt diese Erkenntnis als einen allmählichen Annäherungsprozeß.

40 Vgl. Pontio (1588), Rückseite des Titelblatts: „La presente mia factica ... habbino anco da me coloro, che vogliono di Musica sapere, onde possano con facilità trovare un picciolo lume, che gli mostri la strada".

41 Vgl. l .c., S. 10: „... il vero Musico sarà quello, che dalla Pratica, & della Specolativa havrà cognitione."

stellung der Satz- und Kompositionslehre anschließt.[43] Im „Dialogo" wird von einem der Laien ausdrücklich die Frage gestellt, ob die Kenntnis der Theorie ausreiche, um ein guter Komponist zu werden, und sie wird vom Experten verneint, denn die Theorie lasse zu viele Themen – wie die kompositorische Behandlung von Dissonanzen oder die „invenzione" – außer Acht. Außerdem könne die Theorie gerade das nicht liefern, was ästhetische Qualität ausmache: nämlich die Vielfalt (varietà).[44]

Der Laie soll also die Musik kennenlernen, indem er erfährt, wie sie gemacht ist. Wie sie gemacht ist, ist bei Pontio nun aber nicht mehr von Regeln abhängig, die ihre Legitimation aus ihrer Anschlußfähigkeit an eine wissenschaftliche Theorie der Musik beziehen; es ist eine Frage, die man nur beantworten kann, wenn man auf das schaut, was tatsächlich komponiert wird. Deshalb führt Pontio im „Ragionamento" über 110 Werke von mehr als 35 Komponisten als Beispiele an, von Josquin bis hin zu Cipriano de Rore und Giaches de Wert.[45] Im „Dialogo" wird Pontio noch deutlicher: Das alte Thema der Einstufung von Intervallen als Konsonanzen oder Dissonanzen führt ihn zur Frage nach der beurteilenden Instanz, und seine Antwort – es sei die über lange Zeit geschulte sinnliche Erfahrung – wirft nicht nur die noch von Zarlino hochgehaltene Balance zwischen Verstand und Gehör über den Haufen. Daß es sich um eine *geschulte*, also an der aufgeführten Musik ausgebildete Erfahrung („l'orecchia de' Periti") handelt, heißt, daß auch die Erfahrung keine „natürlichen" Maßstäbe mehr liefert. Die Musik bewegt auch in ihren Fundamenten – dort, wo es um die grundlegenden musikalischen Eigenschaften von Intervallen geht – gänzlich im Bereich dessen, was durch Menschen gemacht wurde.[46] Unter dieser Voraussetzung kann die Theorie keine andere als eine technische Rolle spielen; als Kriterium für ihrer Beurteilung steht einzig ihre praktische Bewährung zur Verfügung.

4. Komposition als Probe auf die Theorie

Wie bereits erwähnt, ist die Unterscheidung zwischen einem wissenschaftlichen und einem technischen Theorieverständnis keine, die die von mir herangezogenen Autoren ausdrücklich vorgenommen haben, und selbst Galilei oder Pontio hätten wohl versucht, sie zu umge-

---

42 l. c., S. 2 (Hervorhebung A.B.) („eine Bewegung der Stimme, gemacht mit Verstand"). Zu Beginn des Gesprächs beschreibt Paolo, der Experte, das Thema des Gesprächs folgendermaßen: Es sind die „cose virtuose, come sonare, cantare, & ragionare di simili …, opera veramente degna di gentilhuomo." Als Paolo zu einer längeren Erörterung der Sphärenharmonie ansetzt, unterbricht ihn Hettore, sein Gesprächspartner: „Non occorre v'affaticate più oltre, che non fa a mio proposito per apprendere la scientia, ch'io cerco, coiè la Musica Artificiale …" (l. c., S. 8).

43 Der Unterschied zwischen „contrapunto" und „compositione" spielt in Pontio (1588) eine große Rolle; vgl. dazu Sachs (1989a). „contrapunto" bezeichnet den einfachen zweistimmigen Satz, mit „compositione" sind drei- und mehrstimmige Sätze gemeint, die auch Anforderungen stellen, die die Anlage einer Komposition als ganze betreffen (beispielsweise Modusregeln).

44 Vgl. Sachs (1989b), S. 132f.

45 Für die Zahlenangaben vgl. Sachs (1989a), S. 145, für eine detailliertere Übersicht das Nachwort von S. Clercx im Nachdruck der „Ragionamenti".

46 Vgl. Sachs (1989b), S. 137f.

hen. Ein wichtiger Grund für diese ambivalente Haltung dürfte darin liegen, daß der Kontext, in dem die bisher betrachteten Dokumente stehen, eine Rechtfertigung dessen forderte, was als musiktheoretisches Wissen präsentiert wurde, und daß eine andere Rechtfertigungsinstanz als die Natur – aufgefaßt nach wie vor in Begriffen, die mehr oder weniger den Vorstellungen des späten Mittelalters entsprachen – in diesem Kontext nicht zur Verfügung stand. Man konnte an die Stelle einer Theorie, die zugleich die Bewegung der Gestirne, das Verhältnis der Seelenteile und den Kirchenchoral zu erklären beanspruchte, nicht einfach nichts setzen, wenn man den Anspruch hatte, Musiktheorie zu betreiben, und nicht lediglich eine Kontrapunktlehre zu schreiben, die ihre Prinzipien (wie die Intervallklassifikation) ohne Begründung voraussetzt.

Es gab im 16. Jahrhundert jedoch zwei Entwicklungen, die die Art und Weise, wie musikalisches Wissen legitimiert wird, selbst betrafen. Die eine dieser Entwicklungen verändert den theoretischen Kontext der Musik von der Mathematik zu dem der Rhetorik; auf sie werde ich im nächsten Abschnitt eingehen. Bei der anderen Entwicklung handelt es sich um die eindeutigste Manifestation dessen, was ich als „technisches" Theorieverständnis bezeichnet habe, denn hier wird die Theorie unmittelbar mit ihrer kompositorischen Anwendung verknüpft. Für diese zweite Richtung steht erstens Nicola Vicentino, dessen Hauptwerk den programmatischen Titel „L'antica musica ridotta alla moderna prattica" trägt und im Entwurf des „Archicembalo" gipfelt, eines Cembalos, auf dem man eine 31-stufige Skala, also eine Skala mit Fünfteltönen, spielen kann. Zweitens gehören hierher eine Reihe von Kompositionen, die man als eine Probe auf temperierte Stimmungen bezeichnen kann.

Temperierte Stimmungen sehen, wie oben bereits erwähnt, u. a. die Teilung von Intervallen in gleich große Abschnitte vor. Dies betrifft auch den Ganzton, der in der modernen temperierten Stimmung halbiert wird; im 16. Jahrhundert kursierten verschiedene Temperierungsverfahren, die nicht nur zur Zweiteilung, sondern auch zur Drittelung oder Fünfteilung des Ganztons führten. In jedem dieser Fälle kann die Teilung des Ganztons zum Ausgangspunkt eines fundamentalen Vorstellungswandels gegenüber der Diatonik werden: Ein Tonsystem wird nun nicht mehr von ausgezeichneten Intervallen (Oktaven, Quinten, Quarten oder Terzen) aus konstruiert, sondern aus elementaren Teilen, den jeweils kleinsten Tonstufen, zusammengesetzt. Folglich gehören alle Teile des Tonsystems, d. h. alle Stufen der Skala in gleichberechtigter Weise zum Tonsystem dazu; es gibt keinen Unterschied mehr zwischen einer *musica vera* und *musica ficta*. Die rein gestimmte diatonische Skala ist geschlossen (oder sollte wenigstens es sein), insofern die ausgezeichneten Intervalle stets die korrekten, „reinen" Proportionen aufweisen. Eine temperierte Skala ist geschlossen, insofern man nach einer festgelegten Anzahl von Schritten eine bestimmte Strecke auf der Tonskala durchmessen hat und zum Oktaväquivalent des Ausgangspunkts gelangt.

Die Gleichwertigkeit aller Stufen und die Geschlossenheit temperierter Tonsysteme waren Eigenschaften, die im Rahmen der mittelalterlichen Musiktheorie keine Rechtfertigung für ein Tonsystem liefern konnten (anders als die Eigenschaft, auf Proportionen der Zahlen von 1 bis 4 zu beruhen). Um temperierte Skalen zu legitmieren, wählten einige Komponisten daher einen anderen Weg: Sie komponierten Werke, die genau diese Eigenschaften temperierter Skalen ausnutzen, indem sie durch alle Stufen hindurch transponieren, so daß jede Stufe in gleicher Weise Ausgangspunkt einer neuen Transposition ist, um schließlich zum Ausgangspunkt zurückzukehren. Wenn eine solche Modulation möglich ist, dann „funktioniert" das Tonsystem: Die Theorie – das Verfahren der Temperierung – wird einer praktischen Bewährungsprobe unterzogen; wenn diese Probe gelingt, dann gilt die Theorie als bestätigt.

Eines der frühesten dieser Werke ist das Duo „Quid non ebrietas" von Adrian Willaert, das bereits 1519 geschrieben wurde, zu Lebzeiten Willaerts aber nur in Expertenkreisen kursierte und erst 1600 von Artusi in seinem Pamphlet gegen Monteverdi gedruckt wurde. Ihm liegt, wie aus mit ihm befaßten zeitgenössischen Briefen hervorgeht, eine 12-stufige Temperatur zugrunde[47], in der der Ganzton in zwei gleiche Hälften geteilt wird. Die Geschlossenheit der temperierten Stimmung macht Willaert auf besondere Weise sinnfällig: Während die eine Stimme des Duos, der Sopran, unverändert bleibt, moduliert der Tenor durch den Quintenzirkel abwärts bis zum Hexachord über eses (mit asas als fa-Stufe); das eses, mit dem der Tenor auch endet, bildet mit dem abschließenden d des Soprans in einer 12-stufiger Temperatur einen Einklang. Ein anderes Beispiel ist die *Chanson spirituelle* „Seigneur Dieu ta pitié" von Guillaume Costeley aus dem Jahr 1558, die auf der mitteltönigen Stimmung mit kleinen Terzen, also einer 19-stufigen Temperatur, und somit, wie Costeley in seinem Vorwort betont, auf der Dreiteilung des Ganztons beruht. Auch hier wird die Geschlossenheit der temperierten Skala zum bestimmenden Moment des Werks: Costeley moduliert von b–fa über die fa-Stufen es–as–des–ges–ces–fes–heses–eses– asas–deses–geses–ceses–feses zu heseses–fa; da jede Erniedrigung ein Drittel des Ganztons ausmacht, ist nach b und heses heseses mit a identisch. Eine textauslegende Funktion der Modulationen liegt in Costeleys Chanson nahe, denn während die Modulationen mit der Schilderung des eigenen Unglücks und der eigenen Erregung verknüpft sind, fällt das Ende der Transpositionskette mit dem Erreichen von heseses mit der Bitte um göttliche Hilfe zusammen; im Schlußteil, einem Lob Gottes, finden sich keine Modulationen mehr. An der Chanson werden übrigens, deutlicher noch als an dem Duo Willaerts, Probleme der Notation offenbar, die darauf zurückzuführen sind, daß das revolutionäre Kompositionsverfahren gänzlich im Rahmen der verfügbaren Musiklehre, vor allem der Hexachordlehre, konzipiert wurde. Die Transposition war für Willaert und Costeley nämlich kein Tonarten-, sondern ein Hexachordwechsel. Die Vorzeichen b und h, die zunächst diatonische Hexachordstufen, später Akzidentien bezeichneten, die im Sinne lokaler Hexachordtranspositionen gedeutet wurden, behielten auch hier ihre Doppelbedeutung als Alterations- und als Hexachordzeichen bei. Folglich weicht gelegentlich die (vermeintlich) notierte Tonhöhe von der gemeinten ab: Ein b vor f bespielsweise bezeichnet immer eine fa-Stufe, bei der es sich jedoch sowohl um f als auch, sofern ces–fa vorausgeht, um fes, und, sofern ceses–fa vorausgeht, um feses handeln kann, denn innerhalb einer Modulationskette bezeichnet ein b immer eine um eine Quint tiefere fa-Stufe als die vorangegangene.

Die Veränderung des Tonsystems, die die Aufwertung der Zwischenstufen zwischen den diatonischen bedeutete, eröffnete außer der Möglichkeit seiner Erprobung durch Transpositionsketten einen weiteren, bis dahin ebenfalls noch nicht beschrittenen Weg der Kompositionstechnik: die lineare Chromatik. Die Aufeinanderfolge von zwei oder mehr Halbtonschritten bzw. kleineren Intervallen ist überhaupt erst denkbar in einem temperierten System, in dem die Intervalle nicht mehr durch ausgezeichnete diatonische Stufen bestimmt sind. Die

---

47 Vgl. E. Lowinsky (1959), S. 14ff. Bei der Erstellung der 12-stufigen Temperatur berief man sich (zu Unrecht) auf die antike Autorität des Aristoxenos, dessen System nicht nur aus Polemiken, sondern auch aus affirmativen Darstellungen venezianischer Kreise (z. B. von Giovanni del Lago) bekannt war (vgl. Lowinsky (1959), S. 7ff.).

einzigen legitimen Halbtonschritte der *musica vera* waren durch die Hexachordstufen mi und fa festgelegt. An dieser Beschränkung änderte auch die *musica ficta* nichts, da auch sie als transponierte Hexachordstufen gedeutet wurde und darüberhinaus Akzidentien jeweils auf eine und nur eine diatonische Stufe bezogen blieben. Demgegenüber sind in einer temperierten Skala die Stufen zwischen den diatonischen Stufen prinzipiell frei und ohne funktionelle Einbindung.

Vicentino ging in seiner „L'antica musica ridotta alla moderna prattica" von der Fünfteilung des Ganztons und einer 31-stufigen Skala aus. Zu Beginn seines Traktats hält er sich nicht lange mit Zahlenproportionen oder Stimmungsproblemen[48] auf. Das kleinste und elementare Intervall, die „Diesis", wird als die Hälfte eines kleinen Halbtons eingeführt. Die Frage, wie man ein Intervall in zwei gleich große Intervalle teilen kann, wird überhaupt nicht angesprochen; der Status der Diesis als kleinster Einheit der Skala, aus der alle anderen Intervalle aufgebaut werden, wird mit dem knappen Hinweis begründet, sie sei das kleinste Intervall, welches das Ohr unterscheiden könnte, wie in der „Prattica" anhand von Beispielen gezeigt werde.[49] Dem gesamten „Libro della Theoria" sind überhaupt nur zwölf Seiten gewidmet. Deutlich mehr Raum nimmt die Vorstellung der verschiedenen neuen Tonschritte in einer Skala mit Fünfteltönen im ersten Buch der „Prattica musicale" ein. Vicentino führt sie am Leitfaden der drei antiken Tongeschlechter, der Diatonik, Chromatik und Enharmonik ein. Das diatonische Tongeschlecht verwendet nur die vertrauten sieben diatonischen Stufen, das chromatische zudem die dazwischenliegenden Halbtonschritte, das enharmonische obendrein jene Stufen, die zwischen den Halbtonschritten liegen. Vicentino präsentiert die resultierenden Skalen, also sein Tonsystem, ausschließlich mit Hilfe von Notenbeispielen. Auf diese Weise stellt er zugleich seine neue Notation vor: das b und das Kreuz, also die ehemaligen Akzidentienzeichen, geben eine Alteration um einen kleinen Halbton oder zwei Diesen an, ein Punkt oberhalb einer Note eine Erhöhung um eine Diesis. Die Stufen von C bis D lauten also folgendermaßen:

C – Ċ – Cis – Des – Dės – D.

Der von Vicentino gewählte Modus der Präsentation ist ein wichtiges Indiz für sein Theorieverständnis. Er führt sein Tonsystem nämlich überhaupt nur in der Form vor, in der es verwendet, in der es in Kompositionen hörbar gemacht werden soll.[50] Die Konstruktion des

---

48 Auf Fragen der Stimmung geht Vicentino erst im Zusammenhang mit seinem „Archicembalo" ein, das er im fünten Buch von „L'antica musica" vorstellt, und zwar indem er ein Verfahren beschreibt, wie man sein Cembalo stimmen soll (vgl. dazu Rippe (1981)). Er geht offensichtlich von einer bereits bestehenden Praxis der temperierten Stimmung aus, etwa wenn er bemerkt, daß die Quinten, die heute verwendet werden, nicht die perfekte Form haben, sondern ein wenig „verschoben" sind (… le Quarte & le Quinte di Boetio sono perfette, & quelle che noi usiamo, sono un poco spontate & scarse nel accordare li strumenti", fol. 13v).

49 Vgl. Vicentino (1555), fol. 6r: das Coma, die Differenz zwischen großem und kleinem Halbton (also eine Diesis), ist „la piu piccola parte di voce che sia atta à essere distinta dal senso dell' udito, come nella prattica ne darò essempi."

50 Auch die während des gesamten Mittelalters und der Renaissance übliche „Demonstration" eines Tonsystems am Monochord, einer einzelnen gespannten Saite, auf der man gemäß den Intervallproportionen Saitenabschnitte festlegen kann, war ein Weg, die Resultate der Theorie hörbar zu machen. Dieser Weg hatte aber nichts mit der musikalischen Praxis zu tun. Vicentinos Präsentationsmodus ist übrigens ein erstes Anzeichen für einen anderen wichtigen Umbruch im Denken über Musik, den Übergang vom Monochord zur Tastatur als dem Mittel, sich ein Tonsystem vorzustellen.

„Archicembalo" ist daher ein integraler Bestandteil seiner gesamten theoretischen Konzeption: Die häufigen Verweise auf sein Instrument und auf den „Sonatore overo Compositore"
erwecken den Eindruck, daß Vicentino seinen Traktat für Leser verfaßt hat, die über ein
Archicembalo verfügen und sich die Beispiele sogleich vorspielen können. Wiederum ist also
das, was man hört, die einzige Instanz, an der sich die Theorie zu bewähren hat. Mit einem
Verfahren der Notation und einem Instrument, auf dem man die notierte Musik spielen kann,
ist für Vicentino offensichtlich alles Notwendige gegeben, um tun zu können, worauf es ihm
eigentlich ankommt: Musik zu komponieren, die die antiken Tongeschlechter der Chromatik
und der Enharmonik wieder hörbar und die wunderbaren Effekte[51] dieser Tongeschlechter
wieder erfahrbar macht. Hierin, in der (erhofften) Wirkung der Musik, liegt Vicentinos eigentliche Rechtfertigung für sein Vorgehen.

Es wird daher auch verständlich, warum Vicentino sein neues Tonsystem nicht zuerst als
ganzes, als 31-stufige Skala, sondern von vornherein in den Kategorien des „genere diatoni-
co", des „genere cromatico" und des „genere enarmonico" einführt. Das System als ganzes
interessiert ihn nicht, sondern nur, was man daraus machen kann. Hierzu bedarf es für Vicentino offensichtlich auch einer Anleitung über mögliche melodische Fortschreitungen; genau das leisten die antiken Tongeschlechter, die insofern eine Aufgabe der Hexachordlehre
übernehmen, die ja gleichfalls die Beziehungen der Töne untereinander und die zulässigen
Bewegungen von einem Ton zu einem anderen festlegte.

Die kompositorische Praxis, die Vicentino in seinem Traktat vorstellt, ist noch auf eine
andere interessante Weise mit der traditionellen Musiklehre verknüpft. Er fügt unter anderem ein von ihm komponiertes Madrigal ein[52], das auf drei Arten gesungen werden könne:
ohne die eingetragenen Vorzeichen, unter Berücksichtigung der chromatischen Vorzeichen
(d. h. b, Kreuz und Auflösungszeichen), und schließlich in der notierten Fassung, also unter
Einschluß der enharmonischen Vorzeichen und Tonschritte. Diese Beliebigkeit ist bemerkenswert: Offensichtlich ändert die Berücksichtigung oder Nichtberücksichtigung der Vorzeichen an der satztechnischen Richtigkeit des Madrigals nichts[53]; der einzige (und entscheidende) Beitrag der Vorzeichen sei ein „großer Gewinn an Harmonie" (gran utile di
armonia).

Im Hintergrund von Vicentinos Kompositionslehre steht die seit dem späten Mittelalter
vertraute Satzlehre. Diese Lehre weist einen bemerkenswerten Freiraum auf, denn sie handelt von Intervallen stets im Sinne von Intervallklassen, d. h. sie unterscheidet nicht zwischen großen und kleinen Intervallen. Satztechnisch konnten daher eine große und eine kleine Terz, ja sogar eine reine und eine übermäßige Quint äquivalent sein (auch wenn im letzte-
ren Fall die erste als eine „relatio harmonica" und die zweite als eine „relatio non harmonica"
galt). Das heißt aber auch, daß man ein Intervall alterieren konnte, ohne daß diese Alteration
ein satztechnischer „Zug" wäre. Alterierungen konnten als *Färbungen* – als Chromatik im

---

51 Vgl. u. a. l. c., fol. 10v.
52 Vicentino, fol. 67v, 68r.
53 Um nur einige Beispiele für die Wirkung der Vorzeichen zu geben: in T. 7 wird ein ganzer Dreiklang um 1/5-
   Ton alteriert; die dabei entstehenden Parallelen wurden nicht als solche verstanden, da sich an der Tonstufe
   nichts ändert. Auch der genaue Aufbau eines Dreiklangs blieb akzidentiell: in T. 8 wird „d-moll" in „D-dur"
   verwandelt (ebenso T. 11). Eine gewisse Gleichgültigkeit gegen die Entstehung von relationes non harmonicae
   durch die Hinzufügung von Akzidentien ist etwa aus T. 7/8, Sopran, T, 9 Sopran und T. 11 Tenor, zu ersehen. Ein
   Beispiel für die Ausnutzung der linearen Chromatik bietet schließlich T. 9/10 Tenor.

urspünglichen Sinne des Wortes – verstanden werden, die den Effekt, aber nicht die Struktur einer Komposition beeinflußten. Genau diese Möglichkeit macht sich Vicentino zu nutze. Die satztechnische Neutralität von Chromatik und Enharmonik schafft den Freiraum, um mit Klängen experimentieren zu können, die erst 500 Jahre später von Komponisten wieder ernsthaft in Erwägung gezogen wurden.[54]

Vicentinos „L'antica musica" ist einer der bemerkenswertesten Kreuzungspunkte in der Entwicklung der Musiktheorie im 16. Jahrhundert. Im Hintergrund seines Werks stehen die Resultate einer der zentralen Debatten der Musiktheorie, nämlich der Entwurf temperierter Stimmungen. In diesen Kontext reiht er sich durch die Abfassung eines Traktats ein. Was für andere ein Tribut an die unvollkommene Instrumentalmusik war, wird für ihn jedoch zu einem Feld völlig neuer kompositorischer Möglichkeiten; Theorie wird zur Technik, zu etwas, das man nutzbringend anwenden kann und darüberhinaus uninteressant ist. Die Legitimation der Theorie ergibt sich allein aus ihrer Verwendung, genauer aus der Wirkung, die die neue Musik hat. Damit bezieht sich Vicentino auf den dritten Bereich des Wissens von der Musik, die ich eingangs unterschieden habe, auf das ästhetische Wissen. Es sind Behauptungen über die „dolcezza" der chromatischen und enharmonischen Musik, auf die er sich letzten Endes beruft, und die anhand der von ihm entwickelten Musik bewiesen werden sollen.[55]

5. Musik und Nachahmung

Musik unter dem Gesichtspunkt der Wirkung, mithin als Gegenstand ästhetischen Wissens zu betrachten, war nichts Neues und gerade im 16. Jahrhundert weit verbreitet, in Folge der Rezeption von Platons „Politeia" und Aristoteles' „Poetik", also zwei Texten, die genau diesen Aspekt der Musik (bzw. der Tragödie) in den Mittelpunkt stellen.[56] Neu war bei Vicentino nur die Legitimationsfunktion des ästhetischen Wissens.

Das ästhetische Wissen von der Musik wurde aber im 16. Jahrhundert in einer wichtigen und neuen Weise weiterentwickelt. Die antike Vorstellung, die verschiedenen Tonarten hät-

---

54 Vgl. dazu Dahlhaus (1982) und (1988b). Vicentinos „kolorierende" Verwendung chromatischer Stufen weicht allerdings insofern von der bisherigen Praxis ab, als Alterationen bisher eine Leittonfunktion hatten und damit der Unterstützung der Satztechnik dienten (so diente die „regola delle terze e seste, vgl. oben, S. , dazu, den Übergang von einer Dissonanz in eine perfekte Konsonanz zu verstärken). Bei Vicentino können Alterationen diese Funktion nicht mehr ausüben, wenn sie satztechnisch neutral sind (vgl. Dahlhaus (1988b), S. 216).

55 Vgl. z. B. l. c., fol. 15r: Wer nicht glaube, daß man heute Tonsysteme mit vielfältigerer Harmonie habe als diejenigen, die Boethius beschreibt, solle doch es doch einfach erfahren („… noi habbiamo molte divisioni, che generano più variata harmonia, che non è quello ch' esso Boetio scrive: e chi non lo crede, ne facci l'esperienza, che ritroverà assai più di quello ch'io ramento e scrivo.").
Ähnliches gilt übrigens für eine andere ob ihrer Radikalität berühmte Figur des 16. Jahrhunderts, für Carlo Gesualdo. Zwar war Gesualdo kein Musiktheoretiker, aber seine Kompositionen zeigen ein ähnliches Verständnis von Chromatik, Kontrapunkt und ästhetischer Legitimation wie bei Vicentino. Für letztere ist bei Gesualdo die Textauslegung durch die Musik ausschlaggebend, die aber auch von Vicentino angeführt wird (vgl. z. B. fol. 27r). Vgl. dazu Dahlhaus (1988a).

56 Aristoteles' „Poetik" wurde erst seit der Übersetzung durch Giorgio Valla im Jahre 1498 breit rezipiert, Platons „Politeia" seit der Übersetzung durch Marsilio Ficino (1485).

ten unterschiedliche Wirkungen auf die Seele, wurde durch das gesamte Mittelalter hindurch als ein Topos der Musiktheorie tradiert, auch dann, wenn man mit den Details nicht mehr viel anfangen konnte. Im 16. Jahrhundert versuchten manche (unter ihnen Vicentino), die Musik zu rekonstruieren, die angeblich jene sagenhaften Wirkungen hatte; andere, wie Zarlino, versuchten, die tatsächlichen Eigenschaften der Kirchentonarten (die, wie man inzwischen wußte, mit den antiken Tonarten nicht identisch waren), zur Erklärung ihrer Wirkung heranzuziehen[57]; wieder andere, wie Galilei, verwarfen diese Form der „Ethos-Lehre" der Musik kurzerhand.[58] Neu ist dagegen, daß das Komponierte selbst zum Gegenstand des ästhetischen Wissens wurde. Es geht jetzt nämlich darum, einzelne Wendungen und Fortschreitungen in Kompositionen zu erklären. Dazu genügt es nicht zu zeigen, daß sie konform zu den Regeln des Kontrapunkts sind. Die chromatischen Werke von Vicentino oder Gesualdo haben, auch wenn es sich um Extremfälle handelt, deutlich gemacht, daß die Regeln des Kontrapunkts einen zu großen Spielraum lassen, als daß sie dasjenige, was die Wirkung erzeugt, wirklich erklären könnten; andere Komponisten wie Marenzio und später Monteverdi demonstrierten, daß man um der Wirkung willen auch Satzregeln verletzen konnte.

Das naheliegendste Beschreibungsmodell für komponierte Musik lieferte die Sprache. Die Grammatik stellte ein Vokabular bereit, das man heranziehen konnte, um etwas, das in gegliederter Weise in der Zeit verläuft, zu erfassen. Entsprechend wird im 16. Jahrhundert der Begriff „Kadenz" in die Musiktheorie aufgenommen und die verschiedenen Ebenen sprachlicher Gliederung durch Punktierung auf die Musik übertragen.[59] Auch hier steht Platon als antike Autorität im Hintergrund, der den Begriff der Musik ja bekanntlich so weit bestimmt hatte, daß auch die Sprache dazuzählt (vgl. z. B. „Politeia" 376e). Die neue Nähe von Musik und Sprache schlägt sich auch in vereinzelten Neuzuordnungen der Musik zu den sprachlichen Disziplinen nieder.[60]

Die bloße Gemeinsamkeit, in der Zeit zu verlaufen, war aber nicht ausreichend, und auch nicht der entscheidende Grund, die Musik unter Rückgriff auf die Sprache zu erklären. Dazu wurde ein anderer Begriff herangezogen: der Begriff der *Nachahmung*. Musik sollte dadurch erklärt werden, daß sie die Sprache nachahmt. Die Nachahmung war zum einen eine intentionale Handlung; Musik in diesem Sinne zu erklären, hieß also zu erkennen, mit welchem Ziel der Komponist sie geschrieben hatte, wie und warum er sie so und nicht anders komponiert hatte. Zum anderen lieferte die Nachahmung eine Erklärung dafür, wie die Musik wirkt – wie die Hörer erfassen, was sie ausdrückt.

Was heißt es aber genau, daß die Musik die Sprache nachahmt? Den meisten Autoren stand als Paradigma die Vokalmusik vor Augen, insbesondere die Madrigale, denen Gedichte der großen italienischen Renaissancedichter wie Petrarca oder Tasso zugrundelagen. Die Nachahmungsbeziehung sollte also zwischen der Musik und dem vertonten Text bestehen. Aus den Diskussionen gehen nun mindestens drei Arten dieser Nachahmung hervor: die Musik konnte entweder die *Deklamation* nachahmen, oder einzelne Worte des Textes bzw. die *concetti*, die diese Worte ausdrücken, oder schließlich das *concetto*, das ein Gedicht insgesamt ausdrückt.

---

57 Zarlino (1573), S. 182.
58 Vgl. Palisca (1985), S. 346.
59 Vgl. Rempp (1989), S. 118 und Palisca (1985), S. 340.
60 Vgl. dazu Palisca (1985), S. 337.

Was ist ein concetto? Concetti sind, generell gesagt, mentale Entitäten. Der Begriff weist eine gewisse Ähnlichkeit zum Terminus „idea" des Empirismus auf, insofern es concetti gibt, die Dingen und Eigenschaften in der Welt repräsentieren, aber auch abstrakte concetti und concetti von seelischen Zuständen. Ferner sind concetti nicht an Worte in der Sprache gebunden und weisen auch keine sprachanalogen Strukturmerkmale auf; sie können durch Worte, aber auch anders, ausgedrückt werden. Die genaue Bestimmung von „concetto" ist hier nicht relevant[61]; wichtiger ist, daß im 16. Jahrhundert bereits die Beziehung eines concetto zu Dingen in der Welt als eine Nachahmungsbeziehung aufgefaßt wurde.[62] Die Stimme ist von der Natur dem Menschen gegeben, concetti auszudrücken. Sie erreicht dies durch die Worte, aber auch durch andere Eigenschaften, wie etwa den Rhythmus oder die Sprachmelodie.[63] Der Ausdruck der concetti durch die Sprache ist wiederum eine Nachahmungsbeziehung.[64] Diese doppelte Nachahmungsbeziehung erklärt nun, wie es der Sprache gelingt, etwas auszudrücken: Wenn die Sprache einem concetto ähnelt, dann ist es dem Hörer möglich, das concetto zu identifizieren; wenn das concetto dem ähnelt, worauf es sich bezieht, dann vermag der Hörer auch den Referenten des concetto zu erfassen. Zumindest für die „musikalischen" Eigenschaften der Sprache, d. h. für den Rhythmus und die Sprachmelodie fällt das Erfassen der Nachahmung mit der Wirkung der Sprache auf die Seele zusammen; kognitives Erfassen und kausaler Effekt gehen ineinander über.[65] Dies ist, in aller Kürze, der theoretische Rahmen, in dem die Nachahmungsbeziehung zum entscheidenden Explanans für die Wirkung der Musik auf die Menschen wird.

Der vertonte Text gibt der Musik nun die concetti vor, die ausgedrückt werden sollen.[66] Die Musik vermag diese concetti nachzuahmen, indem sie diejenigen Eigenschaften der Sprache nachahmt, die das concetto nachahmen; dies gelingt besonders dort, wo es um emotionale concetti geht, denn gerade sie werden in den „musikalischen" Eigenschaften der Sprache ausgedrückt. Girolamo Mei erläutert diesen Zusammenhang im Detail: Es gibt Zustände

---

61 Zum Begriff des „concetto" vgl. Bauer (1970). Bauer zeigt, daß dem concetto im Rahmen der Erkenntnistheorie sowohl mentale wie materiale Eigenschaften zugewiesen wurden und das Erfassen eines concetto daher sowohl ein Akt des Verstehens (also ein kognitiver Akt) wie ein Akt des Fühlens (der kausalen Einwirkung) ist.
62 Vgl. das aufschlußreiche Zitat von Agnolo Segni in Palisca (1985), S. 365.
63 Vgl. Mei (1960), S. 91: „... perche la voce era stata da la natura a gli animanti, et al huomo spezialmente per significazione de concetti intrinsechi, era medesimamente ragionevole che tutte queste diverse qualità di lei, essendo tutte determinatamente distinte fussero appropriate à esprimere ciascuna da per se distintamente l'affezione di alcuni determinanti ... onde l'acuta non potesse acconciamente esprimere l'affezioni dela mezzana ...". Girolamo Mei war eine der entscheidenden Figuren in der Entwicklung solcher Gedanken im Kontext der Musiktheorie, und die wichtigste Inspirationsquelle Galileis. Dieses und die folgenden Zitate stammen aus einem Brief Meis an Galilei; sechs Jahre nach dem Tod Meis wurde aus seinem Nachlaß ein Traktat veröffentlicht (Mei (1602)), der inhaltlich diesem Brief weitgehend entspricht.
64 Vgl. wieder die von Palisca zitierte Passage von Segni (Palisca (1985), S. 365).
65 Beispielsweise wechselt Mei in seiner Beschreibung des Ausdrucks der Sprachmelodie zwischen „dimostrare" (was nahelegt, daß ein Zeichen, etwa aufgrund seiner Ähnlichkeit zum Bezeichneten, erkannt werden soll), und „muovere affezioni", was auf einen kausalen Einfluß hinweist (vgl. Mei 1960, S. 92f.). Für beide Beziehungen galt die Nachahmung als hinreichende Erklärung; dies dürfte ein Grund sein, warum sie nicht deutlich unterschieden wurden. An anderer Stelle erläutert er den Prozeß des Ausdrückens als „rapresentarsene con la loro imitazione al senso del udito, et per mezzo di lui à lo intelletto quasi l'imagine e la verità stessa de le cose imitate" (l. c., S. 116).
66 Vgl. Galilei (1581), S. 88: „l'imitazione de concetti che si trae dalle parole"; dies sei „la piu importante, & principale parte che sia nella Musica" (l. c.).

oder Eigenschaften der Seele, die durch schnelle Bewegung charakterisiert sind. Die Sprache verwendet unterschiedliche Tonhöhen; höhere Töne beruhen auf schnellerer Bewegung der Luft, tiefere Töne auf einer langsameren. Also repräsentiert die Höhe eines Tons ein concetto (bzw. ahmt es nach), das der Ursache des concetto, einer bestimmten Seelenbewegung, gleicht, und der Hörer erfaßt den Ausdruck der Tonhöhe gemäß dieser Ähnlichkeitsbeziehung.[67] Orientiert man sich an einer solchen Erklärung, wird ersichtlich, warum die Musik die Deklamation der Sprache nachahmen soll: Der Melodieverlauf ahmt die Sprachmelodie nach; die Sprachmelodie ahmt das concetto der Emotion, dieses wiederum die Bewegung der Seele in jenem emotionalen Zustand nach; diese Kette von Nachahmungsbeziehungen sorgt dafür, daß die Musik den emotionalen Zustand ausdrückt und als solcher Ausdruck verstanden wird bzw. wirkt. Meis Konzeption war eine der Wurzeln der monodischen Musik, die in der Florentiner Camerata (auf die Mei großen Einfluß hatte) erprobt wurde. Wenn es nämlich darauf ankommt, den Ausdruck der Musik zu erfassen, und wenn der Ausdruck von den rhythmischen und diastematischen Eigenschaften der Melodielinie abhängt, dann kann es nur ablenken, die Kraft des Ausdrucks nur schwächen, wenn mehrere Stimmen gleichzeitig zu hören sind.[68]

Nun bietet die Nachahmungsrelation Raum für mehr und für lockerere Beziehungen, als sie Mei vorschwebten. Die Musik vermag die Sprache auch nachzuahmen, und die concetti, die der Text ausdrückt, auszudrücken, wenn sie Eigenschaften der Sprache (oder der concetti, oder der Referenten der concetti) mit ihren eigenen Mitteln nachahmt. Weist die Sprache beispielsweise eine Folge rauher Laute auf, und taucht im Text das Stichwort „rauh" auf, so soll der Komponist rauhe Klänge wählen.[69] Ist im Text vom Himmel die Rede, so kann die Musik dieses concetto durch hohe Töne nachahmen. Es ist offensichtlich, das gerade das zuletzt genannte Beispiel der Tonmalerei alle Türen öffnet, denn die Ähnlichkeitsbeziehungen, die zwischen Musik, den concetti und dem bestehen, was die concetti repräsentieren, sind nahezu beliebig. Es wundert daher nicht, daß Galilei diese Praxis mit Spott überzog; anstatt jedes einzelne Wort nachzuahmen, sollten sie eher die „concetti dell' animo", die hinter den Worten stehen, auszudrücken versuchen.[70] Das Verfahren, das Galilei hierzu vorschlägt, greift die von Mei propagierte Nachahmung der Deklamation auf: Galilei faßt das auszudrückende „concetto dell' animo" offenbar als einen Zustand der Seele auf, der sich in der Sprechweise und der Gestik niederschlägt und daher durch deren Nachahmung ausgedrückt werden kann. In diesem Sinne vermag übrigens auch die Instrumentalmusik auszudrücken.[71]

---

67 Vgl. die von Palisca (1985), S. 351 zitierte Passage aus einem unveröffentlichten Traktat Meis.

68 Vgl. Mei (1960), S. 93. Mei ging sogar noch einen Schritt weiter, indem er eine wirklich einstimmige Musik propagierte, also auch ohne Baßbegleitung. Im Sinne seiner zitierten Erläuterung war dies konsequent: wenn die Tonhöhe allein den Affekt ausdrückt, dann muß der Zusammenklang eines hohen und eines tiefen Tons den Hörer sehr irritieren.

69 Vgl. Zarlino (1573), S. 419: „Et debbe avvertire di accompagnare quanto potrà in tal maniera ogni parola, che dove ella dinoti asprezza, durezza, crudeltà, amaritudine, & altre cose simili, l'Harmonia sia simile à lei; cioè alquanto dura & aspra; di maniera però che non offendi."

70 Vgl. Galilei (1581), S. 89.

71 Galilei, l. c., S. 89f. Eine ähnliche Differenz zwischen der Wort-für-Wort-Tonmalerei und dem Versuch, der Komposition ein zentrales concetto des vertonten Textes zu unterlegen, findet sich übrigens auch in der Madrigalkomposition des 16. Jahrhunderts. Vgl. dazu Owens (1999); sie zeigt, daß de Wert, im Unterschied zu Maren-

Blickt man auf die weitere Geschichte der Musikästhetik, kann man den Eindruck gewinnen, Mei und Galilei seien die Urheber der Nachahmungsästhetik, die die im 17. und 18. Jahrhundert die mathematische Musiktheorie als beherrschende Theorie ablöste. Wollte man daraus den Schluß ziehen, die beiden Autoren hätten sich selbst als Begründer einer neuen Musiktheorie gesehen, ginge man fehl. Ein Grund dafür mag sein, daß der Nachahmungsbegriff auch in der alten mathematischen Konzeption der Welt seinen Platz hatte, insofern die „musica instrumentalis", die von Menschen erzeugte Musik, die Proportionen der „musica mundana", der Sphärenharmonie, imitierte. Jedenfalls erkannten beide, auch Mei, die mathematische Musiktheorie als diejenige Wissenschaft an, die die Wahrheit ans Licht befördert[72]; die bereits erwähnte scharfe Trennung zwischen Kunst und Natur bei Galilei (die ebenfalls auf Mei zurückgeht) eröffnete nicht nur die Möglichkeit, Musiktheorie als Technik zu begreifen, sie beließ der Theorie auch ihr eigenes Untersuchungsfeld, nämlich die Natur. Mei und Galilei sahen in diesem Nebeneinander offensichtlich keine Probleme. Daß die Trennung von Musik und Natur die Legitimation des alten Wissens über die Musik in Frage stellte, habe ich versucht darzulegen. Daß der Nachahmungsbegriff und der Wechsel zur komponierten Musik als Gegenstand des Wissens eine neue Legitimationsbasis schuf, zeigte sich spätestens in der Auseinandersetzung zwischen Monteverdi und Artusi, in der die Wahrheit des Ausdrucks zur Legitimation für den Regelverstoß wurde.[73]

Literatur

a) Quellen

Benedetti, Giovanni (1585): „De intervallis musicis", in J. Reiß (1925/25): *Joh. Bapt. Benedictus, De intervallis musicis. Zeitschrift für Musikwissenschaft* 7, 13 – 20.

Boethius (1867): *De institutione musica.* Hrsg. von G. Friedlein. Leipzig. (Nachdruck Frankfurt: Minerva 1966.)

De Grocheo, Johannes (1290): *De musica.* Hrsg. von E. Rohloff (Media latinitas musica, Bd. 2). Leipzig: Gebrüder Reinecke 1943.

De Muris, Johannes (1321): *Notitia artis musicae.* Hrsg. von U. Michels (= CSM Bd. 17). American Institute of Musicology 1972.

De Muris, Johannes (1323): „Musica speculativa", in Christoph Falkenroth (Hg.) (1992), *Beihefte zum Archiv für Musikwissenschaft,* Bd. 34. Stuttgart: Steiner.

De Vitry, Philippe (1320): „Ars nova und Ars contrapunctus secundum Philippo de Vitriaco", in E. de Coussemaker (1864-1876), *Scriptorum de musica medii aevi,* Bd. 3. Paris, 13-22.

Fogliani, Lodovico (1529): *Musica theorica.* Venedig: G. A. e fratelli di Sabio. (Nachdruck Bologna: Forni 1970)

---

zio, solche zentralen Ideen entweder durch ein Strukturmerkmal des Madrigals nachgeahmt hat, oder einen Affekt zum dominierenden gemacht hat, von dem auch dann nicht abgewichen wurde, wenn der Text andere Stichworte anbot.

72 Vgl. Mei (1960), S. 103.

73 Vgl. dazu den Beitrag von Matthias Vogel in diesem Band.

Franco von Köln (1280): *Ars cantus mensurabilis*. Hrsg. von G. Reaney und A. Gilles (= CSM Bd. 18). American Institute of Musicology 1974.

Galilei, Vincenzo (1581): *Dialogo della musica antica et della moderna*. Florenz: Giorgio Marescotti. (Reprint New York: Broude Broth 1966.)

Galilei, Vincenzo (1589): *Discorso intorno alle opere die Giseffo Zarlino et altri importanti attenenti alla musica*. Florenz: Giorgio Marescotti. (Nachdruck Mailand 1933.)

Galilei, Vincenzo (o. J.): „Discorso particolare intorno alla diversità delle forme del diapason". Ms. Florenz, Biblioteca Nazionale Centrale, Fondo Anteriori a Galileo III, fol. 44r – 54v. Veröffentlicht in C. Palisca (1989), *The Florentine Camerata*. New Haven: Yale UP.

Mei, Girolamo (1602): *Discorso sopra la musica antica et moderna*. Venedig: G. B. Ciotti. (Nachdruck Bologna: Forni o. J.)

Mei, Girolamo (1960): „Letters to Vincenzo Galilei", in C. Palisca (Hg.), *Musicological Studies & Documents* 3. American Institute of Musicology.

Pontio, Pietro (1588): *Ragionamento di musica*. Parma: Erasmo Viotto. (Nachdruck Kassel: Bärenreiter 1959.)

Vicentino, Nicola (1555): *L'antica musica ridotta alla moderna prattica*. Rom: Antonio Barre. (Nachdruck Kassel: Bärenreiter 1959.)

Zarlino, Gioseffo (1573): *Istitutioni harmoniche*. Venedig: Francesco de i Franceschi. (Nachdruck Ridgewood: The Gregg Press 1966.)

b) Sekundärliteratur

Bauer, Robert (1970): „A Phenomenon of Epistemology in the Renaissance", *Journal of the History of Ideas* 31, 281-288.

Bernhard, Michael (1990): „Das musikalische Fachschrifttum im Mittelalter", in F. Zaminer (Hg.), *Geschichte der Musiktheorie*, Bd. 3. Darmstadt: Wissenschaftliche Buchgesellschaft, 37-104.

Dahlhaus, Carl (1962): „Domenico Belli und der chromatische Kontrapunkt um 1600", *Musikforschung* 15, 315-340.

Dahlhaus, Carl (1963): „Zu Costeleys chromatischer Chanson", *Musikforschung* 16, 253-265.

Dahlhaus, Carl (1968a): *Untersuchungen über die Entstehung der harmonischen Tonalität*. Kassel: Bärenreiter.

Dahlhaus, Carl (1968): „Zur Akzidentiensetzung in den Motetten Josquin des Prez", in Festschrift K. Vötterle, Kassel: Bärenreiter, 206-219.

Dahlhaus, Carl (1969): „Tonsystem und Kontrapunkt um 1500", *Jahrbuch des staatlichen Instituts für Musikforschung Preußischer Kulturbesitz* (1969), 7-18.

Dahlhaus, Carl (1976): „Zur Tonartenlehre des 16. Jahrhunderts", *Musikforschung* 29, 300-303.

Dahlhaus, Carl (1982): „Musikalischer Humanismus als Manierismus", *Musikforschung* 35, 122-129.

Dahlhaus, Carl (1986): „Die Tactus- und Proportionenlehre des 15. bis 17. Jahrhunderts", in F. Zaminer (Hg.), *Geschichte der Musiktheorie*, Bd. 6. Darmstadt: Wissenschaftliche Buchgesellschaft, 333-362.

Dahlhaus, Carl (1988a): „Il cromatismo di Gesualdo", in Fabbri (1988), 207-230.

Dahlhaus, Carl (1988b): „Zur Harmonik des 16. Jahrhunderts", *Musiktheorie* 3, 205-211.

Fabbri, Paolo (Hg.) (1988): *Il madrigale tra Cinque e Seicento*. Bologna: il Mulino.

Gallo, F. Alberto (1984): „Die Notationslehre im 14. und 15. Jahrhundert", in F. Zaminer (Hg.), *Geschichte der Musiktheorie*, Bd. 5. Darmstadt: Wissenschaftliche Buchgesellschaft, 257-356.

Gallo, F. Alberto (1989): „Die Kenntnis der griechischen Theoretikerquellen in der italienischen Renaissance", in F. Zaminer (Hg.), *Geschichte der Musiktheorie*, Bd. 7. Darmstadt: Wissenschaftliche Buchgesellschaft, 7-38.

Groth, Renate (1989): „Italienische Musiktheorie im 17. Jahrhundert", in F. Zaminer (Hg.), *Geschichte der Musiktheorie*, Bd. 7. Darmstadt: Wissenschaftliche Buchgesellschaft, 307-380.

Hortschansky, Klaus (Hg.) (1989): *Zeichen und Struktur in der Musik der Renaissance*. (Musikwissenschaftliche Arbeiten Bd. 28). Kassel: Bärenreiter.

Judd, Cristle C. (2000): *Reading Renaissance Music Theory*. Cambridge: CUP.

Kaufmann, Henry W. (1961): „The motets of Nicola Vicentino", *Musica disciplina* 15, 169-183.

Kaufmann, Henry W. (1963): „Vicentino and the Greek Genera", *Journal of the American Musicological Society* 16, 325-346.

Kaufmann, Henry W. (1966): „A ‚diatonic' and a ‚chromatic' madrigal by Giulio Fiesco", in Festschrift G. Reese. New York.

Levy, Kenneth J. (1955): „Costeley's chromatic chanson", *Annales Musicologiques* 3, 213-263.

Lindley, Mark (1986): „Stimmung und Temperatur", in F. Zaminer (Hg.), *Geschichte der Musiktheorie*, Bd. 6. Darmstadt: Wissenschaftliche Buchgesellschaft, 109-332.

Lowinsky, Edward E. (1946): *Secret chromatic art in the Netherland motet*. New York.

Lowinsky, Edward E. (1948): „On the use of scores by sixteenth-century musicians", *Journal of the American Musicological Society* 1.

Lowinsky, Edward E. (1954): „Music in the culture of the renaissance", *Journal of the History of Ideas* 5, 509-551.

Lowinsky, Edward E. (1956): „Matthaeus Greiters Fortuna: An experiment in chromaticism and musical iconography", *Musical Quarterly* 52, 500-520.

Lowinsky, Edward E. (1959): „Adrian Willaerts chromatic ‚Duo' re-examined", *Tijdschrift voor Muziekwetenschop* 18, 1-36.

Maniates, Maria R. (1975): „Vicentino's ‚Incerta et occulta scientia' Reexamined", *Journal of the American Musicological Society* 28, 335-351.

Moyer, Ann (1997): „Musical Scholarship in Italy at the End of the Renaissance, 1500-1650", in Donald Kelley (Hg.), *History and the Disciplines*. Rochester: The University of Rochester Press, 185-202.

Owens, Jessie (1997): *Composers at Work. The Craft of Musical Composition 1450-1600*. Oxford: OUP.

Owens, Jessie (1999): „Marenzio and Wert read Tasso: a study in contrasting aesthetics", *Early Music* 27, 555-574.

Palisca, Claude V. (1961): „Scientific Empirism in Musical Thought", in H. H. Rhys (Hg.), *Seventeenth Century Science and the Arts*. Princeton: Princeton UP, 91-137.

Palisca, Claude V. (1968): „The Alterati of Florence: Pioneers in the Theory of Dramatic Music", in W. Austin (Hg.), *Festschrift D. G. Grant*. Ithaca: Cornell UP.

Palisca, Claude V. (1974): „Towards an Intrinsically Musical Definition of Mannerism in the Sixteenth Century", *Studi Musicali* 3, 313-346.

Palisca, Claude V. (1985): *Humanism in Italian Renaissance Musical Thought.* New Haven: Yale UP.

Palisca, Claude V. (1989): „Die Jahrzehnte um 1600 in Italien", in F: Zaminer (Hg.), *Geschichte der Musiktheorie,* Bd. 7. Darmstadt: Wissenschaftliche Buchgesellschaft, 221-306.

Rempp, Frieder (1989): „Elementar- und Satzlehre von Tinctoris bis Zarlino", in F. Zaminer (Hg.), *Geschichte der Musiktheorie,* Bd. 7. Darmstadt: Wissenschaftliche Buchgesellschaft, 39-220.

Rippe, Volker (1981): „Nicola Vicentino – sein Tonsystem und seine Instrumente", *Musikforschung* 34, 393-413.

Sachs, Klaus J. (1984): „Die Contrapunctus-Lehre im 14. und 15. Jahrhundert", in F. Zaminer (Hg.), *Geschichte der Musiktheorie,* Bd. 5. Darmstadt: Wissenschaftliche Buchgesellschaft, 161-256.

Sachs, Klaus J. (1989a): „Musikalische ‚Struktur' im Spiegel der Kompositionslehre von Pietro Pontios ‚Ragionamento di musica' (1588)", in Hortschansky (1989), 141-157.

Sachs, Klaus J. (1989b): „‚Theoria e Prattica di Musica' in Pietro Pontios ‚Dialogo' (Parma 1595)", *Musiktheorie* 4, 127-141.

Seidel, Wilhelm (1986): „Französische Musiktheorie im 16. und 17. Jahrhundert", in: F. Zaminer (Hg.), *Geschichte der Musiktheorie,* Bd. 9. Darmstadt: Wissenschaftliche Buchgesellschaft, 1-140.

Walter, Michael (1994): *Grundlagen der Musik des Mittelalters: Schrift – Zeit – Raum.* Stuttgart: Metzler.

Zenck, Hermann (1933): „Nicola Vicentinos ‚L' antica musica'", in *Festschrift Th. Kroyer.* Regensburg, 86-101.

Matthias Vogel

"Truth" in Monteverdi's *"L'incoronazione di Poppea"*

Introduction

Contemporary readers of my talk's title *"Truth" in Monteverdi's "L'incoronazione di Pop-pea"* will normally feel quite uneasy, seeing truth attributed to a piece of music.[1] They will suspect the author to be the victim of a categorial mistake, because truth is said to be a property of declarative sentences and not of musical expressions. Perhaps the (scare) quotes give a little relief, but even the quoted mentioning of truth in connection with music remains at least a provocation. Yet, if this is a provocation, it is not one of those fashionable post-modern ones, but one that is provided by Monteverdi himself, by the way he talks about a fundamental feature of his music.

So, what am I up to?

1. First I try to locate Monteverdi's aesthetical position in the context of the debates that accompany the profound changes in composition and music during his life time.
2. Given that Monteverdi's position has to be distinguished by referring to a concept of truth, I will secondly discuss some philosophical options to reconstruct truth as a property of artworks and then propose a conception of truth of music that allows us to *understand* some possible reasons, that could have motivated Monteverdi to talk of *true music*.
3. Finally I give a very rough sketch of the structures that organize the "Coronation of Pop-pea", leading to an interpretation of the opera that takes advantage of the conception of truth, developed in advance.

1. Monteverdi's aesthetical position

During the lifespan of Monteverdi, some profound developments took place that changed both theory and practice of music. This developments can be understood as the solution of a set of aesthetical problems, which came up with the dramatic changes in the self-understand-ing of 16th century people. While the renaissance mood is optimistic about the human enter-prise, this optimism is disturbed in late renaissance and early baroque. But expressing this changed mood in music challenged the borders of the musical material formed by the accept-ed rules for (good) compositions.

---

1 I would like to thank Michael Kohler for helpful comments on an earlier version of this paper.

These rules grounded in the tradition of those composers, who wrote a vocal polyphonic music, which is organized by integration of voices of equal value, resulting in a stream of thoroughly interwoven lines. The integrity of this textures was protected by a set of compositorial rules that codify a conception of music as an equilibristic unity. Consequently, unlicensed dissonances and certain forms of dependence between the voices, for example consecutive fifth and fourth, were prohibited. These rules thereby prevented composers from falling back into "archaic" forms of early versions of polyphony, and were justified not only by aesthetical claims, but also by nature. In its religious ritual contexts music ought to be the expression of the eternal cosmic harmony. Although this picture is far to simple, especially because it lacks any reference to secular madrigals (for example of Guillaume de Machault), it nonetheless helps to make clear what kind of difficulties composers had to face in expressing human emotions.[2]

In late Renaissance however this situation has changed profoundly. Yet, besides the soloistic voices of the not very well estimated folk-songs only two musical sources directly stood open to Monteverdi and his not very successful precursors for developing the opera as a form of music, that would satisfy the demand for the expression of individual emotions. They only could go back either to the fainting form of *Rappresentatione sacra*, a nonliturgical religious play, in which a lot of text was presented in a monodic form, or to the form of *madrigal-comedy*, a kind of theatre play accompanied by nonmonodic madrigal-like parts.

In the view of the reformers of the Florentine Camerata, which criticized the music of their days for being congealed in counterpoint (as codified by Zarlino), the virtues of Greek theatre should be revitalized. They thought of the ancient drama as accompanied by monophonic music, whose function was to communicate the text and the emotional situation of its protagonist. Vincenzo Galilei, a member of the Camerata and the father of the famous Galileo Galilei, recommended that musicians should visit the theatres in order to study the human emotions. Monteverdi was not much interested in a reconstruction of Greek music (as for example Girolamo Mei) and as far as we know he was not engaged in theoretical debates until his compositions were criticized for involving unlicensed dissonances by Zarlinos' pupil Artusi in 1600.

In his not less than five years delayed reply to Artusi, printed as a foreword of his fifths madrigal-book, Monteverdi coined the slogan of a *seconda pratica*, which he contrasted against the *prima pratica* of the counterpointual orthodoxy.

The aesthetical postulates of the *seconda pratica*, which in fact articulate the principles that are met by figured bass music, are the following:

1. Music has to serve expression! (In the words of Monteverdi's brother: "*che l'oratione sia padrona dell'armonia e non serva.*"[3])
2. The adequate basis for evaluating dissonances are not primarily traditional compositional rules and conceptions of harmony[4] but the capability of dissonances to express the underlying text.

---

2 For a detailed analysis of at least some of the complex developments that took place before Monteverdi approached the scene see Alexander Beckers contribution in this volume.

3 Cf. Guilio Cesare Monteverdi (1607): "Dichiaratione della lettera stampata nel Quinto libro de suoi Madregali," in *Scherzi musicali a tre voci.* Quoted from Ehrmann (1989), p. 131.

3. The expression of the text has to be realized in form of an *imitational* representation. Objects of this imitation are the properties of the recitation of the text (*propria del parlare*) and the affects and passions, which are expressed by the content and the natural articulation of the text.
4. Music has to evoke emotions and affects in its listeners (*muovere l'affetto del animo*).
5. Therefore the polyphonic fabric of music has to be split up into monodic voices and an accompanying instrumental part based on a figured bass. Melody and not harmony is the primary medium of expressing and evoking affects.

Monteverdi closes his reply stressing, that "the modern composer is working on the foundations of truth."[5] But not only his programmatic words lead to the conception of *verità del'arte*[6] (as his brother put it on behalf of him), his concept of imitation as a form of representation seems to involve a relation to truth as well. Insofar representations can be *mis*representations they are connected to a (communicative) function, which has to be rendered in normative vocabulary. Accepting a (proposed) representation could count as taking it to be true accordingly.[7]

2. Is Truth a possible property of Music?

It is evident that the concept of truth is closely linked to nearly any explanation of our understanding of linguistic expressions. And introducing "truth" into the context of music would be a very short way to conceive music as something we can understand. But how could we assign truth values to musical expressions as we do assign them to assertional propositions? What would be truth conditions of musical expressions? Even just asking these questions

---

4 Concerning dissonances Zarlino taught: "That compositions should be made up primarily of consonances, and secondarily and incidentally of dissonances. And although every composition, every counterpoint, and in a word every harmony is made up primarily and principially of consonances, dissonances are used secondarily and incidentially for the sake if greater beauty and elegance. Taken by themselves, these are not very acceptable to the ear ..." See Zarlino (1558), pp. 41f.

5 Cf. Monteverdi (1605): "Preface to Il Quinto Libro de Madrigali a cinque voci," where he writes: "... & credete che il moderno Compositore fabrica sopra li fondamenti della veritá." Quoted from Ehrmann (1989), p. 128.

6 Cf. Monteverdi (1607), p. 133.

7 It is important to see that "imitation" in the list above should be understood as one means of representation among others. Because if Music would *essentially* only be the imitation of the properties of other forms of expression, for instance talk or declamation (as Monteverdi himself sometimes emphasize), this would raise at least two questions. First (in a normative perspective), wouldn't an imitation count as perfect when the imitated and the imitating become indiscernible? And second: Why should we be interested in imitations, if they are only expressions of something, that is already expressed in speech, gesture or mimic. Because I can't imagine any satisfying answers to this questions, I want to suggest, that successful imitation in the required sense is connected to normative conditions which demand more than pure similarity. Rather imitations, must be understood as means of representation, whose conditions of success transcend similarity insofar as they have to represent via those effects they evoke in the listeners. If imitation has to be conceived as a kind of representation, and representations in turn can be correct or incorrect, it follows – under the assumption that correct representations provide a true picture of the world – that imitations are related to truth themselves.

seems to threaten our intellectual health. But nonetheless Monteverdi is claiming that truth is a property which should be realized by musical works of art,[8] and of course he is claiming that his own music is true.

So let me examine some of the common philosophical positions concerning the relation between truth and art.

1. Adopting a pragmatist view we could say: Monteverdi was not a philosopher but a musician and his use of "truth" is more or less metaphorical. Taken literally, the predicate "true" does not apply to musical expressions, but only to propositions. What Monteverdi could only have had in mind, when talking about the truth of music, is an underlying *normative concept* that in some respect is *analogous* to truth. Following this line of thought we have to find some normative property which can be intelligibly attributed to music.

According to pragmatist intuitions, expressed for example by Habermas,[9] we could go on to say: While *assertional sentences* can be *true or false* and *normative sentences* can be *right or wrong* – expressive (speech) acts can be *truthful (wahrhaftig)* or not. But if we accept the idea that in the case of expressive articulations "truth" means "truthfulness", we have to face the problem that this interpretation neither seems to be *adequate* concerning what artists are doing nor *sufficient* to explain musical understanding:

(a) *Inadequate* is a conception of art in terms of truthfulness because it may be just *one aspect* of a work of art to express the artists inner world, or in Hegelian terms, his or her "Innerlichkeit". But very often this is not the whole story. What artists do or intend to do by realizing works of art is simply not restricted to expression and may instead include different kinds of representations (as), investigations of perception, transformations of experience and so on.

(b) *Insufficient* is such an explanation because the normativity of truthfulness cannot play the role of a normative basis which allows for *criticizing* a work of art, because our third-person-perspective on the artist's work does *not* enable us to challenge his or her truthfulness by appeal to intersubjectively accessible facts.

2. According to a *second* strategy to conceptualize "truth" as a property of artworks we could accept the view that truth is a language-specific property, but at the same time conceive truth only as a specification of a broader concept – another specification of which applies to music. As Goodman would have put it: Truth and its aesthetical equivalent just work out as *adequacy* named by different labels.[10] Given this revision of the concept of truth reduced to the fitting of sentences,[11] it is surprising that Goodman never spells out in detail what the aesthetical form of adequacy exactly would be. Instead of doing that he draws attention to the

---

8 I have to leave open the question whether we should suppose an epistemic or semantic notion of truth in this context.

9 Cf. Habermas (1981), pp. 447f.

10 Cf. Goodman (1968), chapter 6.

11 It should be noted, that Goodman's assimilation of truth and its aesthetical analog under the roof of adequacy is based on a well placed conceptual inaccuracy, because it is of course the relation between assertoric sentences and a theory (which they support) and not the relation between these sentences an those parts of reality they describe, which can be characterized as a relation of fitting adequately. Making intelligible that truth is a genus

symbolic character of art, a character, art – again – shares with science, and so he leaves us without any clear criterion to decide whether a work of art is adequate or not. If this is true, I simply can't see what Goodman's contribution to our understanding of truth as a possible property of works of art could be.

3. Following a Hegelian line of thought brings us to a *third* strategy of explaining the truth talk with respect to music or art in general. Discussing possible ends of art, Hegel determines the *genuine and inherent* purpose of art not as teaching, catharsis, moral improvement and so on, but as the *presentation of truth* in form of sensual articulations. Hegel understands art as a product of spirit, a product that articulates the same truth which philosophy is capable to express. But compared to philosophy music is seen as deficient insofar as it lacks *concepts* and therefore is not able to make truth *explicit*. In short: Because the genuine processing of the spirit takes place in a conceptual form, music can only be seen as a preliminary form of expressing the minds inner world.[12] Consequentially, the truth of art is derivative to conceptual truth and therefore does not possess any truth in itself.

I will not go deeper into Adorno's inversion of this Hegelian figure which recognizes art as a *superior* way to articulate truth because this inversion just compensates the consequences of his self-destructive criticism of conceptual thinking as intrinsically violent without giving us any intelligible arguments for conceiving of truth as an intrinsic feature of art. To argue, however, that art is a *superior* form of displaying truth (because it is aware of its status as an appearance) remains hopelessly metaphorical and leaves open the quest for a criterion that ensures the homogenity of a concept of truth, which also is able to cope with aesthetical and conceptual forms of truth.

Consulting philosophers as advocats for Monteverdi's talk of truth seems to end in confused philosophical debates and leaves us with either an insufficient and partly inadequate concept of truthfulness (in the case of Habermas), or with a very broad concept of adequacy (in the case of Goodman), or with the picture of art as a deficient or superior form of nonconceptual truth, which can only be characterized by appeal to trivial differences to its conceptual form (in case of Hegel or Adorno respectively).

But this depressing consequence certainly cannot be the whole story. Rather, as sincere interpreters of Monteverdi's conception of *vertità della musica*, we are obliged to follow the principle of charity, trying to make sense of what he said by maximizing its rationality.

However, there seem to be still one possibility – a postmodern one – in play: Maybe Monteverdi was dealing with a concept of truth completely different from those we nowadays have in mind. He could be engaged in discourses, involving candidates for truth, that disappeared over the centuries. But even if this is the truth about truth – although I didn't find anything in Monteverdi pointing to this possibility – our task would remain the same: Even such a historical relativization does not answer the question what it meant for Monteverdi to take music as a subject of epistemical discourses. Even this strategy would not ease our theoretical burdens. Rather, we have to face the full-blown problem of understanding.

---

of adequacy would instead imply to show how the latter relation, which is fundamental for truth in a naive sense can be understood in terms of adequacy.

12 Cf. Hegel (1840), vol. 13, pp. 82, 141, 142.

In what follows I will try to give a sketch for an alternative approach to the "truth" of a piece of music, which on the one hand addresses features of music, which Monteverdi stresses in his music as well as in his writings about music, and on the other hand tries to reconceptualize the talk of truth (of music) in terms of causes and effects.

Before we get into that, let me distinguish at least two functions, which musical works of art are capable to realize, namely *representation* and *expression*.

Let us first turn to representation. As I will spell out in the last part of this paper, I would like to suggest that the "Coronation of Poppea" can be read as the portrayal of a complex social development, as a depiction that emphasizes the ambivalence of feelings, motives and interests of people as such. I further want to claim that one aspect of the truth of this depiction can be rendered in psychoanalytic terms, because the picture of the acting persons is far too complex to be fully taken into account by everyday psychology. In particular it involves far more ambivalence than common explanations and rationalizations of actions are able to explain. The opera doesn't take the perspective of standard action theory or folk psychology – for what is shown goes beyond the reconstructive resources of reasons, conceived as causes of actions. Rather, it presents human beings as creatures that have to cope with the fact that their behavior is influenced by what are not at all rational sources, directly transparent to their self-understandings. Therefore it is not surprising, that an important aspect of the opera's diagnosis consists in the powerlessness of rationality, embodied in the central figure of Seneca.

So, as a representation or a presentation of *X as Y* – in our case, as a presentation of action *as* desperately influenced by drive, emotion and not very well understood motives – the opera puts forward a *diagnosis* or a *description*, which can be true or false. What is specific for this diagnosis is that it grants access to ambivalence, and its insistence on ambivalence and drive as relevant aspects of being human. It therefore can claim that any description which cannot accommodate ambivalence is not true, because it is *essentially incomplete*. Following this line of understanding "truth", we now see the opera as a specific kind of descriptional representation. This, however, would lead us to the disastrous conclusion that the opera (as such) is superfluous since anything what is said by means of the opera could be said by words, or at least could become sayable with words in the ongoing development of culture.

If we allow works of art to have this representational aspects, what is what I think we should do, the concept of truth which we can easily adopt will depend on these aspects. We then have to face the problem that although representational aspects *can* be properties of works of art, they are not specific for them. So Monteverdi must simply have had something different in mind when he talks of *verità della musica*.

The basic idea of my concluding attempt to explain what could be meant by "true music" is to put weight on the fact, that music is something *expressive*, that has *effects* on us, when we listen to it. Despite the fact that philosophers tend to draw a sharp distinction between the space of reasons and the realm of causes, my suggestion will link truth as a property of music closely to its causal effects on listeners. In order to give back peace of mind to philosophers troubled by this, let me add that my proposal will not challenge the distinction between causes and reasons, between the physical and the mental.

Before spelling that out, let me briefly introduce a concept of aesthetical *communication*, that John Dewey elaborates in his 1934 book *Art as Experience*. Dewey designs a model of aesthetical communication by reconciling the Kantian motive of an individualistic *experien-*

tal aesthetics with the Hegelian motive of a communicative social framework. Its basic structure is the following: P produces a *perceptible* phenomenon A in *anticipation of the reception* of A by potential recipients. And those recipients receive A with regard to its being intentionally produced as a perceptible phenomenon by P.[13] In my view, this analysis, which Dewey differentiates for producers and recipients as involving *active* aspects of designing and interpreting respectively, and more passive aspects of perceiving the relevant properties of the corresponding phenomenon, is of interest for our problem, because it involves a layer of empirical knowledge, that could be seen as backing up the talk of truth.

If it is plausible to understand art as a primarily communicative phenomenon, which I find – *pace* Goodman – strikingly true, then it is a matter of fact that *physical* phenomena are the means of communication. Consequently, any aesthetical experience involves a state of being affected by physical properties of things or processes. Of course, being affected by physical events or states is not sufficient for aesthetical experiences, but it is certainly a *necessary* condition. But if that is correct, then a composer, who tries to anticipate the reception of his composition, will also anticipate effects, that the physical events, which instantiate his music, will have on a listener. In this view, a composition involves an *empirical 'theory'* concerning the effects a piece of music will typically have on its listeners. Of course, it is not necessary for a composer to initiate empirical research about his future listeners. Rather, she can simply take her own reaction to her (imagined) music as a model for her prognosis. In inversion of the figure of self-interpretation as anticipated interpretation by others, she anticipates reactions of others on the basis of her own reactions.

It is important to see that such a theory doesn't imply any form of crude naturalism, because there is no need to understand the effects of music as merely physiological effects. A composer can see himself and his recipients as beings, whose dispositions to react in certain ways to a certain kind of music are deeply influenced by cultural and historical factors. And that would be quite appropriate in the case of Monteverdi, since the awareness of music's historicity originates in his time.[14] But far from reducing musical practices to the production of perceptible phenomena and far from endorsing the reducibility of musical categories to physical (acoustic) or functional categories,[15] we nonetheless have to face the fact, that music involves the production of sensual experience.

If we accept this layer of effects as one level of musical communication among others, we have no problems in admitting that even bodily effects are relevant aspects of our listening to music in general and in our case to "Poppea" in particular. This music is obviously the *cause* of proprioceptively imagined gestures. It manipulates our breath, it makes us experiencing developments of tension and relief, of speeding up and slowing down, in short: It produces bodily states that are typical for *emotional* states.

Suppose, a composer intended to portray certain emotional states by causing reactions, that are typical for those states, with the help of his or her music, a preliminary definition of the truth of a piece of music could be the following:

---

13 Cf. Dewey (1934).
14 Cf. Moyer (1997), pp. 185–202.
15 I try to give an account of this irreducibility as well as of understanding music (and nonlinguistic understanding in general) in terms of a theory of media in my book *Medien der Vernunft*, Vogel (2001).

(TiM) A piece of music *M* is *true* if and only if *M* has the effect that listeners of *M* experience those sensations that are typical for the emotional states (affects) that *M* is intended to present.

The main problem of this definition – a problem that Goodman addresses – is of course that musical expressions do not evoke the same sensations as those we experience, when we have the corresponding emotions in everyday life. If that were the case, the mood of typical listeners would be transformed by music to sanguinarity or loyalty within minutes; it would get them up to enter the stage and intervene in the play they attend. Perhaps one could diminish this problem by postulating the following: Insofar as music is not the cause of everyday affectual experience, music doesn't cause everyday affects as its effects. The idea is to say that music produces emotions that are – as it were – emotions in "brackets".

This of course seems to be just an ad-hoc-theory to get rid of Goodman's sarcasm. But I want to claim that it is an essential feature, may be of art as such, but certainly of opera, to present expressions of emotions that are *marked* – a feature, which is deeply rooted in the basic structures of the affectual communication, in which we were involved as infants. Let me in short point out the basic idea behind this claim, because in my eyes it seems to be also fruitful with regard to another question, any theory of opera has to answer: Why do these people on stage sing to communicate?

As an outcome of research, done by Gergeley and Watson, *parental affect mirroring* can be seen as the motor for the development of emotional self-awareness and self-control in infancy. Gergeley and Watson propose that "infants first become sensitized to their categorial emotion-states through a natural biofeedback process provided by the parents' 'marked' reflection of the baby's emotion displays during affect regulative interactions."[16] As we all can observe, parents often react to the affectual facial displays of their babies by reproducing the appropriate facial display themselves. Thereby it is of fundamental importance that the parents do not reflect the facial expression in the way a mirror would do, because this would simply induce or confirm the corresponding affectual state in the baby, which is not very desirable, say, in case of sadness or fear. So there must be a way for the baby to distinguish between facial expressions that express the emotional states of its parents and those that reflect the babies inner feelings. There must be a specific property of reflecting expressions, that realizes their *marked* character. And, as we all implicitly know, this feature is *exaggeration*. It is the exaggerated version of an affectual behavior that we all could imitate when we try to imitate adults interacting with prelinguistic infants, a behavior that gains the status of an indicator for the emotional state the baby is in.

This as a background – my thesis now is, that *music is a successor of these marked emotional expressions*, and as a successor[17] of an indicator, we would not expect such expressions to produce feelings[18] but to represent and individuate them.

In this perspective, Monteverdi's stylistic means of *genere molle, genere temperato, genere concitato* and so on can be seen as organized forms of sophisticated exaggerations that jointly realize a *stile rappresentativo* – a style that aims to represent by marked expressions.

---

16 Gergeley/Watson (1996), pp. 1181-1212.
17 A successor, by the way, that has precursors in the playing of children. Cf. Fonagy/Target (1996), pp. 217-233.
18 As Goodman supposes for the sake of ridiculing emotions as prominent objects of art.

In this perspective, it is not longer a problem to explain, why people in operas do not talk to each other, but *sing* to communicate. Singing in order to communicate is just one specific form to *mark* a representational expression. Thus, we should reformulate our definition in the following way:

(TiM') A piece of music M is *true* if and only if M is a marked expression which causes those effects in listeners, that enable them to identify the emotional states (affects) that M is intended to present.

Note, that I do not want to propose that we should subscribe to (TiM'). Rather the definition is the utmost I can do to keep truth as a property of music in play in order to understand Monteverdi. The definition only aims to make *intelligible* a notion of truth of music.[19] Besides its ability to allow an account to truth as a musical property, (TiM') solves a problem which, as I mentioned above, is normally associated with the representational account of music, namely the problem that musical representations are substitutable by linguistic ones as long as they represent the same. Because (TiM') talks of effects that music has on listeners, it ensures the untranslatability of musical means[20] insofar our identifying of the expressed and thereby represented emotions may depend on the presence of specific causes. In other words: What can be done by specific musical means cannot necessarily be done by words or concepts.

It should be clear by now that I do not want to claim that this definition provides an overall solution to the problem of the truth-talk in question. However, it opens up a perspective on music that is not reductionist at all, although it is formulated partly by appeal to causal vocabulary. Rather, it points to the possibility that music, insofar as it is an autonomous practice of forming causally effective units, opens up the space for structuring our emotionality and thereby grants access to emotions we otherwise would not have.

Monteverdi was not much interested in a well-founded system of affects. He distinguished three basic affects, namely *ira*, *temperanza* and *humilità*. But of course only *ira* (anger) is – strictly spoken – an affect, whereas temperance and humility are virtues.[21] But unlike his theory his music is a milestone in the process of civilizing emotions, because it offers sophisticated forms of musical articulations by means of which we can learn to look at our emotions, to address and differentiate our passions. In the last section I present a very rough analysis of some of the specific means Monteverdi used in "Coronation of Poppea".

---

19 (TiM') of course doesn't express all conditions that have to be fulfilled by the listeners, above all the ability to structure what they are listening to as music. In the context of an opera one could think that (TiM') should mention criteria ensuring that true music has to relate emotions to their correct owners respectively to adequate objects. But on one hand these criteria would be specific for dramatic music and on the other hand they would involve quasi-predicative relations that cannot be met by music alone.

20 The untranslatablity of music is a widespread intuition (shared for example by Hanslik (1854), p. 79, Webern (1932), p. 46, and Langer (1942), p. 243) that often is expressed by the unintelligible conception of music as an untranslatable language.

21 Cf. Ehrmann (1989).

3. Basic Structures in Monteverdi's "Coronation of Poppea"

Let us take a closer look at the internal structure of the opera. First some basic facts. "The Coronation" is the last of Monteverdi's seven operas, only three of which survived, and it is the first opera in the history of this genre that deals with a historic subject.[22] The opera was composed and first performed in the year 1642 – one year before Monteverdi's death – in the public Opera House of Venice, the city where Monteverdi has been working for 30 years as choirmaster of the world-famous Choir of St. Marco. The libretto was written by Gian Francesco Busenello, a venetian noble, trained as a lawyer, who had been a senior member of the *Accademia degli Incognito*, which had been the "leading intellectual and political force in mid–seicento Venice."[23] The members of this Accademia were involved in many domains of Venetian social life and ran the most successful theatre during the early period of Venetian Opera.

The plot of the opera is quickly told. Busenello used Tacitus and, even more important, a roman play (with the title *Octavia*), which was said to be written by Seneca himself, as basis for his libretto. In short, Busenello is telling a story well known to his audience.

Act One

Nero, the roman emperor is in love with Poppea, a courtesan married to Ottone, who returns to Rome from Lusitania, where he had been serving the empire as praetor. But instead of a lucky reunion, Ottone finds his house guarded by Neros soldiers. He eavesdrops a conversation of these soldiers, who express their commiseration with the fate of the empress Octavia. While she is in despair over the infidelity of her husband, Nero enjoys clandestine nights with his secret love Poppea. The following morning, Nero parts from Poppea promising her to get divorced from Octavia in order to be free to marry her. Poppea praises Love and Fortune and ignores all warnings of her maid servant Arnalta concerning a possible revenge Octavia might take.[24] In the Palace the crestfallen empress meanwhile awaits Seneca, who gives her the advice, not to follow her anger but to stay with the virtues. But Octavia pleads him to make public the adultery. Shortly after that, Seneca gets informed by Nero of his intention to get rid of Octavia in order to marry Poppea. After reminding Nero of the lawlessness of his plan, Seneca has to leave. Poppea very wells knows how to use Nero's anger and induces him to speak a sentence of death over Seneca. After Nero has left, Ottone tries to regain Poppea's affection, but without success. Hurt by this rejection, he tries to find relief in the arms of Poppea's friend Drusilla, whose hand he once turned down in favor of Poppea.

---

22 So the opera leaves behind the well established (and more or less regressive) mythological sujets of the Arcadian wonderland and transcends the self-glorification of music personified by Orfeo, who stands for the musicalization of the world (cf. Katz (1986), pp. 112f.).

23 Cf. Rosand (1985), p. 36.

24 It should be remarked that the secondary characters' function is not only to give the emotions of the main characters clearer shapes by strong contrasts, for example by down-to-earth-commentaries in the scene of Seneca trying to comfort Octavia. In many scenes, this figures can instead be seen as externalized instances of self-reflection, that articulate aspects of the complex psychological situation the main characters are in, aspects which can't be articulated trustworthy in a monological way.

Act Two

A freed slave delivers Nero's order to Seneca that he should commit suicide. Seneca says farewell to his friends, whose efforts to prevent him from carrying out Nero's decree are in vain. Nero feels profound relief by the death of Seneca and indulges in fantasies about soon reaching his goal to marry Poppea. Octavia on the other hand threatens Ottone with torture and death in case he won't slay Poppea, so that Ottone reluctantly agrees. With the help of Drusilla, who lends him some of her clothes, Ottone approaches the sleeping Poppea, ready to murder her. But Amor intervenes and saves the life of Poppea, who sees the attacker fleeing, identifying him as Drusilla.

Act Three

Instead of her joyful expected Ottone, a lictor appears who arrests the supposed murderess Drusilla. When brought before Nero, she admits the murder despite facing painful death. But Ottone confesses, that he has attempted to murder Poppea, forced by Octavia. Upon hearing this, Nero banishes Ottone, allowing Drusilla to go with him, and, now seeing a legal way to get rid of his wife, he expels Octavia as the instigator of an attempted murder. Octavia leaves Rome and Nero crowns Poppea his empress.

Most summaries of the opera sound more or less like this.[25] Given that plot, the opera seems not to be too complex. But "The Coronation" neither is a "Spieloper" nor an opera that lives on intrigues. It simply does not focus on lines of action, but on social constellations between persons, whose actions and relationships are deeply shaped by passions. What is fascinating about "The Coronation" is not the excitement generated by action, but the psychological tensions within the acting characters and between them.[26]

Let me show at least roughly by concentrating on the first act, how this social and psychological configuration[27] is set up by dramatic and musical means:

First of all the characters do not only appear in the context of the story, but also in the light of a competition between Fortuna, Virtu, and Amor, who are involved in a hegemonial struggle throughout the opera. In the light of this struggle, all characters appear as figures on the chessboard of divine – means psychic – powers, who exert their influence on them. As one can see in the schematic Synosis,[28] Act One is structured symmetrically, centered on Scene 7, which shows Seneca, the representant of Virtue, in isolation. This central scene is framed by

---

25 Cf. for example Bleß (1981), or Kloiber (1973), pp. 308-312.

26 In addition to that should be noted that the opera breaks the barriers of what is seemly to be presented on stage. It does not only tell a historical story but a story of murder and crime. Perhaps this has been acceptable for the Venetian audience, because citizens of the Venetian republic were able to see the crime as an outcome of roman decadence.

27 See Fig. 1, p. 314 below.

28 See Fig. 2, p. 316-7 below.

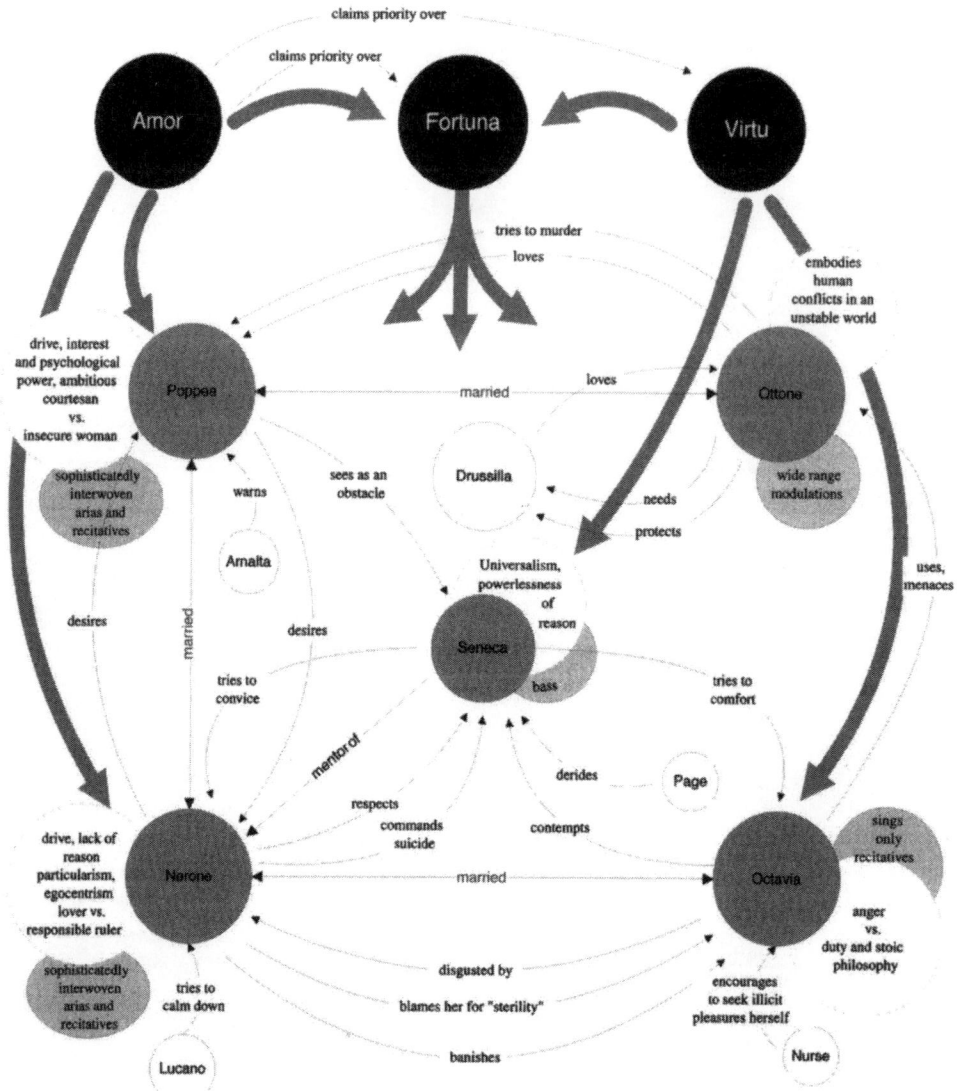

Figure 1: Basic Social Relations in "L'incoronazione di Poppea"

other scenes, in which Seneca talks to Octavia and Nero respectively, the members of the royal couple. These scenes are themselves framed by other scenes presenting the illegitimate pair of Nero and Poppea. All those scenes are in turn embedded in the opening and closing ones in which Ottone appears. But Ottone is not simply introduced to the audience at the beginning of the opera: rather he becomes the witness of the above mentioned conversation between the two soldiers, which indicates to him that something is wrong. What is wrong becomes clear from that conversation, and in this way two main figures of the play are introduced, namely Nero and Poppea. They appear in the next scene as last night's lovers, and it is already here that we encounter Monteverdi's enormous ability to sketch the relationship between the two far beyond what is said by them. From what is *said*, it becomes completely clear that Nero is the emperor. However, the *music* reveals that the person in power really is Poppea. And another aspect comes in play that directly addresses us, the listeners: Although we know, that the love of Poppea is soaked with her will to gain power, we are nonetheless enchanted by the beauty of her singing, precisely like Nero, who obviously becomes the victim of her seductive power. Our own wishful thinking partly puts behind our caution and opens us for the effects of her seductive lines.

In this way the Opera pulls us into the kind of ambivalence it deals with. None of the characters is simply only good or evil. There is an inherent ambivalence in all of them, which is not resolved by reflection but rather expressed through action. In connection with this point it can be claimed that Seneca is assigned the central role in the opera. As the representant of rationality, universalism and stoic virtues, he is not able either to integrate rationally the dispersing aspects of the other characters or to tame them.

This is the reason why Seneca is shown alone and powerless in the middle of Act One and why his death in the very middle of the whole opera is the turning point (*peripetie*) of the drama. After his death the whole machinery is driven exclusively by emotions and particular interest. This is also underwritten by the fact, that the last duet of Poppea and Nero consists of nearly the same musical material as that of Amor and Fortuna in the prologue.

Given this line of interpretation to be adequate, one could say that the opera puts forward a specific picture of human action in terms of the dramatic involvements it displays. This is a picture in which actions are significantly structured also by influences, that are not rational at all. And, of course, such a picture can be true or false. If it is true, however, this kind of truth is not a specific property of the opera, because it can be stated linguistically. That the opera transcends the sort of truth, which can be articulated linguistically, depends on it being the subject of our experience.

It is quite easy to agree with such a linguistic diagnosis. But the opera urges us to discern precisely those emotions in us, which determine the actions of the characters on stage. It does so by appeal to musical means which, against the background of a history of musical socialization, elicit bodily effects in the listeners – effects, which in turn then allow for the identification of those emotions. When he talked about "true" music, Monteverdi perhaps thought of a music, which in this sense successfully uses expressive means in order to achieve representational goals.

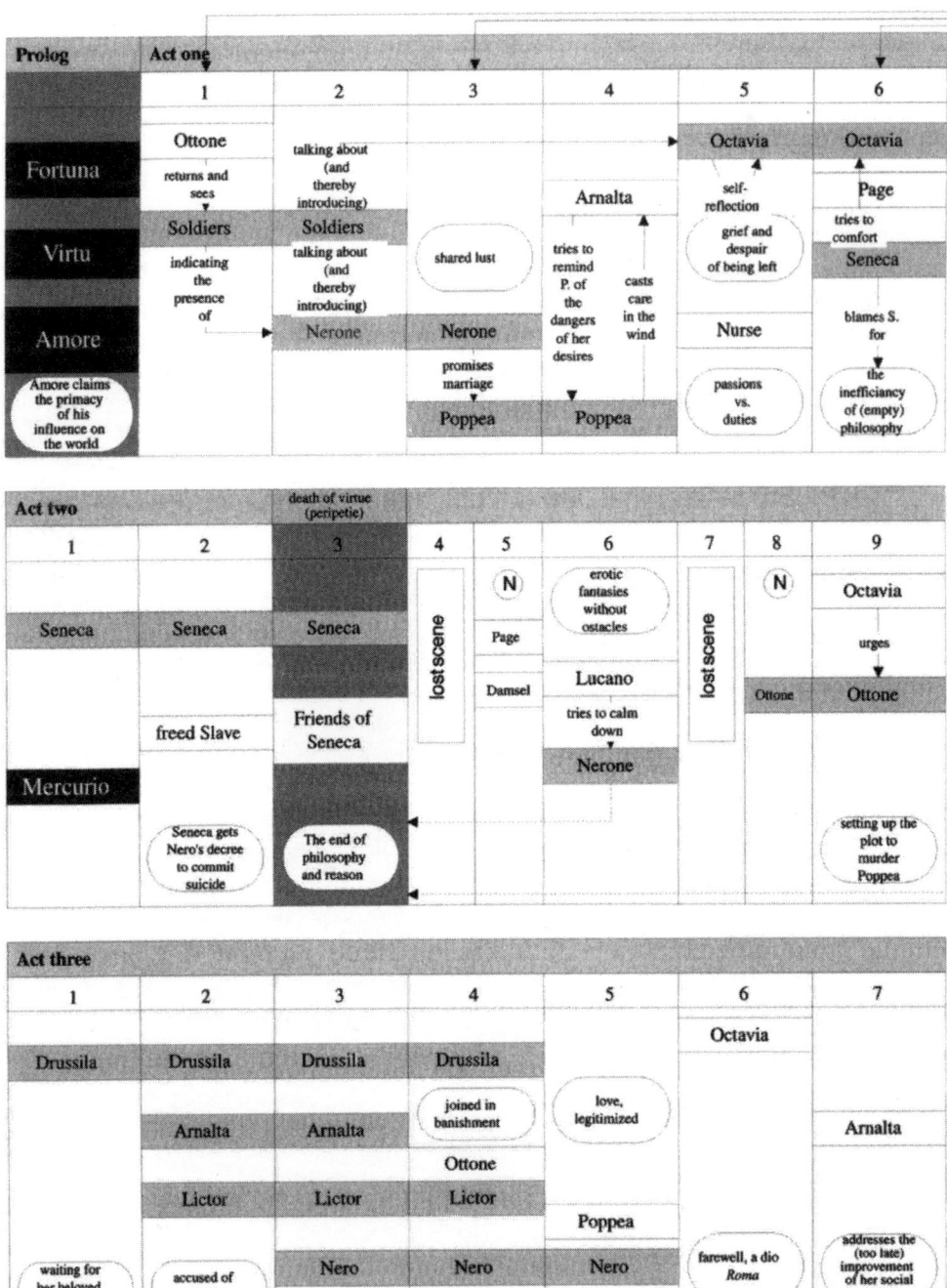

Figure 2: Synopsis of "L'incoronazione di Poppea"

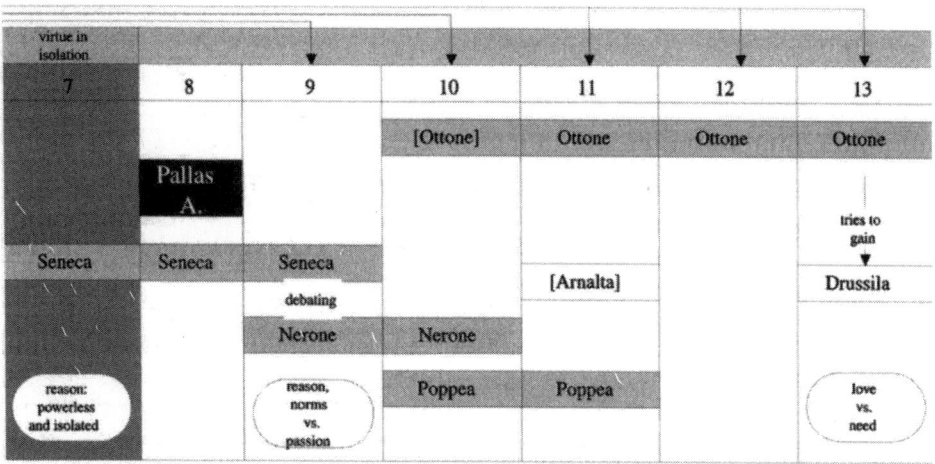

virtue in isolation						
7	8	9	10	11	12	13
			[Ottone]	Ottone	Ottone	Ottone
	Pallas A.					
						tries to gain
Seneca	Seneca	Seneca		[Arnalta]		Drussila
		debating				
		Nerone	Nerone			
reason: powerless and isolated		reason, norms vs. passion	Poppea	Poppea		love vs. need

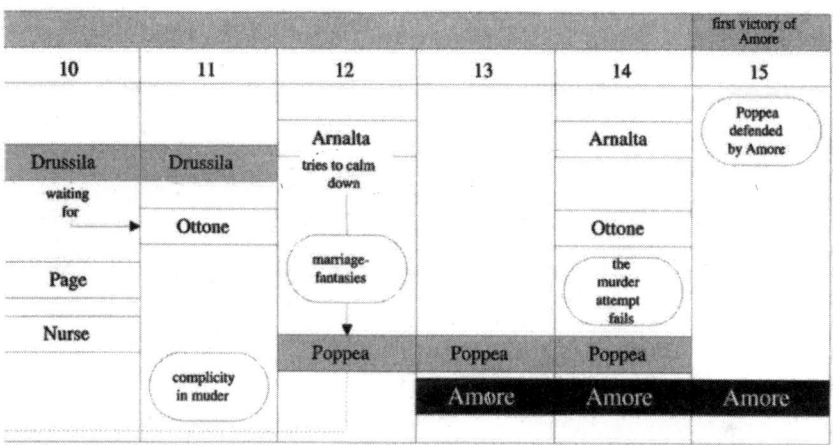

					first victory of Amore
10	11	12	13	14	15
					Poppea defended by Amore
		Arnalta		Arnalta	
Drussila	Drussila	tries to calm down			
waiting for				Ottone	
	Ottone	marriage-fantasies		the murder attempt fails	
Page					
Nurse		Poppea	Poppea	Poppea	
	complicity in muder		Amore	Amore	Amore

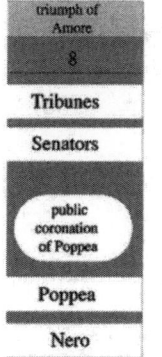

triumph of Amore

8

Tribunes

Senators

public coronation of Poppea

Poppea

Nero

Ottone, husband of Poppea, roman praetor	Soldiers (of Nero's guard)
Poppea, wife of Ottone, later married and crowned by Nerone	Lucano, servant of Nero
	Page (of Octavia), named Valetto
Nerone, husband of Octavia, lover of Poppea, roman emperor	Nurse (of Octavia)
	Freed Slave
Octavia, wife of Nerone, crestfallen empress	Friends of Seneca
Seneca, philosopher, mentor of Nerone, who decrees Seneca's suicide	Lictor (Policemen)
	Tribuns
Drussila, friend of Poppea, whose hand Ottone had once turned out, supposed murdress of Poppea	Consuln
Arnalta, maid servant of Poppea	

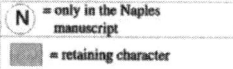 = only in the Naples manuscript

= retaining character

References

Adorno, Theodor W. (1956): "Fragment über Musik und Sprache," in *Gesammelte Schriften, vol. 16: Musikalische Schriften I-III.* Frankfurt/M.: Suhrkamp, 251-256.

Adorno, Theodor W. (1970): *Ästhetische Theorie.* Frankfurt/M.: Suhrkamp 1973.

Beat, Janet E. (1968): "Monteverdi and the Opera Orchestra of his Time," in D. Arnold and N. Fortune (eds.): *The Monteverdi Companion.* London, 277-302.

Bleß, Marion (1981): "Synopsis," in Harnoncourt, Monteverdi: *L'Incoronazione di Poppea.* Hamburg: Teldec.

Chafe, Eric T. (1992): *Monteverdi's Tonal Language.* New York/Toronto: Schirmer Books.

Dahlhaus, Carl (1986): "Second pratica und musikalische Figurenlehre," *Musik-Konzepte* Nr. 88 (1995), 3-12.

Davies, Stephen (1994): *Musical Meaning and Expression.* Ithaca, London: Cornell UP.

Dewey, John (1934): *Art as Experience.* New York: Capricon Books 1958. (German: *Kunst als Erfahrung.* Frankfurt/M.: Suhrkamp 1988.)

Ehrmann, Sabine (1989): *Claudio Monteverdi. Die Grundbegriffe seines musiktheoretischen Denkens.* Pfaffenweiler: Centaurus.

Fischer, Kurt von (1977): "Versuch einer Interpretation von Monteverdis Opern," *Musik-Konzepte* Nr. 88 (1995), 13-19.

Fonagy, P. and Target, M. (1996): "Playing with Reality: I. Theory of Mind and the Normal Development of Psychic Reality," *International Journal of Psycho-Analysis* 77 (1996), 217-233.

Gergeley, G. and Watson, J. (1996): "The Social Biofeedback Theory of Parental Affect Mirroring: The Development of emotional Self-Awareness and Self-Control in Infancy," *International Journal of Psycho-Analysis* 77 (1996), 1181-1212.

Goodman, Nelson (1968): *Languages of Art.* Indianapolis: Hackett Publishing Company, [2]1976. (German: *Sprachen der Kunst.* Frankfurt/M.: Suhrkamp 1995.)

Habermas, Jürgen (1981a): *Theorie des kommunikativen Handelns, vol. 1: Handlungsrationalität und gesellschaftliche Rationalisierung.* Frankfurt/M.: Suhrkamp, [3]1985.

Hanslick, Eduard (1854): *Vom Musikalisch Schönen. Ein Beitrag zur Revision der Tonkunst.* Leipzig 1902.

Hegel, G. W. F (1840): "Vorlesungen über die Ästhetik," in *Werke in zwanzig Bänden*, vols. 13-15. Frankfurt/M.: Suhrkamp 1970.

Kant, Immanuel (1790): "Kritik der Urteilskraft," in *Werkausgabe,* vol. 10 (ed. W. Weischedel). Frankfurt/M.: Suhrkamp 1978.

Katz, Ruth (1986): *Divining the Powers of Music.* New York: Pendragon Press.

Kivy, Peter (1980): *The Corded Shell.* Princeton: Princeton UP.

Kloiber, Rudolf (1973): *Handbuch der Oper,* vol. 1. München: DTV 1978.

Langer, Susanne (1946): *Philosophy in a New Key.* Cambridge/Mass./London: Harvard UP 1956.

Leopold, Silke (1982/93): *Claudio Monteverdi und seine Zeit.* Laaber: Laaber [2]1993.

McClary, Susan (1989): "Konstruktionen des Geschlechts in Monteverdis dramatischer Musik," *Musik-Konzepte* 88 (1995), 34-56.

Monteverdi, Claudio (1966-68): "L'incoronazione di Poppea," in G. F. Malipiero (ed.), *Tutte le opere di Claudio Monteverdi,* vol. XIII. Wien: Eberle, 1-280.

Moyer, Ann (1997): "Musical Scholarship in Italy at the End of the Renaissance, 1500-1650: From Veritas to Verisimilitude," in Donald R. Kelly (ed.), *History and the Disciplines. The Reclassification of Knowledge in Early Modern Europe.* Rochester/New York: University of Rochester Press, 185-202.

Osthoff, Wolfgang (1958): "Neue Beobachtungen zu Quellen und Geschichte von Monteverdis 'Incoronazione di Poppea'," *Die Musikforschung* 11 (1958), 129-138.

Palisca, Claude V. (1968): "The Artusi-Monteverdi Controversy," in D. Arnold/N. Fortune (eds.), *The Monteverdi Companion.* London: The University Press Glasgow, 133-166.

Rosand, Ellen (1985): "Seneca and the Interpretation of L'Incoronazione di Poppea," *Journal of the American Musicological Society* (1985), 34-71.

Siegele, Ulrich (1984): "Wie ist Monteverdis 'seconda pratica' satztechnisch zu verstehen?," *Musik-Konzepte* 83/84 (1994), 31-102.

Strunk, Oliver (ed.) (1952): *Source Readings in Music History,* vol. 2: *The Renaissance.* London, Boston: Faber and Faber 1981.

Titone, Antonio (1995): "'E quai nuovi rumori'. Neben- und komische Rollen in Monteverdis Musiktheater," *Musik-Konzepte* 88 (1995), 20-33.

Vogel, Matthias (2001): *Medien der Vernunft. Studien zu einer Theorie der Rationalität und des Geistes auf Grundlage einer Theorie der Medien.* Frankfurt/M.: Suhrkamp.

Webern, Anton (1932): *Der Weg zur neuen Musik Vorträge 1932/33.* Wien 1960.

Zarlino, Gioseffe (1558): "Istituzioni armoniche," in Oliver Strunk (ed.), *Source Readings in Music History,* vol. 2. London/Boston 1981.

Bernhard Stumpfhaus

Lisez l'histoire et le tableau –
Logik und Poetik im Werk Poussins

Der Maler und gelegentliche Skulpteur Nicolas Poussin gilt seit je als *peintre philosophe*. Dieser Name wird nicht allein damit begründet, dass der Künstler sich nach Art des *pictor doctus* ausgiebig mit bildlichen und schriftlichen Überlieferungen der Antike beschäftigt und diese umfangreich in seinem Werk verarbeitet hat. Vielmehr geben seine Historien und Landschaften allen Anlass, als tiefsinnige Allegorien von bürgerlich-stoischer Moral, geprägt vom Skeptizismus Montaignes, gedeutet zu werden. Poussins *raison* wird hervorgehoben, als eigentliches Produktionsorgan seiner Werke erkannt, deren angemessenes Äquivalent auf Seiten des Betrachters die Apperzeption sei. „Nicht allein unsere Lust (nos appétits) soll urteilen, sondern die Vernunft"[1], mit dieser Bemerkung preist Poussin sich selbst als aufgeklärten Rationalisten.

Die philosophischen Qualitäten seiner Werke werden aber nicht nur in der gelehrten Ikonographie erkannt, sondern auch in der strengen, nach den Regeln der aristotelischen Dichtungslehre organisierten, Struktur der Figurenkompositionen. Besonders im Rahmen der Debatten um seine Werke an der französischen Akademie werden immer wieder die Linearität, Logizität und die Klarheit gelobt, mit denen Poussin seine Bilderzählungen strukturiert, Eigenschaften, die im 17. Jahrhundert allein dem begrifflichen Denken vorbehalten sind. Bemerkungen beispielsweise zur *Landschaft mit einer Schlange*[2] legen eine Analogie zwischen der Ordnung der Historie und der Sequenz eines Satzes nahe[3]. Wie Subjekt, Prädikat, Objekt sich notwendig und linear aneinander reihen, so folgen die einzelnen Handlungsteile, etwa Mord, Entsetzen und Staunen notwendig, linear und irreversibel aufeinander[4]. Das Interesse der frühen Akademiker an der grammatikalischen Struktur und der Logizität einer durch ein Gemälde dargestellten Geschichte interpretiert Puttfarken als ein Interesse weniger an der Malerei als vielmehr am rationalen Gedanken selbst und seiner logischen Verknüpfung. „Linearität, Logizität und Klarheit", so Puttfarken, „wurden jenseits der Reichweite der Malerei gesehen. Sie wurden gewertet als die Grundbedingungen unseres [begrifflichen] Verstehens und unserer Weise, vernünftig [d. h. sprachlich] die Welt zu ordnen"[5]. Historienmalerei wird in dieser Interpretation als ein Organ verstanden, die Welt logisch zu

---

1 „Nos appétits n'en doivent pas juger seulement, mais la raison", Poussin an Chantelou vom 24. Nov. 1647; s. Jouanny (1968) [11911], S. 372; Blunt (1989) [11964], S. 135.

2 1648, London, National Gallery, s. Ausstellungs-Katalog Paris (1994), S. 406-08; Thuillier (1994), S. 259. Im Folgenden nach Puttfarken (1985), S. 12-15.

3 So Puttfarkens Interpretation von Fénelons erdichtetem Dialog zwischen Leonardo und Poussin; in: Œuvres de Fénelon, Archevêque-Duc de Cambrai, Tome neuvième, Dialogues des Morts [11712], Paris 1822, Dialogue LI, Léonard de Vinci et Poussin, S. 263-271.

4 „The sequence of the snake with the dead body, the fleeing man, and the surprised woman is as irreversible as the sequence of subject-verb-object in the French language" (Puttfarken (1985), S. 15).

Abbildung 1: Nicolas Poussins *Mannalese*

repräsentieren, nach den Maßgaben sprachlich geprägter Rationalität. Poussin kann deshalb als *peintre philosophe* gelten, weil er seine Historien am begrifflichen Denken orientiert, seine Kompositionen an der Struktur eines geschriebenen Textes ausrichtet und damit klar und distinkt die ansonsten nur begrifflich adäquat fassbare Ordnung der Welt zur Anschauung bringt. Diese Orientierung auf die begrifflichen Qualitäten Poussins gilt auch dort, wo seine poetischen Fähigkeiten gewürdigt[6], seine künstlerische Imagination und Phantasie diskutiert[7] werden.

---

5 „Linearity, logicality and clarté which, being normally considered as beyond the reach and the limits of painting, were thought of as hall-marks of French esprit and thinking and which were seen as the basic conditions of our understanding and our rational ordering of the world", ebd.

6 So wurden Poussin jüngst für seinen Metapherngebrauch formallogische Grundlagen zugewiesen. Colantuono bringt für die Interpretation des Pariser Selbstporträts Poussins einen rhetorischen Syllogismus gemäß der Definition der aristotelischen Poetik: „J'ai appliqué une forme de logique déductive, qui pourrait s'énoncer comme un syllogisme aristotélicien: a) la toile vierge porte l'image de Nicolas Poussin; b) la seule image visible sur cette toile est une ombre; par conséquent, c) l'image de Poussin est une ombre"; s. Colantuono (1996), Tome II, S. 649.

7 „Even Bellori declared that ‚imagination gives greater wisdom to a painter than imitation'. However, Poussin appears to have held to the view that imagination should not be divorced from rational imitation. For him, invention or ‚fantasy' was founded on reason and judicious imitation, rather than being in contradiction to it"; s. Pace (1999), S. 89.

Die Einschätzung von Poussins Werken als Texte treibt jedoch eigenartige Blüten. Voraussetzung für einen lesbaren Text ist, dass Zeichen vorhanden sind, die gelesen werden können. Dieser Umstand mag in den Diskussionen um Poussins Gemälde der Grund dafür sein, den tatsächlich gegebenen Bildzeichen solche beizufügen, die es gar nicht gibt, die aber nötig sind, dem Geschehen im Bild eine diskursive Ordnung zu geben. In einer der prominentesten Debatten der Kunstgeschichte, in der Sitzung der Akademie Royale[8] zur *Mannalese*[9] (Abb. 1, S. 322), war der Vorwurf zu vernehmen, Poussin halte sich nicht an den Text der Bibel, da er das Manna vom Himmel fallend darstelle[10]. Tatsächlich hat Poussin aber nie vom Himmel fallendes Manna gemalt[11]. Doch ohne das Manna von oben gibt es nicht die von Gott ausgehende, fortfolgende Logik des Geschehens, wie sie im Bild der erhobene Zeigefinger des Moses bedeutet[12].

Auch die neuere Kunstgeschichte verfährt hier nicht zurückhaltender als ihre namhaften Vorgänger. Sheila McTighe, die in ihrer Arbeit Poussins Landschaften als hieroglyphische Bildtexte bestimmt[13], beobachtet im *Sturm*[14] einen Blitz, den es dort nachweislich[15] nie gegeben hat. Sie charakterisiert den dargestellten Augenblick als ein „violant peak of action in the bold of lightning"[16]. Dass sie im Bild einen Blitz sieht, entbehrt nicht der Logik. Sie liegt weniger darin, dass sich McTighe in ihren Betrachtungen auf den Stich Louis Châtillons[17] zu

Es stellt sich für diese Auffassung von Pace die Frage, worin etwa bei dem querliegenden Dampf in Poussins *Orion* (New York, Metropolitan Museum) oder im unnatürlich stillen See der *Gewitterlandschaft mit Pyramus und Thisbe* (Frankfurt am Main, Städelsches Kunstinstitut) die gewissenhafte Imitation liegen mag. Immerhin ist ja gerade die Unwahrscheinlichkeit der ruhigen Seeoberfläche einer schlüssigen Deutung dieser Gewitterlandschaft Poussins im Wege.

8 Le Bruns zur sechsten Konferenz der *Königlichen Akademie der Malerei und Skulptur* am 5. Nov. 1667. Zu der Debatte s. Literatur im Ausst.-Kat. Paris, S. 263; danach vor allem: Schlink (1996); Oy-Marra (1997); Held (2001), bes. S. 92-105, S. 339-67 mit weiterer, aktueller Literatur.

9 Zwischen 1637 und 1639 für Paul Fréart de Chantelou, Öl/Lw, 149 cm x 200 cm, Paris, Louvre, Inv. 7275; am oberen und unteren Rand vergrößert, wodurch die Bewegungsdynamik im Bild geschwächt wird. Wie eine druckgraphische Reproduktion (Guillaume Chasteau 1680) nahe legt, scheint die *Mannalese* an ihrer rechten Seite etwas beschnitten zu sein (Ausstellungs-Katalog Paris (1994), S. 262). Diese Beobachtung ist von Bedeutung, da eine in der Breite vorgenommene Veränderung die Mitte des Bildes verschiebt; s. Ausstellungs-Katalog Paris (1994), Kat.-Nr. 78, S. 263; Thuillier (1994), Kat.-Nr. 135, S. 256.

10 Jouin (1883), S. 63; deutsch: Schlink (1996), S. 46; Held (2001), S. 423-24.

11 In der Reproduktion der *Mannalese* Poussins durch J. Le Maire, in Les Traits de l'histoire universelle von 1769, wird der Mannaregen sichtbar ausformuliert. Der Stich Le Maires zeigt deutlich das, was von den Betrachtern der Akademie in das Bild hineingelesen wurde, das wie Schneeflocken vom Himmel fallende Manna.

12 Selbst die Verteidigung Poussins gibt zu, dass das Manna vom Himmel fällt: „C'est ce qui l'a obligé de représenter cette manne tombant du ciel, parce qu'il ne peut autrement faire connaître que c'est d'où elle vient. Car si on ne la voyoit tomber d'en haut, et que ces hommes et ces femmes la ramassassent seulement à terre, on la pourroit prendre pour une graine ou pour quelque fruit"; s. Jouin (1883), S. 63. Diese Begründung übersieht Poussins eigentliche Strategie, den Ursprung der wunderbaren Nahrung anzuzeigen: Es sind dies der Finger des Moses wie der Rock des Mädchens rechts im Bild. Damit übersieht die Verteidigung einen zentralen Aspekt Poussinscher Malerei, nämlich eine Historie von den Wirkungen, nicht aber von den Ursachen her zu beschreiben.

13 McTighe (1996).

14 1651, Rouen, Musée des Beaux-Arts, Ausstellungs-Katalog Paris (1994), Kat.-Nr. 200, S. 449, Abb. S. 451; Thuillier (1994), Kat. Nr. 200, S. 261.

15 Das Röntgenbild (Rouen, Musée des Beaux-Arts) des *Sturms* zeigt keine Überarbeitungen, die einen ursprünglichen Blitz erahnen lassen; Abb. in: Austellungs-Katalog Frankfurt (1988), Abb. 9, S. 28.

16 McTighe (1996), S. 32, 37.

beziehen scheint. Tatsächlich bringt die zerstörerische Lichterscheinung erst die lineare Folge von kausal aufeinander bezogenen Ereignissen hervor. Der einschlagende Blitz verursacht das Herunterbrechen des Astes, beide Unfälle das Zusammenzucken von Mensch und Tier. Ohne Blitz ist die Szene in ihrer Kausalstruktur völlig offen.

Doch nicht allein die sprachliche Fixierung der Interpreten, deren Deutung sich als Rekonstruktion der pensée, der idée der Komposition versteht, und sich so implizit zum Prätext des Gemäldes erklärt, ist für diese Zurichtung des Bildes verantwortlich. Es ist vor allem die suggestive Kraft von Poussins Figuren, die das zu imaginieren veranlasst, was gar nicht vorhanden ist. Im Fall der *Mannalese* ist es die junge Frau, die wie ein Sterntalermädchen ihren Rock schürzt und damit den Betrachter bewegt, das Manna tatsächlich vom Himmel fallen zu sehen. Im *Sturm* ist es das scheinbar grundlose Abbrechen des Astes über den zusammenzuckenden Bauern, welches die optische Sensation einer Ursache des Geschehens, des Blitzes, hervorruft.

Neben der Beobachtung von der Suggestivkraft der in den Gemälden vorgestellten Bildelemente ist es vor allem die von Poussin aus der Musik in die Malerei transferierte Moduslehre, die Skepsis aufkommen lässt an einer allzu emphatischen Interpretation der Gemälde Poussins als ‚Texte‘. Die Moduslehre betrifft das Gesamt der Gemäldekomposition, wie betont wird[18], und hat die Funktion, eine „union générale"[19] zu konstituieren. Diese vermittelt als expression générale dem Betrachter den Eindruck des Hauptsujets. Das unmittelbare Erfassen des Gesamt einer Stimmung, des Modus, hat jedoch wenig zu tun mit dem sukzessiven Fortschreiten des Betrachterblicks beim Entziffern der einzelnen Bildzeichen, mit der lesenden Rekonstruktion der durch die Verknüpfung der Einzelhandlungen entwickelten Kausalität. Die Forderung der Moduslehre nach einem einheitlichen Charakter der Komposition, der sich ganzheitlich und momentan dem Betrachter erschließt, widerspricht dem Verständnis der Historien Poussins als in der Zeit fortlaufende logische Strukturen, die, analytisch aufgetragen, lesend synthetisiert werden wollen. Dies alles ist für mich Anlass genug, Poussins Historien noch einmal in Hinsicht auf ihren Textcharakter zu diskutieren.

Zwei prinzipiell unterschiedliche Weisen der Textabhängigkeit von Geschichtsmalerei sind in den Debatten der Akademie um die Werke Poussins unterscheidbar. Die eine betrifft die dramatische Struktur des Gemäldes, die andere das Verhältnis zur vorrangigen schriftlichen Überlieferung. Erstere hat zum Ziel, durch den Transfer der Prinzipien der aristotelischen *Poetik* auf die Historienmalerei Gesetzmäßigkeiten von einer sprachgebundenen Gattung, der Tragödie, auf eine bildliche, die Historienmalerei, zu übertragen. Letztere unterliegt der Forderung, genau den Text zu illustrieren, die erwähnten Personen und Umstände, in denen sie sich befinden, ‚wörtlich‘ zu übernehmen und damit die Eigenschaften des Textes auf das Bild zu übertragen[20]. Beide Standpunkte stoßen besonders vehement in der Diskussion um die *Mannalese* aufeinander. Poussin wurde von der einen Partei, vor allem von Philip-

---

17 Stich nach *Der Sturm*, von Louis de Châtillon, 1679, Kupferstich, 30,8 x 43,2 cm Blattmaß, Düsseldorf, Kunstmuseum, Graphische Sammlung; Abb. in: Austellungs-Katalog Frankfurt (1988), Kat.-Nr. P 6, S. 71-73, Abb. S. 72.

18 Montagu (1992), S. 233-48.

19 Félibien (1669), Faksimile in: Boos (1966), Anhang S. 26.

20 Dazu ausführlich Stumpfhaus (2000), [Typoskript] S. 63-64, S. 68, S. 79. Noch Louis Marin steht ganz in dieser Tradition: „Si le peintre a dû lire un livre, un texte, des mots, des phrases pour peindre, pour faire voir, le spectateur doit «lire» le tableau pour voir ce dont le texte parle"; s. Marin (1995), S. 13-14.

pe de Champaigne, vorgeworfen, er halte sich in seiner Komposition nicht an die Bibel, er weiche vom Text ab. Poussin zeige in seiner Fassung des Bibeltextes zuviel Leiden, obwohl doch überliefert sei, dass es am Vorabend bereits Wachteln geregnet habe. Außerdem male er fälschlich das Manna vom Himmel fallend. Die Verteidigung Poussins, Le Brun und Félibien, führt dagegen ins Feld, ein Künstler, zumal als dramatischer Dichter, brauche sich nicht wörtlich an einen Text zu halten, solange er sich regelgerecht verhalte, d. h. gemäß der aristotelischen Forderung alle drei Handlungsteile einer Geschichte aufführe und nach der Wahrscheinlichkeit zu einer Einheit verknüpfe[21]. Die Kritiker richten sich an der Textnähe aus, die Verteidiger an der Schrift des Aristoteles, beide aber an Produkten der Sprache, um das Bild zu evaluieren. Beiden ist diese Orientierung gleichzeitig Kriterium für die Wahrheit des Historienbildes. Die Kritiker verstehen die Texttreue einer Komposition als Ausweis ihrer Wahrheit, die Verteidiger sehen sie in Entfaltung der Handlung nach einer kausalen Ordnung begründet.

Schauen wir nun ins Bild, um zu sehen, welche der beiden Parteien die Sache besser trifft: Schon allein die kaum übersehbare Fülle der Figuren lässt Zweifel aufkommen, ob sich der Künstler an die Bibel hält. Tatsächlich soll, so Félibien in besagter Debatte, Poussin auch auf den säkularen Text des Josephus Flavius zurückgegriffen haben[22], eine Erwähnung, die von den Kunstgeschichtlern, die sich zur *Mannalese* geäußert haben, meist überlesen worden ist[23], denn man geht einheitlich von 2 Moses 16, 13-15 als Vorlage aus[24].

Betrachten wir einige Besonderheiten in diesem Gemälde: links die *Caritas*-Gruppe, rechts die Hilfe schickende Mutter, hinter ihr die balgenden Knaben, der Felsen links, eine kleine Gruppe vor den Bäumen rechts, einen aktiven, die Erde mit einer Hacke behauenden jungen Mann in Gelb und einen kontemplierenden Greis in Rot, beide hervorgehoben durch ihre Vereinzelung im Raum. Auffällig ist natürlich auch das fehlende Wachtelwunder, sowie der unsichtbare Mannaregen.

In 4 Moses 11 wird eine Geschichte vom Mannawunder erzählt, welche das Unmaß und die Unzufriedenheit des israelischen Volkes mit Gottes Wundern zum Thema hat. Hier wird berichtet, dass das Mannawunder vor dem Wachtelregen geschehe[25], denn das Volk murrte, weil es nur Brot zu essen bekomme, nicht aber auch Fleisch und Fisch. Daraufhin zürnt Gott seinem auserwählten Volk, und Moses klagt: „Habe denn ich dieses ganze Volk empfangen, oder habe ich es geboren, dass du zu mir sagst: Trage es an deinem Busen, wie eine Wärterin den Säugling trägt, in das Land, das du deinen Vätern zugeschworen hast? Woher nehme ich Fleisch für dieses ganze Volk?"[26] Die Textstelle ist für das Bild von zentraler Bedeutung,

---

21 Jouin (1883), S. 64. Félibien ist über diesen Punkt ausführlicher in seinem Vorwort zu seiner Veröffentlichung der Conférences; s. Boos (1966), Anhang S. 17-18.
22 Jouin (1883), ebd.
23 Jüngst machte Held darauf aufmerksam (Held (2001), S. 102). Allerdings ist für Held der Verweis auf Flavius lediglich eine Frage der Argumentationsstrategie in der Debatte der Akademiker, weniger ein Hinweis auf mögliche Quellen für die *Mannalese* Poussins.
24 Z. B. Ausstellungs-Katalog Paris (1994), Kat.-Nr. 78, S. 263: „Le sujet est tiré de L'Exode (16: 13-15), durant la traversée du désert." Ebenso der Ausstellungs-Katalog London 1995, Kat. Nr. 37, S. 216: „The subject is from Exodus XVI, 11-18, and depicts the miraculous fall of the manna to the Israelites in the wildernes."
25 Diese Erwähnung in der Bibel ist für die *Mannalese* von eminenter Bedeutung, denn in dieser Variation der Erzählung regnet es erst Wachteln, *nachdem* die Israeliten Brot, Manna, zu essen bekamen. Wollte man sich virtuell in den Streit der Akademiker einmischen, so könnte die Kritik mit dem Verweis auf *diese* Bibelstelle berichtigt werden.

denn sie liefert die Motive, die auch die Komposition prägen. Die menschliche Ohnmacht, das Volk zu sättigen, findet seine Entsprechung in der *Caritas*-Gruppe links. Der Hunger des Knaben bleibt ungestillt, trotz der ,wunderbaren' Hilfsbereitschaft seiner Mutter. Das von Moses angesprochene Muttermotiv gibt dem Gemälde die Hauptfigur. Mütter beherrschen das Bild. Abgesehen von der *Caritas*-Gruppe links und der weisenden Frau rechts finden wir eine dritte inmitten der Komposition: eine Frau in Gelb trägt ein Kind und spricht mit einer Frau in Blau, eine Szene, wie wir sie von der Begegnung Marias mit Elisabeth als *Heimsuchung*[27] kennen. Aber auch im Bericht des Josephus Flavius[28] erfahren wir einiges zum Mannawunder. In einer Textpassage kontrastiert Flavius Bedürftigkeit und gieriges Einsammeln. Um Letzteres zu mäßigen, sollen die Starken, so schreibt Flavius, nur eine bestimmte Menge auflesen, damit auch die Schwachen zu dem Ihren kommen. Die Gier etwa ist im Bild deutlich mit den beiden um das Manna streitenden Knaben ausgedrückt. In 5 Moses 8, 16-17 wird daran erinnert, dass die Mannagabe nicht selbst erworben sei, „dass du nicht bei dir selber sprichst: ,Meine Kraft und die Stärke meiner Hand hat mir diesen Reichtum erworben'". Es ist augenscheinlich, dass der Jüngling rechts im Bild, winzig vor der gewaltigen Erscheinung der Bäume, umsonst die Erde bebaut.

Poussin hält sich offenbar *nicht* an *den* Text der Bibel, 2 Moses 16, vielmehr kompiliert er einzelne Episoden aus verschiedenen biblischen wie außerbiblischen Berichten. Abgesehen davon, dass er durch seine Auswahl Bibel und Säkulartext gleichrangig behandelt und damit die Bibel in ihrem ausgezeichneten Status profaniert, verzichtet Poussin ebenfalls darauf, anders als er es selbst behauptet, sich eng an der Schrift zu orientieren. Das auf Textgenauigkeit gegründete Wahrheitskriterium entfällt.

Die Verteidigung scheint genauer zu treffen, wie die Komposition zu verstehen sei. Poussin stellt, wie von Le Brun und Félibien hervorgehoben, alle drei Teile der Geschichte dar, die anfängliche Not links im Bild, das erlösende Ereignis des Mannawunders rechts, den abschließenden Dank im Mittelgrund, und er verknüpft sie in zeitlicher Linearität unumkehrbar. Die geforderte Organisation einer dramatischen Handlung durch die Malerei ist mustergültig eingehalten. Der Betrachterblick wird durch ein komplexes Verweissystem aus Fingern und Blicken von einer Gruppe zur nächsten geführt. Er ,liest' das Bild, die Geschichte. Dabei ist die Stringenz der Handlungsführung so zwingend, dass sich dem Betrachter die repräsentierte Geschichte allein aus der Beobachtung der einzelnen Gruppen – auch ohne die Kenntnis der biblischen Erzählung – erschließt[29].

---

26 4 Moses 11, 12-13.

27 Beispielsweise Piero di Cosimo, *Heimsuchung mit dem Hl. Nikolaus und dem Hl. Antonius*, Holz, 184 x 189 cm, 1490-1500, Washington, National Gallery of Art, Kress Coll.; s. Austellungs-Katalog Washington (1995) (11984), Kat.-Nr. 70, S. 102.

28 Josephus Flavius, *Antiquitates Judaeorum*, lib. III, Kap. 1, Abs. 6: „Es [das maßvolle Auflesen des Mannas] geschah aus Vorsicht, damit nicht die Stärkeren, die mehr zu sammeln vermochten, den Schwächeren das Einsammeln ihrer Nahrung erschweren konnten," zitiert nach der Übersetzung Clementz (1959), S. 141.

29 Dieser hohe Grad an kompositioneller Stringenz ist insofern bemerkenswert, als kein Handlungselement eliminierbar, keine Figur umdeutbar ist. Ein solches Gefüge wäre demnach nicht durch z. B. politische Propaganda ,ausbeutbar'.
Damit ist ein weiteres Argument für die These gegeben, dass sich die Fraktion der Akademiker, welche sich um die Unabhängigkeit der Kunst, sei es von den Einmischungen der Kirche, sei es von den Interventionen der Propagandisten des Königs, besorgten (s. Kirchner (1991), S. 10-28). Gerade mit seiner erzählerischen Stringenz macht Poussin seine Geschichten zu Ereignissen, die nicht politisch verwertbar sind. Dieser Umstand ist

Allerdings, so plausibel die Ausführungen Le Bruns und Félibiens die Handlungsstruktur der Komposition mit der Lehre des Aristoteles in Verbindung zu bringen und damit mögliche Abweichungen des Dichters von der Bibel zu rechtfertigen wissen, so wenig ist ihre Erklärung der Handlungsstruktur in der Lage, den Vorwurf an Poussin, er stelle besonders mit der alten Frau, die sich stillen lasse, zuviel Leiden dar, außer Kraft zu setzen[30]. So bleibt zu fragen, warum Poussin seine *Mannalese* angesichts des von ihm selbst in Gestalt der Heimsuchungsszene inmitten des Gemäldes alludierten typologischen Verweises auf den erlösenden Opfertod Christi – den göttlichen Triumph über Tod und Hinfälligkeit – als ein vorwiegend leidvolles Geschehen schildert[31]; der Künstler selbst betont auch in seinen Briefen die Seite des Mangels.

Die leidvolle Düsternis der *Mannalese* ist zweifellos zu rechtfertigen mit dem alttestamentlichen Gedanken der Prüfung, dem gemäß das Maß für die sittliche Gesinnung des auserwählten Volkes die Größe seines Leidens ist: „Und du sollst gedenken des ganzen Weges, den dich der Herr, dein Gott, nun vierzig Jahre lang geführt hat in der Wüste, um dich zu erproben, auf dass er erkenne, wie du gesinnt seiest"[32]. Erst durch das tugendhafte Ertragen der Not zeigt sich, ob Israel gottgefällig und der Wunder Gottes würdig ist. Poussin forciert in seiner Interpretation gegen das göttliche Wunder auf der rechten Seite der Komposition die selbstgewählte Tugendtat der stillenden Mutter, indem er sie als das Erste im Bild zeigt und sie in überall herrschende Agonie und Hilflosigkeit einbettet. Er setzt sie an den Anfang der Geschichte und impliziert, dass diese Tugend das folgende Gnadenwunder begründet, insofern sich das Volk erst durch eine solche Tat der weiteren Gnade Gottes würdig erweist.

Diese Betonung des Leidens lässt sich zwanglos verknüpfen mit stoischer Philosophie[33]. Der Gedanke, dass Gott die Tugendstärke durch zugeteiltes Leid prüft, gehört zu den zentralen Gedanken der Stoiker, etwa Senecas[34]. Bei ihm liegt das Maß für die Stärke der Tugend

---

insofern von Interesse, da in Frankreich nach der Fronde die ideologisch besetzte Mythologie besonders in den Historien Mode wurde – berühmtes Beispiel, die an Cortonas florentinische und römische Arbeiten orientierten Dekorationen von Vaux-le-Vicomte von 1658-61. „Durch die Allegorisierung büßten Mythos und Geschichte Autonomie, Narrativität und Vieldeutigkeit ein, um gänzlich zu propagandistischen Instrumenten zu werden" (Germer (1997), S. 147). Solchen Instrumentalisierungen stehen Poussins Geschichtsdarstellungen entgegen.

30 Le Brun stimmt dem Vorwurf sogar insoweit zu, dass der ärgste Hunger des Volkes tatsächlich schon am Vorabend durch ein Wachtelwunder gestillt worden sei. Zur Verteidigung Poussins führt er jedoch an: „Toutefois, comme ce peuple étoit en grand nombre et répandu dans une ample étendue de pays, il n'est pas hors d'apparence qu'il n'y en eût plusieurs qui n'eussent pas encore appris la promesse qui leur avoit été fait ; ... quelques-uns n'ajoutassent pas foi aux promesses de Moise, puisqu'ils étoient naturellement fort incrédules"; s. Jouin (1883), S. 63. Dass sich Le Brun überhaupt auf die Ebene der Textexegese einlässt, lässt den Verdacht aufkommen, er selbst trumpfe mit seinen gelehrten Ausführungen nur auf, um die – berechtigte – Kritik zum schweigen zu bringen. Das Protokoll schliesst mit der apodiktischen Bemerkung: „De sorte que, bien loin de trouver à redire à tout ce que M. Poussin a peint dans ce tableau, on doit plutôt admirer de quelle manière il s'est conduit dans la représentation d'un sujet si grand et si difficile"; s. ebd. S. 65.

31 Während Poussin noch im ersten Brief an Jacques Stella (dat. 1637) beide Grundstimmungen, das Leiden, wie die Freude des Volkes Israel, gleichwertig aufführt, hebt er, nachdem er kurz vorher (19. März 1639) sein Bild als beendet erklärt hat, im Brief vom 28. April 1639 an Chantelou die hungernde Gruppe links im Bild hervor: „Car les sept premières figures à main gauche vous diront tout ce qui est ici écrit et tout le reste est de même étoffe"; s. Jouanny (1968), S. 21; Lettres et propos 1989, S. 45.

32 5 Moses 8, 2.

33 Prominent zur stoischen Prägung Poussins: Blunt (1995) (11967), hier bes. S. 174.

34 Etwa Seneca (1992), S. 6-45.

ebenfalls im Ertragen des Unglücks, das dem Menschen zustößt. Bedeutsam in diesem Zusammenhang ist die Annahme, dass eine gute Tat im Unglück sehr viel schwieriger auszuführen ist als im Glück, weil sie stärkere Widrigkeiten zu überwinden hat. Es wird eine spezifische Proportion von Leid und Tugend angenommen: je stärker das Leiden, desto stärker muss die Tugend sein, es zu überwinden. Zudem macht erst das Unglück die Tugend offenbar, weil sie ohne Hilfe von außen auskommen muss, also ganz ein Ergebnis des tugendhaften Charakters selbst darstellt. Deshalb zeichnet sich der Gerechte erst im Leiden als ein Mann Gottes aus. „Die also, die Gott anerkennt und die er liebt, die stählt, prüft und schleift er"[35]. Im Unterschied jedoch zum Alten Testament weiß sich der tugendhafte Stoiker, die Gerechtigkeit des Schicksals erkennend, Gott verwandt, ja sogar ihm ähnlich.

Altes Testament und Stoa geben uns Kriterien zur Hand, die Betonung der Not in der *Mannalese* zu erklären. Die Düsternis und die Not, die im Gemälde trotz der Erlösungssituation überwiegen, betonen die glanzvollen Taten der Mütter und mit ihnen die Qualität der Tugenden, die das Volk Israel zu leisten vermag. Die beiden Mütter sind im Bild gleich groß und überragen optisch alle anderen Figuren im Bild[36]. Sie tun dies, weil sie sich den Extremsituationen, denen sie ausgeliefert sind, dem Mangel links, wie dem Überfluss rechts, nicht überlassen. Da sich die Gesinnung nach dem Maß der Prüfung äußert, ist jedoch die Tugend der stillenden Mutter links höher zu werten als die der weisende rechts. Denn es ist schwieriger, im Mangel tugendhaft zu sein, als im Überfluss. Tugend in der Not ohne die sichtbare, himmlische Unterstützung ist der stärkste Beweis für Sittlichkeit. Damit wird deutlich, warum Poussin auf die Gruppe links in besonderer Weise aufmerksam macht. Die dort geschilderte Tugend ist die vollkommenere[37]. Sie zeigt an, was Tugend ist und was sie leistet. Eine solche Tugend hebt den Menschen über den Zwang des Schicksals und den der menschlichen Natur hinaus und kann dadurch den Charakter einer Metapher für die Barmherzigkeit Gottes selbst annehmen[38].

Bemerkenswert ist, dass uns Poussin in den beiden Bildhälften seiner *Mannalese* eine Wertabwägung von verschiedenen Formen der Tugend präsentiert: links zeigt der Künstler eine gute Tat umgeben von Zeichen der äußersten Not, rechts diejenige inmitten des Gott gegebenen Überflusses, begleitet von menschlichen Schwächen, etwa der Habsucht, in Gestalt der balgenden Knaben[39], oder der Selbstsucht, in Gestalt der für sich sammelnden Männern und Frauen. Welche von beiden Tugenden Poussin im Bild den Vorzug gibt, bleibt

---

35 „Hos itaque deus, quos probat, quos amat, indurat, recognoscit, exercet; eos autem, quibus indulgere videtur, quibus pacere, mollis venturis malis servat"; ebd., S. 28-29.

36 Le Brun bezeichnet sie jeweils als "la principale partie du groupe"; Jouin 1883, S. 52.

37 Ähnlich Mildner-Flesch (1983), S. 10. Mit dem oben im Text getroffenen Befund kann eine negative Wertung der *Caritas-Romana*, wie sie Imdahl (Imdahl (1985), S. 153-54) und auch heute noch Held (Held (2001), S. 160; Held spricht von einer Beraubung des Kindes) vornehmen, widerlegt werden. Das Kriterium für die Bewertung der Barmherzigkeit der stillenden Mutter als unvollkommen ist der Erfolg. Da das Kind ungestillt bleibt, ist die Tugend nicht vollkommen. Nach stoischer Lehre ist Tugend jedoch keinem weiteren Zweck, etwa der vollständigen Sättigung der Familie, untergeordnet. Tugend – nicht nur im stoischen Sinne – verbürgt die Glückseligkeit des weisen Menschen und ist damit Selbstzweck. Sie trägt den höchsten Wert in sich (s. Hossenfelder (1995), S. 54 ff.). Auch die katholische Vorstellung von den guten Werken als Verdienste vor Gott rechnet nicht mit dem materiellen Erfolg einer Tat. Ein Werk der gläubigen Liebe kann auch ohne vollständige Bedürfnisbefriedigung vor Gott ein gerechtes sein; s. LThK, Bd. 10, SW. Verdienst, Sp. 675 ff.

38 In diesem Sinne ist Imdahl zuzustimmen, dass die stillende Mutter der *Caritas Romana* Gruppe das *Subjekt* der dargestellten Geschichte sei; s. Imdahl (1985), S. 147-48.

unklar. Die linke Seite bezeichnet die menschlich vollkommenste Tugend, die rechte göttliche Barmherzigkeit. Die eine Tat hebt er durch einen Betrachter hervor, die andere ist kostbarer geschmückt und erscheint in glänzendem Licht. Dass beide Gruppen als Antagonisten zu verstehen sind, das zeigt ihre kontrastive Inszenierung[40]. Die Beobachtung der Konfrontation zweier Tugenden auf den beiden Seiten der Komposition ist für die Beurteilung sowohl des Bildsinns, wie auch der Einschätzung der Historienkomposition als sukzessivem Text folgenreich.

Zum einen zeichnet sich ein fundamentaler Widerspruch zur tridentinischen Gnadenlehre ab, welche menschliche Tugend ohne den Einfluss Gottes schlichtweg ablehnt[41], in der Figuration der *Caritas-Romana* von Poussin aber deutlich formuliert wird[42]. Zwar nimmt im Bild die Erfahrbarkeit göttlichen Wirkens in der Geschichte von links nach rechts bis in den Mittelgrund zu[43]. Der Wert der vom Menschen selbst geübten Tugend steigert sich jedoch in umgekehrter Richtung. Als eine Tat der unerlösten Antike, oder auch der unerlösten Zeit des Alten Testaments, kommt sie ohne die vorausgehende Gnade durch Christi Erlösungstat aus, und erhebt sich über die Natur durch ihre eigene, selbstverantwortete Tat. Diese Tat der Barmherzigkeit rechtfertigt sich selbst[44].

Poussin spielt in der *Mannalese* die menschliche Willensfreiheit gegen die Gnadenvorstellung des Tridentinums aus und nutzt dafür die Zweidimensionalität der Gemäldefläche.

---

39 Ich kann hier nicht Helds auf Le Bruns Worten basierenden Interpretation der sich balgenden Knaben zustimmen, dass sie ihre spontanen Ausbrüche in toleranten Grenzen hielten; s. Held (2001), S. 157. Le Bruns Deutung, die Buben balgten sich nicht, um sich etwas Böses zu tun, „mais seulement que l'un empêche l'autre d'avoir ce qu'ils voyent tous deux leur être si nécessaire" (Jouin (1883), S. 58) verwischt den Umstand, dass die balgenden Jungs im Kontrast zur Caritas der weisenden Mutter und erst recht zu der der stillenden als selbstsüchtig desavouiert werden.

40 „Que cèst là qu'on trouve ce contraste judicieux"; s. Jouin (1883), S. 52.

41 Tridentinum, 6. Sektion, 13.01.1547, § V.

42 S. hierzu Held (2001), S. 104–05. Nach Held zeigt das ikonographische Motiv der Caritas Romana die interesselose Liebe der Mutter, „die gegen ihre natürlichen Neigungen handelt". „Sie stellt zugleich eine Geschichte des Mangels dar, die die Gnadentat hervorruft, die allen natürlichen Gefühlen zuwiderläuft." Die Abwendung der Mutter von ihrem Kind unter gleichzeitiger Zuwendung zur alten Frau neben ihr stellt für Held „einen Bruch zwischen der natürlichen Liebe und der göttlich inspirierten Liebe" dar, welche „allein im Willen ohne Eigeninteresse" wurzelt, „der eine Gabe der Gnade ist". Allerdings scheint mir Held bei dieser Bewertung der Caritas Romana-Gruppe zu übersehen, dass Poussin beide Gruppen so gegeneinander setzt, dass menschliche Tugend unter der Gnade (rechts im Bild) und außerhalb ihrer (links) konfrontiert wird. In der Caritas Romana-Gruppe deutet nichts auf das Einwirken göttlicher Gnade.

43 Die im Bild formulierte Aussage geht weit über das im Alten Testament geschilderte Ereignis hinaus. Mit dem die Heimsuchung alludierenden Frauenpaar und der sie flankierenden Naturzeichen für Christus, der durchlichtete Felsen und die mächtigen Bäume, formuliert Poussin den gesamten Verlauf der Menschheit. Es sind die großen christlichen Zeitalter angesprochen: Die Zeit des Alten Testaments und die der gleichzeitigen Antike im Vordergrund (sub lege) und die Zeit von Christi Wirken im Mittelgrund (sub gratia).

44 Die Gegenüberstellung der zwei Gnadenauffassungen, von der menschlichen Freiheit und Tugendfähigkeit auch ohne Gottes Gnaden einerseits, und ihrer Gegenposition, dass menschliche Tugend ohne göttliches Zutun nicht möglich sei, andererseits, legt eine zumindest intellektuelle Auseinandersetzung Poussins mit den Glaubensstreitigkeiten in Frankreich, zwischen der jansenistisch orientierten Abtei Port-Royal des Champs und den Jesuiten, nahe. Seit den zwanziger Jahren befand sich das Kloster unter Abbé de St. Cyran, Vertreter einer streng gefassten Gnadenlehre, im Streit mit den Jesuiten, die der Überzeugung waren, dass sich der sündhafte Mensch aus eigener Kraft, Gott zuwenden könne; ein Streit, der letztlich zur politischen Isolierung des Klosters führte (Schmidt-Biggemann (1999), S. 24-27). Diesen Hinweis verdanke ich Barbara Bauer.

Er inszeniert seine *Mannalese* als einen *Vergleich* zwischen den Tugenden links und rechts, ohne den damit formulierten Wertekonflikt einer Lösung zuzuführen. In diesem Sinne übersetzt Poussin die Abfolge der verschiedenen Handlungsteile, Not, Erlösung und Dank, in die Simultaneität des Gemäldes und verlässt die Linearität und Sukzessivität der vorbildlich niedergeschriebenen Geschichte. Poussin inszeniert die Abfolge von Not und Glück, von menschlicher Caritas und göttlicher Gnade weniger im Sinne der Dramatik des Geschehens, sondern vor allem als einen Wertekonflikt, oder, ästhetisch ausgedrückt, als einen Kontrast von dunkler Not und labender Erlösung. Der Betrachterblick kann nun vergleichend zwischen den Gruppen hin und her wandern. Poussin löst die Sukzessivität des kausal geordneten Handlungsablaufes auf in den ästhetisch komponierten Kontrast und setzt seine Betrachtung ganz der ins Belieben des Zuschauers gesetzten Zeitlichkeit des *Vergleichs* aus[45]. Dieser Strategie, die beiden Tugendtaten unentschieden nebeneinander zu stellen, ist auch nicht mit dem von Le Brun und Félibien für ein Historienbild geforderten Wahrheitsanspruch beizukommen. Keinesfalls basiert sie auf der logischen Ordnung der Handlungen. Der simultan gegebene Vergleich zweier verschiedener Handlungsmodelle steht jenseits der Linearität einer einfachen, in ihrem Ursache-Wirkungsverhältnis sukzessiven Kausalfolge.

Überwunden wird der Wertkonflikt zwischen den beiden Tugendmodellen allein artistisch. Denn die von Poussin vorgenommene Kontrastierung von Leiden und Überfluss, Stabilität und Bewegung, Willensfreiheit und Gnade, fundiert eine genuin ästhetische Qualität, die sich im Gesamtlob Le Bruns ausspricht, Poussin habe seine Figuren und die Landschaft so in Kontrastverhältnisse gesetzt, dass eine *belle harmonie* entstehe[46]. Offensichtlich übersetzt Poussin nicht nur einfach eine Textstruktur in eine von schriftähnlichen Bildzeichen, er transformiert den Text – die Texte –, so, dass die einzelnen Figuren und Gruppen nach den Gesetzen der Malerei auf eine Leinwand nebeneinander einen schönen Kontrast bilden und damit ästhetischen Maßgaben folgen. Diese harmoniestiftende Antithetik kann nur durch künstlerische Mittel hergestellt werden. Zu diesen Mitteln zählen nicht nur die kontrastive Situierung der beiden Hauptgruppen auf der Fläche der Leinwand, ihre künstlerische Ausarbeitung durch Größe, Licht und Farbe, sondern auch die Proportionierung der Gruppen so, dass sie im Sinne eines Gegensatzes gleichwertig und gleichgewichtig aufeinander bezogen sind. Die optische Plausibilität des in den beiden Hauptgruppen links und rechts der Komposition formulierten Wertkonflikts ist allein eine Frage der künstlerischen Gestaltung und entzieht sich jeglicher begriffsbezogenen Diskursivität.

Allerdings wäre es zu früh geschlossen, Poussin als bloßen Ästheten im Sinne des l'art pour l'art zu verstehen. Denn das Ziel des Kontrastes ist nicht nur die schöne Harmonie, sondern vor allem eine Verstärkung des Ausdrucks. „Nichts kann unsere Seele stärker bewegen als Kontraste"[47], so schrieb schon Monteverdi in dem Vorwort zu den *Madrigali guerrie-*

---

45 Ein reflektierendes Hin und Her des Betrachters dürfte wohl gemeint sein, wenn Le Brun davon spricht, dass der Kontrast „sert à donner du mouvement"; = Fortsetzung des Zitats von Anm. 40.

46 „Que c'est là qu'on trouve ce contraste judicieux qui sert à donner du movement, et qui provient des différentes dispositiones des figures qui la composent, dont la situation, l'aspect et les mouvements étant conformes à l'histoire engrendent cette unité d'action et cette belle harmonie qu'on voit dans ce tableau"; ebd., S. 52.
Dazu auch Held (2001), S. 98. Anders als Held sehe ich den Kontrast nicht allein auf die Lebensalter, das Geschlecht, die körperliche Haltung oder mentale Einstellung der Figuren bezogen. Der Kontrast ist *die* die Komposition – die belle harmonie als ihr Ziel – konstituierende Hauptsache und umfasst alle Teile der Malerei (auch die Farben, das Licht- und Schattensiel etc.), und nicht nur die des decorums.

ri ed amorosi. Im Kontrast entfaltet eine Komposition die stärkste Wirkkraft auf den Zuhörer. In gleichem Sinn wird das Kontrastverfahren auch in den Akademiedebatten um Poussins Werke debattiert. So führt Philippe de Champaigne am Schluss der Darstellung[48] von Poussins *Rebekka und Eliezer am Brunnen*[49] aus: „Um die Tugend noch liebenswürdiger und leuchtender erscheinen zu lassen, muss man sie der Untugend entweder entgegensetzen oder sie mit ihr vergleichen. Daraus folgt, dass rechtschaffene Menschen solange nicht in ihrem eigentlichen Glanz erscheinen, bis sie niederträchtigen gegenübergestellt werden"[50]. So, wie die Schönheit einer Komposition des Schattens bedarf, so bedarf die Tugend, um besonders zu leuchten, des Schattens der Untugend[51]. Das Kontrastverfahren hat also nicht nur ästhetische Ambitionen, sondern vor allem ethische. Da der Kontrast im Dienste der Ausdrucksintensität steht, macht Le Brun zur Verteidigung Poussins korrekt darauf aufmerksam, dass es der Not bedarf, um das „Überwältigende des Wunders"[52] zu zeigen. Wir haben es in der *Mannalese* also weniger mit der Übersetzung einer Geschichte in ein Bild zu tun, sei es im Sinne eines ‚wörtlichen' Transfers, sei es im Sinne einer Strukturangleichung von Bild und Textbericht, sondern wesentlich mit der Inszenierung einer begrifflich schwer fassbaren emotionalen Qualität, dem „Überwältigenden des Wunders".

Poussin löst sich, wie wir gesehen haben, in zweierlei Hinsicht von der Autorität des vorrangigen Textes. Zum einen formuliert er nicht eine einzige Schriftstelle in Malerei um, sondern kompiliert eine Reihe von verschiedenen Texten, so dass es sich hier weniger um einen künstlerischen Kommentar zur Erzählung des Mannawunders handelt, als um eine eigene Erfindung des Künstlers. Damit löst er die unmittelbaren Bande von Bild und Text. Durch die Ordnung der Komposition in Kontrasten löscht er auch mittelbar den Textcharakter seiner Malerei. Nicht die Sukzession einer dramatischen Erzählung nachzuahmen, nicht die Einhaltung des aristotelischen Regelwerks ist Leitfaden seiner Malerei, sondern die Fundierung ästhetischer Werte, einer schönen Harmonie, deren Ziel die – gänzlich unbegriffliche – Ausdruckssteigerung ist. Poussin formuliert die Wandlungen eines Textes um zu einem aus Gegensätzen gefügten Ganzen. Seine angemessene Rezeption ist der Vergleich[53], nicht die diskursive Erschließung eines fortfolgenden Ereignisses.

Doch es lässt sich noch auf einer dritten Ebene die Textbezüglichkeit Poussins problematisieren, auf jener der semantischen Zeichen, die erst einen lesbaren Text als solchen fundie-

---

47 Monteverdi, *Quinto libri de' madrigali*, 1605, S. 299 s. LeCoat (1975), S. 128.

48 Achte Konferenz, Jouin (1883), S. 87-99.

49 Um 1648, Paris, Musée du Louvre ; Ausstellungs-Katalog Paris (1994), Kat. Nr. 166, S. 382-85; Thuillier (1994), Kat. Nr. 173, S. 259.

50 „En effet, si, pour rendre la vertu plus aimable et plus éclatante, il la falloit opposer ou comparer au vice, il s'ensuivroit que les hommes de probité ne pourroient jamais être dans une véritable splendeur que lorsqu'il seroient confrontés à des scélérats"; Jouin (1883), S. 96.

51 „Et que la vertu seroit redevable au vice de tout ce qu'elle a de brilliant. Et quand, pour soutenir le même axiome, M. de Champaigne a dit que M. Poussin n'auroit pu former les beautés de tableau de Rébecca sans le secours des ombres, il a dit vrai, et n'a rien prouvé qui lui soit favorable, car les ombres et les jours sont des parties relatives et réciproques, toutes deux essentielles dans la peinture," ebd.

52 „La grandeur de ce miracle", Jouin (1883), S. 63. Deshalb ist es nicht verwunderlich, wenn Le Brun, wie Schlink moniert, das Volk Israel in der Vordergrund seiner Abhandlung stellt, und nicht die Manna*lese*.

53 Dieser Befund ist wichtig für die Offenheit der Wirkung einer Komposition. Es ist ganz dem Betrachter anheim gestellt, für welche Seite er sich als die wertvollere entscheidet. Poussin selbst schreibt ja an Chantelou (im Modusbrief vom 24. Nov. 1647), dass die Wirkung eines Bildes nicht allein vom Sujet, sondern auch von der Disposition des Betrachters abhänge; s. Jouanny (1968), S. 372; Lettres et propos, S. 134.

Abbildung 2: Nicolas Poussins *Gewitterlandschaft mit Pyramus und Thisbe*

ren. Welche das für die Malerei sein mögen, können wir von Poussin selbst erfahren: „Wie die 24 Buchstaben des Alphabetes dazu dienen, unsere Worte zu formen und unsere Gedanken auszudrücken", so zitiert Félibien Poussin, „auf die selbe Weise transportieren die Umrisse des menschlichen Körpers die verschiedenen Passionen der Seele, um das im Außen erscheinen zu lassen, was man im Geiste hat"[54]. Poussin beruft sich hier offensichtlich auf das für die bildende Kunst verbindliche Regelsystem der Rhetorik. Gesten und Mimik vermitteln bedeutungsvoll eine im Geiste und in der Seele formulierte Botschaft, die von der Zuhörerschaft, bzw. von den Betrachtern, gelesen werden kann[55]. Voraussetzung der Adäquanz von Ausdruck und seelischer Bewegung ist die Vorstellung von der Kongruenz von Körper und Seele. Prominentestes Beispiel für ein so fundiertes Ausdruckssystem dürfte Le Bruns Lehre sein[56]. Die von Poussin vorgenommene Analogie von Buchstaben und Aus-

---

54 „En parlant de la peinture, [Poussin] dit que de même que les vingt-quatre lettres de l'alphabet servent à former nos paroles et exprimer nos pensées, de même les linéaments du corps humain à exprimer les diverses passions de l'âme pour faire paraître au-dehors ce que l'on a dans l'esprit"; Lettres et propos, S. 196-97.

55 In diesem Sinne versieht Sebastian Schütze Poussins Personal mit dem Attribut „icon of pathos". „Poussin", so resümiert Schütze seine Analyse zu Poussins frühen Ausdrucksfiguren, „l'utilisait [le recours à des exemples illustres] pour présenter sous une forme figurée des notions générales et universelles"; cf. Schütze (1996), bes. S. 581-82.

56 Diese Lehre ist in diesem Zusammenhang nicht ganz unwichtig. Le Brun führt nicht nur sein semantisches Ausdruckssystem vor, er begründet es auch, so dass wir sehen können, welche Voraussetzungen gegeben sein

drucksfiguren legt nahe, bei Poussin ein ähnliches Alphabet von Mimik und Gestik zu erwarten. Am Beispiel der Hauptfigur der Frankfurter *Gewitterlandschaft mit Pyramus und Thisbe*[57] (Abb. 2, S. 332) möchte ich jedoch einige Argumente gegen diese Annahme vorbringen.

Dargestellt ist die dramatische Auffindung des sterbenden Pyramus durch seine Geliebte, Thisbe, während über beiden unerbittlich ein Gewitter tost. Das Gemälde wird in der Forschung nahezu einheitlich als ein moralisches exemplum malum[58] gedeutet. Durch rasende Liebe, den entfesselten Sturm der Leidenschaften, liefert sich ein junges Paar hoffnungslos dem Tode aus[59]. Diese, ganz stoischer Tradition verpflichtete, Deutung stützt sich vor allem auf die Beobachtung der Figur Thisbes. Ihre Haltung ist geprägt vom Lauf gegen den Wind. Die am Leib klebenden Kleider, ihre rudernden Arme wie ihr zum Schreien aufgerissener Mund werden nicht nur verstanden als Folge des Sturms, sondern auch als Ausdruck ihrer verwerflichen Affektgebundenheit, wobei ihr äußerer Kampf gegen das Unwetter gleichgesetzt wird mit der zuchtlosen Hingabe an den inneren Sturm der Leidenschaft[60]. Mit ihrer Haltlosigkeit besiegelt Thisbe ihr eigenes Schicksal. Die von den Interpreten vorgenommene Gleichsetzung von innerer und äußerer Natur[61] fußt einerseits auf der Mehrdeutigkeit des Wortes *fortuna*, welches Sturm und gleichzeitig Schicksal bedeutet. Zum anderen erhält ihre Deutung Nahrung durch den Befund, dass Poussin mit der Figur Thisbes Leonardos Studie eines gegen den Sturm laufendes Mannes[62] rezipiert. So scheint die Ausdrucksfigur Thisbe sowohl begrifflich als auch künstlerisch allein von ‚fortuna‘ geprägt zu sein. Es ist bisher kein ernsthafter Einspruch gegen dieses Verständnis vorgebracht worden.

---

müssen, um im Außen das sehen zu können, was uns im Innern bewegt. Bei Le Bruns Ausdruckschema handelt es sich trotz aller anatomischen und philosophischen Beschreibungen um ein semantisches System von Nähe und Ähnlichkeiten, von Analogien, um es mit Foucault zu sagen. Die Bewegungen im Gesicht sind aus zwei Gründen lesbar. Sie befinden sich zum einen in direkter Nähe zum Gehirn, so dass sich dessen Dynamik unmittelbar dem Gesicht mitteilen kann. Zum anderen bestehen Formanalogien zwischen Gesicht uns Herz: Weil das Herz bedrückt ist, sind auch die Augenbrauen gesenkt; s. Le Brun (1994). Für Le Brun gilt dieses Raster nicht nur hinsichtlich des auf den Moment bezogenen Ausdrucks von Emotionen sondern auch der physiognomischen Charakterbestimmungen des Menschen. Diese werden, abhängig von della Portas Charakterstudien, fest in Paracelsischer Tradition wurzelnd, in Hinsicht auf ihre Ähnlichkeiten und Analogien zu dem Reich der Tiere getroffen; s. Bryson (1981), S. 43-47.

57 Öl/Lw, 192,5 x 273,5 cm, Frankfurt am Main, Städelsches Kunstinstitut, Inv.Nr. 1849; Ausstellungs-Katalog Paris (1994), Kat. Nr. 203, S. 453-456; Thuillier (1994), Kat.-Nr. 202, S. 261. Zur Datierung, Restaurierung und zur Provenienz s. Aussellungs-Katalog Frankfurt (1988). Die kunstgeschichtliche Literatur bis 1988 wird im selben Katalog von Michael Maek-Gérard, S. 57-62, unter Kat.-Nr. 1, besprochen.

58 „Stultitia amoris", wie Brandt es nennt: Brandt (1989), S. 249.

59 „Ce sujet, tiré d'Ovide, montre Thisbé découvrant le corps de Pyrame qui, la croyant tuée par un lion, vient de se suicider : épisode théâtral fort célèbre et maintes fois traité par les dramaturgues et les peintres, mais qui est ici mis en scène dans un grand paysage d'orage où l'aveuglement de la nature s'accorde au tragique des erreurs humaines", Thuillier (1994), S. 261.

60 Diese findet sich bei Brandt am deutlichsten formuliert: „Das Nachvornstürzen der Thisbe ist also für stoische Vorstellungen der körperliche Ausdruck einer hemmungslosen Triebbestimmtheit"; Brandt (1989), S. 248.

61 „Sie ist von Äußerlichem bestimmt, so wie ihre stürzende Bewegung, die vom Affekt der Trauer ausgelöst wird, zugleich eine Bewegung gegen den äußeren Sturm bildet"; ebd.

62 S. dazu ausführlicher mit Abbildungen und Literatur: Rosenberg/Prat (1994), Bd. 1, Nr. 129.24, S. 250 (Nr. 129, 129 R1 – 129.29, S. 240-51). Bereits Kate Steinitz geht jedoch ausführlich auf die stilistischen Unterschiede zwischen Leonardo und Poussin ein. Poussin ‚petrifiziert‘, so Steinitz, Leonardos Bewegungsstudien; Steinitz (1953).

Thisbe kann damit in ihrem Ausdruck als eindeutig gelten. Doch bereits eine erste Vergegen-
wärtigung der Situation, in welcher sie dargestellt ist, verunklärt die Eindeutigkeit wieder.
So ist nicht klar, ob sie die Arme ausbreitet als Folge ihrer Instabilität im Wind oder, um
ihren Geliebten zu umarmen. Im ersten Fall wird ihre Haltung allein durch den Sturm motiviert, im zweiten durch ihre Liebe zu Pyramus. Werfen wir nun einen vergleichenden Blick
in andere Werke des Künstlers, so verliert sich die Herleitung ihrer Haltung allein aus dem
Anlaufen gegen den Sturm. Stattdessen finden wir ähnliche Figuren in ganz anderen Zusam-
menhängen. Im Gemälde *Landschaft mit einem von einer Schlange getöteten Mann*[63] kann
man einen Mann entdecken, der in gleicher Richtung wie Thisbe, erschreckt von dem, was
er sieht, davonläuft[64]. Eine weitere, Thisbe ähnliche Figur finden wir im *Orpheus*[65] Poussins. Hier sehen wir eine junge Frau, die halb kniend, von einer Schlange aufgeschreckt,
ängstlich von der Stelle strebend, so ihre Arme nach vorne streckt, wie wir es von Thisbe her
kennen. In keiner dieser Landschaften weht ein Wind. Die Gemeinsamkeit all dieser Figuren
liegt allein in ihrer Begegnung mit dem Tod. Poussin stellt mit der Geste Thisbes scheinbar
nicht nur ein Rudern gegen den Wind, sondern eine von Todesfurcht durch und durch affi-
zierte Person dar. Diese These lässt sich erhärten durch einen Blick auf die Allegorie Jan
Swart van Groningens zur Vergänglichkeit der Welt. In seinem *Triumph des Todes*[66] können
wir im Mittelgrund eine Frau beobachten, die vor dem riesigen Pfeil des hinter ihr schweben-
den Todes mit ausgestreckten Armen flieht[67]. Von der Schrittstellung, den unterschiedlich
vor der Brust ausgestreckten Armen bis hin zum kapuzenähnlichen Tuchwulst auf den Schultern gleicht alles an ihr Thisbe. Poussin stellt in seiner tragischen Hauptperson weniger ein
Laufen gegen den Sturm dar, als vielmehr Todesschrecken[68].

Noch eine weitere – wahrscheinlichere – Vorlage lässt sich für die Figur Thisbes benen-
nen. Wir finden Thisbe nahezu identisch wieder in der Personifikation der *volontà* aus Cesa-

---

63 1648, London, National Gallery; s. o. Anm. 2.
64 Die gemeinsamen Merkmale sind das Laufen wie die von der Brust ausgestreckten Arme. Diesen Ausdrucks-
gestus läßt Fénelon seinen Poussin wie folgt beschreiben: „Il lève un bras en haut, l'autre tombe en bas ; mais les
deux mains s'ouvrent, elles marquent la surprise et l'horreur", Fénelon (1832), S. 266. Dieser halbnackte Mann
mit seinem wehenden Mantel und seiner gesenkten Rechten hat meiner Meinung nach mehr mit Leonardos
Figur gemeinsam als Thisbe, obwohl er gegen keinen Sturm anzurennen hat.
65 Vor 1660, Paris, Louvre; Ausstellungs-Katalog Paris (1994), Kat. Nr. 180, S. 409 ff.; Thuillier (1994), Kat.-
Nr. 179, S. 259.
66 Dat. Ende 1. Hälfte 16. Jh., 33 x 54 cm, Öl/Eichenholz, Prag, Národní galerie v Praze; s. Sternbersky (1992),
Nr. 41, S. 74.
67 Die leichte Gewandung dieser flüchtenden Frauenfigur verweist auf die Antike. Wir haben es offenbar mit
einer Umdeutung des antiken Typus der ekstatischen Mänade zu tun. Sie ist, ähnlich wie die sog. *Ignudo della
paura* (Ausstellungs-Katalog Frankfurt (1986), Kat.-Nr. 84 ff., S. 389 ff.), zu werten als eine Inversionsfigur im
Warburgschen Sinn.
68 Wer an einem direkten künstlerischen Einfluss Leonardos auf Poussins Thisbe festhalten möchte, wird einen
solchen in der kleinen Sturmstudie Leonardos entdecken können, die er unterhalb seiner Bemerkungen *Über
eine Figur, die sich gegen den Wind bewegt* gesetzt hat (C.U. 137 r) sowie in der Studie *Verfolgung einer Frau
durch einen Mann* (Windsor Castle, Royal Library, Inv. 12708 r). Obwohl die zuerst genannte Figur als Bewe-
gungsstudie einer gegen den Wind laufenden Person gewertet wird, die Verfolgte jedoch in einer Fluchtbewe-
gung ohne Wind begriffen ist, gleichen sie sich. Poussin verfährt in seinem Gemälde umgekehrt zu seinen Illu-
strationen von Leonardos Trattato. Gibt er hier die Bewegungsfiguren Leonardos statuarisch in antikisierender
Proportion, so entfällt dieses Moment bei Thisbe. Sie kommt den Figürchen Leonardos wesentlich näher als ihr
Pendant, die männliche gegen einen Sturm laufende Figur; cf. Barone (2001).

re Ripas *Iconologia*[69]. Ripa beschreibt die Figur als mit Flügeln begabt, blind und mit zwei erhobenen Armen versehen, „so als wolle sie sich festhalten". Analog zum Willen, der an sich blind ist, aber doch zwischen Recht und Unrecht zu unterscheiden vermag, ordnet Ripa seiner Personifikation zwei Wertungen zu. Zum einen bleibt sie rastlos und wendet sich zur Erde, zum anderen aber findet sie Ruhe und erstarkt, ist sie gen Himmel orientiert. Läuft sie hinter den Sinnen her, ist sie blind, schwach und unedel, dem Verstand folgend aber tapfer und wertvoll. Schauen wir auf die *volontà* bei Ripa, so bestehen viele Gemeinsamkeiten mit der Haltung Thisbes. In beiden Fällen handelt es sich um eine voranschreitende Frau, die Arme nach vorn gestreckt, die rechte Hand über die linke erhoben. Selbst der zwischen den Beinen wehende Rock stimmt überein[70]. Poussin folgt in seiner ‚Sturmfigur' den Angaben Ripas und gestaltet den Ausdruck Thisbes ambivalent. Zum einen zeigt sie die der Vernunft folgende Variante des Willens, denn sie stemmt sich mutig, sehenden Auges, gegen den Sturm. Zum anderen hat sie den Kopf gegen ihren sterbenden Geliebten geneigt und spielt damit auf den sinnlichen Teil des Willen an.

Es zeigt sich nach diesen Vergleichen, dass Thisbe sowohl von ihrem künstlerischen Herkommen, als auch im Vergleich mit ihre Verwandtschaft in anderen Kompositionen Poussins mitnichten der eindeutige Ausdruck eignet, der ihr zugeschrieben wird. Bei Thisbe ist unklar, ob der Künstler sie der Sturmfigur Leonardos entlehnt, Todesfurcht im Sinne der *Vanitas* Swart van Groningens ausdrückt oder aber von der *volontà* Ripas inspiriert ist. Es ist unklar, was im Bild Thisbes Haltung motiviert, der Sturm, ihr Todesschrecken oder ihr Wille. Bei dieser Mehrdeutigkeit der Ausdrucksfigur fragt sich, wie Poussin ihren Charakter bestimmt.

Er tut dies durch die Strukturierung seiner Komposition. Wir sehen das von links kommende Unwetter, welches die Welt umwälzt. Blitze schlagen ein in Burg und Baum, der Sturm biegt die Bäume und weht Tiere wie Menschen vor sich her – allerdings scheinen sie auch vor einem Löwenangriff zu fliehen. Unten verdichtet sich das Geschehen zur dramatischen Auffindung des sterbenden Pyramus durch seine geliebte Thisbe. Die Dramatik des Unwetters und seiner Folgen entfaltet sich von links nach rechts und motiviert, wie gesagt, alle Bewegungen, die des Himmels, der Bäume, der Tiere und des Menschen[71]. Poussin gelingt dies, weil er das Geschehen konsequent kausal ordnet[72]. Der Wind kann als Ursache aller äußeren Bewegungen gelten. Diese Ordnung ist sogar Poussins ausdrückliches Ziel[73].

---

69 *Volontà* „Una giovane mal vestita di rosso, & giallo, haverà ali alle spalle, & a' piedi; serà cieca, sporgendo ambedue le mani avanti una più dell'altra in atto di volersi appiare ad alcuna cosa. Il color rosso, & giall, cagionati presso al Sole per l'abbondanza della luce, potranno in questo luogo, secondo quella corrispondenza dimostrar le verità, che è chiarezza, e splendore dell'intelletto. Si dipinge con l'ali, perche si domanda col nome di volontà, & perche con un perpetuo volo discorrendo inquieta se stessa per cercar la quiete, la qual non ritrovando, con volo ordinario vicino alla terra, ingagliardisce, il suo moto in verso il Cielo. La cecità le conviene, perche non vedendo per se stessa cosa alcuna, và quasi tentone dietro al senso, se è debole, & ignobile, ò dietro alla ragione, se è gagliarda, e di prezzo", Ripa (1618).

70 Auch die schriftliche Ausführung wird von Poussin, wenngleich in Variation, alludiert. Thisbe ist, wie Ripa es fordert, in grelles Licht gerückt, allerdings nicht in das der Sonne, sondern des Blitzes. Damit mag Poussin das Thema der Semele anspielen; s. dazu Bätschmann (1987), S. 60-62.

71 Wir können hier eine Ordnung nach Cicero *De Officiis*, II, 6, 19-20, annehmen. Cicero gibt drei Weisen an, durch welche das Schicksal das menschliche Leben bestimmt: durch die unbeseelte Natur, das sind Stürme, Schiffbrüche, Katastrophen, durch wilde Tiere, die treten, beißen und hinterrücks angreifen, und zuletzt durch den Menschen selbst; s. Verdi (1982), S. 648.

Die Pointe dabei ist, dass er den Sturm nicht an sich selbst charakterisiert, sondern über seine Wirkungen, denn die Ursache all des Durcheinanders im Bild, der Wind, ist an sich unsichtbar[74]. Alle Folgen sind dazu angetan, auf verschiedenste Weise den Sturm zu erklären. So zeigt der kontrastreiche Wetterumschwung die Schnelligkeit des Sturms, die zerzausten Bäume, wie die wallenden Rücken der Tiere seine physische Stärke. Auch die Menschen geben deutlich Zeugnis von der Kraft des Windes. So zeigt etwa der Eselreiter ganz rechts, wie schwer es ist, sogar auf einem starken Lasttier gegen den Sturm zu bestehen. Zudem hält er sich die Augen mit der Hand zu und bedeutet mit dieser Geste, dass die Luft wohl staubig ist[75]. Der Sturm hat auch soziale Folgen, insofern er die Menschen von einander trennt. Sie laufen alle, gleich dem Vieh[76], in verschiedene Richtungen, ohne dass einer vom Unglück des anderen Notiz nimmt. Kurzum, Poussin gelingt es über die Konzentration auf die Wirkungen des Sturmes, diesen in seinen Kräften und Qualitäten komplex zu charakterisieren. Allein Thisbe widersteht dieser allgemeinen Dynamik. Sie ist die einzige, die ohne weitere Unterstützung, sehenden Auges gegen den Sturm ankämpft, hin zu ihrem Geliebten Pyramus. Ihre Gegenläufigkeit und Andersartigkeit betont Poussin durch einen nahezu spiegelbildlich formulierten Hirten, der, ihr gegenüber, einsam sich seiner Flucht ergibt und der in seinem rückwärts gewandten Blick Thisbes Unglück übersieht. Schon hier wird deutlich, dass der Künstler im Rahmen seiner Komposition Thisbe nicht als die närrisch Liebende wertet. Im Gegenteil, sie ist die einzige, die sich gegen das Unwetter und seine Wirkungen stemmt.

Allerdings sind dies nur äußere Gegebenheiten. Bisher haben wir Thisbes Status abhängig von den gleichsam physikalischen Folgen des Sturmes betrachtet. Poussin beschränkt sich jedoch nicht allein auf die externe Schilderung des Dramas. Er weiß auch ihre psychischen Konstellationen im Bild zu entfalten. Hat der Künstler zur Darstellung der äußeren Dynamik auf eine kausallogische Strukturierung der Handlung zurückgegriffen, so bedient er sich zur Entfaltung der inneren Disposition der Analogie. Diese veranschaulicht er durch ein pikturales Mittel, durch parallele Zuordnungen. Eine solche stellt die Beziehung von

---

72 Poussin folgt Leonardos Anliegen, die Natur von ihren Wirkungen her zu begreifen. Künstlerisch geht er jedoch insofern über Leonardo hinaus, als es diesem nicht gelungen ist, eine einheitliche Bewegung sowohl der Menschen als auch der Natur in einem Sturm zu komponieren; s. Stumpfhaus (2000), [Typoskript] S. 142-52.

73 „J'ay essayé de représenter une tempeste sur terre, imitant le mieux que j'ay pû l'effet d'un vent impétueux, d'un air rempli d'obscurité, de pluye, d'éclairs et de foudres qui tombent en plusieurs endroits, non sans y faire du désordre. Toutes les figures qu'on y voit joüent leur personnage selon le temps qu'il fait", Poussin an Jacques Stella, 1651; Jouanny (1968), S. 425; Lettres et propos, S. 160; deutsch: Bätschmann (1987), S. 19. Das Brieffragment an Jacques Stella wird überliefert von Félibien, in: Félibien (1725) (11685), S. 160, in Pace (1981), S. 149.

74 Bätschmanns komplexe Analyse, wie Malerei qua Indizien das Unsichtbare [i. e. Bacchus als mythische Figuration für die Wandelbarkeit der Welt und das Schicksal] erschließt (Bätschmann, ebd. S. 77), kann ich nicht teilen. Poussins Malerei bezieht sich zwar auf Unsichtbares, hier der Wind, doch nur um den Blick auf das Sichtbare, die Welt, zu lenken. Poussin nimmt im Laufe seines Schaffens zunehmend Abstand, auf irgendwelche unsichtbaren Transzendenzen zu verweisen. Seine Definition der Malerei lautet am Ende seines Lebens: „C'est une imitation faite avec lignes et couleurs en quelque superficie de tout ce qui se voit dessous le soleil", Poussin an Chambray, 1. März 1665; Jouanny (1968), S. 462; Lettres et propos, S. 174.

75 Er hat dieselbe Funktion wie das Mädchen mit dem geschürzten Rock in der *Mannalese*; s. o. S. 322.

76 Das Leben einzelgängerischer und vereinsamter Menschen entspräche, so Scaliger in Anlehnung an die Nikomachische Ethik des Aristoteles, entweder demjenigen Gottes oder eines Tieres; s. Scaliger (1561) III, 1, 82b, in: Scaliger (1994), Bd. 2, S. 76-77.

Blitz und Löwenangriff dar. Offensichtlich sind sie metaphorisch aufeinander bezogen, denn beide Ereignisse stehen auf einer Vertikalen übereinander, sind aber nicht kausal voneinander abhängig. Sie haben die gleiche Bewegungsrichtung, nicht nur für sich selbst, sondern auch in Bezug auf ihre Folgen. Ja, Poussin setzt die Folgen dieser beiden Begebenheiten sogar in eins. Mensch und Tier fliehen sowohl vor dem Unwetter als auch vor dem Löwen. Damit explizieren sich Unwetter und Löwenangriff gegenseitig. So unvermutet und zerstörerisch der Blitz in den Baum einschlägt, so plötzlich und Tod bringend greift der Löwe die Reiter an. So unerbittlich der Sturm Mensch und Tier vor sich her in die Vereinzelung treibt, so heillos fliehen sie vor dem furchtbaren Löwen. Die Zusammenführung von Gewitter und Löwenangriff in ihrer Wirkung, der Flucht, ist insofern bedeutungsvoll, als hier physikalische und psychische Zustände gleichgesetzt werden[77]. Wie die Menschen und Tiere von dem Wind verweht werden, so reißt sie ihre Furcht vor dem Löwen mit sich. Äußere Bewegung und Affektdynamik werden in der Fluchtgruppe des Mittelgrundes in eins gesetzt. Damit visualisiert Poussin nun nicht nur allein den Umstand der allgemeinen Bewegungen, dass Mensch und Tier fliehen, sondern er veranschaulicht gleichzeitig ihre Furcht, genauer, die *Macht* ihrer Furcht. Der Sturm gibt metaphorisch das Maß der Gewalt an, mit welcher das Entsetzen Mensch und Tier gepackt hat. Umgekehrt ist darin der Widerstand zu sehen, den Thisbe zu überwinden hat, um den Geliebten zu erreichen. Im Sturm sehen wir die Gewalt der Angst, der sich ihre Liebe widersetzt. Damit wiederum haben wir ein Maß für die Macht ihres Willens und ihrer Liebe, sich Flucht, Furcht und Vereinzelung entgegenzustemmen. Der Gegenstand der Darstellung ist somit vor allem eine Gefühlsintensität, ein Gegenstand, der sich begrifflich nur schwer einholen lässt.

Was heißt das aber für die Bewertung Thisbes? Thisbe wird, wie gesehen, erklärt durch ihre Stellung im Gefüge der Komposition. Es dient dazu, nicht allein die äußeren Bedingungen ihres Unglücks zu schildern, sondern auch die Intensität ihrer seelischen Bewegung, ihres Willen, ihrer zu überwindenden Ängste und Nöte. Poussin gibt sich nicht damit zufrieden, allein vermittels Zeichen, mit Mimik, Gestik und Körperhaltung, abstrakt einen Affekt zu bezeichnen – dazu ist Thisbe auch viel zu klein gegeben[78]. Vielmehr erhalten wir durch die Gesamtkomposition einen Eindruck von der *Macht und Stärke* der Thisbe bewegenden seelischen Regungen. Poussin inszeniert mit seinem Unwetter weniger ein exemplum malum haltloser Begierde und verwerflicher Liebe, als eher eine heroische Leidenschaft.

Das ganze Bild spricht vom überwältigenden Zwang der Natur und von der Macht der Gefühle, derer es bedarf, sich ihr zu widersetzen. Von hier aus wird verständlich, was mit dem durch den Modus vermittelten Gesamteindruck gemeint ist, der gleich und ganzheitlich dem Betrachter ausdrückt, wovon die Historie handelt. Denn das riesige Gewitter im Vergleich zu den kleinen Figuren, deren größte sich ihm im Vordergrund widersetzt, zeigt in seiner überwältigenden Düsternis die Begegnung mit dem Tod und die Schrecken, die er verursacht. Erst im weiteren Hinsehen erschließt sich peu à peu der genauere Hergang der Geschichte, lernen wir die Gefahren, ihre Wirkungen und Gegenkräfte kennen, allerdings ohne dass sich das Sujet ändert. Es erfährt lediglich eine Nuancierung und Vertiefung. Wenn

---

[77] Was die Eindeutigkeit der Kausalität im Bild verunklärt. Es ist an den Figuren selbst nicht ablesbar, ob ihre Haltung durch die physische Gewalt des Unwetters motiviert wird, oder durch die psychische ihrer Furcht. Ihr Ausdruck ist ambivalent.

[78] So sieht es Brandt: „Der Betrachter soll kein Mitleid empfinden, sondern nur aus der Gestik den Affekt dechiffrieren, der Thisbe bewegt"; Brandt (1989), S. 249.

an der Vorstellung vom Lesen eines Bildes festgehalten werden soll, so liegt es allenfalls auf der Ebene der Durchsicht des Gemäldes nach Bezüglichkeiten der Bildelemente und der Art ihres formalen Zusammenhangs. Dieser jedoch ist nicht vergleichbar mit der grammatikalischen Struktur eines Satzes. Wir können nach der Diskussion so verschiedener Kompositionen wie der *Mannalese* und der *Sturmlandschaft mit Pyramus und Thisbe* konstatieren, dass Poussin seine Werke weder in Hinsicht auf eine enge Bindung an einer vorrangigen Schrift orientiert noch strukturell einem Text angleicht, auch funktionieren sie nicht nach den Regeln harter Begrifflichkeit. Poussin begnügt sich nicht damit, den Ausdruck seiner Hauptrollen allein zeichenhaft lesbar zu vermitteln, weshalb ein Lesen der Affekte, ein bloß ikonographisch angeleiteter Zugang in die Irre führen muss. Eine sprachlich angeleitete Diskursivität hat dort ein Ende, wo es um die Einsicht der im Bild entfalteten Gefühlsintensitäten geht. Sie obliegt allein der Intuition, die das eigentliche Wahrnehmungsorgan der gegebenen Metaphern und Proportionen stellt.

Die Entäußerung der Darstellung von Emotionen, weg von der eigentlichen Ausdrucksfigur hin in die Struktur der Gesamtkomposition, legt den Verdacht nahe, dass Poussin die Deutlichkeit und Wahrheit von Mimik und Gestik bezweifelt. Er vertraut nicht, so meine These, der Kongruenz von Seele und Körper – Voraussetzung für ein auf Zeichen und Lesbarkeit ausgerichtetes Ausdruckssystem in Mimik und Gestik. Das adäquate Zusammenspiel von Körperoberfläche und seiner seelischen Tiefe ist ihm nicht selbstverständlich, wie sonst ließe sich eine so komplexe Ausdrucksstruktur außerhalb der dargestellten Figuren erklären?

Poussin scheint mit seinen Kompositionen auf Erfahrungen der Zeit zu reagieren, die eben jene Kongruenz von Leib und Seele grundsätzlich in Frage stellen. Immerhin beklagt sich der Künstler des öfteren darüber, dass man den Menschen leider nicht ins Herz schauen könne[79]. Dieser Seufzer entfährt ihm zum ersten Mal am französischen Hofe, und wir können in ihm wohl eine kritische Auseinandersetzung mit dem Ideal der *Honnêteté*, der „maîtrise de soi", erkennen[80]. Offensichtlich distanziert sich der Künstler von dieser Fähigkeit, sein Äußeres zu beherrschen, um es undurchschaubar zu machen[81], und nimmt die Honnêteté-Kritik etwa La Bruyères vom Ende des Jahrhunderts vorweg[82].

---

79 „Qu'il aurait souhaité, de même que faisait autrefois un philosophe, qu'on pût voir ce qui se passe dans l'homme, parce que non seulement on y découvrirait le vice et la vertu, mais aussi les sciences et les bonnes disciplines; ce qui serait d'un grand avantage pour les personnes savantes, desquelles on pourrait mieux connaître le mérite : mais comme la nature en a usé d'une autre sorte, il est aussi difficile de bien juger de la capacité des personnes dans les sciences et dans les arts que de leurs bonnes ou de leurs mauvaises inclinations dans les moeurs", Poussin an Sublet de Noyers, o. D., in: Félibien (1725) (11685), S. 40, in: Pace (1981), S. 119; Lettres et propos, S. 70. Poussin polemisiert hier offensichtlich auch gegen das seit Horaz (*Ars poetica*, V. 102-3) gültige Klischee, dass ein Künstler, der ein wirkungsvolles Werk zu präsentieren wünscht, selbst alle darzustellenden Emotionen zu erleben habe.

80 Galle (1985), S. 39.

81 Poussin steht mit der Skepsis an der Aussagefähigkeit des menschlichen Äußeren nicht allein. Auch Descartes äußert in seinem 1649 herausgegebenen Werk *Die Leidenschaften der Seele* Zweifel an der Verlässlichkeit der Mimik: „Et generalement toutes les actions, tant du visage que des yeux, peuvent estre changées par l'ame, lors que, voulant cacher sa passion, elle en imagine fortement une contraire: en sorte qu'on s'en peut aussi bien servir à dissimuler ses passions, qu'à les declarer"; Descartes (1993), § 113, S. 172/74; deutsch: Hammacher, ebd., S. 173/75. Descartes führt mehrere Argumente an, die erklären, warum es schwer ist, eindeutig in einem Gesicht zu lesen. Neben dem oben im Text genannten Grund bemerkt der Philosoph die nicht immer unterscheidbare Ähnlichkeit verschiedener auch konträrer Affektäußerungen im Gesicht. Darüber hinaus seien diese Äußerungen so differenziert und kleinteilig in ihrer Zusammensetzung, dass es schwer fällt, sie getrennt wahr zu

Doch Poussin reagiert mit seinen Kompositionen nicht nur auf soziale Bewegungen. Auch Philosophie und Naturwissenschaft formulieren ihren grundsätzlichen Zweifel an der Einheit von Körper und Seele. So wird mit Descartes' substanziellen Unterscheidung von Körper und Seele die für die Kunst so zentrale Voraussetzung der Übereinstimmung von Körper und Seele prinzipiell problematisch[83]. Zum anderen zeigt die Erfindung des Mikroskops, dass der bloße Gebrauch der Sinne niemals den letzten Grund der Dinge zu schauen vermag[84]. Diese Erfindung Mitte der zwanziger Jahre des 17. Jahrhunderts im Kreis der Barberini[85], eines Kreises, dem Poussin seit seinem ersten Jahr in Rom, 1624, zugehörte, eröffnet dem Auge neue Welten, die mit dem, was sich auf der Oberfläche abzeichnet, wenig zu tun haben. Wie dieses Instrument dem Blick eine unermessliche, nicht mehr überschaubare Welt eröffnet, die weit entfernt ist, sich auf einer Außenseite adäquat abzuzeichnen, so führt Poussin die seelischen Bewegungen als ein eigenes Reich auf, das in seiner Vielfalt und Dynamik nicht ohne weiteres auf einfache Oberflächensignaturen übertragbar ist.

---

nehmen. Ein unzweideutiger – jedoch nicht wirklich verlässlicher – Indikator der inneren Bewegungen ist Descartes die Gesichtsfarbe, denn sie zeigt abhängig vom Grad der Hautdurchblutung unmittelbarer den Zustand des Herzens an; Descartes ebd., §§ 114ff. Diesen Hinweis verdanke ich Ulrich Pfarr.

Das „dissimuler" bereitet auch Poussin Schwierigkeiten. So schreibt er am 3. Nov. 1647 an Chantelou: „Il est difficile de connaître les personnes dissimulées si se n'est avec un longtemps. Quand à moi je vous jure que j'ai eté trompé"; Jouanny (1986), S. 367. Hier ist es weniger die Wissenschaft und Philosophie, die das Innere des Menschen erkennen lassen, sondern allein der Umgang mit ihnen, die Erfahrung.

82 Bruyère kritisiert die undurchdringliche Maskerade der Selbstbeherrschung der Höflinge und wertet das Raffinement ihres Verhaltens als schlechten Dienst und als Untugend; s. hierzu ausführlich Galle (1985), S. 39ff. Obwohl Poussin in seinen Briefen auf die Taktik des ‚dissimuler' zurückgreift, um „so offen wie möglich, so bedeckt wie nötig zu erscheinen" (Bruhn (2000), S. 47), übt er doch entschieden Kritik an dieser Taktik; s. o. Anm. 64; 66. Für seine Historien jedoch gilt zweifellos das Ideal der Naturnähe und – in komplexer Weise – der Echtheit des Ausdrucks, die Poussins Personal einheitlich zugeschrieben wird; das sind Qualitäten, die der Sincérité zugeordnet sind.

83 Es sei an dieser Stelle erwähnt, dass sich die Ausdruckslehren Le Bruns und Testelins zwar ausführlich an Descartes orientieren, von ihnen selbst aber auch als ungenügend wahrgenommen werden. So äußert Testelin Zweifel an der Möglichkeit, alle Zeichen der verschiedenen Leidenschaften erfassen zu können; s. Larsson (1990), S. 180. Des weiteren ist es bezeichnend, dass Le Bruns Affekt- und Charakterlehren „nur in allegorischen Gestalten" ihren Niederschlag fanden; s. ebd., S. 187.

84 Die Erfindung des Mikroskops weckte harsche Kritik an einer humanistisch angeleiteten Vorstellung von der Natur als Symbolschrift Gottes: „One cannot achieve insight into the secrets of nature by studying Hieroglyphs"; s. Wilson (1988), S. 94. Freundl. Hinweis von Claus Zittel.

Zu einem ähnlichen Ergebnis kommt Freedberg, dass das Mikroskop dem auf der Figur der Analogie beruhenden humanistisch-christlichen Weltbild neue Ordnungskategorien abverlangt, vor allem deshalb, weil es dem Blick des Menschen Welten eröffnet, die wenig mit dem zu tun haben, was der makroskopische Bereich sehen lässt: „There are limitations beyond the limitations of sight itself, and beyond the impossibility of ever seeing everything that goes into the constitution of natural bodies"; Freedberg (1998), S. 286.

85 Poussin wurde schon früh, bereits in seinem ersten Jahr in Rom, 1624, von Marino über Marcello Sacchetti mit dem Kardinal Barberini, Neffe von Urban VIII. bekannt gemacht (s. Félibien (1725), S. 10); in Pace (1981), S. 111). Der erste Druck einer mikroskopischen Beobachtung, von einer Biene, dem Wappentier der Barberini, erschien 1625 zu Ehren des neu gewählten Papstes.

Literatur

Austellungs-Katalog Frankfurt (1988): Nicolas Poussin: Zu den Bildern im Städel. Hrsg. von Claude Lorrain. Frankfurt/M.

Ausstellungs-Katalog Paris (1994): *Nicolas Poussin 1594-1665*. Galeries nationales du Grand Palais.

Austellungs-Katalog Washington (1995): National Gallery of Art. Hrsg. von John Walker. New York.

Barone, Juliana (2001): „Illustrations of figures by Nicolas Poussin and Stefano della Balla in Leonardo's *Trattato*", *Gazette des Beaux-Arts, 6ᵉ période*, Tome CXXXVIII, 143ᵉ Année (2001), 1-14.

Bätschmann, Oskar (1987): *Nicolas Poussin Landschaft mit Pyramus und Thisbe. Das Liebesunglück und die Grenzen der Malerei*. Frankfurt/M.

Blunt, Anthony (1995): „Nicolas Poussin", in Ausst.-Kat. Washington (1995), 157-76.

Blunt, Anthony (Hg.) (1989): *Lettres et propos sur l'art* (1964). Paris.

Boos, Manfred (1966): *Französische Kunstliteratur 1648 und 1669*. Radolfzell/Bodensee.

Brandt, Reinhard (1989): „Pictor philosophus: Nicolas Poussin, Gewitterlandschaft mit Pyramus und Thisbe", *Städel-Jahrbuch* 12 (1989), 249.

Bruhn, Matthias (2000): *Nicolas Poussin. Bilder und Briefe*. Berlin.

Bryson, Norman (1981): *Word and image, French painting of the ancien regime*. Cambridge.

Clementz, Heinrich (1959): *Des Flavius Josephus Jüdische Altertümer.* Köln.

Colantuono, Anthony (1996) : „Interpréter Poussin. Métaphore, similarité et ‚maniera magnifica‘", in *Nicolas Poussin (1594-1665), Actes du Colloque organisé au Musée du Louvre par le Service culturel du 19 au 21 Octobre 1994*. Paris.

Descartes, René (1996): *Les Passions de l'Ame* (1649). *Die Leidenschaften der Seele*. Hrsg. und übersetzt von Klaus Hammacher. Hamburg.

Félibien (1669): *Conferences de l'Academie Royale de Peinture et de Sculpture pendant l'année 1667*. Paris.

Félibien (1725) : „8ᵉ Entretien" (1685) , in Claire Pace (1981).

Freedberg, David (1998): „Iconography between the History of Art and the History of Sience: Art, Science, and the Case of the Urban Bee", in Caroline A. Jones und Peter Galison (1998), *Picturing Science – Producing Art*. New York/London.

Galle, Roland (1985): „Honnêteté und Sincérité", in Fritz Nies und Karlheinz Stierle (Hg.), *Französische Klassik. Theorie-Literatur-Malerei*. München.

Germer, Stefan (1997): *Kunst – Macht – Diskurs. Die intellektuelle Karriere des André Félibien im Frankreich von Louis XIV*. München.

Held, Jutta (2001): *Französische Kunsttheorie des 17. Jahrhunderts und der absolutistische Staat*. Berlin

Hossenfelder, Malte (1995): „Die Philosophie der Antike. Stoa, Epikureismus und Skepsis", in Wolfgang Röd (Hg.), *Geschichte der Philosophie*, Bd. 3. München, 54 ff.

Imdahl, Max (1985): „Caritas und Gnade. Zur ikonischen Zeitstruktur in Poussins Mannalese", in F. Nies und K. Stierle (Hg.), *Französische Klassik: Theorie-Literatur-Malerei*. München, 153-54.

Jouanny, Ch. (Hg.) (1968): *Correspondance de Nicolas Poussin* (1911). Paris.

Jouin, Henry (Hg.) (1883): *Conférences de l'Académie Royale de Peinture et de Sculpture*. Paris.

Kirchner, Thomas (1991): *L'expression des passions*. Mainz.

Larsson, Lars Olof (1990): „Der Maler als Erzähler: Gebärdensprache und Mimik in der französischen Malerei und Kunsttheorie des 17. Jahrhunderts am Beispiel Charles Le Bruns", in Volker Kapp (Hg.), *Die Sprache der Zeichen und Bilder: Rhetorik und nonverbale Kommunikation in der frühen Neuzeit*. Marburg.

Le Brun, Charles (1994): *L'expression des passions et autres conférences*. Hrsg. von Philippe Julien. Maisonneuve et Larose.

Marin, Louis (1995): „Lire un tableau en 1639 d'après une lettre de Poussin", in *Sublime Poussin*. Paris, 13-14. (Zuerst in Chartier, R. (Hg.) (1983): Pratiques de la lecture. Marseille/ Rivages, 101-124.)

McTighe, Sheila (1996): *Nicolas Poussin's Landscape Allegories*. Cambridge u. a.

Mildner-Flesch, Ursula (1983): *Das Decorum. Herkunft, Wesen und Wirkung des Sujetstils am Beispiel Nicolas Poussin*. St. Augustin.

Montagu, Jennifer (1992): „The Theory of the musical modes in the Académie royale de peinture et de sculpture", *Journal of the Warburg and Courtauld Institutes*, LV (1992), 233-48.

Oy-Marra, Elisabeth (1997): „Poussins ‚Mannalese': Zur Debatte um Zeitlichkeit in der Historienmalerei", *Marburger Jahrbuch für Kunstwissenschaft*, Bd. 24 (1997), 20-12.

Pace, Claire (1981): *Félibien's Life of Poussin*. London.

Pace, Claire (1999): „Nicolas Poussin: ‚peintre-poète'?", in *Commemorating Poussin. Reception and Interpretation of the Artist*. Cambridge University Press.

Puttfarken, Thomas (1985): *Roger de Piles' theory of art*. New Haven/London.

Ripa, Cesare (1618): „Nova Iconologia del Cavalier Cesare Ripa Perugino", in Cesare Ripa (1992), *Iconologia*. Hrsg. von Piero Buscaroli. Mailand, SW *Volontà*, 485-86.

Rosenberg, Pierre und Prat, Louis-Antoine (Hg.) (1994): *Nicolas Poussin 1594-1665. Catalogue raisonné des dessins*, 2 Bde. Genève.

Scaliger, Iulius C. (1994): *Poetices libri septem. Die sieben Bücher über die Dichtkunst*, bisher 5 Bde. Hrsg. von Luc Deitz und Gregor Vogt-Spira. Stuttgart-Bad Cannstatt.

Schütze, Sebastian (1996): „Aristide de Thèbes, Raphaël et Poussin. La représentation des affetti dans les grands tableaux d'histoire de Poussin des années 1620-30", in *Nicolas Poussin. 1594-1665, Actes du Colloque organisé au Musée du Louvre par le Service culturel du 19 au 21 Octobre 1994*, Tome II. Paris, 571-602;

Schlink, Wilhelm (1996): *Ein Bild ist kein Tatsachenbericht. Le Bruns Akademierede von 1667 über Poussins* Mannawunder. Freiburg im Breisgau.

Schmidt-Biggemann, Wilhelm (1999): *Blaise Pascal*. München.

Seneca, Lucius A. (1992): „*De Providentia*", in Lucius A. Seneca, *Die kleinen Dialoge*, 2 Bde. Hrsg. und übersetzt von Gerhard Fink. München u.a., 6-45.

Steinitz, Kate (1953): „Poussin, Illustrator of Leonardo da Vinci and the Problem of Replicas in Poussin's Studio", *The Art Quarterly* XVI (1953), 40-55.

Sternbersky, Palac (1992): Staré evropské umení. Národni galerie v Prage. Prag.

Stumpfhaus, Bernhard (2000): *Modus – Affekt – Allegorese bei Poussin. Ein Beitrag zur Emotionsforschung in der Französischen Malerei des 17. Jahrhunderts*. Frankfurt/M.

Thuillier, Jacques (1994): *Nicolas Poussin*. Paris.

Verdi, Richard (1982): „Poussin and the ‚Tricks of Fortune'", *The Burlington Magazine* CXXIV, 956 (Nov. 1982), 648.

Wilson, Catherine (1988): „Visual surface and visual symbol. The Microscope and the occult in early modern Europe", *Journal of the history of ideas* 1988.

Contributors

Alexander Becker studied musicology, philosophy, and history at the University of Frankfurt and at the Warburg Institute, London. Ph. D. 1998 in philosophy with a thesis on "Understanding and Consciousness" (published Paderborn: mentis, 2000). He works currently in the research project "Wissenskultur und gesellschaftlicher Wandel" at the University of Frankfurt am Main on the dialogue form and the transfer of knowledge in the Platonic dialogues.

Mario Biagioli is Professor of the History of Science at Harvard University. He is author of *Galilei Courtier* (1993) and editor of *The Science Studies Reader* (1999).

Laurence Brockliss is Reader in Modern History at the University of Oxford and Fellow and Tutor of Magdalen College. His publications include *French Higher Education in the Seventeenth and Eighteenth Centuries: A Cultural History* (Oxford, 1987); (with Colin Jones) *The Medical World of Early Modern France* (Oxford, 1997); and Calvet's Web: Enlightenment and the Republic of Letters in Eighteenth-Century France (Oxford, 2002).

Wolfgang Detel studied classics, mathematics, and philosophy at the universities of Tübingen, Mannheim, and Hamburg. He is currently professor of philosophy (chair for ancient philosophy and philosophy of science) at the universitiy of Frankfurt/Main, Germany. As a visiting professor he taught at the universities of Princeton, Pittsburgh, and Columbia/New York. He is author of books on Plato's theory of false statements (*Platons Beschreibung des falschen Satzes im "Theätet" und "Sophistes"*) (1974), on Gassendi´s physics (*"Scientia Rerum Natura Occultarum." Methodologische Studien zur Physik Pierre Gassendis)* (1978), on Foucault's *History and Sexuality, vol.2 (Macht, Moral, Wissen. Foucault und die klassische Antike)* (1998), an introduction, translation, and commentary on Aristotle's *Posterior Analytics* (in 2 volumes) (*Aristoteles: Analytica Posteriora. Einleitung, Übersetzung und Kommentar von W. D. Aristoteles, Werke (Hrg. v. H. Flashar 3,II2))* (1993) as well as many articles on different topics. His main research areas are currently Plato, modern semantics, and perspectives of Critical Theory.

Daniel Garber is Lawrence Kimpton Distinguished Service Professor in Philosophy and in the Committee on the Conceptual Foundation of Science at the University of Chicago. He is *inter alia* the author of *Descartes' Metaphysical Physics* (Chicago 1992), *Descartes Embodied* (Cambridge 2001) and co-editor of *The Cambridge History of Seventeenth-Century Philosophy.*

Peter Machamer is a professor of History and Philosophy of Science at the University of Pittsburgh, and associate director of the Center for Philosophy of Science. He has been visiting professor in Greece (University of Athens), Italy (Universities of Undine and Pisa) and Bogazici University in Istanbul, Turkey. He has published widely in diverse fields, including 17th Century science and philosophy, philosophy of psychology and neuroscience, value theory, and epistemology. He was the wine columnist for the Pittsburgh Post Gazette for 15 years. He recently edited *A Cambridge Companion to Galileo* (Cambridge), *Scientific Controversies* (Oxford), *Theories and Methods in the Neurosciences* (Pittsburgh) and *The Blackwell Guide to Philosophy of Science* (Blackwell).

Wolfgang Neuber, geboren 1956, studierte Anglistik, Germanistik und Philosophie. Promotion 1980 in Wien, Habilitation 1988 in Wien, ebd. Associate Professor seit 1989, Professor an der Universität Frankfurt/M. 1995-2000, seither Professor an der FU Berlin. Publikationen (in Auswahl): *Nestroys Rhetorik. Wirkungspoetik und Altwiener Volkskomödie im 19. Jahrhundert* (Bonn 1987); *Fremde Welt im europäischen Horizont. Zur Topik der deutschen Amerika-Reiseberichte der Frühen Neuzeit* (Berlin 1991); Er ist Herausgeber bzw. Mitherausgeber u. a. von: *Ars memorativa.* Zur kulturgeschichtlichen Bedeutung der Gedächtniskunst 1400-1750 (Tübingen 1993); *Intertextualität in der Frühen Neuzeit. Studien zu ihren theoretischen und praktischen Perspektiven* (Frankfurt a. M. u. a. 1994); DOCUMENTA MNEMONICA, 7 Bde (Tübingen 1998); *Daphnis.* Zeitschrift für Mittlere Deutsche Literatur und Kultur der Frühen Neuzeit (1400-1750) 29- (2000-); *Chloe.* Beihefte zum Daphnis (ab September 1999); *Seelenmaschinen. Gattungstraditionen, Funktionen und Leistungsgrenzen der Mnemotechniken vom späten Mittelalter bis zum Beginn der Moderne* (Wien/Köln/Weimar 2000).

Eileen Reeves is Assistant Professor for Comparative Literature at Princeton University. She is author of *Painting the Heavens: Art and Science in the Age of Galileo* (Princeton 1997).

Klaus Reichert ist seit 1975 Professor für Anglistik/Amerikanistik an der Universität Frankfurt. Gastprofessuren in Italien und den USA. Seit 1993 geschäftsführender Direktor des neugegründeten Zentrums zur Erforschung der Frühen Neuzeit. Wichtigste Veröffentlichungen: „Neue Formen des Geheimen am Beginn der Moderne", in: *Das Geheimnis am Beginn der europäischen Moderne. Zeitsprünge* 6 (2002), Heft 1-4, S. 12-20; „Shakespeares mimetische Rivalen", in *Das Opfer – aktuelle Kontroversen: religionspolitischer Diskurs im Kontext der mimetischen Theorie*, hrsg. v. Bernhard Diekmann, Münster u. a.: LIT 2000, S. 207-224. „Von der Wissenschaft zur Magie: John Dee", in *Der Magus. Seine Ursprünge und seine Geschichte in veschiedenen Kulturen*, hrsg. v. Anthony Grafton u. Moshe Idel, Berlin: Akademie Verlag, 2001, S. 87 – 106; „Zeitsprünge. Von Jahrhundertenden, Jahrtausendenden, Enden der Welt", in Forschung Frankfurt 1/2000, S. 6-15; „Zur Geschichte der christlichen Kabbala", in *Kabbala und die Literatur der Romantik*, hrsg. v. Eveline Goodman-Thau, Gert Mattenklott, Christoph Schulte, Max Niemeyer Verlag 1999, S. 1-16; „Friede am Himmel wie auf Erden: Giordano Brunos Spaccio de la Bestia Trionfante", in *Zeitsprünge. Forschungen zur Frühen Neuzeit*, Band 3 (1999), Heft 1/2, S. 49-64; *Der fremde Shakespeare*, München: Hanser 1998; „Pico della Mirandola and the Beginning of Christian Kabbala", in

Mysticism, Magic and Kabbalah in Ashkenazi Judaism, Berlin, New York: de Gruyter, 1995, S. 195-207; „Joseph Glanvill's Plus Ultra and Beyond: Or How to Delay the Rise of Modern Science" in *Technology, Pessimism, and Postmodernism*, Dordrecht: Kluwer Academic Publishers, 1993, S. 39-51.; „The Two Faces of Scientific Progress; or Institution as Utopia", in *Utopian Vision, Technological Innovation and Poetic Imagination*, ed. K. L. Berghahn, R. Grimm, Heidelberg: Winter Universitätsverlag, 1990, S. 11-28; *Vielfacher Schriftsinn. Gesammelte Essays zu Finnegans Wake*, Frankfurt: edition suhrkamp, 1989. *Fortuna oder die Beständigkeit des Wechsels*, Frankfurt: Suhrkamp, 1985.

Friedrich Steinle has, after his studies of physics, worked in history and philosophy of science, and is currently research fellow at the Max Planck Institute for the History of Science in Berlin. He has published on early modern natural philosophy, focussing on the formation of Newton's mechanical and optical concepts (cf. his book on Newton's manuscript *De gravitatione*), and on the rise of the concept of laws of nature. He has many articles, moreover, on the history and philosophy of experiment, with a focus on the history of electricity and of color theory. In his book-length Habilitationsschrift (2000), he studied research practice in early electromagnetism – a crucial period for the development of 19th century physics – and analyzed the particular type of exploratory experimentation.

Bernhard Stumpfhaus: Freischaffender Kunsthistoriker; Promotion 2001 Universität Frankfurt am Main; 1999 Stipendiat des Graduiertenkollegs „Psychische Energien bildender Kunst"; 1997-98 wiss. Mitarbeiter am SFP der DFG: Pierre Puget; organisiert Ausstellungen (u. a. zwei Retrospektiven zur sowjetischen Oppositionskunst in Oxford und Frankfurt am Main). Forschungsschwerpunkte: Malerei und Skulptur des 17. Jahrhunderts, Kunst der Moderne, zeitgenössische Kunst, Geschichte der Kunstgeschichte; Publikationen (in Auswahl): *Transrealismus. Kunst zur Zeit der Perestroika* (März 1995), *Modus-Affekt-Allegorese bei Poussin. Ein Beitrag zur Emotionsforschung in der französischen Malerei des 17. Jahrhunderts* (Heidelberg 2002).

Matthias Vogel studied philosophy and musicology in Hamburg and is currently Assistant Professor at the Philosophy Department of the Johann Wolfgang Goethe-Universität Frankfurt/Main. Recent Publications: *Medien der Vernunft* (Frankfurt/Main, Suhrkamp 2001); *Wissen zwischen Entdeckung und Konstruktion* (Frankfurt/Main , Suhrkamp 2002) (ed. together with Lutz Wingert).

Catherine Wilson: Professor of Philosophy, University of British Columbia. She has publications in the history of philosophy and the history and philosophy of 17th and 18th century science, including *Leibniz's Metaphysics* (1989) and The *Invisible World: Early Modern Philosophy and the Invention of the Microscope* (1995).

Claus Zittel ist wissenschaftlicher Mitarbeiter im DFG-Sonderforschungsbereich „Wissenskultur und gesellschaftlicher Wandel" an der J. W. Goethe-Universität Frankfurt am Main (Teilprojekt: Wissen und Wissenschaft in der antiken und frühneuzeitlichen Philosophie). Publikationen zur Philosophie des 17. und 19. Jahrhunderts, Wissenschaftstheorie und zur Ästhetik, darunter: *Selbstaufhebungsfiguren bei Nietzsche* (Würzburg 1995); *Das ästheti-*

sche Kalkül von Friedrich Nietzsches ‚Also sprach Zarathustra' (Würzburg 2000). *Mirabilis scientiae fundamenta.* Die Philosophie des jungen Descartes (1619-1628). In: *Seelenmaschinen. Gattungstraditionen, Funktionen und Leistungsgrenzen der Mnemotechniken vom späten Mittelalter bis zum Beginn der Moderne.* Hrsg. von Jörg J. Berns und Wolfgang Neuber. Wien 2000, S. 309-262. Er ist Herausgeber eines weiteren Bandes in der gleichen Reihe mit dem Titel: *Wissen und soziale Konstruktion* (Berlin 2002).

Index

Picture sources

P. 42: Galileo's 1609 washdrawings of the Moon, Ms. Gal 48, fol. 28r, courtesy of the Biblioteco Nazionale Centrale, Florence.

P. 43: Galileo's engravings of the Moon in *Sidereus muncius* (1610), courtesy of Owen Gingerich.

P. 44: An example of Galileo's maps of Jupiter and its satellites in *Sidereus nuncius* (1610),
courtesy of Owen Gingerich.

P. 46: Map R from Hevelius' *Selenographia* (1647), courtesy of Owen Gingerich.

P. 67: Foldout of sunspots illustrations from Scheiner *Tres epistolae* (1612). This reproduction is taken from the 1613 Roman edition, courtesy of the William Andrews Clark Memorial
Library, UCLA.

P. 70/71: Galileo's illustrations of sunspots for June 26-9 from his *Istoria e dimonstrationi*
(1613), courtesy of the William Andrews Clark Memorial Library, UCLA.

P. 78/79: Sunspots illustrations from Scheiner's *Accuratior disquisitio* (1612) as reproduced
in the Roman 1613 edition, courtesy of the William Andrews Clark Memorial Library, UCLA.

P. 314: Basic Social Relations in "L'incoronazione di Poppea," by Matthias Vogel.

P. 316/317: Synopsis of "L'incoronazione di Poppea," by Matthias Vogel.

P. 322: Nicolas Poussins *Mannalese,* in: Christopher Wright (1985): *Poussin. A catalogue
raisonné.* Buccaneer Books.

P. 332: Nicolas Poussin: *Gewitterlandschaft mit Pyramus und Thisbe,* in: Christopher Wright
(1985): *Poussin. A catalogue raisonné.* Buccaneer Books.